CONVEX ANALYSIS

PRINCETON LANDMARKS
IN MATHEMATICS AND PHYSICS

Convex Analysis

BY

R. TYRRELL ROCKAFELLAR

PRINCETON, NEW JERSEY
PRINCETON UNIVERSITY PRESS

Published by Princeton University Press, 41 William Street,
Princeton, New Jersey 08540
In the United Kingdom: Princeton University Press,
Chichester, West Sussex

First Published in the Princeton Mathematical Series in 1970; tenth printing
and first paperback printing in the Princeton Landmarks in Mathematics
and Physics series, 1997

ISBN 0-691-08069-0 (cloth)
ISBN 0-691-01586-4 (paperback)
L. C. Card 68-56318

Princeton University Press books are printed on acid-free paper and meet the
guidelines for permanence and durability of the Committee on Production
Guidelines for Book Longevity of the Council on Library Resources

http://pup.princeton.edu

Printed in the United States of America

17 19 20 18

ISBN-13: 978-0-691-01586-6 (pbk.)

This book is dedicated to
WERNER FENCHEL

Preface

Convexity has been increasingly important in recent years in the study of extremum problems in many areas of applied mathematics. The purpose of this book is to provide an exposition of the theory of convex sets and functions in which applications to extremum problems play the central role.

Systems of inequalities, the minimum or maximum of a convex function over a convex set, Lagrange multipliers, and minimax theorems are among the topics treated, as well as basic results about the structure of convex sets and the continuity and differentiability of convex functions and saddle-functions. Duality is emphasized throughout, particularly in the form of Fenchel's conjugacy correspondence for convex functions.

Much new material is presented. For example, a generalization of linear algebra is developed in which "convex bifunctions" are the analogues of linear transformations, and "inner products" of convex sets and functions are defined in terms of the extremal values in Fenchel's Duality Theorem. Each convex bifunction is associated with a generalized convex program, and an adjoint operation for bifunctions that leads to a theory of dual programs is introduced. The classical correspondence between linear transformations and bilinear functionals is extended to a correspondence between convex bifunctions and saddle-functions, and this is used as the main tool in the analysis of saddle-functions and minimax problems.

Certain topics which might properly be regarded as part of "convex analysis," such as fixed-point theorems, have been omitted, not because they lack charm or applications, but because they would have required technical developments somewhat outside the mainstream of the rest of the book.

In view of the fact that economists, engineers, and others besides pure mathematicians have become interested in convex analysis, an attempt has been made to keep the exposition on a relatively elementary technical level, and details have been supplied which, in a work aimed only at a mathematical in-group, might merely have been alluded to as "exercises." Everything has been limited to R^n, the space of all n-tuples of real numbers, even though many of the results can easily be formulated in a broader setting of functional analysis. References to generalizations and extensions are collected along with historical and bibliographical comments in a special section at the end of the book, preceding the bibliography itself.

As far as technical prerequisites are concerned, the reader should be able to get by, for the most part, with a sound knowledge of linear algebra and elementary real analysis (convergent sequences, continuous functions, open and closed sets, compactness, etc.) as pertains to the space R^n. Nevertheless, while no actual familiarity with any deeper branch of abstract mathematics is required, the style does presuppose a certain "mathematical maturity" on the part of the reader.

A section of remarks at the beginning of the book describes the contents of each part and outlines a selection of material which would be appropriate for an introduction to the subject.

This book grew out of lecture notes from a course I gave at Princeton University in the spring of 1966. In a larger sense, however, it grew out of lecture notes from a similar course given at Princeton fifteen years earlier by Professor Werner Fenchel of the University of Copenhagen. Fenchel's notes were never published, but they were distributed in mimeographed form, and they have served many researchers long and well as the main, and virtually the only, reference for much of the theory of convex functions. They have profoundly influenced my own thinking, as evidenced, to cite just one aspect, by the way conjugate convex functions dominate much of this book. It is highly fitting, therefore, that this book be dedicated to Fenchel, as honorary co-author.

I would like to express my deep thanks to Professor A. W. Tucker of Princeton University, whose encouragement and support has been a mainstay since student days. It was Tucker in fact who suggested the title of this book. Further thanks are due to Dr. Torrence D. Parsons, Dr. Norman Z. Shapiro, and Mr. Lynn McLinden, who looked over the manuscript and gave some very helpful suggestions. I am also grateful to my students at Princeton and the University of Washington, whose comments on the material as it was taught led to many improvements of the presentation, and to Mrs. Janet Parker for her patient and very competent secretarial assistance.

Preparation of the 1966 Princeton lecture notes which preceded this book was supported by the Office of Naval Research under grant NONR 1858(21), project NR-047-002. The Air Force Office of Scientific Research subsequently provided welcome aid at the University of Washington in the form of grant AF-AFOSR-1202-67, without which the job of writing the book itself might have dragged on a long time, beset by interruptions.

R. T. R.

Contents

PART I: BASIC CONCEPTS

PART II: TOPOLOGICAL PROPERTIES

PART III: DUALITY CORRESPONDENCES

PART IV: REPRESENTATION AND INEQUALITIES

PART V: DIFFERENTIAL THEORY

PART VI: CONSTRAINED EXTREMUM PROBLEMS

PART VII: SADDLE-FUNCTIONS AND MINIMAX THEORY

PART VIII: CONVEX ALGEBRA

Introductory Remarks: A Guide for the Reader

This book is not really meant to be read from cover to cover, even if there were anyone ambitious enough to do so. Instead, the material is organized as far as possible by subject matter; for example, all the pertinent facts about relative interiors of convex sets, whether of major or minor importance, are collected in one place (§6) rather than derived here and there in the course of other developments. This type of organization may make it easier to refer to basic results, at least after one has some acquaintance with the subject, yet it can get in the way of a beginner using the text as an introduction. Logical development is maintained as the book proceeds, but in many of the earlier sections there is a mass of lesser details toward the end in which one could get bogged down.

Nevertheless, this book can very well be used as an introduction if one makes an appropriate selection of material. The guidelines are given below, where it is described just which results in each section are really essential and which can safely be skipped over, at least temporarily, without causing a gap in proof or understanding.

Part I: Basic Concepts

Convex sets and convex functions are defined here, and relationships between the two concepts are discussed. The emphasis is on establishing criteria for convexity. Various useful examples are given, and it is shown how further examples can be generated from these by means of operations such as addition or taking convex hulls.

The fundamental idea to be understood is that the convex functions on R^n can be identified with certain convex subsets of R^{n+1} (their epigraphs), while the convex sets in R^n can be identified with certain convex functions on R^n (their indicators). These identifications make it easy to pass back and forth between a geometric approach and an analytic approach. Ordinarily, in dealing with functions one thinks geometrically in terms of the graphs of the functions, but in the case of convex functions pictures of epigraphs should be kept in mind instead.

Most of the material, though elementary, is basic to the rest of the book, but some parts should be left out by a reader who is encountering the subject for the first time. Although only linear algebra is involved in §1

(Affine Sets), the concepts may not be entirely familiar; §1 should therefore be perused up through the definition of barycentric coordinate systems (preceding Theorem 1.6) as background for the introduction of convexity. The remainder of §1, concerning affine transformations, is not crucial to a beginner's understanding. All of §2 (Convex Sets and Cones) is essential and the first half of §3, but the second half of §3, starting with Theorem 3.5, deals with operations of minor significance. Very little should be skipped in §4 (Convex Functions) except some of the examples. However, the end of §5 (Functional Operations), following Theorem 5.7, is not needed in any later section.

Part II: Topological Properties

The properties of convexity considered in Part I are primarily algebraic: it is shown that convex sets and functions form classes of objects which are preserved under numerous operations of combination and generation. In Part II, convexity is considered instead in relation to the topological notions of interior, closure, and continuity.

The remarkably uncomplicated topological nature of convex sets and functions can be traced to one intuitive fact: if a line segment in a convex set C has one endpoint in the interior of C and the other endpoint on the boundary of C, then all the intermediate points of the line segment lie in the interior of C. A concept of "relative" interior can be introduced, so that this fact can be used as a basic tool even in situations where one has to deal with configurations of convex sets whose interiors are empty. This is discussed in §6 (Relative Interiors of Convex Sets). The principal results which every student of convexity should know are embodied in the first four theorems of §6. The rest of §6, starting with Theorem 6.5, is devoted mainly to formulas for the relative interiors of convex sets constructed from other convex sets in various ways. A number of useful results are established (particularly Corollaries 6.5.1 and 6.5.2, which are cited often in the text, and Corollary 6.6.2, which is employed in the proof of an important separation theorem in §11), but these can all be neglected temporarily and referred to as the need arises.

In §7 (Closures of Convex Functions) the main topic is lower semi-continuity. This property is in many ways more important than continuity in the case of convex functions, because it relates directly to epigraphs: a function is lower semi-continuous if and only if its epigraph is closed. A convex function which is not already lower semi-continuous can be made so simply by redefining its values (in a uniquely determined manner) at certain boundary points of its effective domain. This leads to the notion of the closure operation for convex functions, which corresponds to the closure operation for epigraphs (as subsets of R^{n+1}) when the functions are proper. All of §7, with the exception of Theorem 7.6, is essential if one is to understand what follows.

All of §8 (Recession Cones and Unboundedness) is also needed in the long run, although the need is not as ubiquitous as in the case of §6 and §7. The first half of §8 elucidates the idea that unbounded convex sets are just like bounded convex sets, except that they have certain "points at infinity." The second half of §8 applies this idea to epigraphs to obtain results about the growth properties of convex functions. Such properties are important in formulating a number of basic existence theorems scattered throughout the book, the first ones occurring in §9 (Some Closedness Criteria).

The question which §9 attempts to answer is this: when is the image of a closed convex set under a linear transformation closed? It turns out that this question is fundamental in investigations of the existence of solutions to various extremum problems. The principal results of §9 are given in Theorems 9.1 and 9.2 (and their corollaries). The reader would do well, however, to skip §9 entirely on the first encounter and return to it later, if desired, in connection with applications in §16.

Only the first theorem of §10 (Continuity of Convex Functions) is basic to convex analysis as a whole. The fancier continuity and convergence theorems are a culmination in themselves. They are used only in §24 and §25 to derive continuity and convergence theorems for subdifferentials and gradient mappings of convex functions, and in §35 to derive similar results in the case of saddle-functions.

Part III: Duality Correspondences

Duality between points and hyperplanes has an important role to play in much of analysis, but nowhere perhaps is the role more remarkable than in convex analysis. The basis of duality in the theory of convexity is, from a geometric point of view, the fact that a closed convex set is the intersection of all the closed half-spaces which contain it. From the point of view of functions, however, it is the fact that a closed convex function is the pointwise supremum of all the affine functions which minorize it. These two facts are equivalent when regarded in terms of epigraphs, and a geometric formulation is usually preferable for the sake of intuition, but in this case both formulations are important. The second formulation of the basis of duality has the advantage that it leads directly to a symmetric one-to-one duality correspondence among closed convex functions, the conjugacy correpsondence of Fenchel.

Conjugacy contains, as a special case in a certain sense, a symmetric one-to-one correspondence among closed convex cones (polarity), but it has no symmetric counterpart in the class of general closed convex sets. The analogous correspondence in the latter context is between convex sets on the one hand and positively homogeneous convex functions (their support functions) on the other. For this reason it is often better in applications, as far as duality is concerned, to express a given situation in terms of convex functions, rather than convex sets. Once this is done, geometric reasoning can still be applied, of course, to epigraphs.

The foundations for the theory of duality are laid in §11 (Separation Theorems). All of the material in this section, except Theorem 11.7, is essential. In §12 (Conjugates of Convex Functions), the conjugacy correspondence is defined, and a number of examples of corresponding functions are given. Theorems 12.1 and 12.2 are the fundamental results which should be known; the rest of §12 is dispensible.

Conjugacy is applied in §13 (Support Functions) to produce results about the duality between convex sets and positively homogeneous convex functions. The support functions of the effective domain and level sets of a convex function f are calculated in terms of the conjugate function f^* and its recession function. The main facts are stated in Theorems 13.2, 13.3, and 13.5, the last two presupposing familiarity with §8. The other theorems, as well as all the corollaries, can be skipped over and referred to if and when they are needed.

In §14 (Polars of Convex Sets), the conjugacy correspondence for convex functions is specialized to the polarity correspondence for convex cones, whereupon the latter is generalized to the polarity correspondence for arbitrary closed convex sets containing the origin. Polarity of convex cones has several applications elsewhere in this book, but the more general polarity is not mentioned subsequently, except in §15 (Polars of Convex Functions), where its relationship with the theory of norms is discussed. The purpose of §15, besides the development of Minkowski's duality correspondence for norms and certain of its generalizations, is to provide (in Theorem 15.3 and Corollary 15.3.1) further examples of conjugate convex functions. However, of all of §14 and §15, it would suffice, as long as one was not specifically interested in approximation problems, to read merely Theorem 14.1.

The theorems of §16 (Dual Operations) show that the various functional operations in §5 fall into dual pairs with respect to the conjugacy correspondence. The most significant result is Theorem 16.4, which describes the duality between addition and infimal convolution of convex functions. This result has important consequences for systems of inequalities (§21) and the calculus of subgradients (§23), and therefore for the theory of extremum problems in Part VI. The second halves of Theorems 16.3, 16.4, and 16.5 (which give conditions under which the respective minima are attained and the closure operation is not needed in the duality formulas) depend on §9. This much of the material could be omitted on a first reading of §16, along with Lemma 16.2 and all corollaries.

Part IV: Representation and Inequalities

The objective here is to obtain results about the representation of convex sets as convex hulls of sets of points and directions, and to apply these results to the study of systems of linear and nonlinear inequalities. Most of the material concerns refinements of convexity theory which take special advantage of dimensionality or the presence of some degree of linearity. The reader could skip Part IV entirely without jeopardizing his understanding of the remainder of this book. Or, as a compromise, only the more fundamental material in Part IV, as indicated below, could be covered.

The role of dimensionality in the generation of convex hulls is explored in §17 (Carathéodory's Theorem), the principal facts being given in Theorems 17.1 and 17.2. Problems of representing a given convex set in terms of extreme points, exposed points, extreme directions, exposed directions, and tangent hyperplanes are taken up in §18 (Extreme Points and Faces of Convex Sets). All of §18 is put to use in §19 (Polyhedral Convexity); applications also occur in the study of gradients (§25) and in the maximization of convex functions (§32). The most important results in §19 are Theorems 19.1, 19.2, 19.3, and their corollaries.

In §20 (Some Applications of Polyhedral Convexity), it is shown how certain general theorems of convex analysis can be strengthened in the case where some, but not necessarily all, of the convex sets or functions involved are polyhedral. Theorems 20.1 and 20.2 are used in §21 to establish relatively difficult refinements of Helly's Theorem and certain other results which are applicable in §27 and §28 to the existence of Lagrange multipliers and optimal solutions to convex programs. Theorem 20.1 depends on §9, although Theorem 20.2 does not. However, it is possible to understand the fundamental results of §21 (Helly's Theorem and Systems of Inequalities) and their proofs without knowledge of §20, or even of §18 or §19. In this case one should simply omit Theorems 21.2, 21.4, and 21.5.

Finite systems of equations and linear inequalities, weak or strict, are the topic in §22 (Linear Inequalities). The results are special, and they are not invoked anywhere else in the book. At the beginning, various facts are stated as corollaries of fancy theorems in §21, but then it is demonstrated that the same special facts can be derived, along with some improvements, by an elementary and completely independent method which uses only linear algebra and no convexity theory.

Part V: Differential Theory

Supporting hyperplanes to convex sets can be employed in situations where tangent hyperplanes, in the sense of the classical theory of smooth surfaces, do not exist. Similarly, subgradients of convex functions, which correspond to supporting hyperplanes to epigraphs rather than tangent hyperplanes to graphs, are often useful where ordinary gradients do not exist.

The theory of subdifferentiation of convex functions, expounded in §23 (Directional Derivatives and Subgradients), is a fundamental tool in the analysis of extremum problems, and it should be mastered before proceeding. Theorems 23.6, 23.7, 23.9, and 23.10 may be omitted, but one should definitely be aware of Theorem 23.8, at least in the non-polyhedral case for which an alternative and more elementary proof is given. Most of §23 is independent of Part IV.

The main result about the relationship between subgradients and ordinary gradients of convex functions is established in Theorem 25.1, which can be read immediately following §23. No other result from §24, §25, or §26 is specifically required elsewhere in the book, except in §35, where analogous theorems are proved for saddle-functions. The remainder of Part V thus serves its own purpose.

In §24 (Differential Continuity and Monotonicity), the elementary theory of left and right derivatives of closed proper convex functions of a single variable is developed. It is shown that the graphs of the subdifferentials of such functions may be characterized as "complete non-decreasing curves." Continuity and monoticity properties in the one-dimensional case are then generalized to the n-dimensional case.

Aside from the theorem already referred to above, §25 (Differentiability of Convex Functions) is devoted mainly to proving that, for a finite convex function on an open set, the ordinary gradient mapping exists almost everywhere and is continuous. The question of when the gradient mapping comprises the entire subdifferential mapping, and when it is actually one-to-one, is taken up in §26 (The Legendre Transformation). The central purpose of §26 is to explain the extent to which conjugate convex functions can, in principle, be calculated in a classical manner by inverting a gradient mapping. The duality between smoothness and strict convexity is also discussed. The development in §25 and §26 depends to some extent on §18, but not on any sections of Part IV following §18.

Part VI: Constrained Extremum Problems

The theory of extremum problems is, of course, the source of motivation for many of the results in this book. It is in §27 (The Minimum of a Convex

Function) that applications to this theory are begun in a systematic way. The stage is set by Theorem 27.1, which summarizes some pertinent facts proved in earlier sections. All the theorems of §27 concern the manner in which a convex function attains its minimum relative to a given convex set, and all should be included in a first reading, except perhaps for refinements which take advantage of polyhedral convexity.

Problems in which a convex function is minimized subject to a finite system of convex inequalities are studied in §28 (Ordinary Convex Programs and Lagrange Multipliers). The emphasis is on the existence, interpretation, and characterization of certain vectors of Lagrange multipliers, called Kuhn-Tucker vectors. The text may be simplified somewhat by deleting the provisions for linear equation constraints, and Theorem 28.2 may be replaced by its special case Corollary 28.2.1 (which has a much easier proof), but beyond this nothing other than examples ought to be omitted.

The theory of Lagrange multipliers is broadened and in some ways sharpened in §29 (Bifunctions and Generalized Convex Programs). The concept of a convex bifunction, which can be regarded as an extension of that of a linear transformation, is used to construct a theory of perturbations of minimization problems. Generalized Kuhn-Tucker vectors measure the effects of the perturbations. Theorems 29.1, 29.3, and their corollaries contain all the facts needed in the sequel.

In §30 (Adjoint Bifunctions and Dual Programs) the duality theory of extremum problems is set forth. Practically everything up through Theorem 30.5 is fundamental, but the remainder of §30 consists of examples and may be truncated as desired. Duality theory is continued in §31 (Fenchel's Duality Theorem). The primary purpose of §31 is to furnish additional examples interesting for their applications. Later sections do not depend on the material in §31, except for §38.

Results of a rather different character are described in §32 (The Maximum of a Convex Function). The proofs of these results involve none of the preceding sections of Part VI, but familiarity with §18 and §19 is required. No subsequent reference is made to §32.

Part VII: Saddle-functions and Minimax Theory

Saddle-functions are functions which are convex in some variables and concave in others, and the extremum problems naturally associated with them involve "minimaximization," rather than simple minimization or maximization. The theory of such minimax problems can be developed by much the same approach as in the case of minimization of convex functions. It turns out that the general minimax problems for (suitably regularized) saddle-functions are precisely the Lagrangian saddle-point problems associated with generalized (closed) convex programs. Understandably, therefore, convex bifunctions are central to the discussion of saddle-functions, and the reader should not proceed without already being familiar with the basic ideas in §29 and §30.

Saddle-functions on $R^m \times R^n$ correspond to convex bifunctions from R^m to R^n in much the same way that bilinear functions on $R^m \times R^n$ correspond to linear transformations from R^m to R^n. This is the substance of §33 (Saddle-functions). In §34 (Closures and Equivalence Classes), certain closure operations for saddle-functions similar to the one for convex functions are studied. It is shown that each finite saddle-function defined on a product of convex sets in $R^m \times R^n$ determines a unique equivalence class of closed saddle-functions defined on all of $R^m \times R^n$, but one does not actually have to read up on the latter fact (embodied in Theorems 34.4 and 34.5) before passing to minimax theory itself.

The results about saddle-functions proved in §35 (Continuity and Differentiability) are mainly analogues or extensions of results about convex functions in §10, §24, and §25, and they are not a prerequisite for what follows.

Saddle-points and saddle-values are discussed in §36 (Minimax Problems). It is then explained how the study of these can be reduced to the study of convex and concave programs dual to each other. Existence theorems for saddle-points and saddle-values are then derived in §37 (Conjugate Saddle-functions and Minimax Theorems) in terms of a conjugacy correspondence for saddle-functions and the "inverse" operation for bifunctions.

Part VIII: Convex Algebra

The analogy between convex bifunctions and linear transformations, which features so prominently in Parts VI and VII, is pursued further in §38 (The Algebra of Bifunctions). "Addition" and "multiplication" of bifunctions are studied in terms of a generalized notion of inner product based on Fenchel's Duality Theorem. It is a remarkable and non-trivial fact that such natural operations for bifunctions are preserved, as in linear algebra, when adjoints are taken.

The results about bifunctions in §38 are specialized in §39 (Convex Processes) to a class of convex-set-valued mappings which are even more analogous to linear transformations.

Part 1 · Basic Concepts

SECTION 1

Affine Sets

Throughout this book, R denotes the real number system, and R^n is the usual vector space of real n-tuples $x = (\xi_1, \ldots, \xi_n)$. Everything takes place in R^n unless otherwise specified. The inner product of two vectors x and x^* in R^n is expressed by

$$\langle x, x^* \rangle = \xi_1 \xi_1^* + \cdots + \xi_n \xi_n^*$$

The same symbol A is used to denote an $m \times n$ real matrix A and the corresponding linear transformation $x \to Ax$ from R^n to R^m. The transpose matrix and the corresponding adjoint linear transformation from R^m to R^n are denoted by A^*, so that one has the identity

$$\langle Ax, y^* \rangle = \langle x, A^*y^* \rangle.$$

(In a symbol denoting a vector, $*$ has no operational significance; all vectors are to be regarded as column vectors for purposes of matrix multiplication. Vector symbols involving $*$ are used from time to time merely to bring out the familiar duality between vectors considered as points and vectors considered as the coefficient n-tuples of linear functions.) The end of a proof is signalled by $\|$.

If x and y are different points in R^n, the set of points of the form

$$(1 - \lambda)x + \lambda y = x + \lambda(y - x), \qquad \lambda \in R,$$

is called the *line through x and y*. A subset M of R^n is called an *affine set* if $(1 - \lambda)x + \lambda y \in M$ for every $x \in M$, $y \in M$ and $\lambda \in R$. (Synonyms for "affine set" used by other authors are "affine manifold," "affine variety," "linear variety" or "flat.")

The empty set $\emptyset$ and the space R^n itself are extreme examples of affine sets. Also covered by the definition is the case where M consists of a solitary point. In general, an affine set has to contain, along with any two different points, the entire line through those points. The intuitive picture is that of an endless uncurved structure, like a line or a plane in space.

The formal geometry of affine sets may be developed from the theorems of linear algebra about subspaces of R^n. The exact correspondence between affine sets and subspaces is described in the two theorems which follow.

3

THEOREM 1.1. *The subspaces of R^n are the affine sets which contain the origin.*

PROOF.   Every subspace contains 0 and, being closed under addition and scalar multiplication, is in particular an affine set.

Conversely, suppose M is an affine set containing 0. For any $x \in M$ and $\lambda \in R$, we have

$$\lambda x = (1 - \lambda)0 + \lambda x \in M,$$

so M is closed under scalar multiplication. Now, if $x \in M$ and $y \in M$, we have

$$\tfrac{1}{2}(x + y) = \tfrac{1}{2}x + (1 - \tfrac{1}{2})y \in M,$$

and hence

$$x + y = 2(\tfrac{1}{2}(x + y)) \in M.$$

Thus M is also closed under addition and is a subspace. ‖

For $M \subset R^n$ and $a \in R^n$, the *translate* of M by a is defined to be the set

$$M + a = \{x + a \mid x \in M\}.$$

A translate of an affine set is another affine set, as is easily verified.

An affine set M is said to be *parallel* to an affine set L if $M = L + a$ for some a. Evidently "M is parallel to L" is an equivalence relation on the collection of affine subsets of R^n. Note that this definition of parallelism is more restrictive than the everyday one, in that it does not include the idea of a line being parallel to a plane. One has to speak of a line which is parallel to another line within a given plane, and so forth.

THEOREM 1.2. *Each non-empty affine set M is parallel to a unique subspace L. This L is given by*

$$L = M - M = \{x - y \mid x \in M, y \in M\}.$$

PROOF.   Let us show first that M cannot be parallel to two different subspaces. Subspaces L_1 and L_2 parallel to M would be parallel to each other, so that $L_2 = L_1 + a$ for some a. Since $0 \in L_2$, we would then have $-a \in L_1$, and hence $a \in L_1$. But then $L_1 \supset L_1 + a = L_2$. By a similar argument $L_2 \supset L_1$, so $L_1 = L_2$. This establishes the uniqueness. Now observe that, for any $y \in M$, $M - y = M + (-y)$ is a translate of M containing 0. By Theorem 1.1 and what we have just proved, this affine set must be the unique subspace L parallel to M. Since $L = M - y$ no matter which $y \in M$ is chosen, we actually have $L = M - M$. ‖

The *dimension* of a non-empty affine set is defined as the dimension of the subspace parallel to it. (The dimension of $\emptyset$ is -1 by convention.) Naturally, affine sets of dimension 0, 1 and 2 are called *points*, *lines* and *planes*, respectively. An $(n - 1)$-dimensional affine set in R^n is called a *hyperplane.* Hyperplanes are very important, because they play a role dual to the role of points in n-dimensional geometry.

Hyperplanes and other affine sets may be represented by linear functions and linear equations. It is easy to deduce this from the theory of orthogonality in R^n. Recall that, by definition, $x \perp y$ means $\langle x, y \rangle = 0$. Given a subspace L of R^n, the set of vectors x such that $x \perp L$, i.e. $x \perp y$ for every $y \in L$, is called the *orthogonal complement* of L, denoted $L^\perp$. It is another subspace, of course, and

$$\dim L + \dim L^\perp = n.$$

The orthogonal complement $(L^\perp)^\perp$ of $L^\perp$ is in turn L. If $b_1, \ldots, b_m$ is a basis for L, then $x \perp L$ is equivalent to the condition that $x \perp b_1, \ldots, x \perp b_m$. In particular, the $(n-1)$-dimensional subspaces of R^n are the orthogonal complements of the one-dimensional subspaces, which are the subspaces L having a basis consisting of a single non-zero vector b (unique up to a non-zero scalar multiple). Thus the $(n-1)$-dimensional subspaces are the sets of the form $\{x \mid x \perp b\}$, where $b \neq 0$. The hyperplanes are the translates of these. But

$$\{x \mid x \perp b\} + a = \{x + a \mid \langle x, b \rangle = 0\}$$
$$= \{y \mid \langle y - a, b \rangle = 0\} = \{y \mid \langle y, b \rangle = \beta\},$$

where $\beta = \langle a, b \rangle$. This leads to the following characterization of hyperplanes.

THEOREM 1.3. *Given $\beta \in R$ and a non-zero $b \in R^n$, the set*

$$H = \{x \mid \langle x, b \rangle = \beta\}$$

is a hyperplane in R^n. Moreover, every hyperplane may be represented in this way, with b and β unique up to a common non-zero multiple.

In Theorem 1.3, the vector b is called a *normal* to the hyperplane H. Every other normal to H is either a positive or a negative scalar multiple of b. A good interpretation of this is that every hyperplane has "two sides," like one's picture of a line in R^2 or a plane in R^3. Note that a plane in R^4 would *not* have "two sides," any more than a line in R^3 has.

The next theorem characterizes the affine subsets of R^n as the solution sets to systems of simultaneous linear equations in n variables.

THEOREM 1.4. *Given $b \in R^m$ and an $m \times n$ real matrix B, the set*

$$M = \{x \in R^n \mid Bx = b\}$$

is an affine set in R^n. Moreover, every affine set may be represented in this way.

PROOF. If $x \in M$, $y \in M$ and $\lambda \in R$, then for $z = (1 - \lambda)x + \lambda y$ one has

$$Bz = (1 - \lambda)Bx + \lambda By = (1 - \lambda)b + \lambda b = b,$$

so $z \in M$. Thus the given M is affine.

On the other hand, starting with an arbitrary non-empty affine set M other than R^n itself, let L be the subspace parallel to M. Let $b_1, \ldots, b_m$ be a basis for $L^\perp$. Then

$$L = (L^\perp)^\perp = \{x \,|\, x \perp b_1, \ldots, x \perp b_m\}$$
$$= \{x \,|\, \langle x, b_i \rangle = 0, \quad i = 1, \ldots, m\} = \{x \,|\, Bx = 0\},$$

where B is the $m \times n$ matrix whose rows are $b_1, \ldots, b_m$: Since M is parallel to L, there exists an $a \in R^n$ such that

$$M = L + a = \{x \,|\, B(x - a) = 0\} = \{x \,|\, Bx = b\},$$

where $b = Ba$. (The affine sets R^n and $\emptyset$ can be represented in the form in the theorem by taking B to be the $m \times n$ zero matrix, say, with $b = 0$ in the case of R^n and $b \neq 0$ in the case of $\emptyset$.) ‖

Observe that in Theorem 1.4 one has

$$M = \{x \,|\, \langle x, b_i \rangle = \beta_i, i = 1, \ldots, m\} = \bigcap_{i=1}^m H_i,$$

where b_i is the ith row of B, β_i is the ith component of b, and

$$H_i = \{x \,|\, \langle x, b_i \rangle = \beta_i\}.$$

Each H_i is a hyperplane ($b_i \neq 0$), or the empty set ($b_i = 0$, $\beta_i \neq 0$), or R^n ($b_i = 0$, $\beta_i = 0$). The empty set may itself be regarded as the intersection of two different parallel hyperplanes, while R^n may be regarded as the intersection of the empty collection of hyperplanes of R^n. Thus:

COROLLARY 1.4.1. *Every affine subset of R^n is an intersection of a finite collection of hyperplanes.*

The affine set M in Theorem 1.4 can be expressed in terms of the vectors $b_1', \ldots, b_n'$ which form the columns of B by

$$M = \{x = (\xi_1, \ldots, \xi_n) \,|\, \xi_1 b_1' + \cdots + \xi_n b_n' = b\}.$$

Obviously, the intersection of an arbitrary collection of affine sets is again affine. Therefore, given any $S \subset R^n$ there exists a unique smallest affine set containing S (namely, the intersection of the collection of affine sets M such that $M \supset S$). This set is called the *affine hull* of S and is denoted by aff S. It can be proved, as an exercise, that aff S consists of all the vectors of the form $\lambda_1 x_1 + \cdots + \lambda_m x_m$, such that $x_i \in S$ and $\lambda_1 + \cdots + \lambda_m = 1$.

A set of $m + 1$ points $b_0, b_1, \ldots, b_m$ is said to be *affinely independent* if aff $\{b_0, b_1, \ldots, b_m\}$ is m-dimensional. Of course

$$\text{aff } \{b_0, b_1, \ldots, b_m\} = L + b_0,$$

where

$$L = \text{aff } \{0, b_1 - b_0, \ldots, b_m - b_0\}.$$

By Theorem 1.1, L is the same as the smallest subspace containing $b_1 - b_0, \ldots, b_m - b_0$. Its dimension is m if and only if these vectors are linearly independent. Thus $b_0, b_1, \ldots, b_m$ are affinely independent if and only if $b_1 - b_0, \ldots, b_m - b_0$ are linearly independent.

All the facts about linear independence can be applied to affine independence in the obvious way. For instance, any affinely independent set of $m + 1$ points in R^n can be enlarged to an affinely independent set of $n + 1$ points. An m-dimensional affine set M can be expressed as the affine hull of $m + 1$ points (translate the points which correspond to a basis of the subspace parallel to M).

Note that, if $M = \text{aff } \{b_0, b_1, \ldots, b_m\}$, the vectors in the subspace L parallel to M are the linear combinations of $b_1 - b_0, \ldots, b_m - b_0$. The vectors in M are therefore those expressible in the form

$$x = \lambda_1(b_1 - b_0) + \cdots + \lambda_m(b_m - b_0) + b_0,$$

i.e. in the form

$$x = \lambda_0 b_0 + \lambda_1 b_1 + \cdots + \lambda_m b_m, \qquad \lambda_0 + \lambda_1 + \cdots + \lambda_m = 1.$$

The coefficients in such an expression of x are unique if and only if $b_0, b_1, \ldots, b_m$ are affinely independent. In that event, $\lambda_0, \lambda_1, \ldots, \lambda_m$, as parameters, define what is called a *barycentric coordinate system* for M.

A single-valued mapping $T: x \to Tx$ from R^n to R^m is called an *affine transformation* if

$$T((1 - \lambda)x + \lambda y) = (1 - \lambda)Tx + \lambda Ty$$

for every x and y in R^n and $\lambda \in R$.

THEOREM 1.5.   *The affine transformations from R^n to R^m are the mappings T of the form $Tx = Ax + a$, where A is a linear transformation and $a \in R^m$.*

PROOF.   If T is affine, let $a = T0$ and $Ax = Tx - a$. Then A is an affine transformation with $A0 = 0$. A simple argument resembling the one in Theorem 1.1 shows that A is actually linear.

Conversely, if $Tx = Ax + a$ where A is linear, one has

$$T((1 - \lambda)x + \lambda y) = (1 - \lambda)Ax + \lambda Ay + a = (1 - \lambda)Tx + \lambda Ty.$$

Thus T is affine. $\|$

The inverse of an affine transformation, if it exists, is affine.

As an elementary exercise, one can demonstrate that if a mapping T from R^n to R^m is an affine transformation the image set $TM = \{Tx \mid x \in M\}$ is affine in R^m for every affine set M in R^n. In particular, then, affine transformations preserve affine hulls:

$$\text{aff}\,(TS) = T(\text{aff } S).$$

THEOREM 1.6. *Let* $\{b_0, b_1, \ldots, b_m\}$ *and* $\{b_0', b_1', \ldots, b_m'\}$ *be affinely independent sets in* R^n. *Then there exists a one-to-one affine transformation* T *of* R^n *onto itself, such that* $Tb_i = b_i'$ *for* $i = 0, \ldots, m$. *If* $m = n$, T *is unique.*

PROOF. Enlarging the given affinely independent sets if necessary, we can reduce the question to the case where $m = n$. Then, as is well known in linear algebra, there exists a unique one-to-one linear transformation A of R^n onto itself carrying the basis $b_1 - b_0, \ldots, b_n - b_0$ of R_n onto the basis $b_1' - b_0', \ldots, b_n' - b_0'$. The desired affine transformation is then given by $Tx = Ax + a$, where $a = b_0' - Ab_0$. ‖

COROLLARY 1.6.1. *Let* M_1 *and* M_2 *be any two affine sets in* R^n *of the same dimension. Then there exists a one-to-one affine transformation* T *of* R^n *onto itself such that* $TM_1 = M_2$.

PROOF. Any m-dimensional affine set can be expressed as the affine hull of an affinely independent set of $m + 1$ points, and affine hulls are preserved by affine transformations. ‖

The graph of an affine transformation T from R^n to R^m is an affine subset of R^{n+m}. This follows from Theorem 1.4, for if $Tx = Ax + a$ the graph of T consists of the vectors $z = (x, y)$, $x \in R^n$ and $y \in R^m$, such that $Bz = b$, where $b = -a$ and B is the linear transformation $(x, y) \to Ax - y$ from R^{n+m} to R^m.

In particular, the graph of a linear transformation $x \to Ax$ from R^n to R^m is an affine set containing the origin of R^{n+m}, and hence it is a certain subspace L of R^{n+m} (Theorem 1.1). The orthogonal complement of L is then given by

$$L^\perp = \{(x^*, y^*) \mid x^* \in R^n, y^* \in R^m, x^* = -A^*y^*\},$$

i.e. $L^\perp$ is the graph of $-A^*$. Indeed, $z^* = (x^*, y^*)$ belongs to $L^\perp$ if and only if

$$0 = \langle z, z^* \rangle = \langle x, x^* \rangle + \langle y, y^* \rangle$$

for every $z = (x, y)$ with $y = Ax$. In other words, $(x^*, y^*) \in L^\perp$ if and only if

$$0 = \langle x, x^* \rangle + \langle Ax, y^* \rangle = \langle x, x^* \rangle + \langle x, A^*y^* \rangle = \langle x, x^* + A^*y^* \rangle$$

for every $x \in R^n$. That means $x^* + A^*y^* = 0$, i.e. $x^* = -A^*y^*$.

Any non-trivial affine set can be represented in various ways as the graph of an affine transformation. Let M be an n-dimensional affine set in R^N with $0 < n < N$. First of all, one can express M as the set of vectors $x = (\xi_1, \ldots, \xi_N)$ whose coordinates satisfy a certain linear system of equations,

$$\beta_{i1}\xi_1 + \cdots + \beta_{iN}\xi_N = \beta_i, \qquad i = 1, \ldots, k.$$

This is always possible, according to Theorem 1.4. The n-dimensionality of M means that the coefficient matrix $B = (\beta_{ij})$ has nullity n and rank $m = N - n$. One can therefore solve the system of equations for $\xi_{\overline{n+1}}, \ldots, \xi_{\overline{N}}$ in terms of $\xi_{\overline{1}}, \ldots, \xi_{\overline{n}}$, where $\overline{1}, \ldots, \overline{N}$ is some permutation of the indices $1, \ldots, N$. One obtains then a system of the special form

$$\xi_{\overline{n+i}} = \alpha_{i1}\xi_{\overline{1}} + \cdots + \alpha_{in}\xi_{\overline{n}} + \alpha_i, \qquad i = 1, \ldots, m,$$

which again gives a necessary and sufficient condition for a vector $x = (\xi_1, \ldots, \xi_N)$ to belong to M. This system is called a *Tucker representation* of the given affine set. It expresses M as the graph of a certain affine transformation from R^n to R^m. There are only finitely many Tucker representations of M (at most $N!$, corresponding to the various ways m of the coordinate variables ξ_i of vectors in M can be expressed in terms of the other n coordinate variables in some particular order).

Often a theorem involving an affine set can be interpreted as a theorem about "linear systems of variables," in the sense that the affine set may be given a Tucker representation. This is important, for example, in certain results in the theory of linear inequalities (Theorems 22.6 and 22.7) and in certain applications of Fenchel's Duality Theorem (Corollary 31.4.2).

The Tucker representations of a subspace L are, of course, of the homogeneous form

$$\xi_{\overline{n+1}} = \alpha_{i1}\xi_{\overline{1}} + \cdots + \alpha_{in}\xi_{\overline{n}}, \qquad i = 1, \ldots, m.$$

Given such a representation of L as the graph of a linear transformation, we know, as pointed out above, that $L^\perp$ corresponds to the graph of the negative of the adjoint transformation. Thus $x^* = (\xi_1^*, \ldots, \xi_N^*)$ belongs to $L^\perp$ if and only if

$$-\xi_{\overline{j}}^* = \xi_{\overline{n+i}}^*\alpha_{1j} + \cdots + \xi_{\overline{n+m}}^*\alpha_{mj}, \qquad j = 1, \ldots, n.$$

This furnishes a Tucker representation of $L^\perp$. Thus there is a simple and useful one-to-one correspondence between the Tucker representations of a given subspace and those of its orthogonal complement.

SECTION 2

Convex Sets and Cones

A subset C of R^n is said to be *convex* if $(1 - \lambda)x + \lambda y \in C$ whenever $x \in C$, $y \in C$ and $0 < \lambda < 1$. All affine sets (including $\emptyset$ and R^n itself) are convex. What makes convex sets more general than affine sets is that they only have to contain, along with any two distinct points x and y, a certain portion of the line through x and y, namely

$$\{(1 - \lambda)x + \lambda y \mid 0 \leq \lambda \leq 1\}.$$

This portion is called the *(closed) line segment between x and y*. Solid ellipsoids and cubes in R^3, for instance, are convex but not affine.

Half-spaces are important examples of convex sets. For any non-zero $b \in R^n$ and any $\beta \in R$, the sets

$$\{x \mid \langle x, b \rangle \leq \beta\}, \qquad \{x \mid \langle x, b \rangle \geq \beta\},$$

are called *closed half-spaces*. The sets

$$\{x \mid \langle x, b \rangle < \beta\}, \qquad \{x \mid \langle x, b \rangle > \beta\},$$

are called *open half-spaces*. All four sets are plainly non-empty and convex. Notice that the same quartet of half-spaces would appear if b and β were replaced by λb and $\lambda \beta$ for some $\lambda \neq 0$. Thus these half-spaces depend only on the hyperplane $H = \{x \mid \langle x, b \rangle = \beta\}$ (Theorem 1.3). One may speak unambiguously, therefore, of the open and closed half-spaces corresponding to a given hyperplane.

THEOREM 2.1. *The intersection of an arbitrary collection of convex sets is convex.*

PROOF. Elementary. ‖

COROLLARY 2.1.1. *Let $b_i \in R^n$ and $\beta_i \in R$ for $i \in I$, where I is an arbitrary index set. Then the set*

$$C = \{x \in R^n \mid \langle x, b_i \rangle \leq \beta_i, \forall i \in I\}$$

is convex.

PROOF. Let $C_i = \{x \mid \langle x, b_i \rangle \leq \beta_i\}$. Then C_i is a closed half-space or R^n or $\emptyset$ and $C = \bigcap_{i \in I} C_i$. ‖

The conclusion of the corollary would still be valid, of course, if some of the inequalities $\leq$ were replaced by $\geq$, $>$, $<$ or $=$. Thus, given any system of simultaneous linear inequalities and equations in n variables, the set C of solutions is a convex set in R^n. This is a significant fact both in theory and in applications.

Corollary 2.1.1 will be generalized by Corollary 4.6.1.

A set which can be expressed as the intersection of *finitely* many closed half spaces of R^n is called a *polyhedral* convex set. Such sets are considerably better behaved than general convex sets, mostly because of their lack of "curvature." The special theory of polyhedral convex sets will be treated briefly in §19. It is applicable, of course, to the study of *finite* systems of simultaneous linear equations and weak linear inequalities.

A vector sum

$$\lambda_1 x_1 + \cdots + \lambda_m x_m$$

is called a *convex combination* of $x_1, \ldots, x_m$ if the coefficients λ_i are all non-negative and $\lambda_1 + \cdots + \lambda_m = 1$. In many situations where convex combinations occur in applied mathematics, $\lambda_1, \ldots, \lambda_m$ can be interpreted as probabilities or proportions. For instance, if m particles with masses $\alpha_1, \ldots, \alpha_m$ are located at points $x_1, \ldots, x_m$ of R^3, the center of gravity of the system is the point $\lambda_1 x_1 + \cdots + \lambda_m x_m$, where $\lambda_i = \alpha_i/(\alpha_1 + \cdots + \alpha_m)$. In this convex combination, λ_i is the proportion of the total weight which is at x_i.

THEOREM 2.2. *A subset of R^n is convex if and only if it contains all the convex combinations of its elements.*

PROOF. Actually, by definition, a set C is convex if and only if $\lambda_1 x_1 + \lambda_2 x_2 \in C$ whenever $x_1 \in C$, $x_2 \in C$, $\lambda_1 \geq 0$, $\lambda_2 \geq 0$ and $\lambda_1 + \lambda_2 = 1$. In other words, the convexity of C means that C is closed under taking convex combinations with $m = 2$. We must show that this implies C is also closed under taking convex combinations with $m > 2$. Take any $m > 2$, and make the induction hypothesis that C is closed under taking all convex combinations of fewer than m vectors. Given a convex combination $x = \lambda_1 x_1 + \cdots + \lambda_m x_m$ of elements of C, at least one of the scalars λ_i differs from 1 (since otherwise $\lambda_1 + \cdots + \lambda_m = m \neq 1$); let it be λ_1 for convenience. Let

$$y = \lambda_2' x_2 + \cdots + \lambda_m' x_m, \quad \lambda_i' = \lambda_i/(1 - \lambda_1).$$

Then $\lambda_i' \geq 0$ for $i = 2, \ldots, m$, and

$$\lambda_2' + \cdots + \lambda_m' = (\lambda_2 + \cdots + \lambda_m)/(\lambda_2 + \cdots + \lambda_m) = 1.$$

Thus y is a convex combination of $m - 1$ elements of C, and $y \in C$ by induction. Since $x = (1 - \lambda_1)y + \lambda_1 x_1$, it now follows that $x \in C$. $\|$

The intersection of all the convex sets containing a given subset S of R^n is called the *convex hull* of S and is denoted by conv S. It is a convex set by Theorem 2.1, the unique smallest one containing S.

THEOREM 2.3. *For any* $S \subset R^n$, *conv* S *consists of all the convex combinations of the elements of S.*

PROOF. The elements of S belong to conv S, so all their convex combinations belong to conv S by Theorem 2.2. On the other hand, given two convex combinations $x = \lambda_1 x_1 + \cdots + \lambda_m x_m$ and $y = \mu_1 y_1 + \cdots + \mu_r y_r$, where $x_i \in S$ and $y_j \in S$. The vector

$$(1 - \lambda)x + \lambda y$$

$$= (1 - \lambda)\lambda_1 x_1 + \cdots + (1 - \lambda)\lambda_m x_m + \lambda_1 \mu_1 y_1 + \cdots + \lambda_r \mu_r y_r,$$

where $0 \leq \lambda \leq 1$, is another convex combination of elements of S. Thus the set of convex combinations of elements of S is itself a convex set. It contains S, so it must coincide with the smallest such convex set, conv S. ‖

Actually, it suffices in Theorem 2.3 to consider convex combinations involving $n + 1$ or fewer elements at a time. This important refinement, known as Carathéodory's Theorem, will be proved in §17. Another refinement of Theorem 2.3 will be given in Theorem 3.3.

COROLLARY 2.3.1. *The convex hull of a finite subset* $\{b_0, \ldots, b_m\}$ *of* R^n *consists of all the vectors of the form* $\lambda_0 b_0 + \cdots + \lambda_m b_m$, *with* $\lambda_0 \geq 0, \ldots, \lambda_m \geq 0, \lambda_0 + \cdots + \lambda_m = 1$.

PROOF. Every convex combination of elements selected from $\{b_0, \ldots, b_m\}$ can be expressed as a convex combination of $b_0, \ldots, b_m$ by including the unneeded vectors b_i with zero coefficients. ‖

A set which is the convex hull of finitely many points is called a *polytope*. If $\{b_0, b_1, \ldots, b_m\}$ is affinely independent, its convex hull is called an *m-dimensional simplex*, and $b_0, \ldots, b_m$ are called the *vertices* of the simplex. In terms of barycentric coordinates on aff $\{b_0, b_1, \ldots, b_m\}$, each point of the simplex is *uniquely* expressible as a convex combination of the vertices. The point $\lambda_0 b_0 + \cdots + \lambda_m b_m$ with $\lambda_0 = \cdots = \lambda_m = 1/(1 + m)$ is called the *midpoint* or *barycenter* of the simplex. When $m = 0, 1, 2$ or 3, the simplex is a *point*, *(closed) line segment*, *triangle* or *tetrahedron*, respectively.

In general, by the *dimension* of a convex set C one means the dimension of the affine hull of C. Thus a convex disk is two-dimensional, no matter what the dimension of the space in which it is embedded. (The dimension of an affine set or simplex as already defined agrees with its dimension as a convex set.) The following fact will be used in §6 in proving that a non-empty convex set has a non-empty relative interior.

THEOREM 2.4. *The dimension of a convex set C is the maximum of the dimensions of the various simplices included in C.*

PROOF. The convex hull of any subset of C is included in C. The maximum dimension of the various simplices included in C is thus the largest m such that C contains an affinely independent set of $m + 1$ elements. Let $\{b_0, b_1, \ldots, b_m\}$ be such a set with m maximal, and let M be its affine hull. Then dim $M = m$ and $M \subset$ aff C. Furthermore $C \subset M$, for if $C \setminus M$ contained an element b, the set of $m + 2$ elements $b_0, \ldots, b_m$, b in C would be affinely independent, contrary to the maximality of m. (Namely, aff $\{b_0, \ldots, b_m, b\}$ would include M properly and hence would be more than m-dimensional.) Since aff C is the smallest affine set which includes C, it follows that aff $C = M$ and hence that dim $C = m$. ‖

A subset K of R^n is called a *cone* if it is closed under positive scalar multiplication, i.e. $\lambda x \in K$ when $x \in K$ and $\lambda > 0$. Such a set is a union of half-lines emanating from the origin. The origin itself may or may not be included. A *convex cone* is a cone which is a convex set. (Note: many authors do not call K a convex cone unless, in addition, K contains the origin. Thus for these authors a convex cone is a *non-empty* convex set which is closed under *non-negative* scalar multiplication.)

One should not necessarily think of a convex cone as being "pointed." Subspaces of R^n are in particular convex cones. So are the open and closed half-spaces corresponding to a hyperplane through the origin.

Two of the most important convex cones are the *non-negative orthant* of R^n,

$$\{x = (\xi_1, \ldots, \xi_n) \mid \xi_1 \geq 0, \ldots, \xi_n \geq 0\},$$

and the *positive orthant*

$$\{x = (\xi_1, \ldots, \xi_n) \mid \xi_1 > 0, \ldots, \xi_n > 0\}.$$

These cones are useful in the theory of inequalities. It is customary to write $x \geq x'$ if $x - x'$ belongs to the non-negative orthant, i.e. if

$$\xi_j \geq \xi_j' \quad \text{for} \quad j = 1, \ldots, n.$$

In this notation, the non-negative orthant consists of the vectors x such that $x \geq 0$.

THEOREM 2.5. *The intersection of an arbitrary collection of convex cones is a convex cone.*

PROOF. Elementary. ‖

COROLLARY 2.5.1. *Let $b_i \in R^n$ for $i \in I$, where I is an arbitrary index set. Then*

$$K = \{x \in R^n \mid \langle x, b_i \rangle \leq 0, i \in I\}$$

is a convex cone.

PROOF.   As in Corollary 2.1.1.   ‖

Of course, ≤ 0 may be replaced by $\geq$, $>$, $<$ or $=$ in Corollary 2.5.1. Thus the set of solutions to a system of linear inequalities is a convex cone, rather than merely a convex set, if the inequalities are homogeneous.

The following characterization of convex cones highlights an analogy between convex cones and subspaces.

THEOREM 2.6.   *A subset of R^n is a convex cone if and only if it is closed under addition and positive scalar multiplication.*

PROOF.   Let K be a cone. Let $x \in K$ and $y \in K$. If K is convex, the vector $z = (1/2)(x + y)$ belongs to K, and hence $x + y = 2z \in K$. On the other hand, if K is closed under addition, and if $0 < \lambda < 1$, the vectors $(1 - \lambda)x$ and λy belong to K, and hence $(1 - \lambda)x + \lambda y \in K$. Thus K is convex if and only if it is closed under addition.   ‖

COROLLARY 2.6.1.   *A subset of R^n is a convex cone if and only if it contains all the positive linear combinations of its elements (i.e. linear combinations $\lambda_1 x_1 + \cdots + \lambda_m x_m$ in which the coefficients are all positive).*

COROLLARY 2.6.2.   *Let S be an arbitrary subset of R^n, and let K be the set of all positive linear combinations of S. Then K is the smallest convex cone which includes S.*

PROOF.   Clearly K is closed under addition and positive scalar multiplication, and $K \supset S$. Every convex cone including S must, on the other hand, include K.   ‖

A simpler description is possible when S is convex, as follows.

COROLLARY 2.6.3.   *Let C be a convex set, and let*

$$K = \{\lambda x \mid \lambda > 0, x \in C\}.$$

Then K is the smallest convex cone which includes C.

PROOF.   This follows from the preceding corollary. Namely, every positive linear combination of elements of C is a positive scalar multiple of a convex combination of elements of C and hence is an element of K.   ‖

The convex cone obtained *by adjoining the origin* to the cone in Corollary 2.6.2 (or Corollary 2.6.3) is known as the *convex cone generated by S* (or C) and is denoted by cone S. (Thus the convex cone generated by S is not, under our terminology, the same as the smallest convex cone containing S, unless the latter cone happens to contain the origin.) If $S \neq \emptyset$, cone S consists of all *non-negative* (rather than positive) linear combinations of elements of S. Clearly

$$\text{cone } S = \text{conv (ray } S),$$

where ray S is the union of the origin and the various *rays* (half-lines of the form $\{\lambda y \mid \lambda \geq 0\}$) generated by the non-zero vectors $y \in S$.

Just as an elliptical disk can be regarded as a certain cross-section of a solid circular cone, so can every convex set C in R^n be regarded as a cross-section of some convex cone K in R^{n+1}. Indeed, let K be the convex cone generated by the set of pairs $(1, x)$ in R^{n+1} such that $x \in C$. Then K consists of the origin of R^{n+1} and the pairs $(\lambda, \lambda x)$ such that $\lambda > 0$, $x \in C$. The intersection of K with the hyperplane $\{(\lambda, y) \mid \lambda = 1\}$ can be regarded as C. This fact makes it possible, if one so chooses, to deduce many general theorems about convex sets from the corresponding (usually simpler) theorems about convex cones.

A vector x^* is said to be *normal* to a convex set C at a point a, where $a \in C$, if x^* does not make an acute angle with any line segment in C with a as endpoint, i.e. if $\langle x - a, x^* \rangle \leq 0$ for every $x \in C$. For instance, if C is a half-space $\{x \mid \langle x, b \rangle \leq \beta\}$ and a satisfies $\langle a, b \rangle = \beta$, then b is normal to C at a. In general, the set of all vectors x^* normal to C at a is called the *normal cone* to C at a. The reader can verify easily that this cone is always convex.

Another easily verified example of a convex cone is the *barrier cone* of a convex set C. This is defined as the set of all vectors x^* such that, for some $\beta \in R$, $\langle x, x^* \rangle \leq \beta$ for every $x \in C$.

Each convex cone containing 0 is associated with a pair of subspaces as follows.

THEOREM 2.7. *Let K be a convex cone containing 0. Then there is a smallest subspace containing K, namely*

$$K - K = \{x - y \mid x \in K, y \in K\} = \text{aff } K,$$

and there is a largest subspace contained within K, namely $(-K) \cap K$.

PROOF. By Theorem 2.6, K is closed under addition and positive scalar multiplication. To be a subspace, a set must further contain 0 and be closed under multiplication by -1. Clearly $K - K$ is the smallest such set containing K, and $(-K) \cap K$ is the largest such set contained within K. The former must coincide with aff K, since the affine hull of a set containing 0 is a subspace by Theorem 1.1. ‖

The Algebra of Convex Sets

The class of convex sets is preserved by a rich variety of algebraic operations.

For instance, if C is a convex set in R^n then so is every translate $C + a$ and every *scalar multiple* λC, where

$$\lambda C = \{\lambda x \mid x \in C\}.$$

In geometric terms, if $\lambda > 0$, λC is the image of C under the transformation which expands (or contracts) R^n by the factor λ with the origin fixed.

The *symmetric reflection* of C across the origin is $-C = (-1)C$. A convex set is said to be *symmetric* if $-C = C$. Such a set (if non-empty) must contain the origin, since it must contain along with each vector x, not only $-x$, but the entire line segment between x and $-x$. The non-empty convex cones which are symmetric are the subspaces (Theorem 2.7).

THEOREM 3.1. *If C_1 and C_2 are convex sets in R^n, then so is their sum $C_1 + C_2$, where*

$$C_1 + C_2 = \{x_1 + x_2 \mid x_1 \in C_1, x_2 \in C_2\}.$$

PROOF. Let x and y be points in $C_1 + C_2$. There exist vectors x_1 and y_1 in C_1 and x_2 and y_2 in C_2, such that

$$x = x_1 + x_2, \qquad y = y_1 + y_2.$$

For $0 < \lambda < 1$, one has

$$(1 - \lambda)x + \lambda y = [(1 - \lambda)x_1 + \lambda y_1] + [(1 - \lambda)x_2 + \lambda y_2],$$

and by the convexity of C_1 and C_2

$$(1 - \lambda)x_1 + \lambda y_1 \in C_1, \qquad (1 - \lambda)x_2 + \lambda y_2 \in C_2.$$

Hence $(1 - \lambda)x + \lambda y$ belongs to $C_1 + C_2$. ‖

To illustrate, if C_1 is any convex set and C_2 is the non-negative orthant, then

$$C_1 + C_2 = \{x_1 + x_2 \mid x_1 \in C_1, x_2 \geq 0\}$$
$$= \{x \mid \exists x_1 \in C_1, x_1 \leq x\}.$$

The latter set is thus convex by Theorem 3.1 when C_1 is convex.

The convexity of a set C means by definition that

$$(1 - \lambda)C + \lambda C \subset C, \qquad 0 < \lambda < 1.$$

We shall see in a moment that equality actually holds for convex sets. A set K is a convex cone if and only if $\lambda K \subset K$ for every $\lambda > 0$, and $K + K \subset K$ (Theorem 2.6).

If $C_1, \ldots, C_m$ are convex sets, then so is the linear combination

$$C = \lambda_1 C_1 + \cdots + \lambda_m C_m.$$

Naturally, this C is called a *convex combination* of $C_1, \ldots, C_m$ when $\lambda_1 \geq 0, \ldots, \lambda_m \geq 0$ and $\lambda_1 + \cdots + \lambda_m = 1$. In that case, it is appropriate to think of C geometrically as a sort of mixture of $C_1, \ldots, C_m$. For instance, let C_1 and C_2 be a triangle and a circular disk in R^2. As λ progresses from 0 to 1,

$$C = (1 - \lambda)C_1 + \lambda C_2$$

changes from a triangle to a triangle with rounded corners. The roundness dominates more and more, until ultimately there is just a circular disk.

For the sake of geometric intuition, it is sometimes helpful to regard $C_1 + C_2$ as the union of all the translates $x_1 + C_2$ as x_1 varies over C_1.

What algebraic laws are valid for the addition and scalar multiplication of sets? Trivially, even without convexity being involved, one has

$$C_1 + C_2 = C_2 + C_1,$$

$$(C_1 + C_2) + C_3 = C_1 + (C_2 + C_3),$$

$$\lambda_1(\lambda_2 C) = (\lambda_1 \lambda_2)C,$$

$$\lambda(C_1 + C_2) = \lambda C_1 + \lambda C_2.$$

The convex set consisting of 0 alone is the identity element for the addition operation. Additive inverses do not exist for sets containing more then one point; the best one can say in general is that $0 \in [C + (-C)]$ when $C \neq \emptyset$.

There is at least one important law of set algebra which does depend on convexity, as is shown in the next theorem. The satisfaction of this distributive law is in fact *equivalent* to the convexity of the set C, since the law implies that $\lambda C + (1 - \lambda)C$ is included in C when $0 \leq \lambda \leq 1$.

THEOREM 3.2.   *If C is a convex set and $\lambda_1 \geq 0$, $\lambda_2 \geq 0$, then*

$$(\lambda_1 + \lambda_2)C = \lambda_1 C + \lambda_2 C.$$

PROOF.   The inclusion $\subset$ would be true whether C were convex or not.

The reverse inclusion follows from the convexity relation

$$C \supset (\lambda_1/(\lambda_1 + \lambda_2))C + (\lambda_2/(\lambda_1 + \lambda_2))C$$

upon multiplying through by $\lambda_1 + \lambda_2$, provided $\lambda_1 + \lambda_2 > 0$. If λ_1 or λ_2 is 0, the assertion of the theorem is trivial. ‖

It follows from this theorem, for instance, that $C + C = 2C$, $C + C + C = 3C$, and so forth, when C is convex.

Given any two convex sets C_1 and C_2 in R^n, there is a unique largest convex set included in both C_1 and C_2, namely $C_1 \cap C_2$, and a unique smallest convex set including both C_1 and C_2, namely conv $(C_1 \cup C_2)$. The same is true starting, not just with a pair, but with an arbitrary family $\{C_i, i \in I\}$. In other words, the collection of all convex subsets of R^n is a *complete lattice* under the natural partial ordering corresponding to inclusion.

THEOREM 3.3. *Let* $\{C_i \mid i \in I\}$ *be an arbitrary collection of non-empty convex sets in* R^n, *and let* C *be the convex hull of the union of the collection. Then*

$$C = \bigcup \{\textstyle\sum_{i \in I} \lambda_i C_i\},$$

where the union is taken over all finite convex combinations (i.e. over all non-negative choices of the coefficients λ_i *such that only finitely many are non-zero and these add up to* 1).

PROOF. By Theorem 2.3, C is the set of all convex combinations $x = \mu_1 y_1 + \cdots + \mu_m y_m$, where the vectors $y_1, \ldots, y_m$ belong to the union of the sets C_i. Actually, we can get C just by taking those combinations in which the coefficients are non-zero and vectors are taken from different sets C_i. Indeed, vectors with zero coefficients can be omitted from the combination, and if two of the vectors with positive coefficients belong to the same C_i, say y_1 and y_2, then the term $\mu_1 y_1 + \mu_2 y_2$ can be replaced by μy, where $\mu = \mu_1 + \mu_2$ and

$$y = (\mu_1/\mu)y_1 + (\mu_2/\mu)y_2 \in C_i.$$

Thus C is the union of the finite convex combinations of the form

$$\mu_1 C_{i_1} + \cdots + \mu_m C_{i_m},$$

where the indices $i_1, \ldots, i_m$ are all different. Except for notation, this is the same as the union described in the theorem. ‖

Given any linear transformation A from R^n to R^m, we define

$$AC = \{Ax \mid x \in C\} \quad \text{for} \quad C \subset R^n,$$
$$A^{-1}D = \{x \mid Ax \in D\} \quad \text{for} \quad D \subset R^m,$$

as is customary. We call AC the *image* of C under A and $A^{-1}D$ the *inverse image* of D under A. It turns out that convexity is preserved when such images are taken. (The notation $A^{-1}D$ here is not meant to imply, of course, that the inverse linear transformation exists as a *single-valued* mapping.)

THEOREM 3.4. *Let A be a linear transformation from R^n to R^m. Then AC is a convex set in R^m for every convex set C in R^n, and $A^{-1}D$ is a convex set in R^n for every convex set D in R^m.*

PROOF. An elementary exercise. ‖

COROLLARY 3.4.1. *The orthogonal projection of a convex set C on a subspace L is another convex set.*

PROOF. The orthogonal projection mapping onto L is a linear transformation, the one which assigns to each point x the unique $y \in L$ such that $(x - y) \perp L$. ‖

One interpretation of the convexity of $A^{-1}D$ in Theorem 3.4 is that, as y ranges over a convex set, the solutions x to the system of simultaneous linear equations expressed by $Ax = y$ will range over a convex set too. If $D = K + a$, where K is the non-negative orthant of R^m and $a \in R^m$, then $A^{-1}D$ is the set of vectors x such that $Ax \geq a$, i.e. the solution set to a certain linear inequality system in R^n. If C is the non-negative orthant of R^n, then AC is the set of vectors $y \in R^m$ such that the equation $Ax = y$ has a solution $x \geq 0$.

THEOREM 3.5. *Let C and D be convex sets in R^m and R^p, respectively. Then*

$$C \oplus D = \{x = (y, z) \mid y \in C, z \in D\}$$

is a convex set in R^{m+p}.

PROOF. Trivial. ‖

The set $C \oplus D$ in Theorem 3.5 is called the *direct sum* of C and D. The same name is also applied to an ordinary sum $C + D$, $C \subset R^n$, $D \subset R^n$, if each vector $x \in C + D$ can be expressed *uniquely* in the form $x = y + z$, $y \in C$, $z \in D$. This happens if and only if the symmetric convex sets $C - C$ and $D - D$ have only the zero vector of R^n in common. (It can be shown that then R^n may be expressed as a direct sum of two subspaces, one containing C and the other containing D.)

THEOREM 3.6. *Let C_1 and C_2 be convex sets in R^{m+p}, and let C be the set of vectors $x = (y, z)$ (where $y \in R^m$ and $z \in R^p$) such that there exist vectors z_1 and z_2 with $(y, z_1) \in C_1$, $(y, z_2) \in C_2$ and $z_1 + z_2 = z$. Then C is a convex set in R^{m+p}.*

PROOF.   Let $(y, z) \in C$, with z_1 and z_2 as indicated. Likewise (y', z'), z_1' and z_2'. Then, for $0 \leq \lambda \leq 1$, $y'' = (1 - \lambda)y + \lambda y'$ and $z'' = (1 - \lambda)z + \lambda z'$, we have

$$(y'', (1 - \lambda)z_1 + \lambda z_1') = (1 - \lambda)(y, z_1) + \lambda(y', z_1') \in C_1,$$

$$(y'', (1 - \lambda)z_2 + \delta z_2') = (1 - \lambda)(y, z_2) + \lambda(y', z_2') \in C_2,$$

$$z'' = (1 - \lambda)(z_1 + z_2) + \lambda(z_1' + z_2')$$

$$= ((1 - \lambda)z_1 + \lambda z_1') + ((1 - \lambda)z_2 + \lambda z_2').$$

Thus the vector

$$(1 - \lambda)(y, z) + \lambda(y', z') = (y'', z'')$$

belongs to C.   ‖

Observe that Theorem 3.6 describes a certain *commutative* and *associative* operation for convex sets in R^{m+p}. Now there are infinitely many ways of introducing a linear coordinate system on R^n and then representing every vector as a pair of components $y \in R^m$ and $z \in R^p$ relative to the co-ordinates. Each of these ways yields an operation of the type in Theorem 3.6. (The operations are different if the corresponding decompositions of R^n into a direct sum of two subspaces are different.) An operation of this type will be called a *partial addition*. Ordinary addition (i.e. the operation of forming $C_1 + C_2$) can be regarded as the extreme case corresponding to $m = 0$ in Theorem 3.6, while intersection (i.e. the operation of forming $C_1 \cap C_2$) corresponds similarly to $p = 0$. Between these extremes are infinitely many partial additions for the collection of all convex sets in R^n, and each is a commutative, associative binary operation.

The infinitely many operations just mentioned seem rather arbitrary in character. But, by more special considerations, we can single out four of these operations as the "natural" ones. Recall that, corresponding to each convex set C in R^n, there is a certain convex cone K in R^{n+1} containing the origin and having a cross-section identifiable with C, namely the convex cone generated by $\{(1, x) \mid x \in C\}$. The correspondence is one-to-one. The class of cones K forming the range of the correspondence consists precisely of the convex cones which have only $(0, 0)$ in common with the half-space $\{(\lambda, x) \mid \lambda \leq 0\}$. An operation which preserves this class of cones in R^{n+1} corresponds to an operation for convex sets in R^n. The decomposition of R^{n+1} into pairs (λ, x) focuses our attention on four partial additions in R^{n+1}. These are the operation of adding in the x argument alone, the operation of adding in the λ argument alone, and the two extreme cases of partial addition, namely the operations of adding in both λ and x, and of adding in neither. All four operations clearly do preserve the special class of convex cones K in question.

Let us see what four operations for convex sets these partial additions amount to. Suppose K_1 and K_2 correspond to the convex sets C_1 and C_2 respectively. If we perform partial addition in the x argument alone on K_1 and K_2, $(1, x)$ will belong to the resulting K if and only if $x = x_1 + x_2$ for some $(1, x_1) \in K_1$ and $(1, x_2) \in K_2$. Thus the convex set corresponding to K will be $C = C_1 + C_2$. If we perform partial addition in both arguments, $(1, x)$ will belong to K if and only if $x = x_1 + x_2$ and $1 = \lambda_1 + \lambda_2$ for some $(\lambda_1, x_1) \in K_1$ and $(\lambda_2, x_2) \in K_2$. Thus C will be the union of the sets $\lambda_1 C_1 + \lambda_2 C_2$ over $\lambda_1 \geq 0$, $\lambda_2 \geq 0$, $\lambda_1 + \lambda_2 = 1$, and this is conv $(C_1 \cup C_2)$ by Theorem 3.3. Adding in neither λ nor x is the same as intersecting K_1 and K_2, which obviously corresponds to forming $C_1 \cap C_2$. The remaining operation is addition in λ alone. Here $(1, x) \in K$ if and only if $(\lambda_1, x) \in K_1$ and $(\lambda_2, x) \in K_2$ for some $\lambda_1 \geq 0$ and $\lambda_2 \geq 0$ with $\lambda_1 + \lambda_2 = 1$. Thus

$$C = \bigcup \{\lambda_1 C_1 \cap \lambda_2 C_2 \mid \lambda_i \geq 0, \lambda_1 + \lambda_2 = 1\}$$
$$= \bigcup \{(1 - \lambda)C_1 \cap \lambda C_2 \mid 0 \leq \lambda \leq 1\}.$$

We shall denote this set by $C_1 \# C_2$. The operation $\#$ will be called *inverse addition*.

THEOREM 3.7. *If C_1 and C_2 are convex sets in R^n, then so is their inverse sum $C_1 \# C_2$.*

PROOF. By the preceding remarks. ‖

Inverse addition is a commutative, associative binary operation for the collection of all convex sets in R^n. It resembles ordinary addition in that it can be expressed in terms of a pointwise operation. To show this, we note first that $C_1 \# C_2$ consists of all the vectors x which can be expressed in the form

$$x = \lambda x_1 = (1 - \lambda)x_2, \quad 0 \leq \lambda \leq 1, \quad x_1 \in C_1, \quad x_2 \in C_2.$$

Such an expression requires that x_1, x_2 and x lie along some common ray $\{\alpha e \mid \alpha \geq 0\}$, $e \neq 0$. Then, in fact, for some $\alpha_1 \geq 0$ and $\alpha_2 \geq 0$, one has $x_1 = \alpha_1 e$, $x_2 = \alpha_2 e$ and

$$x = [\alpha_1 \alpha_2 / (\alpha_1 + \alpha_2)]e = (\alpha_1^{-1} + \alpha_2^{-1})^{-1}e.$$

(The last coefficient may be interpreted as 0 if $\alpha_1 = 0$ or $\alpha_2 = 0$.) Here x actually depends only on x_1 and x_2, not on the choice of e. We might call it the inverse sum of x_1 and x_2 and denote it by $x_1 \# x_2$. Inverse addition of vectors is commutative and associative to the extent that it is defined, which is only for vectors on a common ray. We have

$$C_1 \# C_2 = \{x_1 \# x_2 \mid x_1 \in C_1, x_2 \in C_2\}$$

in parallel with the formula for $C_1 + C_2$.

All the operations we have been discussing clearly preserve the class of all convex cones in R^n, except for the operation of translation. Thus the sets $K_1 + K_2$, $K_1 \# K_2$, conv $(K_1 \cup K_2)$, $K_1 \cap K_2$, $K_1 \oplus K_2$, AK, $A^{-1}K$ and λK are convex cones when K_1, K_2 and K are convex cones. Positive scalar multiplication is a trivial operation for cones: one has $\lambda K = K$ for every $\lambda > 0$. For this reason, addition and inverse addition reduce essentially to the lattice operations in the case of cones.

THEOREM 3.8.   *If K_1 and K_2 are convex cones containing the origin, then*

$$K_1 + K_2 = \text{conv } (K_1 \cup K_2),$$
$$K_1 \# K_2 = K_1 \cap K_2.$$

PROOF.   By Theorem 3.3, conv $(K_1 \cup K_2)$ is the union over $\lambda \in [0, 1]$ of $(1 - \lambda)K_1 + \lambda K_2$. The latter set is $K_1 + K_2$ when $0 < \lambda < 1$, K_1 when $\lambda = 0$, and K_2 when $\lambda = 1$. Since $0 \in K_1$ and $0 \in K_2$, $K_1 + K_2$ includes both K_1 and K_2. Thus conv $(K_1 \cup K_2)$ coincides with $K_1 + K_2$. Similarly, $K_1 \# K_2$ is the union over $\lambda \in [0, 1]$ of $(\lambda K_1) \cap (1 - \lambda)K_2$. The latter set is $K_1 \cap K_2$ when $0 < \lambda < 1$, and it is $\{0\} \subset K_1 \cap K_2$ when $\lambda = 0$ or $\lambda = 1$. Thus $K_1 \# K_2 = K_1 \cap K_2$. $\|$

There is one other interesting construction which we would like to mention here. Given two different points x and y in R^n, the half-line $\{(1 - \lambda)x + \lambda y \mid \lambda \geq 1\}$ might be thought of as the "shadow of y cast by a light source at x." The union of these half-lines as y ranges over a set C would be the "shadow of C." This suggests that we define the *umbra* of C with respect to S, for any disjoint subsets C and S of R^n, as

$$\bigcap_{x \in S} \bigcup_{\lambda \ge 1} \{(1 - \lambda)x + \lambda C\}$$

and the *penumbra* of C with respect to S as

$$\bigcup_{x \in S} \bigcup_{\lambda \ge 1} \{(1 - \lambda)x + \lambda C\}.$$

We leave it to the reader to show, as an exercise, that the umbra is convex if C is convex, and that the penumbra is convex if both S and C are convex.

SECTION 4

Convex Functions

Let f be a function whose values are real or $\pm\infty$ and whose domain is a subset S of R^n. The set

$$\{(x, \mu) \mid x \in S, \mu \in R, \mu \geq f(x)\}$$

is called the *epigraph* of f and is denoted by epi f. We define f to be a *convex* function on S if epi f is convex as a subset of R^{n+1}. A *concave* function on S is a function whose negative is convex. An *affine* function on S is a function which is finite, convex and concave.

The *effective domain* of a convex function f on S, which we denote by dom f, is the projection on R^n of the epigraph of f:

$$\text{dom} f = \{x \mid \exists \mu, (x, \mu) \in \text{epi} f\} = \{x \mid f(x) < +\infty\}.$$

This is a convex set in R^n, since it is the image of the convex set epi f under a linear transformation (Theorem 3.4). Its dimension is called the *dimension* of f. Trivially, the convexity of f is equivalent to that of the restriction of f to dom f. All the interest really centers on this restriction, and S itself has little role of its own.

There are weighty reasons, soon apparent, why one does not want to consider merely the class of all convex functions having a certain fixed C as their common effective domain. Two good technical approaches remain. One could limit attention to functions which are nowhere $+\infty$, so that S would always coincide with dom f (but would vary with f). Or one could limit attention to functions given on all of R^n, since a convex function f on S can always be extended to a convex function on all of R^n by setting $f(x) = +\infty$ for $x \notin S$.

The second approach will be taken in this book. *Thus by a "convex function" we shall henceforth always mean a "convex function with possibly infinite values which is defined throughout the space R^n," unless otherwise specified.* This approach has the advantage that technical nuisances about effective domains can be suppressed almost entirely. For example, when a convex function f is constructed according to certain formulas, the same formulas specify the effective domain of f implicitly, because they specify where $f(x)$ is or is not $+\infty$. In the other approach, one would always have to describe the effective domain of f explicitly before the values of f on that domain could be given.

The approach taken here does, however, lead to arithmetic calculations involving $+\infty$ and $-\infty$. The rules we adopt are the obvious ones:

$$\alpha + \infty = \infty + \alpha = \infty \quad \text{for} \quad -\infty < \alpha \leq \infty,$$

$$\alpha - \infty = -\infty + \alpha = -\infty \quad \text{for} \quad -\infty \leq \alpha < \infty,$$

$$\alpha\infty = \infty\alpha = \infty, \qquad \alpha(-\infty) = (-\infty)\alpha = -\infty \quad \text{for} \quad 0 < \alpha \leq \infty,$$

$$\alpha\infty = \infty\alpha = -\infty, \qquad \alpha(-\infty) = (-\infty)\alpha = \infty \quad \text{for} \quad -\infty \leq \alpha < 0,$$

$$0\infty = \infty 0 = 0 = 0(-\infty) = (-\infty)0, \qquad -(-\infty) = \infty,$$

$$\inf \emptyset = +\infty, \qquad \sup \emptyset = -\infty.$$

The combinations $\infty - \infty$ and $-\infty + \infty$ are undefined and are avoided. Under these rules, the familiar laws of arithmetic:

$$\alpha_1 + \alpha_2 = \alpha_2 + \alpha_1, \qquad (\alpha_1 + \alpha_2) + \alpha_3 = \alpha_1 + (\alpha_2 + \alpha_3),$$

$$\alpha_1\alpha_2 = \alpha_2\alpha_1, \qquad (\alpha_1\alpha_2)\alpha_3 = \alpha_1(\alpha_2\alpha_3),$$

$$\alpha(\alpha_1 + \alpha_2) = \alpha\alpha_1 + \alpha\alpha_2,$$

are still valid, provided that none of the indicated binary sums $\alpha + \beta$ is the forbidden $\infty - \infty$ (or $-\infty + \infty$). This can be verified directly by testing all possible combinations of finite versus infinite values for the α's.

Avoiding $\infty - \infty$ naturally requires some cautious attention, like avoiding division by zero. In practice, one or the other of the infinities is usually excluded automatically from a given calculation by the hypothesis, so no complications arise.

A convex function f is said to be *proper* if its epigraph is non-empty and contains no vertical lines, i.e. if $f(x) < +\infty$ for at least one x and $f(x) > -\infty$ for every x. Thus f is proper if and only if the convex set $C = \operatorname{dom} f$ is non-empty and the restriction of f to C is finite. Put another way, a proper convex function on R^n is a function obtained by taking a finite convex function f on a non-empty convex set C and then extending it to all of R^n by setting $f(x) = +\infty$ for $x \notin C$.

A convex function which is not proper is *improper*. Proper convex functions are the real object of study, but improper functions do arise from proper ones in many natural situations, and it is more convenient to admit them than to exclude them laboriously from consideration. An example of an improper convex function which is not simply identically $+\infty$ or $-\infty$ is the function f on R defined by

$$f(x) = \begin{cases} -\infty & \text{if} \quad |x| < 1, \\ 0 & \text{if} \quad |x| = 1, \\ +\infty & \text{if} \quad |x| > 1. \end{cases}$$

Convex functions have an important interpolation property. By definition, f is convex on S if and only if

$$(1 - \lambda)(x, \mu) + \lambda(y, \nu) = ((1 - \lambda)x + \lambda y, (1 - \lambda)\mu + \lambda\nu)$$

belongs to epi f whenever (x, μ) and (y, ν) belong to epi f and $0 \leq \lambda \leq 1$. In other words, one is to have $(1 - \lambda)x + \lambda y \in S$ and

$$f((1 - \lambda)x + \lambda y) \leq (1 - \lambda)\mu + \lambda\nu,$$

whenever $x \in S$, $y \in S$, $f(x) \leq \mu \in R$, $f(y) \leq \nu \in R$ and $0 \leq \lambda \leq 1$. This condition can be expressed in several different ways. The following two variants are especially useful.

THEOREM 4.1. *Let f be a function from C to $(-\infty, +\infty]$, where C is a convex set (for example $C = R^n$). Then f is convex on C if and only if*

$$f((1 - \lambda)x + \lambda y) \leq (1 - \lambda)f(x) + \lambda f(y), \qquad 0 < \lambda < 1,$$

for every x and y in C.

THEOREM 4.2. *Let f be a function from R^n to $[-\infty, +\infty]$. Then f is convex if and only if*

$$f((1 - \lambda)x + \lambda y) < (1 - \lambda)\alpha + \lambda\beta, \qquad 0 < \lambda < 1,$$

whenever $f(x) < \alpha$ and $f(y) < \beta$.

Another useful variant can be deduced by applying Theorem 2.2 to epigraphs.

THEOREM 4.3 (Jensen's Inequality). *Let f be a function from R^n to $(-\infty, +\infty]$. Then f is convex if and only if*

$$f(\lambda_1 x_1 + \cdots + \lambda_m x_m) \leq \lambda_1 f(x_1) + \cdots + \lambda_m f(x_m)$$

whenever $\lambda_1 \geq 0, \ldots, \lambda_m \geq 0, \lambda_1 + \cdots + \lambda_m = 1$.

PROOF. An elementary exercise. ‖

Concave functions, of course, satisfy the opposite inequalities under similar hypotheses. Affine functions satisfy the inequalities as equations. Thus the affine functions on R^n are the affine transformations from R^n to R.

The inequality in Theorem 4.1 is often taken as the *definition* of the convexity of a function f from a convex set C to $(-\infty, +\infty]$. This approach causes difficulties, however, when f can have both $+\infty$ and $-\infty$ among its values, since the expression $\infty - \infty$ could arise. Of course, the condition in Theorem 4.2 could be used as the definition of convexity in the general case, but the definition given at the beginning of this section seems preferable because it emphasizes the geometry which is fundamental to the theory of convex functions.

Some classical examples of convex functions on the real line are obtained from the following theorem.

THEOREM 4.4.   *Let f be a twice continuously differentiable real-valued function on an open interval (α, β). Then f is convex if and only if its second derivative f'' is non-negative throughout (α, β).*

PROOF.   Suppose first that f'' is non-negative on (α, β). Then f' is non-decreasing on (α, β). For $\alpha < x < y < \beta, 0 < \lambda < 1$ and $z = (1 - \lambda)x + \lambda y$, we have

$$f(z) - f(x) = \int_x^z f'(t)\, dt \leq f'(z)(z - x),$$

$$f(y) - f(z) = \int_z^y f'(t)\, dt \geq f'(z)(y - z).$$

Since $z - x = \lambda(y - x)$ and $y - z = (1 - \lambda)(y - x)$, we have

$$f(z) \leq f(x) + \lambda f'(z)(y - x),$$
$$f(z) \leq f(y) - (1 - \lambda)f'(z)(y - x).$$

Multiplying the two inequalities by $(1 - \lambda)$ and λ respectively and adding them together, we get

$$(1 - \lambda)f(z) + \lambda f(z) \leq (1 - \lambda)f(x) + \lambda f(y).$$

The left side is just $f(z) = f((1 - \lambda)x + \lambda y)$, so this proves the convexity of f on (α, β) by Theorem 4.1. As for the converse assertion of the theorem, suppose that f'' were not non-negative on (α, β). Then f'' would be negative on a certain subinterval (α', β') by continuity. By the obvious argument, exactly parallel to the one just given, on (α', β') we would have

$$f(z) - f(x) > f'(z)(z - x),$$
$$f(y) - f(z) < f'(z)(y - z),$$

and hence

$$f((1 - \lambda)x + \lambda y) > (1 - \lambda)f(x) + \lambda f(y).$$

Thus f would not be convex on (α, β).   ‖

Theorem 4.4 will be generalized in Theorems 24.1 and 24.2.

Here are some functions on R whose convexity is a consequence of Theorem 4.4.

1. $f(x) = e^{\alpha x}$, where $-\infty < \alpha < \infty$;

2. $f(x) = x^p$ if $x \geq 0, f(x) = \infty$ if $x < 0$, where $1 \leq p < \infty$.

3. $f(x) = -x^p$ if $x \geq 0, f(x) = \infty$ if $x < 0$, where $0 \leq p \leq 1$;

4. $f(x) = x^p$ if $x > 0, f(x) = \infty$ if $x \leq 0$, where $-\infty < p \leq 0$;

5. $f(x) = (\alpha^2 - x^2)^{-1/2}$ if $|x| < \alpha, f(x) = \infty$ if $|x| \geq \alpha$, where $\alpha > 0$;

6. $f(x) = -\log x$ if $x > 0, f(x) = \infty$ if $x \leq 0$.

In the multidimensional case, it is trivial from Theorem 4.1 that every function of the form

$$f(x) = \langle x, a \rangle + \alpha, \quad a \in R^n, \quad \alpha \in R$$

is convex on R^n, in fact affine. Every affine function on R^n is actually of this form (Theorem 1.5). A *quadratic* function

$$f(x) = \tfrac{1}{2}\langle x, Qx \rangle + \langle x, a \rangle + \alpha,$$

where Q is a symmetric $n \times n$ matrix, is convex on R^n if and only if Q is *positive semi-definite*, i.e.

$$\langle z, Qz \rangle \geq 0 \quad \text{for every} \quad z \in R^n.$$

That is immediate from the following multidimensional version of Theorem 4.4.

THEOREM 4.5. *Let f be a twice continuously differentiable real-valued function on an open convex set C in R^n. Then f is convex on C if and only if its Hessian matrix*

$$Q_x = (q_{ij}(x)), \qquad q_{ij}(x) = \frac{\partial^2 f}{\partial \xi_i \, \partial \xi_j} (\xi_1, \ldots, \xi_n),$$

is positive semi-definite for every $x \in C$.

PROOF. The convexity of f on C is equivalent to the convexity of the restriction of f to each line segment in C. This is the same as the convexity of the function $g(\lambda) = f(y + \lambda z)$ on the open real interval $\{\lambda \mid y + \lambda z \in C\}$ for each $y \in C$ and $z \in R^n$. A straightforward calculation shows that

$$g''(\lambda) = \langle z, Q_x z \rangle, \qquad x = y + \lambda z.$$

Thus, by Theorem 4.4, g is convex for each $y \in C$ and $z \in R^n$ if and only if $\langle z, Q_x z \rangle \geq 0$ for every $z \in R^n$ and $x \in C$. ∥

An interesting function on R^n whose convexity may be verified by Theorem 4.5 is the negative of the geometric mean:

$$f(x) = f(\xi_1, \ldots, \xi_n) = \begin{cases} -(\xi_1 \xi_2 \cdots \xi_n)^{1/n} & \text{if} \quad \xi_1 \geq 0, \ldots, \xi_n \geq 0, \\ +\infty \quad \text{otherwise.} \end{cases}$$

Direct computation shows that

$$\langle z, Q_x z \rangle = n^{-2} f(x) \left[\left(\sum_{j=1}^{n} \zeta_j / \xi_j \right)^2 - n \sum_{j=1}^{n} (\zeta_j / \xi_j)^2 \right]$$

for $z = (\zeta_1, \ldots, \zeta_n)$, $x = (\xi_1, \ldots, \xi_n)$, $\xi_1 > 0, \ldots, \xi_n > 0$. This quantity is non-negative, because $f(x) < 0$ and

$$(\alpha_1 + \cdots + \alpha_n)^2 \leq n(\alpha_1^2 + \cdots + \alpha_n^2)$$

(inasmuch as $2\alpha_j\alpha_k \leq \alpha_j^2 + \alpha_k^2$) for any real numbers α_j.

One of the most important convex functions on R^n is the *Euclidean norm*

$$|x| = \langle x, x \rangle^{1/2} = (\xi_1^2 + \cdots + \xi_n^2)^{1/2}.$$

This is, of course, just the absolute value function when $n = 1$. The convexity of the Euclidean norm follows from the familiar laws

$$|x + y| \leq |x| + |y|, \qquad |\lambda x| = \lambda |x| \quad \text{for} \quad \lambda \geq 0.$$

There are several useful correspondences between convex sets and convex functions. The simplest associates with each set C in R^n the *indicator function* $\delta(\cdot \mid C)$ of C, where

$$\delta(x \mid C) = \begin{cases} 0 & \text{if} \quad x \in C, \\ +\infty & \text{if} \quad x \notin C. \end{cases}$$

The epigraph of the indicator function is a "half-cylinder with cross-section C." Clearly C is a convex set if and only if $\delta(\cdot \mid C)$ is a convex function on R^n. Indicator functions play a fundamental role in convex analysis similar to the role of characteristic functions of sets in other branches of analysis.

The *support function* $\delta^*(\cdot \mid C)$ of a convex set C in R^n is defined by

$$\delta^*(x \mid C) = \sup \{\langle x, y \rangle \mid y \in C\}.$$

The *gauge* $\gamma(\cdot \mid C)$ is defined by

$$\gamma(x \mid C) = \inf \{\lambda \geq 0 \mid x \in \lambda C\}, \qquad C \neq \emptyset.$$

The (Euclidean) *distance function* $d(\cdot, C)$ is defined by

$$d(x, C) = \inf \{|x - y| \mid y \in C\}.$$

The convexity of these functions on R^n could be verified now directly, but we shall wait until the next section, where the convexity can be shown to follow from general principles.

Convex functions give rise to convex sets in an important way.

THEOREM 4.6. *For any convex function f and any $\alpha \in [-\infty, +\infty]$, the level sets $\{x \mid f(x) < \alpha\}$ and $\{x \mid f(x) \leq \alpha\}$ are convex.*

PROOF. In the case of strict inequality the result is immediate from Theorem 4.2, with $\beta = \alpha$. The convexity of $\{x \mid f(x) \leq \alpha\}$ then follows from the fact that it is the intersection of the convex sets $\{x \mid f(x) < \mu\}$ for

$\mu > \alpha$. A more geometric way of seeing this convexity is to observe that $\{x \mid f(x) \leq \alpha\}$ is the projection on R^n of the intersection of epi f and the horizontal hyperplane $\{(x, \mu) \mid \mu = \alpha\}$ in R^{n+1}, so that $\{x \mid f(x) \leq \alpha\}$ can be regarded as a horizontal cross-section of epi f. ‖

COROLLARY 4.6.1. *Let f_i be a convex function on R^n and α_i be a real number for each $i \in I$, where I is an arbitrary index set. Then*

$$C = \{x \mid f_i(x) \leq \alpha_i, \forall i \in I\}$$

is a convex set.

PROOF. Like Corollary 2.1.1. ‖

Taking f to be a quadratic convex function in Theorem 4.6, we can conclude that the set of points satisfying a quadratic inequality

$$\tfrac{1}{2}\langle x, Qx \rangle + \langle x, a \rangle + \alpha \leq 0$$

is convex when Q is positive semi-definite (Theorem 4.5). Sets of this form include all "solid" ellipsoids and paraboloids, and in particular spherical balls like $\{x \mid \langle x, x \rangle \leq 1\}$.

Theorem 4.6 and Corollary 4.6.1 have a clear significance for the theory of systems of nonlinear inequalities. But convexity enters into the analysis of other aspects of the theory of inequalities too, because various classical inequalities can be regarded as special cases of Theorem 4.3. For example, take f on R to be the negative of the logarithm, as in example 6 above. For a convex combination of positive numbers $x_1, \ldots, x_m$, we have

$$-\log (\lambda_1 x_1 + \cdots + \lambda_m x_m) \leq -\lambda_1 \log x_1 - \cdots - \lambda_m \log x_m$$

by Theorem 4.3. Multiplying by -1 and taking the exponential of both sides, we have

$$\lambda_1 x_1 + \cdots + \lambda_m x_m \geq x_1^{\lambda_1} \cdots x_m^{\lambda_m}.$$

In particular, for $\lambda_1 = \cdots = \lambda_m = 1/m$,

$$(x_1 + \cdots + x_m)/m \geq (x_1 \cdots x_m)^{1/m}.$$

This is the famous inequality between the arithmetic mean and geometric mean of a family of positive numbers.

Sometimes a non-convex function can be transformed into a convex one by a nonlinear change of variables. An outstanding example is the class of (positive) algebraic functions on the positive orthant of R^n which are sums of terms of the form

$$g(x) = g(\xi_1, \ldots, \xi_n) = \beta \xi_1^{\alpha_1} \cdots \xi_n^{\alpha_n},$$

where $\beta > 0$ and the exponents α_j are arbitrary real numbers. (Such functions occur in an important application at the end of §30.) A particular function in this class would be

$$f(\xi_1, \xi_2) = \xi_1^{-2} + (\xi_1\xi_2)^{1/3} + 2\xi_2^4, \ \xi_1 > 0, \ \xi_2 > 0.$$

The substitution $\zeta_j = \log \xi_j$ converts the general term g into

$$h(z) = h(\zeta_1, \ldots, \zeta_n) = \beta e^{\alpha_1 \zeta_1} \cdots e^{\alpha_n \zeta_n} = \beta e^{\langle a, z \rangle},$$

where $a = (\alpha_1, \ldots, \alpha_n)$. It will be seen in the next section that h, and any sum of functions of the form of h, is convex. Notice that the same change of variables transforms the set $\{x \mid g(x) = \alpha\}$ into a hyperplane

$$\{z \mid h(z) = \alpha\} = \{z \mid \langle a, z \rangle = \log (\alpha/\beta)\}.$$

A function f on R^n is said to be *positively homogeneous* (of degree 1) if for every x one has

$$f(\lambda x) = \lambda f(x), \qquad 0 < \lambda < \infty.$$

Obviously, positive homogeneity is equivalent to the epigraph being a cone in R^{n+1}. An example of a positively homogeneous convex function which is not simply a linear function is $f(x) = |x|$.

THEOREM 4.7.   *A positively homogeneous function f from R^n to $(-\infty, +\infty]$ is convex if and only if*

$$f(x + y) \leq f(x) + f(y)$$

for every $x \in R^n$, $y \in R^n$.

PROOF.   This is implied by Theorem 2.6, because the subadditivity condition on f is equivalent to epi f being closed under addition.   ‖
COROLLARY 4.7.1.   *If f is a positively homogeneous proper convex function, then*

$$f(\lambda_1 x_1 + \cdots + \lambda_m x_m) \leq \lambda_1 f(x_1) + \cdots + \lambda_m f(x_m)$$

whenever $\lambda_1 > 0, \ldots, \lambda_m > 0$.
COROLLARY 4.7.2.   *If f is a positively homogeneous proper convex function, then $f(-x) \geq -f(x)$ for every x.*
PROOF. $f(x) + f(-x) \geq f(x - x) = f(0) \geq 0$.   ‖

THEOREM 4.8.   *A positively homogeneous proper convex function f is linear on a subspace L if and only if $f(-x) = -f(x)$ for every $x \in L$. This is true if merely $f(-b_i) = -f(b_i)$ for all the vectors in some basis $b_1, \ldots, b_m$ for L.*

PROOF. Assume the latter. Then $f(\lambda_i b_i) = \lambda_i f(b_i)$ for every $\lambda_i \in R$, not just for $\lambda_i > 0$. For any $x = \lambda_1 b_1 + \cdots + \lambda_m b_m \in L$ we have

$$f(\lambda_1 b_1) + \cdots + f(\lambda_m b_m) \geq f(x) \geq -f(-x)$$
$$\geq -(f(-\lambda_1 b_1) + \cdots + f(-\lambda_m b_m)) = f(\lambda_1 b_1) + \cdots + f(\lambda_m b_m)$$

(Theorem 4.7 and Corollary 4.7.2), and hence

$$f(x) = f(\lambda_1 b_1) + \cdots + f(\lambda_m b_m) = \lambda_1 f(b_1) + \cdots + \lambda_m f(b_m).$$

Thus f is linear on L, and in particular $f(-x) = -f(x)$ for $x \in L$. ‖

Certain positively homogeneous convex functions will be characterized in §13 as support functions of convex sets and in §15 as gauge functions of convex sets (including norms). Convex functions which are "positively homogeneous of degree $p > 1$" will be considered in Corollaries 15.3.1 and 15.3.2.

Functional Operations

How can new convex functions be obtained from functions already known to be convex? There are many operations which preserve convexity, as it turns out. Some of the operations, like pointwise addition of functions, are familiar from ordinary analysis. Others, like taking the convex hull of a collection of functions, are geometrically motivated. Often the constructed function is expressed as a constrained infimum, thereby suggesting applications to the theory of extremum problems.

Familiarity with the operations below is helpful, especially, when one has to prove that a given function with a complicated formula is a convex function.

THEOREM 5.1. *Let f be a convex function from R^n to $(-\infty, +\infty]$, and let φ be a convex function from R to $(-\infty, +\infty]$ which is non-decreasing. Then $h(x) = \varphi(f(x))$ is convex on R^n (where one sets $\varphi(+\infty) = +\infty$).*

PROOF. For x and y in R^n and $0 < \lambda < 1$, we have

$$f((1 - \lambda) x + \lambda y) \leq (1 - \lambda)f(x) + \lambda f(y)$$

(Theorem 4.1). Applying φ to both sides of this inequality, we get

$$h((1 - \lambda)x + \lambda y) \leq \varphi((1 - \lambda)f(x) + \lambda f(y)) \leq (1 - \lambda)h(x) + \lambda h(y).$$

Thus h is convex (Theorem 4.1). ‖

It follows from Theorem 5.1 that $h(x) = e^{f(x)}$ is a proper convex function on R^n if f is. Also, $h(x) = f(x)^p$ is convex for $p > 1$ when f is convex *and non-negative*. This is proved by taking

$$\varphi(\xi) = \begin{cases} \xi^p & \text{if } \xi \geq 0, \\ 0 & \text{if } \xi < 0. \end{cases}$$

In particular, $h(x) = |x|^p$ is convex on R^n for $p \geq 1$ ($|x|$ being the Euclidean norm). If g is a concave function, then $h(x) = 1/g(x)$ is convex on $C = \{x \mid g(x) > 0\}$. To see this, apply to the convex function $f = -g$ the function φ defined by

$$\varphi(\xi) = \begin{cases} -1/\xi & \text{if } \xi < 0, \\ +\infty & \text{if } \xi \geq 0. \end{cases}$$

Taking φ to be an affine function on R with positive slope λ, we get the important fact that $\lambda f + \alpha$ is a proper convex function when f is a proper convex function and λ and α are real numbers, $\lambda \geq 0$. Further examples based on Theorem 5.1 will be found in Theorem 15.3.

THEOREM 5.2. *If f_1 and f_2 are proper convex functions on R^n, then $f_1 + f_2$ is convex.*

PROOF. Evident from Theorem 4.1. ‖

Notice that $(f_1 + f_2)(x) < \infty$ if and only if $f_1(x) < \infty$ and $f_2(x) < \infty$. Thus the effective domain of $f_1 + f_2$ is the intersection of the effective domains of f_1 and f_2, which might be empty, in which case $f_1 + f_2$ would be improper. The properness in the hypothesis of Theorem 5.2 is for the sake of avoiding $\infty - \infty$ when $f_1 + f_2$ is formed.

A linear combination $\lambda_1 f_1 + \cdots + \lambda_m f_m$ of proper convex functions with non-negative coefficients is convex.

If f is a finite convex function, say, and C is a non-empty convex set, then

$$f(x) + \delta(x \mid C) = \begin{cases} f(x) & \text{if } x \in C, \\ +\infty & \text{if } x \notin C, \end{cases}$$

where $\delta(\cdot \mid C)$ is the indicator function of C. Thus adding an indicator function to f amounts to restricting the effective domain of f.

A common device for constructing convex functions on R^n is to construct a convex set F in R^{n+1} and then take the function whose graph is the "lower boundary" of F in the sense of the following theorem.

THEOREM 5.3. *Let F be any convex set in R^{n+1}, and let*

$$f(x) = \inf \{\mu \mid (x, \mu) \in F\}.$$

Then f is a convex function on R^n.

PROOF. Evident from Theorem 4.2. (Notice the usefulness here of the convention that an infimum over the empty set of real numbers is $+\infty$.) ‖

As the first application of the device in Theorem 5.3, we introduce the functional operation which corresponds to the addition of epigraphs as sets in R^{n+1}.

THEOREM 5.4. *Let $f_1, \ldots, f_m$ be proper convex functions on R^n, and let*

$$f(x) = \inf \{f_1(x_1) + \cdots + f_m(x_m) \mid x_i \in R^n, x_1 + \cdots + x_m = x\}.$$

Then f is a convex function on R^n.

PROOF. Let $F_i = \text{epi } f_i$ and $F = F_1 + \cdots + F_m$. Then F is a convex set in R^{n+1}. By definition, $(x, \mu) \in F$ if and only if there exist $x_i \in R^n$,

$\mu_i \in R$, such that $\mu_i \geq f(x_i)$, $\mu = \mu_1 + \cdots + \mu_m$ and $x = x_1 + \cdots + x_m$. Thus the f defined in the theorem is the convex function obtained from F by the construction in Theorem 5.3. ‖

The function f in Theorem 5.4 will be denoted by $f_1 \,\square\, f_2 \,\square\, \cdots \,\square\, f_m$. The operation $\square$ is called *infimal convolution*. This terminology arises from the fact that, when only two functions are involved, $\square$ can be expressed by

$$(f \,\square\, g)(x) = \inf_y \{f(x - y) + g(y)\},$$

and this is analogous to the classical formula for integral convolution. Infimal convolution is dual to the operation of addition of convex functions in a sense to be explained in §16.

If $g = \delta(\cdot \,|\, a)$ for a certain point $a \in R^n$ (where $\delta(x \,|\, a) = \infty$ if $x \neq a$ and $\delta(a \,|\, a) = 0$), then $(f \,\square\, g)(x) = f(x - a)$. Thus $f \,\square\, \delta(\cdot \,|\, a)$ is the function whose graph is obtained by translating the graph of f horizontally by a. For an arbitrary g and for $h(y) = f(-y)$, the infimal convolute $f \,\square\, g$ expresses the infimum over R^n of g plus the translate $h \,\square\, \delta(\cdot \,|\, x)$, as a function of the translation x. The effective domain of $f \,\square\, g$ is the sum of dom f and dom g.

Taking f to be the Euclidean norm and g to be the indicator function of a convex set C, we get

$$(f \,\square\, g)(x) = \inf_y \{|x - y| + \delta(y \,|\, C)\} = \inf_{y \in C} |x - y| = d(x, C).$$

This establishes the convexity of the distance function $d(\cdot, C)$.

Other examples of infimal convolution will be found following Corollary 9.2.2.

Properness of convex functions is not always preserved by infimal convolution, since the infimum in the formula in Theorem 5.4 may be $-\infty$. Nor is infimal convolution of *improper* functions defined by this formula, because of the rule of avoiding $\infty - \infty$. However, $f_1 \,\square\, f_2$ can be defined for any functions f_1 and f_2 from R^n to $[-\infty, +\infty]$ directly in terms of addition of epigraphs:

$$(f_1 \,\square\, f_2)(x) = \inf \{\mu \,|\, (x, \mu) \in (\mathrm{epi}\, f_1 + \mathrm{epi}\, f_2)\}.$$

As an operation on the collection of all functions from R^n to $[-\infty, +\infty]$, infimal convolution is commutative, associative and convexity-preserving. The function $\delta(\cdot \,|\, 0)$ acts as the identity element for this operation.

It has already been pointed out that the operation of non-negative *left* scalar multiplication preserves convexity, where

$$(\lambda f)(x) = \lambda(f(x)).$$

There is also a useful operation of *right* scalar multiplication, which corresponds to scalar multiplication of epigraphs. For any convex function f on R^n and any λ, $0 \le \lambda < \infty$, we define $f\lambda$ to be the convex function obtained from Theorem 5.3 with $F = \lambda(\text{epi } f)$. Thus

$$(f\lambda)(x) = \lambda f(\lambda^{-1}x), \qquad \lambda > 0,$$

while for $\lambda = 0$ we have

$$(f0)(x) = \delta(x \mid 0), \qquad f \not\equiv +\infty.$$

(Trivially $f0 = f$ if $f \equiv +\infty$). A function f is positively homogeneous if and only if $f\lambda = f$ for every $\lambda > 0$.

Let h be any convex function in R^n, and let F be the convex cone in R^{n+1} generated by epi h. The function obtained by applying Theorem 5.3 to F has as its epigraph a convex cone in R^{n+1} containing the origin. It is the greatest of the positively homogeneous convex functions f such that $f(0) \le 0$ and $f \le h$. Naturally, we shall call this f the *positively homogeneous convex function generated* by h. Since F consists of the origin and the union of the sets $\lambda(\text{epi } h)$ for $\lambda > 0$, we have

$$f(x) = \inf \{(h\lambda)(x) \mid \lambda \ge 0\}$$

when $h \not\equiv +\infty$. Of course, $\lambda = 0$ can be omitted from the infimum if $x \ne 0$ or if $h(0) < +\infty$.

For any proper convex function f on R^n, the function g on R^{n+1} defined by

$$g(\lambda, x) = \begin{cases} (f\lambda)(x) & \text{if } \lambda \ge 0, \\ +\infty & \text{if } \lambda < 0, \end{cases}$$

is a positively homogeneous proper convex function, the positively homogeneous convex function generated by

$$h(\lambda, x) = \begin{cases} f(x) & \text{if } \lambda = 1, \\ +\infty & \text{if } \lambda \ne 1. \end{cases}$$

In particular, then, $\varphi(\lambda) = (f\lambda)(x)$ is a proper convex function of $\lambda \ge 0$ for any $x \in \text{dom } f$.

The gauge of a non-empty convex set C in R^n is the positively homogeneous convex function generated by $\delta(\cdot \mid C) + 1$. Indeed, for $h(x) = \delta(x \mid C) + 1$ we have $(h\lambda)(x) = \delta(x \mid \lambda C) + \lambda$, so that

$$\inf \{(h\lambda)(x) \mid \lambda \ge 0\} = \inf \{\lambda \ge 0 \mid x \in \lambda C\} = \gamma(x \mid C).$$

THEOREM 5.5. *The pointwise supremum of an arbitrary collection of convex functions is convex.*

PROOF. This follows from the fact that the intersection of a collection of convex sets is convex. Indeed, if

$$f(x) = \sup \{f_i(x) \,|\, i \in I\},$$

the epigraph of f is the intersection of those of the functions f_i. ‖

The convexity of the support function $\delta^*(\cdot \,|\, C)$ of a set C in R^n is implied by Theorem 5.5, because this function is by definition the pointwise supremum of a certain collection of linear functions, namely the functions $\langle \cdot, y \rangle$ as y ranges over C.

As a further illustration, consider the function f which assigns to each $x = (\xi_1, \dots, \xi_n)$ the greatest of the components ξ_j of x. This f is convex by Theorem 5.5, because it is the pointwise supremum of the linear functions $\langle x, e_j \rangle$, $j = 1, \dots, n$, where e_j is the vector forming the jth row of the $n \times n$ identity matrix. Observe that f is also positively homogeneous; in fact f is the support function of the simplex

$$C = \{y = (\eta_1, \dots, \eta_n) \,|\, \eta_j \geq 0, \eta_1 + \cdots + \eta_n = 1\}.$$

The convexity of the function

$$k(x) = \max \{|\xi_j| \,|\, j = 1, \dots, n\},$$

which is called the *Tchebycheff norm* on R^n, can be seen similarly from Theorem 5.5. The latter function is the support function of the convex set

$$D = \{y = (\eta_1, \dots, \eta_n) \,|\, |\eta_1| + \cdots + |\eta_n| \leq 1\}$$

and at the same time the gauge of the n-dimensional cube

$$E = \{x = (\xi_1, \dots, \xi_n) \,|\, -1 \leq \xi_j \leq 1, j = 1, \dots, n\}.$$

(Any non-negative support function is the gauge of some closed convex set containing the origin, and conversely, as will be explained in §14.)

The *convex hull* of a non-convex function g is the function $f = \text{conv } g$ obtained from Theorem 5.3 with

$$F = \text{conv (epi } g).$$

It is the greatest convex function majorized by g. By Theorem 2.3, a point (x, μ) belongs to F if and only if it can be expressed as a convex combination

$$(x, \mu) = \lambda_1(x_1, \mu_1) + \cdots + \lambda_m(x_m, \mu_m)$$

$$= (\lambda_1 x_1 + \cdots + \lambda_m x_m, \lambda_1 \mu_1 + \cdots + \lambda_m \mu_m),$$

where $(x_i, \mu_i) \in \text{epi } g$ (i.e. $g(x_i) \leq \mu_i \in R$). Thus

$$f(x) = \inf \{\lambda_1 g(x_1) + \cdots + \lambda_m g(x_m) \,|\, \lambda_1 x_1 + \cdots + \lambda_m x_m = x\},$$

where the infimum is taken over all expressions of x as a convex combination of points of R^n (provided g does not take on the value $-\infty$, so that the summation is unambiguous).

The convex hull of an arbitrary collection of functions $\{f_i \mid i \in I\}$ on R^n is denoted by

$$\text{conv } \{f_i \mid i \in I\}.$$

It is the convex hull of the pointwise infimum of the collection, i.e. it is the function f obtained via Theorem 5.3 from the convex hull F of the union of the epigraphs of the functions f_i. It is the greatest convex function f (not necessarily proper) on R^n such that $f(x) \leq f_i(x)$ for every $x \in R^n$ and every $i \in I$.

THEOREM 5.6. *Let* $\{f_i \mid i \in I\}$ *be a collection of proper convex functions on* R^n, *where I is an arbitrary index set, and let f be the convex hull of the collection. Then*

$$f(x) = \inf \{\textstyle\sum_{i \in I} \lambda_i f_i(x_i) \mid \sum_{i \in I} \lambda_i x_i = x\},$$

where the infimum is taken over all representations of x as a convex combination of elements x_i, such that only finitely many coefficients λ_i are nonzero. (The formula is also valid if one actually restricts x_i to lie in dom f_i.)

PROOF. By definition $f(x)$ is the infimum of the values of μ such that $(x, \mu) \in F$, where F is the convex hull of the union of non-empty convex sets $C_i = \text{epi} f_i$. By Theorem 3.3, $(x, \mu) \in F$ if and only if (x, μ) can be expressed as a finite convex combination of the form

$$(x, \mu) = \textstyle\sum_{i \in I} \lambda_i(x_i, \mu_i) = (\sum_{i \in I} \lambda_i x_i, \sum_{i \in I} \lambda_i \mu_i),$$

where $(x_i, \mu_i) \in C_i$ (only finitely many of the coefficients being non-zero). Thus $f(x)$ is the infimum of $\sum_{i \in I} \lambda_i \mu_i$ over all expressions of x as a finite convex combination $\sum_{i \in I} \lambda_i x_i$ with $\mu_i \geq f_i(x_i)$ for every i. This is the same as the infimum in the theorem. $\|$

A useful case of Theorem 5.6 occurs when all the functions f_i are of the form

$$f_i(x) = \delta(x \mid a_i) + \alpha_i = \begin{cases} \alpha_i & \text{if } x = a_i, \\ +\infty & \text{if } x \neq a_i, \end{cases}$$

a_i and α_i being fixed elements of R^n and R, respectively. Then f is the greatest convex function satisfying

$$f(a_i) \leq \alpha_i, \qquad \forall i \in I,$$

and we have

$$f(x) = \inf \{\textstyle\sum_{i \in I} \lambda_i \alpha_i \mid \sum_{i \in I} \lambda_i a_i = x\},$$

where the infimum is taken over all representations of x as a convex combination of the a_i (with only finitely many non-zero coefficients).

A somewhat stronger version of Theorem 5.6 will be given in §17 as a consequence of Carathéodory's Theorem.

The formula in Theorem 5.6 can also be expressed by infimal convolution. For simplicity of notation, let us assume $I = \{1, \ldots, m\}$. Then f is obtained via Theorem 5.3 from the set

$$F = \text{conv} \{C_1, \ldots, C_m\} = \cup \{\lambda_1 C_1 + \cdots + \lambda_m C_m\},$$

where the union is taken over all convex combinations of the sets $C_i = \text{epi} f_i$ (Theorem 3.3). But $f_1 \lambda_1 \,\square\, \cdots \,\square\, f_m \lambda_m$ is the function obtained via Theorem 5.3 from the convex set $\lambda_1 C_1 + \cdots + \lambda_m C_m$ in R^{n+1}. Taking a union of epigraphs in R^{n+1} amounts to taking the pointwise infimum of the corresponding functions. Therefore $f = \text{conv} \{f_1, \ldots, f_m\}$ is also given by

$$f(x) = \inf\{(f_1 \lambda_1 \,\square\, \cdots \,\square\, f_m \lambda_m)(x) \mid \lambda_i \geq 0, \lambda_1 + \cdots + \lambda_m = 1\}$$

when $f_1, \ldots, f_m$ are proper convex functions.

The collection of all convex functions on R^n, regarded as a partially ordered set relative to the pointwise ordering (where $f \leq g$ if and only if $f(x) \leq g(x)$ for every x), is a complete lattice. The greatest lower bound of a family of convex functions f_i is $\text{conv} \{f_i \mid i \in I\}$ (relative to this particular partially ordered set!), while the least upper bound is $\sup \{f_i \mid i \in I\}$.

Constructions involving a linear transformation are considered in the next theorem.

THEOREM 5.7. *Let A be a linear transformation from R^n to R^m. Then, for each convex function g on R^m, the function gA defined by*

$$(gA)(x) = g(Ax)$$

is convex on R^n. For each convex function h on R^n, the function Ah defined by

$$(Ah)(y) = \inf \{h(x) \mid Ax = y\}$$

is convex on R^m.

PROOF. Direct verification is elementary using the criterion in Theorem 4.2. The convexity of $f = Ah$ also follows from applying Theorem 5.3 to the image F of the epigraph of h under the linear transformation $(x, \mu) \to (Ax, \mu)$ from R^{n+1} to R^{m+1}. ‖

The function Ah in Theorem 5.7 is called the *image of h under A*, while gA is called the *inverse image of g under A*. This terminology is suggested by the case where g and h are indicators of convex sets.

As an important example of the operation $h \to Ah$, we mention the case where A is a projection. For

$$A : x = (\xi_1, \ldots, \xi_m, \xi_{m+1}, \ldots, \xi_n) \to (\xi_1, \ldots, \xi_m),$$

say, we have

$$(Ah)(\xi_1, \ldots, \xi_m) = \inf_{\xi_{m+1}, \ldots, \xi_n} h(\xi_1, \ldots, \xi_m, \xi_{m+1}, \ldots, \xi_n).$$

This is convex in $y = (\xi_1, \ldots, \xi_m)$, when h is convex, according to the theorem.

When A is non-singular, $Ah = hA^{-1}$.

Partial addition of epigraphs can be used to define infinitely many commutative, associative binary operations for the collection of all convex functions on R^n. An example is the *partial infimal convolution*

$$h(y, z) = \inf_u \{f(y, z - u) + g(y, u)\},$$

where $x = (y, z)$ with $y \in R^m$, $z \in R^p$, $m + p = n$.

In the case of convex sets, there is a "natural" set of four commutative associative binary operations, and these reduce to only two operations when the sets are cones containing the origin. These operations are obtained from partial additions of convex cones of the form

$$K = \{(\lambda, x) \mid \lambda \geq 0, x \in \lambda C\} \subset R^{n+1}$$

corresponding to convex sets C in R^n; see the discussion following Theorem 3.6. One is led similarly to *eight* "natural" commutative associative binary operations in the collection of all convex functions on R^n, when the sets C are replaced by epigraphs. Specifically, we associate with each convex function f the convex cone K which is the epigraph of the positively homogeneous convex function on R^{n+1} generated by h, where $h(\lambda, x) = f(x) + \delta(\lambda \mid 1)$. If f is not identically $+\infty$,

$$K = \{(\lambda, x, \mu) \mid \lambda \geq 0, x \in R^n, \mu \geq (f\lambda)(x)\} \subset R^{n+2}.$$

(If $f = +\infty$, K is the non-negative μ-axis.) There are eight partial additions which arise from adding sets in R^{n+2} in various combinations of the three arguments λ, x and μ. In each case, we take the partial sum K of the cones K_1 and K_2 corresponding to two convex functions f_1 and f_2 on R^n. We then apply Theorem 5.3 to

$$F = \{(x, \mu) \mid (1, x, \mu) \in K\}$$

to get f. The resulting operation $(f_1, f_2) \to f$ is evidently commutative and associative. Four of the operations defined in this manner turn out to be among those previously defined. Namely, adding in μ alone forms $f_1 + f_2$. Adding in x and μ forms $f_1 \square f_2$. Adding in λ, x and μ forms conv $\{f_1, f_2\}$. Adding in none of the arguments forms the pointwise maximum of f_1 and f_2. The remaining four operations are described in the theorem below. (Here max $\{\alpha_1, \ldots, \alpha_m\}$ denotes, of course, the greatest of the m real numbers $\alpha_1, \ldots, \alpha_m$.)

THEOREM 5.8.   Let $f_1, \ldots, f_m$ be proper convex functions on R^n. Then the following are convex functions also:

$$f(x) = \inf \{\max \{f_1(x_1), \ldots, f_m(x_m)\} \mid x_1 + \cdots + x_m = x\},$$
$$g(x) = \inf \{(f_1\lambda_1)(x) + \cdots + (f_m\lambda_m)(x) \mid \lambda_i \geq 0, \lambda_1 + \cdots + \lambda_m = 1\},$$
$$h(x) = \inf \{\max \{(f_1\lambda_1)(x), \ldots, (f_m\lambda_m)(x)\} \mid \lambda_i \geq 0, \lambda_1 + \cdots + \lambda_m = 1\},$$
$$k(x) = \inf \{\max \{\lambda_1 f_1(x_1), \ldots, \lambda_m f_m(x_m)\}\},$$

where the last infimum is taken over all representations of x as a convex combination $x = \lambda_1 x_1 + \cdots + \lambda_m x_m$.

PROOF.   In the sense of the preceding discussion, adding in x alone yields f. Adding in λ and μ yields g. Adding in λ alone yields h. Adding in λ and x yields k.   ‖

The first operation in Theorem 5.8 can be expressed in "convolution" form when $m = 2$:

$$f(x) = \inf_y \max \{f_1(x - y), f_2(y)\}.$$

Observe that, with this operation,

$$\{x \mid f(x) < \alpha\} = \{x \mid f_1(x) < \alpha\} + \{x \mid f_2(x) < \alpha\},$$

for any α. The third operation amounts to inverse addition of epigraphs.

Part II · Topological Properties

Relative Interiors of Convex Sets

The *Euclidean distance* between two points x and y in R^n is by definition

$$d(x, y) = |x - y| = \langle x - y, x - y \rangle^{1/2}$$

The function d, the *Euclidean metric*, is convex as a function on R^{2n}. (This follows from the fact that d is obtained by composing the Euclidean norm $f(z) = |z|$ with the linear transformation $(x, y) \to x - y$ from R^{2n} to R^n.) The familiar topological concepts of closed set, open set, closure, and interior in R^n are usually introduced in terms of convergence of vectors with respect to the Euclidean metric. But such convergence is, of course, equivalent to the convergence of a sequence of vectors in R^n component by component.

The topological properties of convex sets in R^n are notably simpler than those of arbitrary sets, as we shall see below.

Convex functions are one important source of open and closed convex sets. Any continuous real-valued function f on R^n gives rise to a family of open level sets $\{x \mid f(x) < \alpha\}$ and closed level sets $\{x \mid f(x) \leq \alpha\}$, and these sets are convex if f is convex (Theorem 4.6).

Throughout this section, we shall denote by B the *Euclidean unit ball* in R^n:

$$B = \{x \mid |x| \leq 1\} = \{x \mid d(x, 0) \leq 1\}.$$

This is a closed convex set (a level set of the Euclidean norm, which is continuous and convex). For any $a \in R^n$, the *ball with radius* $\varepsilon > 0$ *and center* a is given by

$$\{x \mid d(x, a) \leq \varepsilon\} = \{a + y \mid |y| \leq \varepsilon\} = a + \varepsilon B.$$

For any set C in R^n, the set of points x whose distance from C does not exceed ε is

$$\{x \mid \exists y \in C, d(x, y) \leq \varepsilon\} = \bigcup \{y + \varepsilon B \mid y \in C\} = C + \varepsilon B.$$

The *closure* cl C and *interior* int C of C can therefore be expressed by the formulas

$$\text{cl } C = \bigcap \{C + \varepsilon B \mid \varepsilon > 0\},$$
$$\text{int } C = \{x \mid \exists \varepsilon > 0, x + \varepsilon B \subset C\}.$$

In the case of convex sets, the concept of interior can be absorbed into a more convenient concept of relative interior. This concept is motivated by the fact that a line segment or triangle embedded in R^3 does have a natural interior of sorts which is not truly an interior in the sense of the whole metric space R^3. The *relative interior* of a convex set C in R^n, which we denote by ri C, is defined as the interior which results when C is regarded as a subset of its affine hull aff C. Thus ri C consists of the points $x \in$ aff C for which there exists an $\varepsilon > 0$, such that $y \in C$ whenever $y \in$ aff C and $d(x, y) \leq \varepsilon$. In other words,

$$\text{ri } C = \{x \in \text{aff } C \mid \exists \varepsilon > 0, (x + \varepsilon B) \cap (\text{aff } C) \subset C\}.$$

Needless to say,

$$\text{ri } C \subset C \subset \text{cl } C.$$

The set difference $(\text{cl } C) \setminus (\text{ri } C)$ is called the *relative boundary* of C. Naturally, C is said to be *relatively open* if ri $C = C$.

For an n-dimensional convex set, aff $C = R^n$ by definition, so ri $C = $ int C.

A pitfall to be noted is that, while the inclusion $C_1 \supset C_2$ implies cl $C_1 \supset$ cl C_2 and int $C_1 \supset$ int C_2, it does *not* in general imply ri $C_1 \supset$ ri C_2. For example, if C_1 is a cube in R^3 and C_2 is one of the faces of C_1, ri C_1 and ri C_2 are both non-empty but disjoint.

An affine set is relatively open by definition. Every affine set is at the same time closed. This is clear from the fact that an affine set is an intersection of hyperplanes (Corollary 1.4.1), and every hyperplane H can be expressed as a level set of a continuous function (Theorem 1.3):

$$H = \{x = (\xi_1, \ldots, \xi_n) \mid \beta_1 \xi_1 + \cdots + \beta_n \xi_n = \beta\}.$$

Observe that

$$\text{cl } C \subset \text{cl } (\text{aff } C) = \text{aff } C$$

for any C. Thus any line through two different points of cl C lies entirely in aff C.

Closures and relative interiors are preserved under translations and more generally under any one-to-one affine transformation of R^n onto itself. Indeed, such a transformation preserves affine hulls and is continuous in both directions (since the components of the image of a vector x under an affine transformation are linear or affine functions of the components ξ_j of x). One should keep this in mind as a useful device for simplifying proofs. For example, if C is an m-dimensional convex set in R^n, there exists by Corollary 1.6.1 a one-to-one affine transformation T of R^n onto itself which carries aff C onto the subspace

$$L = \{x = (\xi_1, \ldots, \xi_m, \xi_{m+1}, \ldots, \xi_n) \mid \xi_{m+1} = 0, \ldots, \xi_n = 0\}.$$

This L can be regarded as a copy of R^m. It is often possible in this manner to reduce a question about general convex sets to the case where the convex set is of full dimension, i.e. has the whole space as its affine hull.

The following property of closures and relative interiors of convex sets is fundamental.

THEOREM 6.1.   *Let C be a convex set in R^n. Let $x \in$ ri C and $y \in$ cl C. Then $(1 - \lambda)x + \lambda y$ belongs to ri C (and hence in particular to C) for $0 \leq \lambda < 1$.*

PROOF.   In view of the preceding remark, we can limit attention to the case where C is n-dimensional, so that ri $C =$ int C. Let $\lambda \in [0, 1)$. We must show that $(1 - \lambda)x + \lambda y + \varepsilon B$ is contained in C for some $\varepsilon > 0$. We have $y \in C + \varepsilon B$ for every $\varepsilon > 0$, because $y \in$ cl C. Hence for every $\varepsilon > 0$

$$(1 - \lambda)x + \lambda y + \varepsilon B \subset (1 - \lambda)x + \lambda(C + \varepsilon B) + \varepsilon B$$

$$= (1 - \lambda)[x + \varepsilon(1 + \lambda)(1 - \lambda)^{-1}B] + \lambda C.$$

The latter set is contained in $(1 - \lambda)C + \lambda C = C$ when ε is sufficiently small, since $x \in$ int C by hypothesis.   ‖

The next two theorems describe the most important properties of the operations "cl" and "ri" on the collection of all convex sets in R^n.

THEOREM 6.2.   *Let C be any convex set in R^n. Then cl C and ri C are convex sets in R^n having the same affine hull, and hence the same dimension, as C. (In particular, ri $C \neq \emptyset$ if $C \neq \emptyset$.)*

PROOF.   The set $C + \varepsilon B$ is convex for any ε, because it is a linear combination of convex sets. The intersection of the collection of these sets for $\varepsilon > 0$ is cl C. Hence cl C is convex. The affine hull of cl C is at least as large as that of C, and since cl $C \subset$ aff C it must actually coincide with aff C. The convexity of ri C is a corollary of the preceding theorem (take y to be in ri C). To complete the proof, it is enough now to show that, in the case where C is n-dimensional with $n > 0$, the interior of C is not empty. An n-dimensional convex set contains an n-dimensional simplex (Theorem 2.4). We shall show that such a simplex S has a non-empty interior. Applying an affine transformation if necessary, we can assume that the vertices of S are the vectors $(0, 0, \ldots, 0)$, $(1, 0, \ldots, 0)$, $\ldots$,

$(0, \ldots, 0, 1)$:

$$S = \{(\xi_1, \ldots, \xi_n) \mid \xi_j \geq 0, \xi_1 + \cdots + \xi_n \leq 1\}.$$

But this simplex does have a non-empty interior, namely

$$\text{int } S = \{(\xi_1, \ldots, \xi_n) \mid \xi_j > 0, \xi_1 + \cdots + \xi_n < 1\}.$$

Hence int $C \neq \emptyset$ as claimed. ‖

For any set C in R^n, convex or not, the laws

$$\text{cl (cl } C) = \text{cl } C, \qquad \text{ri (ri } C) = \text{ri } C,$$

are valid. The following complementary laws are valid in the presence of convexity.

THEOREM 6.3. *For any convex set C in R^n, cl (ri C) = cl C and ri (cl C) = ri C.*

PROOF. Trivially, cl (ri C) is contained in cl C, since ri $C \subseteq C$. On the other hand, given any $y \in$ cl C and any $x \in$ ri C (such an x exists by the last theorem when $C \neq \emptyset$), the line segment between x and y lies entirely in ri C except perhaps for y (Theorem 6.1). Thus $y \in$ cl (ri C). This proves cl (ri C) = cl C. The inclusion ri (cl C) $\supset$ ri C holds, since cl $C \supset C$ and the affine hulls of cl C and C coincide.

Now let $z \in$ ri (cl C). We shall show $z \in$ ri C. Let x be any point of ri C. (We can suppose $x \neq z$, for otherwise $z \in$ ri C trivially.) Consider the line through x and z. For values $\mu > 1$ with $\mu - 1$ sufficiently small, the point

$$y = (1 - \mu)x + \mu z = z - (\mu - 1)(x - z)$$

on this line still belongs to ri (cl C) and hence to cl C. For such a y, we can express z in the form $(1 - \lambda)x + \lambda y$ with $0 < \lambda < 1$ (specifically with $\lambda = \mu^{-1}$). By Theorem 6.1, $z \in$ ri C. ‖

COROLLARY 6.3.1. *Let C_1 and C_2 be convex sets in R^n. Then cl $C_1 =$ cl C_2 if and only if ri $C_1 =$ ri C_2. These conditions are equivalent to the condition that ri $C_1 \subseteq C_2 \subseteq$ cl C_1.*

COROLLARY 6.3.2. *If C is a convex set in R^n, then every open set which meets cl C also meets ri C.*

COROLLARY 6.3.3. *If C_1 is a convex subset of the relative boundary of a non-empty convex set C_2 in R^n, then dim $C_1 <$ dim C_2.*

PROOF. If C_1 had the same dimension as C_2, it would have interior points relative to aff C_2. But such points could not be in cl (ri C_2), since ri C_2 is disjoint from C_1, and hence they could not be in cl C_2. ‖

The following characterization of relative interiors is frequently helpful.

THEOREM 6.4. *Let C be a non-empty convex set in R^n. Then $z \in$ ri C if and only if, for every $x \in C$, there exists a $\mu > 1$ such that $(1 - \mu)x + \mu z$ belongs to C.*

PROOF. The condition means that every line segment in C having z as one endpoint can be prolonged beyond z without leaving C. This is certainly true if $z \in$ ri C. Conversely, suppose z satisfies the condition. Since ri $C \neq \emptyset$ by Theorem 6.2, there exists a point $x \in$ ri C. Let y be the corresponding point $(1 - \mu)x + \mu z$ in C, $\mu > 1$, whose existence is hypothesized. Then $z = (1 - \lambda)x + \lambda y$, where $0 < \lambda = \mu^{-1} < 1$. Hence $z \in$ ri C by Theorem 6.1. ‖

COROLLARY 6.4.1. *Let C be a convex set in R^n. Then $z \in$ int C if and only if, for every $y \in R^n$, there exists some $\varepsilon > 0$ such that $z + \varepsilon y \in C$.*

We turn now to the question of how relative interiors behave under the common operations performed on convex sets.

THEOREM 6.5. *Let C_i be a convex set in R^n for $i \in I$ (an index set). Suppose that the sets ri C_i have at least one point in common. Then*

$$\text{cl} \bigcap \{C_i \mid i \in I\} = \bigcap \{\text{cl } C_i \mid i \in I\}.$$

If I is finite, then also

$$\text{ri} \bigcap \{C_i \mid i \in I\} = \bigcap \{\text{ri } C_i \mid i \in I\}.$$

PROOF. Fix any x in the intersection of the sets ri C_i. Given any y in the intersection of the sets cl C_i, the vector $(1 - \lambda)x + \lambda y$ belongs to every ri C_i for $0 \leq \lambda < 1$ by Theorem 6.1, and y is the limit of this vector as $\lambda \uparrow 1$. It follows that

$$\bigcap_i \text{cl } C_i \subset \text{cl} \bigcap_i \text{ri } C_i \subset \text{cl} \bigcap_i C_i \subset \bigcap_i \text{cl } C_i.$$

This establishes the closure formula in the theorem, and it proves at the same time that $\bigcap_i$ ri C_i and $\bigcap_i C_i$ have the same closure. By Corollary 6.3.1, these last two sets must also have the same relative interior. Therefore

$$\text{ri} \bigcap_i C_i \subset \bigcap_i \text{ri } C_i.$$

Assuming I is finite, we now demonstrate the opposite inclusion. Take any $z \in \bigcap_i$ ri C_i. By Theorem 6.4, any line segment in $\bigcap_i C_i$ with z as endpoint can be prolonged slightly beyond z in each of the sets C_i. The intersection of these prolonged segments, since there are only finitely many of them, is a prolongation in $\bigcap_i C_i$ of the original segment. Thus $z \in$ ri $\bigcap_i C_i$ by the criterion of Theorem 6.4. ‖

The formulas in Theorem 6.5 can fail when the sets ri C_i do not have a point in common, as is shown by the case where $I = \{1, 2\}$, C_1 is the positive orthant in R^2 with the origin adjoined, and C_2 is the "horizontal axis" of R^2. The finiteness of I in the second formula is also necessary: the intersection of the real intervals $[0, 1 + \alpha]$ for $\alpha > 0$ is $[0, 1]$, but the intersection of the intervals ri $[0, 1 + \alpha]$ for $\alpha > 0$ is not ri $[0, 1]$.

COROLLARY 6.5.1. *Let C be a convex set, and let M be an affine set (such as a line or a hyperplane) which contains a point of* ri C. *Then*

$$\text{ri } (M \cap C) = M \cap \text{ri } C, \qquad \text{cl } (M \cap C) = M \cap \text{cl } C.$$

PROOF. ri $M = M = $ cl M for an affine set. ∥

COROLLARY 6.5.2. *Let C_1 be a convex set. Let C_2 be a convex set contained in* cl C_1 *but not entirely contained in the relative boundary of C_1. Then* ri $C_2 \subset$ ri C_1.

PROOF. The hypothesis implies ri C_2 has a point in common with ri (cl C_1) = ri C_1, for otherwise the relative boundary cl $C_1 \setminus$ ri C_1, which is a closed set, would contain ri C_2 and its closure cl C_2. Hence

$$\text{ri } C_2 \cap \text{ri } C_1 = \text{ri } C_2 \cap \text{ri } (\text{cl } C_1) = \text{ri } (C_2 \cap \text{cl } C_1) = \text{ri } C_2,$$

i.e. ri $C_2 \subset$ ri C_1. ∥

THEOREM 6.6. *Let C be a convex set in R^n, and let A be a linear transformation from R^n to R^m. Then*

$$\text{ri } (AC) = A(\text{ri } C), \qquad \text{cl } (AC) \supset A(\text{cl } C).$$

PROOF. The closure inclusion merely reflects the fact that a linear transformation is continuous; it does not depend on C being convex. To prove the formula for relative interiors, we argue first that

$$\text{cl } A(\text{ri } C) \supset A(\text{cl } (\text{ri } C)) = A(\text{cl } C) \supset AC \supset A(\text{ri } C).$$

This implies that AC has the same closure as $A(\text{ri } C)$, and hence also the same relative interior by Corollary 6.3.1. Therefore ri $(AC) \subset A(\text{ri } C)$. Suppose now that $z \in A$ (ri C). We shall use Theorem 6.4 to show that $z \in$ ri (AC). Let x be any point of AC. Choose any elements $z' \in$ ri C and $x' \in C$, such that $Az' = z$ and $Ax' = x$. There exists some $\mu > 1$ such that the vector $(1 - \mu)x' + \mu z'$ belongs to C. The image of this vector under A is $(1 - \mu)x + \mu z$. Thus, for the same $\mu > 1$, $(1 - \mu)x + \mu z$ belongs to AC. Therefore $z \in$ ri (AC). ∥

The possible discrepancy in Theorem 6.6 between cl (AC) and $A(\text{cl } C)$, and how to ensure against it, will be discussed in §9.

COROLLARY 6.6.1. *For any convex set C and any real number λ,* ri $(\lambda C) = \lambda$ ri C.

PROOF. Take $A: x \rightarrow \lambda x$. ∥

It is elementary that, for the direct sum $C_1 \oplus C_2$ in R^{m+p} of convex sets

$C_1 \subset R^m$ and $C_2 \subset R^p$, one has

$$\text{ri } (C_1 \oplus C_2) = \text{ri } C_1 \oplus \text{ri } C_2,$$
$$\text{cl } (C_1 \oplus C_2) = \text{cl } C_1 \oplus \text{cl } C_2.$$

When this is combined with Theorem 6.6, we get the following fact.

COROLLARY 6.6.2. *For any convex sets C_1 and C_2 in R^n,*

$$\text{ri } (C_1 + C_2) = \text{ri } C_1 + \text{ri } C_2,$$
$$\text{cl } (C_1 + C_2) \supset \text{cl } C_1 + \text{cl } C_2.$$

PROOF. $C_1 + C_2 = A(C_1 \oplus C_2)$, where A is the addition linear transformation from R^{2n} to R^n, i.e. $A : (x_1, x_2) \rightarrow x_1 + x_2$. ‖

Corollary 6.6.2 will be sharpened in Corollaries 9.1.1 and 9.1.2.

THEOREM 6.7. *Let A be a linear transformation from R^n to R^m. Let C be a convex set in R^m such that $A^{-1}(\text{ri } C) \neq \emptyset$. Then*

$$\text{ri } (A^{-1}C) = A^{-1}(\text{ri } C), \qquad \text{cl } (A^{-1}C) = A^{-1}(\text{cl } C).$$

PROOF. Let $D = R^n \oplus C$, and let M be the graph of A. Then M is an affine set (in fact a subspace as explained in §1), and M contains a point of ri D. Let P be the projection $(x, y) \rightarrow x$ from R^{n+m} to R^n. Then $A^{-1}C = P(M \cap D)$. Calculating with the rules in Theorem 6.6 and Corollary 6.5.1, we get

$$\text{ri } (A^{-1}C) = P(\text{ri } (M \cap D)) = P(M \cap \text{ri } D) = A^{-1}(\text{ri } C),$$
$$\text{cl } (A^{-1}C) \supset P(\text{cl } (M \cap D)) = P(M \cap \text{cl } D) = A^{-1}(\text{cl } C).$$

The remaining inclusion cl $(A^{-1}C) \subset A^{-1}(\text{cl } C)$ is implied by the continuity of A. ‖

A counterexample for Theorem 6.7, in the case where the relative interior condition is violated, is obtained when $m = n = 2$, C is the positive orthant of R^2 with the origin adjoined, and A maps (ξ_1, ξ_2) onto $(\xi_1, 0)$.

The class of relatively open convex sets is preserved under finite intersections, scalar multiplication, addition, and taking images or inverse images under linear (or affine) transformations, according to the results above.

THEOREM 6.8. *Let C be a convex set in R^{m+p}. For each $y \in R^m$, let C_y be the set of vectors $z \in R^p$ such that $(y, z) \in C$. Let $D = \{y \mid C_y \neq \emptyset\}$. Then $(y, z) \in \text{ri } C$ if and only if $y \in \text{ri } D$ and $z \in \text{ri } C_y$.*

PROOF. The projection $(y, z) \rightarrow y$ carries C onto D, and hence ri C onto ri D by Theorem 6.6. For a given $y \in \text{ri } D$ and the affine set $M = \{(y, z) \mid z \in R^p\}$, the points of ri C projecting onto y are the points of

$$M \cap \text{ri } C = \text{ri } (M \cap C) = \{(y, z) \mid z \in \text{ri } C_y\}.$$

The first inequality in the latter formula is justified by Corollary 6.5.1. Thus, for any given $y \in$ ri D, we have $(y, z) \in$ ri C if and only if $z \in$ ri C_y, and this proves the result. ‖

COROLLARY 6.8.1. *Let C be a non-empty convex set in R^n, and let K be the convex cone in R^{n+1} generated by $\{(1, x) \mid x \in C\}$. Then* ri K *consists of the pairs (λ, x) such that $\lambda > 0$ and $x \in \lambda$* ri C.

PROOF. Apply the theorem with $R^m = R$, $R^p = R^n$. ‖

The reader can show as a simple exercise that, more generally, the relative interior of the convex cone in R^n generated by a non-empty convex set C consists of the vectors of the form λx with $\lambda > 0$ and $x \in$ ri C. A formula for the closure of this cone will be given in Theorem 9.8.

Observe that the relative interior and the closure of a convex cone are always convex cones too. This is immediate from Corollary 6.6.1, because a convex set C is a convex cone if and only if $\lambda C = C$ for every $\lambda > 0$.

THEOREM 6.9. *Let $C_1, \ldots, C_m$ be non-empty convex sets in R^n, and let $C_0 = $ conv $(C_1 \cup \cdots \cup C_m)$. Then*

$$\text{ri } C_0 = \bigcup \{\lambda_1 \text{ ri } C_1 + \cdots + \lambda_m \text{ ri } C_m \mid \lambda_i > 0, \lambda_1 + \cdots + \lambda_m = 1\}.$$

PROOF. Let K_i be the convex cone in R^{n+1} generated by $\{(1, x_i) \mid x_i \in C_i\}$, $i = 0, 1, \ldots, m$. Then

$$K_0 = \text{conv } (K_1 \cup \cdots \cup K_m) = K_1 + \cdots + K_m$$

(Theorem 3.8), and hence by Corollary 6.6.2

$$\text{ri } K_0 = \text{ri } K_1 + \cdots + \text{ri } K_m.$$

By Corollary 6.8.1, ri K_i consists of the pairs (λ_i, x_i) such that $\lambda_i > 0$, $x_i \in \lambda_i$ ri C_i. Thus $x_0 \in$ ri C_0 is equivalent to $(1, x_0) \in$ ri K_0, and that is equivalent in turn to

$$x_0 \in (\lambda_1 \text{ ri } C_1 + \cdots + \lambda_m \text{ ri } C_m)$$

some choice of $\lambda_1 > 0, \ldots, \lambda_m > 0$ with $\lambda_1 + \cdots + \lambda_m = 1$. ‖

The closure of the C_0 in Theorem 6.9 will be considered in Theorem 9.8.

Closures of Convex Functions

The continuity of a linear function is a consequence of an algebraic property, linearity. With convex functions, things are not quite so simple, but still a great many topological properties are implied by convexity alone. These can be deduced by applying the theory of closures and relative interiors of convex sets to the epigraphs or level sets of convex functions. One of the principal conclusions which can be reached is that lower semi-continuity is a "constructive" property for convex functions. It will by demonstrated below, namely, that there is a simple closure operation which makes any proper convex function lower semi-continuous merely by redefining it at certain relative boundary points of its effective domain.

Recall that, by definition, an extended-real-valued function f given on a set $S \subset R^n$ is said to be *lower semi-continuous* at a point x of S if

$$f(x) \leq \lim_{i \to \infty} f(x_i)$$

for every sequence $x_1, x_2, \ldots$, in S such that x_i converges to x and the limit of $f(x_1), f(x_2), \ldots$, exists in $[-\infty, +\infty]$. This condition may be expressed as:

$$f(x) = \lim_{y \to x} \inf f(y) = \lim_{\varepsilon \downarrow 0} (\inf \{f(y) \mid |y - x| \leq \varepsilon\}).$$

Similarly, f is said to be *upper semi-continuous* at x if

$$f(x) = \lim_{y \to x} \sup f(y) = \lim_{\varepsilon \downarrow 0} (\sup \{f(y) \mid |y - x| \leq \varepsilon\}).$$

The combination of lower and upper semi-continuity at x is ordinary continuity at x.

The natural importance of lower semi-continuity in the study of convex functions is apparent from the following result.

THEOREM 7.1. *Let f be an arbitrary function from R^n to $[-\infty, +\infty]$. Then the following conditions are equivalent:*

(a) *f is lower semi-continuous throughout R^n;*
(b) *$\{x \mid f(x) \leq \alpha\}$ is closed for every $\alpha \in R$;*
(c) *The epigraph of f is a closed set in R^{n+1}.*

PROOF. Lower semi-continuity at x can be reexpressed as the condition that $\mu \geq f(x)$ whenever $\mu = \lim \mu_i$ and $x = \lim x_i$ for sequences $\mu_1, \mu_2, \ldots,$ and $x_1, x_2, \ldots,$ such that $\mu_i \geq f(x_i)$ for every i. But this condition is the same as (c). It also implies (b) (take $\alpha = \mu = \mu_1 = \mu_2 = \cdots$). On the other hand, suppose (b) holds. Suppose x_i converges to x and $f(x_i)$ converges to μ. For every real $\alpha > \mu$, $f(x_i)$ must ultimately be less than α, and hence

$$x \in \text{cl} \{y \mid f(y) \leq \alpha\} = \{y \mid f(y) \leq \alpha\}.$$

Hence $f(x) \leq \mu$. This proves (b) implies (a). ‖

Given any function f on R^n, there exists a greatest lower semi-continuous function (not necessarily finite) majorized by f, namely the function whose epigraph is the closure in R^{n+1} of the epigraph of f. In general, this function is called the *lower semi-continuous hull* of f.

The *closure* of a convex function f is defined to be the lower semi-continuous hull of f if f nowhere has the value $-\infty$, whereas the closure of f is defined to be the constant function $-\infty$ if f is an improper convex function such that $f(x) = -\infty$ for some x. Either way, the closure of f is another convex function; it is denoted by $\text{cl} f$. (The purpose of the exception in the definition of $\text{cl} f$ is to make the formula $f^{**} = \text{cl} f$ in Theorem 12.2 valid even when f is improper, as is often convenient, especially in the theory of saddle-functions.)

A convex function is said to be *closed* if $\text{cl} f = f$. *For a proper convex function, closedness is thus the same as lower semi-continuity.* But the only closed improper convex functions are the constant functions $+\infty$ and $-\infty$.

If f is a proper convex function such that dom f is closed and f is continuous relative to dom f, then f is closed by criterion (b) of Theorem 7.1. However, a convex function can be closed without its effective domain being closed, for example the function on R given by $f(x) = 1/x$ when $x > 0$, $f(x) = \infty$ when $x \leq 0$.

Suppose f is a proper convex function. Then

$$\text{epi} \, (\text{cl} f) = \text{cl} \, (\text{epi} f)$$

by definition. It is clear from this and the proof of Theorem 7.1 that $\text{cl} f$ can be expressed by the formula

$$(\text{cl} f)(x) = \liminf_{y \to x} f(y).$$

Alternatively, $(\text{cl} f)(x)$ can be regarded as the infimum of values of μ such that x belongs to cl $\{x \mid f(x) \leq \mu\}$. Thus

$$\{x \mid (\text{cl} f)(x) \leq \alpha\} = \bigcap_{\mu > \alpha} \text{cl} \{x \mid f(x) \leq \mu\}.$$

In any case $\operatorname{cl} f \leq f$, and $f_1 \leq f_2$ implies $\operatorname{cl} f_1 \leq \operatorname{cl} f_2$. The functions f and $\operatorname{cl} f$ plainly have the same infimum on R^n.

To get a good idea of what the closure operation is like, consider the convex function f on R defined by $f(x) = 0$ for $x > 0, f(x) = \infty$ for $x \leq 0$. Here $\operatorname{cl} f$ agrees with f everywhere except at the origin, where its value is 0 instead of $+\infty$. For another example, take any circular disk C in R^2. Let $f(x)$ be 0 in the interior of C and $+\infty$ outside of C, and assign *arbitrary* values in $[0, \infty]$ to f on the boundary of C. Then f is a proper convex function on R^n. The closure of f is obtained by redefining $f(x)$ to be 0 on the boundary of C.

These examples suggest that the closure operation is a reasonable normalization which makes convex functions more regular by redefining their values at certain points where there are unnatural discontinuities. This is the secret of the great usefulness of the operation in theory and in applications. It usually enables one to reduce a given situation, without significant loss of generality, to the case where the convex functions in the situation are closed. The functions then have the three important properties in Theorem 7.1.

We proceed now with the detailed comparison of $\operatorname{cl} f$ and f in the general case. It is expedient to treat improper convex functions first. For this we need the following fact, which is really the chief theorem that can be proved about improper convex functions.

THEOREM 7.2. *If f is an improper convex function, then $f(x) = -\infty$ for every $x \in \operatorname{ri}(\operatorname{dom} f)$. Thus an improper convex function is necessarily infinite except perhaps at relative boundary points of its effective domain.*

PROOF. If the effective domain of f contains any points at all, it contains (by the definition of "improper") points where f has the value $-\infty$. Let u be such a point, and let $x \in \operatorname{ri}(\operatorname{dom} f)$. By Theorem 6.4, there exists a $\mu > 1$ such that $y \in \operatorname{dom} f$, where $y = (1 - \mu)u + \mu x$. We have $x = (1 - \lambda)u + \lambda y$, where $0 < \lambda = \mu^{-1} < 1$. Hence by Theorem 4.2

$$f(x) = f((1 - \lambda)u + \lambda y) < (1 - \lambda)\alpha + \lambda\beta$$

for any $\alpha > f(u)$ and $\beta > f(y)$. Since $f(u) = -\infty$ and $f(y) < +\infty, f(x)$ must be $-\infty$. ‖

COROLLARY 7.2.1. *A lower semi-continuous improper convex function can have no finite values.*

PROOF. The set of points x where $f(x) = -\infty$ must include $\operatorname{cl}(\operatorname{ri}(\operatorname{dom} f))$ by lower semi-continuity, and

$$\operatorname{cl}(\operatorname{ri}(\operatorname{dom} f)) = \operatorname{cl}(\operatorname{dom} f) \supset \operatorname{dom} f$$

by Theorem 6.3. ‖

COROLLARY 7.2.2. *Let f be an improper convex function. Then* cl f *is a closed improper convex function which agrees with f on* ri (dom f).

According to these results, the closure of a convex function f which has the value $-\infty$ somewhere is not so drastically different from the lower semi-continuous hull $\check{f}$ of f as might have been gathered from the seeming arbitrariness of the definition. Indeed, $\check{f}(x)$ is $-\infty$ on cl (dom f) and $+\infty$ outside cl (dom f), whereas $(\text{cl} f)(x)$ is $-\infty$ everywhere, for such a function f.

We would like to point out another consequence of Theorem 7.2 in passing, even though it has nothing to do with the main topic of this section, lower semi-continuity.

COROLLARY 7.2.3. *If f is a convex function whose effective domain is relatively open (for instance if* dom $f = R^n$), *then either $f(x) > -\infty$ for every x or $f(x)$ is infinite for every x.*

As a typical application of this corollary (and therefore of the theory of improper convex functions), consider any finite convex function f on R^2. The function

$$g(\xi_1) = \inf_{\xi_2} f(\xi_1, \xi_2)$$

is convex (see the comment after Theorem 5.7), and its effective domain is R. We may conclude that the infimum is finite for every ξ_1 or it is $-\infty$ for every ξ_1. Thus, if f is bounded below along just one of the lines parallel to the ξ_2-axis, it is bounded below along *every* such line.

The most important topological property of convex sets in R^n is the intimate relationship between their closures and relative interiors. Since closing a proper convex function f amounts to closing epi f, the relative interior of epi f will understandably be important in the analysis of cl f.

LEMMA 7.3. *For any convex function f,* ri (epi f) *consists of the pairs (x, μ) such that $x \in$ ri (dom f) and $f(x) < \mu < \infty$.*

PROOF. This result is the special case of Theorem 6.8 where $m = n$, $p = 1$ and $C = $ epi f, and it can easily be deduced directly from Theorems 6.4 and 6.1. However, we shall also furnish an alternative proof. It suffices actually to show that

$$\text{int (epi} f) = \{(x, \mu) \mid x \in \text{int (dom} f), f(x) < \mu < \infty\}.$$

The inclusion $\subset$ is obvious, so only $\supset$ needs verification. Let $\bar{x} \in$ int (dom f), and let $\bar{\mu}$ be a real number such that $\bar{\mu} > f(\bar{x})$. Let $a_1, \ldots, a_r$ be points of dom f such that $\bar{x} \in$ int P, where

$$P = \text{conv} \{a_1, \ldots, a_r\},$$

and let

$$\alpha = \max \{f(a_i) \mid i = 1, \ldots, r\}.$$

Given any $x \in P$, we can express x as a convex combination

$$x = \lambda_1 a_1 + \cdots + \lambda_r a_r, \qquad \lambda_i \geq 0, \qquad \lambda_1 + \cdots + \lambda_r = 1,$$

and therefore

$$f(x) \leq \lambda_1 f_1(a_1) + \cdots + \lambda_r f_r(a_r) \leq (\lambda_1 + \cdots + \lambda_r)\alpha = \alpha.$$

Hence the open set

$$\{(x, \mu) \mid x \in \text{int } P, \, \alpha < \mu < \infty\}$$

is included in epi f. In particular, for every $\mu > \alpha$ we have

$$(\bar{x}, \mu) \in \text{int (epi} f),$$

and it follows that $(\bar{x}, \bar{\mu})$ can be viewed as a relative interior point of a "vertical" line segment in epi f which meets int (epi f). This implies

$$(\bar{x}, \bar{\mu}) \in \text{int (epi} f)$$

by Theorem 6.1. ‖

COROLLARY 7.3.1.   *Let α be a real number, and let f be a convex function such that, for some x, $f(x) < \alpha$. Then actually $f(x) < \alpha$ for some $x \in \text{ri (dom} f)$.*

PROOF.   If the open half-space $\{(x, \mu) \mid x \in R^n, \mu < \alpha\}$ in R^{n+1} meets epi f, then it also must meet ri (epi f) (Corollary 6.3.2). ‖

COROLLARY 7.3.2.   *Let f be a convex function, and let C be a convex set such that $\text{ri } C \subset \text{dom } f$. Let α be a real number such that $f(x) < \alpha$ for some $x \in \text{cl } C$. Then actually $f(x) < \alpha$ for some $x \in \text{ri } C$.*

PROOF.   Let $g(x) = f(x)$ for $x \in \text{cl } C$, $g(x) = +\infty$ for $x \notin \text{cl } C$. Then

$$\text{ri } C \subset \text{dom } g \subset \text{cl } C,$$

and hence ri (dom g) = ri C. By hypothesis, there is an x such that $g(x) < \alpha$. Then $g(x) < \alpha$ for some $x \in \text{ri (dom} g)$ by the preceding corollary. In other words, $f(x) < \alpha$ for some $x \in \text{ri } C$. ‖

COROLLARY 7.3.3.   *Let f be a convex function on R^n, and let C be a convex set on which f is finite. If $f(x) \geq \alpha$ for every $x \in C$, then also $f(x) \geq \alpha$ for every $x \in \text{cl } C$.*

PROOF.   This is obvious from the preceding corollary. ‖

Another easy consequence of Lemma 7.3 is the fact that the closure of a convex function f is completely determined by the restriction of f to ri (dom f):

COROLLARY 7.3.4.   *If f and g are convex functions on R^n such that*

$$\text{ri (dom} f) = \text{ri (dom} g),$$

and f and g agree on the latter set, then $\text{cl } f = \text{cl } g$.

PROOF. The hypothesis implies that

$$\text{ri (epi} f) = \text{ri (epi} g)$$

and hence by Theorem 6.3 that

$$\text{cl (epi} f) = \text{cl (epi} g).$$

This relation says precisely that $\text{cl } f = \text{cl } g$, at least if f and g are proper. In the case of improper functions, the conclusion follows trivially from Theorem 7.2. ‖

The most important theorem about $\text{cl } f$ is the following.

THEOREM 7.4. *Let f be a proper convex function on R^n. Then $\text{cl } f$ is a closed proper convex function. Moreover, $\text{cl } f$ agrees with f except perhaps at relative boundary points of $\text{dom } f$.*

PROOF. Since $\text{epi (cl } f) = \text{cl (epi} f)$, and $\text{epi} f$ is convex, $\text{epi (cl } f)$ is a closed convex set in R^{n+1} and $\text{cl } f$ is a lower semi-continuous convex function. The properness of $\text{cl } f$, and hence also its closedness, will follow from the last assertion of the theorem in view of Corollary 7.2.1, because f is finite on $\text{dom } f$. Given any $x \in \text{ri (dom} f)$, consider the vertical line $M = \{(x, \mu) \mid \mu \in R\}$. This M meets $\text{ri (epi} f)$ by Lemma 7.3. Hence

$$M \cap \text{cl (epi} f) = \text{cl } (M \cap \text{epi} f) = M \cap \text{epi} f$$

by Corollary 6.5.1 (or by an argument directly based on Theorem 6.1). This says $(\text{cl } f)(x) = f(x)$. Now suppose on the other hand that $x \notin \text{cl (dom} f)$. From the "lim inf" formula for $\text{cl } f$ we have

$$\text{cl (dom} f) \supset \text{dom (cl } f) \supset \text{dom } f,$$

and hence $(\text{cl } f)(x) = \infty = f(x)$. ‖

COROLLARY 7.4.1. *If f is a proper convex function, then $\text{dom (cl } f)$ differs from $\text{dom } f$ at most by including some additional relative boundary points of $\text{dom } f$. In particular, $\text{dom (cl } f)$ and $\text{dom } f$ have the same closure and relative interior, as well as the same dimension.*

COROLLARY 7.4.2. *If f is a proper convex function such that $\text{dom } f$ is an affine set (which is true in particular if f is finite throughout R^n), then f is closed.*

PROOF. Here $\text{dom } f$ has no relative boundary points, so $\text{cl } f$ agrees with f everywhere. ‖

Theorems 7.2 and 7.4 imply that a convex function f is always lower semi-continuous except perhaps at relative boundary points of $\text{dom } f$. We shall see in §10 that f is actually continuous relative to $\text{ri (dom } f)$.

Various formulas for the closures of convex functions constructed by the operations in §5 will be derived in §9.

The closure operation for convex functions has been described in terms of a "lim inf". We can now show that much simpler limits really suffice for calculating cl f from f.

THEOREM 7.5. *Let f be a proper convex function, and let $x \in$ ri (dom f).* *Then*

$$(cl f)(y) = \lim_{\lambda \uparrow 1} f((1 - \lambda)x + \lambda y)$$

for every y. (The formula is also valid when f is improper and $y \in$ cl (dom f).)

PROOF. Since cl f is lower semi-continuous and cl $f \leq f$, we have

$$(cl f)(y) \leq \liminf_{\lambda \uparrow 1} f((1 - \lambda)x + \lambda y).$$

We only need to show that

$$(cl f)(y) \geq \limsup_{\lambda \uparrow 1} f((1 - \lambda)x + \lambda y)$$

as well. Assume β is any real number such that $\beta \geq (cl f)(y)$. Take any real number $\alpha > f(x)$. Then

$$(y, \beta) \in \text{epi } (cl f) = cl \text{ (epi } f),$$

while $(x, \alpha) \in$ ri (epi f) by Lemma 7.3. Therefore

$$(1 - \lambda)(x, \alpha) + \lambda(y, \beta) \in \text{ri (epi } f), \qquad 0 \leq \lambda < 1$$

(Theorem 6.1), so that

$$f((1 - \lambda)x + \lambda y) < (1 - \lambda)\alpha + \beta, \qquad 0 \leq \lambda < 1.$$

Consequently

$$\limsup_{\lambda \uparrow 1} f((1 - \lambda)x + \lambda y) \leq \limsup_{\lambda \uparrow 1} [(1 - \lambda)\alpha + \lambda\beta] = \beta,$$

which is the desired conclusion. The formula also holds when f is improper and $y \in$ cl (dom f), because then $f((1 - \lambda)x + \lambda y) = -\infty$ for $0 \leq \lambda < 1$ by Theorem 6.1 and Theorem 7.2. ‖

COROLLARY 7.5.1. *For a closed proper convex function f, one has*

$$f(y) = \lim_{\lambda \uparrow 1} f((1 - \lambda)x + \lambda y)$$

for every $x \in$ dom f and every y.

PROOF. Let $\varphi(\lambda) = f((1 - \lambda)x + \lambda y)$. Then φ is a proper convex function on R with $\varphi(0) = f(x) < \infty$ and $\varphi(1) = f(y)$. Moreover, φ is lower semi-continuous by Theorem 7.1, since $\{\lambda \mid \varphi(\lambda) \leq \alpha\}$ is the inverse image of the closed set $\{z \mid f(z) \leq \alpha\}$ under the continuous transformation $\lambda \to (1 - \lambda)x + \lambda y = z$. The effective domain of φ is a certain interval.

If interior points of the interval lie between 0 and 1, then the limit of $\varphi(\lambda)$ as $\lambda \uparrow 1$ is $(\operatorname{cl} \varphi)(1) = \varphi(1)$ by the theorem. Otherwise the limit and $\varphi(1)$ are both trivially $+\infty$.　‖

Theorem 7.5 and Corollary 7.5.1 will be extended in Theorems 10.2 and 10.3.

Sometimes Theorem 7.5 is useful in showing that a given function is convex. For instance, let $f(x) = -(1 - |x|^2)^{1/2}$ for $|x| \leq 1$, $f(x) = +\infty$ for $|x| > 1 (x \in R^n)$. The effective domain of f is the unit ball $B = \{x \mid |x| \leq 1\}$. On the interior of B, the convexity of f can be proved from the second partial derivative condition (Theorem 4.5). Since the values of f on the boundary of B are the limits of its values along radii, Theorem 7.5 then implies f is a closed proper convex function.

In the theory of inequalities and elsewhere, level sets of the form $\{x \mid f(x) \leq \alpha\}$ are naturally important. The advantage of being able to arrange, by means of the closure operation for convex functions, that such sets are closed, is clear enough. The relative interiors of such sets, likewise, are conveniently obtained from the function f itself, as we now show.

THEOREM 7.6.　*Let f be any proper convex function, and let $\alpha \in R$, $\alpha > \inf f$. The convex level sets $\{x \mid f(x) \leq \alpha\}$ and $\{x \mid f(x) < \alpha\}$ then have the same closure and the same relative interior, namely*

$$\{x \mid (\operatorname{cl} f)(x) \leq \alpha\}, \qquad \{x \in \operatorname{ri}(\operatorname{dom} f) \mid f(x) < \alpha\},$$

respectively. Furthermore, they have the same dimension as $\operatorname{dom} f$ *(and f).*

PROOF.　Let M be the horizontal hyperplane $\{(x, \alpha) \mid x \in R^n\}$ in R^{n+1}. By Corollary 7.3.1 and Lemma 7.3, M meets $\operatorname{ri}(\operatorname{epi} f)$. We are concerned with the closure and relative interior of

$$M \cap \operatorname{epi} f = \{(x, \alpha) \mid f(x) \leq \alpha\}.$$

By Corollary 6.5.1, these are $M \cap \operatorname{cl}(\operatorname{epi} f)$ and $M \cap \operatorname{ri}(\operatorname{epi} f)$, respectively. Of course, $\operatorname{cl}(\operatorname{epi} f) = \operatorname{epi}(\operatorname{cl} f)$. Therefore

$$\operatorname{cl} \{x \mid f(x) \leq \alpha\} = \{x \mid (\operatorname{cl} f)(x) \leq \alpha\},$$
$$\operatorname{ri} \{x \mid f(x) \leq \alpha\} = \{x \in \operatorname{ri}(\operatorname{dom} f) \mid f(x) < \alpha\}.$$

The latter formula implies that

$$\operatorname{ri} \{x \mid f(x) \leq \alpha\} \subset \{x \mid f(x) < \alpha\} \subset \{x \mid f(x) \leq \alpha\},$$

and hence that $\{x \mid f(x) < \alpha\}$ has the same closure and relative interior as $\{x \mid f(x) \leq \alpha\}$ (Corollary 6.3.1). The dimensions of these sets are equal by Theorem 6.2. They coincide in fact with the dimension of $M \cap \operatorname{ri}(\operatorname{epi} f)$, which is obviously one less than the dimension of $\operatorname{ri}(\operatorname{epi} f)$. The latter dimension is one more than the dimension of $\operatorname{dom} f$.　‖

COROLLARY 7.6.1. *If f is a closed proper convex function whose effective domain is relatively open (in particular if dom f is an affine set), then for* $\inf f < \alpha < +\infty$ *one has*

$$\text{ri} \{x \,|\, f(x) \leq \alpha\} = \{x \,|\, f(x) < \alpha\},$$
$$\text{cl} \{x \,|\, f(x) < \alpha\} = \{x \,|\, f(x) \leq \alpha\}.$$

PROOF. Here $\text{cl} f = f$ and $\text{ri} (\text{dom} f) = \text{dom} f$. ‖

The relationship in the corollary depends on the convexity of f, not just on the convexity of the level sets. For example, consider the non-convex function f on R defined by

$$f(x) = \begin{cases} 0 & \text{if } |x| \leq 1, \\ 1 & \text{if } |x| > 1. \end{cases}$$

All the level sets $\{x \,|\, f(x) \leq \alpha\}$ and $\{x \,|\, f(x) < \alpha\}$ of this function are convex. Moreover, f is lower semi-continuous (by condition (b) of Theorem 7.1), and its "effective domain" is relatively open, being all of R. But $\{x \,|\, f(x) < 1\}$ is not the relative interior of $\{x \,|\, f(x) \leq 1\}$, nor is $\{x \,|\, f(x) \leq 1\}$ the closure of $\{x \,|\, f(x) < 1\}$.

All the closure and relative interior formulas in Theorem 7.6 and Corollary 7.6.1 are trivially valid also when $\alpha < \inf f$, because all the sets in question are empty in that case. The formulas can fail when $\alpha = \inf f$ since then $\{x \,|\, f(x) < \alpha\}$ is empty but $\{x \,|\, f(x) \leq \alpha\}$ might not be empty.

Recession Cones and Unboundedness

Closed bounded subsets of R^n are usually easier to work with than unbounded ones. When the sets are convex, however, the difficulties with unboundedness are very much less, and that is fortunate, since so many of the sets we need to consider, like epigraphs, are unbounded by their nature.

Unbounded closed convex sets have a simple behavior "at infinity," according to one's intuition. Suppose that C is such a set and x is a point of C. It seems that C must actually contain some entire half-line starting at x, or the unboundedness would be contradicted. The *directions* of such half-lines seem not to depend on x: the half-lines in C starting at a different point y are apparently just the translates of those starting at x. These directions in which C recedes indefinitely might possibly be thought of as ideal points of C lying at infinity, "horizon points," after the fashion of projective geometry. The half-lines in C starting at x could then be interpreted as the segments joining x with such ideal points of C.

The objective below is to put these intuitive notions on a sound mathematical foundation and to apply them to the study of convex functions.

Let us first see how the concept of "direction" can be formalized. Each closed half-line in R^n should have a "direction," and two should have the same "direction" if and only if they are translates of each other. We therefore define a *direction* of R^n simply to be an equivalence class of the collection of all closed half-lines of R^n under the equivalence relation "half-line L_1 is a translate of half-line L_2." The direction of the half-line $\{x + \lambda y \mid \lambda \geq 0\}$, where $y \neq 0$, is then by definition the set of all translates of the half-line, and that is independent of x. We shall also call this the *direction of y*. Two vectors in R^n have the same direction if and only if they are positive scalar multiples of each other. The zero vector has no direction. It is clear what one would mean by the *opposite* of a given direction.

Under the natural correspondence between points of R^n and points of the hyperplane $M = \{(1, x) \mid x \in R^n\}$ in R^{n+1}, a point $x \in R^n$ can be represented by the ray $\{\lambda(1,x) \mid \lambda \geq 0\}$. The directions of R^n can then be represented by the rays $\{\lambda(0, y) \mid \lambda \geq 0\}$, $y \neq 0$, lying in the hyperplane parallel to M through the origin of R^{n+1}. This suggests referring to the directions of R^n alternatively as *points of R^n at infinity*. (This usage differs from that of projective geometry, where a point at infinity is an equivalence class of parallel *lines;* each such point of projective geometry would correspond to a pair of opposite points at infinity in our sense.) Forming the convex hull of two rays in R^{n+1} which intersect M corresponds to forming the line segment between the points of R^n they represent. If one of the rays represents a point at infinity, one gets, instead of a line segment, a half-line with a certain endpoint and direction.

Let C be a non-empty convex set in R^n. We shall say that C *recedes in the direction D* if C includes all the half-lines in the direction D which start at points of C. In other words, C recedes in the direction of y, where $y \neq 0$, if and only if $x + \lambda y \in C$ for every $\lambda \geq 0$ and $x \in C$. The set of all vectors $y \in R^n$ satisfying the latter condition, including $y = 0$, will be called the *recession cone* of C. The recession cone of C will be denoted by 0^+C, for reasons to be explained shortly. Directions in which C recedes will also be referred to as *directions of recession* of C.

The recession cone of cl C has elsewhere been called the *asymptotic cone* of C. We shall not adopt that terminology here, since it does not really agree with other uses of "asymptote" and "asymptotic" and might be misleading.

THEOREM 8.1. *Let C be a non-empty convex set. The recession cone 0^+C is then a convex cone containing the origin. It is the same as the set of vectors y such that $C + y \subset C$.*

PROOF. Each $y \in 0^+C$ has the property that $x + y \in C$ for every $x \in C$, i.e. $C + y \subset C$. On the other hand, if $C + y \subset C$ then

$$C + 2y = (C + y) + y \subset C + y \subset C$$

and so forth, implying $x + my \in C$ for every $x \in C$ and positive integer m. The line segments joining the points $x \in C$, $x + y$, $x + 2y$, ..., are then all contained in C by convexity, so that $x + \lambda y \in C$ for every $\lambda \geq 0$. Thus $y \in 0^+C$. Since positive scalar multiplication does not change directions, 0^+C is truly a cone. It remains to be shown that 0^+C is convex. If y_1 and y_2 are vectors in 0^+C and $0 \leq \lambda \leq 1$, we have

$$(1 - \lambda)y_1 + \lambda y_2 + C = (1 - \lambda)(y_1 + C) + \lambda(y_2 + C)$$

$$\subset (1 - \lambda)C + \lambda C = C$$

(using the distributive law in Theorem 3.2). Hence $(1 - \lambda)y_1 + \lambda y_2$ is in 0^+C. ‖

As examples of recession cones of convex sets in R^2, for

$$C_1 = \{(\xi_1, \xi_2) \mid \xi_1 > 0, \xi_2 \geq 1/\xi_1\},$$
$$C_2 = \{(\xi_1, \xi_2) \mid \xi_2 \geq \xi_1^2\},$$
$$C_3 = \{(\xi_1, \xi_2) \mid \xi_1^2 + \xi_2^2 \leq 1\},$$
$$C_4 = \{(\xi_1, \xi_2) \mid \xi_1 > 0, \xi_2 > 0\} \cup \{(0, 0)\},$$

one has

$$0^+C_1 = \{(\xi_1, \xi_2) \mid \xi_1 \geq 0, \xi_2 \geq 0\},$$
$$0^+C_2 = \{(\xi_1, \xi_2) \mid \xi_1 = 0, \xi_2 \geq 0\},$$
$$0^+C_3 = \{(\xi_1, \xi_2) \mid \xi_1 = 0 = \xi_2\} = \{(0, 0)\}.$$
$$0^+C_4 = \{(\xi_1, \xi_2) \mid \xi_1 > 0, \xi_2 > 0\} \cup \{(0, 0)\} = C_4.$$

The recession cone of a non-empty affine set M is, of course, the subspace L parallel to M. If C is the set of solutions to a system of weak linear inequalities on R^n,

$$C = \{x \mid \langle x, b_i \rangle \geq \beta_i, \forall i \in I\} \neq \emptyset,$$

the recession cone of C is given by the corresponding system of homogeneous inequalities, as is easily verified:

$$0^+C = \{x \mid \langle x, b_i \rangle \geq 0, \forall i \in I\}.$$

When points of R^n are represented by rays in R^{n+1} in the manner described above, a non-empty convex set C is represented as the union of the rays representing its points. This union is the convex cone

$$K = \{(\lambda, x) \mid \lambda \geq 0, x \in \lambda C\},$$

which, except for the origin, lies entirely in the open half-space $\{(\lambda, x) \mid \lambda > 0\}$. Let us consider how K might be enlarged to a convex cone of the form $K \cup K_0$, where K_0 is a cone lying in the hyperplane $\{(0, x) \mid x \in R^n\}$. Since K is already a convex cone, for $K \cup K_0$ to be a convex cone it is necessary and sufficient that K_0 be convex and $K + K_0 \subset K \cup K_0$ (Theorem 2.6). We will have $K + K_0 \subset K \cup K_0$ if and only if each $(0, x) \in K_0$ has the property that $(1, x') + (0, x)$ belongs to K for every $(1, x') \in K$. This property means that $x' + x \in C$ for every $x' \in C$, and hence by Theorem 8.1 that $x \in 0^+C$. It follows that there exists a unique *largest* convex cone K' in the half-space $\{(\lambda, x) \mid \lambda \geq 0\}$ whose intersection with the half-space $\{(\lambda, x) \mid \lambda > 0\}$ is $K \setminus \{(0, 0)\}$, namely

$$K' = \{(\lambda, x) \mid \lambda > 0, x \in \lambda C\} \cup \{(0, x) \mid x \in 0^+C\}.$$

In this sense, 0^+C can be regarded as what happens to λC as $\lambda \to 0^+$, whence the notation.

THEOREM 8.2. *Let C be a non-empty closed convex set in R^n. Then 0^+C is closed, and it consists of all possible limits of sequences of the form $\lambda_1 x_1, \lambda_2 x_2, \ldots,$ where $x_i \in C$ and $\lambda_i \downarrow 0$. In fact, for the convex cone K in R^{n+1} generated by $\{(1, x) \mid x \in C\}$ one has*

$$\text{cl } K = K \cup \{(0, x) \mid x \in 0^+C\}.$$

PROOF. The hyperplane $M = \{(1, x) \mid x \in R^n\}$ must intersect ri K (e.g. by Corollary 6.8.1), so

$$M \cap \text{cl } K = \text{cl } (M \cap K) = M \cap K = \{(1, x) \mid x \in C\}$$

by the closure rule in Corollary 6.5.1. The cone K' defined just prior to the theorem must therefore contain cl K, because of its maximality property. On the other hand, since K' is contained in the half-space $H = \{(\lambda, x) \mid \lambda \geq 0\}$ and meets int H, ri K' must be entirely contained in int H (Corollary 6.5.2). Hence ri $K' \subset K$, and we have

$$\text{cl } K \subset K' \subset \text{cl } (\text{ri } K') \subset \text{cl } K.$$

This proves the formula cl $K = K'$ asserted in the theorem. The set $\{(0, x) \mid x \in 0^+C\}$ is the intersection of cl K with $\{(0, x) \mid x \in R^n\}$, so it is closed and consists of the limits of sequences of the form $\lambda_1(1, x_1),$ $\lambda_2(1, x_2), \ldots,$ where $x_i \in C$ and $\lambda_i \downarrow 0$. ‖

The fact that 0^+C can fail to be closed when C is not closed is shown by the set C_4 above.

Suppose that C is a closed convex set and z is a point such that, for some $x \in C$, the relative interior of the line segment between x and z lies in C. Then $z \in C$, so that the same property holds for *every* $x \in C$. The next theorem may be interpreted as a generalization of this fact to the case where z is a point at infinity.

THEOREM 8.3. *Let C be a non-empty closed convex set, and let $y \neq 0$. If there exists even one x such that the half-line $\{x + \lambda y \mid \lambda \geq 0\}$ is contained in C, then the same thing is true for every $x \in C$, i.e. one has $y \in 0^+C$. Moreover, then $\{x + \lambda y \mid \lambda \geq 0\}$ is actually contained in ri C for each $x \in$ ri C, so that $y \in 0^+(\text{ri } C)$.*

PROOF. Let $\{x + \lambda y \mid \lambda \geq 0\}$ be contained in C. Then y is the limit of the sequence $\lambda_1 x_1, \lambda_2 x_2, \ldots,$ where $\lambda_k = 1/k$ and $x_k = x + ky \in C$. Hence $y \in 0^+C$ by Theorem 8.2. The last assertion of the theorem is immediate from the fact that any line segment in C which meets ri C must have its relative interior in ri C (Theorem 6.1). ‖

COROLLARY 8.3.1. *For any non-empty convex set C, one has $0^+(\text{ri } C) = 0^+(\text{cl } C)$. In fact, given any $x \in$ ri C, one has $y \in 0^+(\text{cl } C)$ if and only if $x + \lambda y \in C$ for every $\lambda > 0$.*

COROLLARY 8.3.2.   *If C is a closed convex set containing the origin, then*

$$0^+C = \{y \mid \varepsilon^{-1}y \in C, \forall \varepsilon > 0\} = \bigcap_{\varepsilon > 0} \varepsilon C.$$

COROLLARY 8.3.3.   *If $\{C_i \mid i \in I\}$ is an arbitrary collection of closed convex sets in R^n whose intersection is not empty, then*

$$0^+(\bigcap_{i \in I} C_i) = \bigcap_{i \in I} 0^+C_i.$$

PROOF.   Let x be any point in the closed convex set $C = \bigcap_{i \in I} C_i$. The direction of a given vector y is a direction in which C recedes, if and only if the half-line $\{x + \lambda y \mid \lambda \geq 0\}$ is contained in every C_i. But the latter means that every C_i recedes in the direction of y.   ‖

COROLLARY 8.3.4.   *Let A be a linear transformation from R^n to R^m, and let C be a closed convex set in R^m such that $A^{-1}C \neq \emptyset$. Then $0^+(A^{-1}C) = A^{-1}(0^+C)$.*

PROOF.   Since A is continuous and C is closed, $A^{-1}C$ is closed. Take any $x \in A^{-1}C$. We have $y \in 0^+(A^{-1}C)$ if and only if, for every $\lambda \geq 0$, C contains $A(x + \lambda y) = Ax + \lambda Ay$. The latter means $Ay \in 0^+C$, i.e. $y \in A^{-1}(0^+C)$.   ‖

The first assertion of Theorem 8.3 is not valid when C is not closed: the C_4 above contains the half-line consisting of all points of the form $(1, 1) + \lambda(1, 0)$, but $(1, 0)$ does not belong to 0^+C_4. Observe also, in connection with Corollary 8.3.1, that $0^+(\text{ri } C_4)$ is properly larger than 0^+C_4.

An unbounded closed convex set contains at least one point at infinity, i.e. recedes in at least one direction, according to the next theorem. Its unboundedness, therefore, is really of the simplest sort that can be hoped for.

THEOREM 8.4.   *A non-empty closed convex set C in R^n is bounded if and only if its recession cone 0^+C consists of the zero vector alone.*

PROOF.   If C is bounded, it certainly contains no half-lines, so that $0^+C = \{0\}$. If C is unbounded, on the other hand, it contains a sequence of non-zero vectors $x_1, x_2, \ldots$, whose Euclidean norms $|x_i|$ increase without bound. The vectors $\lambda_i x_i$, where $\lambda_i = 1/|x_i|$, all belong to the unit sphere $S = \{x \mid |x| = 1\}$. Since S is a closed bounded subset of R^n, some subsequence of $\lambda_1 x_1, \lambda_2 x_2, \ldots$, will converge to a certain $y \in S$. This y is a non-zero vector of 0^+C by Theorem 8.2.   ‖

COROLLARY 8.4.1.   *Let C be a closed convex set, and let M be an affine set such that $M \cap C$ is non-empty and bounded. Then $M' \cap C$ is bounded for every affine set M' parallel to M.*

PROOF.   We have $0^+M' = 0^+M$ by definition of "parallel." Assuming $M' \cap C$ is not actually empty, we have

$$0^+(M' \cap C) = 0^+M' \cap 0^+C = 0^+M \cap 0^+C = 0^+(M \cap C)$$

by the intersection rule in Corollary 8.3.3. Since $M \cap C$ is bounded, this implies $0^+(M' \cap C) = 0$, and hence $M' \cap C$ is bounded. ‖

If C is a non-empty convex set, the set $(-0^+C) \cap 0^+C$ is called the *lineality space* of C. It consists of the zero vector and all the non-zero vectors y such that, for every $x \in C$, the line through x in the direction of y is contained in C. The directions of the vectors y in the lineality space are called *directions in which C is linear*. Of course, if C is closed and contains a certain line M, then all the lines parallel to M through points of C are contained in C. (This is a special case of Theorem 8.3.) The lineality space is the same as the set of vectors y such that $C + y = C$; this may be proved as an elementary exercise.

The lineality space of C is a subspace, the largest subspace contained in the convex cone 0^+C (Theorem 2.7). Its dimension is called the *lineality* of C.

Consider, for example, the cylinder

$$C = \{(\xi_1, \xi_2, \xi_3) \mid \xi_1^2 + \xi_2^2 \leq 1\} \subset R^3.$$

The lineality space of C is the ξ_3-axis, so that C has lineality 1. Here C is the direct sum of a line and a circular disk.

In general, if C is a non-empty convex set with a non-trivial lineality space L, one can obviously express C as the direct sum

$$C = L + (C \cap L^\perp),$$

where $L^\perp$ is the orthogonal complement of L. The lineality of the set $C \cap L^\perp$ in this expression is 0. The dimension of $C \cap L^\perp$, which is the dimension of C minus the lineality of C, is called the *rank* of C. It is a measure of the nonlinearity of C.

The convex sets of rank 0 are the affine sets. The rank of a closed convex set coincides with its dimension if and only if the set contains no lines.

In the case where

$$C = \{x \mid \langle x, b_i \rangle \geq \beta_i, \forall i \in I\},$$

the lineality space L of C is given by a system of equations:

$$L = \{x \mid \langle x, b_i \rangle = 0, \forall i \in I\}.$$

We turn now to the application of the above results to convex functions. Let f be a convex function on R^n not identically $+\infty$. The epigraph of f, as a non-empty convex set in R^{n+1}, has a recession cone $0^+(\mathrm{epi}\, f)$. By definition, $(y, \nu) \in 0^+(\mathrm{epi}\, f)$ if and only if

$$(x, \mu) + \lambda(y, \nu) = (x + \lambda y, \mu + \lambda \nu) \in \mathrm{epi}\, f$$

for every $(x, \mu) \in \mathrm{epi}\, f$ and $\lambda \geq 0$. This means that

$$f(x + \lambda y) \leq f(x) + \lambda \nu$$

for every x and every $\lambda \geq 0$. Actually, by Theorem 8.1, the latter inequality holds for every x and every $\lambda \geq 0$ if it merely holds for every x with $\lambda = 1$. At all events, for a given y, the values of ν for which $(y, \nu) \in 0^+(\text{epi } f)$ will form a closed interval of R unbounded above, or the empty interval. Thus $0^+(\text{epi } f)$ is the epigraph of a certain function. We call this function the *recession function* of f, and we denote it by $f0^+$. By definition, then,

$$\text{epi } (f0^+) = 0^+(\text{epi } f).$$

Thus the $f0^+$ notation is in line with our previous notation of right scalar multiplication in §5.

THEOREM 8.5. *Let f be a proper convex function. The recession function $f0^+$ of f is then a positively homogeneous proper convex function. For every vector y, one has*

$$(f0^+)(y) = \sup \{f(x + y) - f(x) \mid x \in \text{dom} f\}.$$

If f is closed, $f0^+$ is closed too, and for any $x \in \text{dom} f$, $f0^+$ is given by the formula

$$(f0^+)(y) = \sup_{\lambda > 0} \frac{f(x + \lambda y) - f(x)}{\lambda} = \lim_{\lambda \to \infty} \frac{f(x + \lambda y) - f(x)}{\lambda} .$$

PROOF. The first formula is a consequence of the observations just made. The condition $\nu \geq (f0^+)(y)$ also means that

$$\nu \geq \sup_{\lambda > 0} \{[f(x + \lambda y) - f(x)]/\lambda\}, \qquad \forall x \in \text{dom} f.$$

(Note from this that $(f0^+)(y)$ cannot be $-\infty$.) For any fixed $x \in \text{dom} f$, the supremum gives the smallest real ν (if any) such that epi f contains the half-line in the direction of (y, ν) with endpoint $(x, f(x))$. If f is closed, epi f is closed, and by Theorem 8.3 this ν is independent of x. This proves the second supremum formula in the theorem. The supremum is the same as the limit as $\lambda \to \infty$, because the difference quotient $[f(x + \lambda y) - f(x)]/\lambda$ is a non-decreasing function of λ by the convexity of f (see Theorem 23.1). The epigraph $0^+(\text{epi } f)$ is a non-empty convex cone, closed if f is closed; therefore $f0^+$ is a positively homogeneous proper convex function, closed if f is closed. ‖

COROLLARY 8.5.1. *Let f be a proper convex function. Then $f0^+$ is the least of the functions h such that*

$$f(z) \leq f(x) + h(z - x), \forall z, \forall x.$$

The recession function of f can be viewed in terms of a closure construction, when f is a closed proper convex function. Let g be the positively homogeneous convex function generated by h, where

$$h(\lambda, x) = f(x) + \delta(\lambda \mid 1).$$

In other words,

$$g(\lambda, x) = \begin{cases} (f\lambda)(x) & \text{if } \lambda \geq 0, \\ +\infty & \text{if } \lambda < 0. \end{cases}$$

It is immediate from Theorem 8.2 and the definition of $f0^+$ that

$$(\text{cl } g)(\lambda, x) = \begin{cases} (f\lambda)(x) & \text{if } \lambda > 0, \\ (f0^+)(x) & \text{if } \lambda = 0, \\ +\infty & \text{if } \lambda < 0. \end{cases}$$

COROLLARY 8.5.2. *If f is any closed proper convex function, one has*

$$(f0^+)(y) = \lim_{\lambda \downarrow 0} (f\lambda)(y)$$

for every $y \in \text{dom } f$. If $0 \in \text{dom } f$, this formula actually holds for every $y \in R^n$.

PROOF. If $0 \in \text{dom } f$, the last formula in Theorem 8.5 yields

$$(f0^+)(y) = \lim_{\lambda \uparrow \infty} [f(\lambda y) - f(0)]/\lambda = \lim_{\lambda \downarrow 0} \lambda f(\lambda^{-1} y).$$

Even if $0 \notin \text{dom } f$, we have (for g as above)

$$(\text{cl } g)(0, y) = \lim_{\lambda \downarrow 0} (\text{cl } g)(\lambda, y)$$

by Corollary 7.5.1 when (λ, y) belongs to dom $(\text{cl } g)$ for some $\lambda > 0$. The latter condition is certainly met when $y \in \text{dom } f$.   ‖

To illustrate, consider

$$f_1(x) = (1 + \langle x, Qx \rangle)^{1/2},$$

where Q is a symmetric $n \times n$ positive semi-definite matrix. (The convexity of f_1 may be deduced from Theorem 5.1 and the convexity of $f_0(x) = \langle x, Qx \rangle^{1/2}$, which follows easily by diagonalizing Q.) By Corollary 8.5.2,

$$(f_1 0^+)(y) = \lim_{\lambda \downarrow 0} \lambda f_1(\lambda^{-1} y)$$

$$= \lim_{\lambda \downarrow 0} (\lambda^2 + \langle y, Qy \rangle)^{1/2} = \langle y, Qy \rangle^{1/2}.$$

On the other hand, for

$$f_2(x) = \langle x, Qx \rangle + \langle a, x \rangle + \alpha$$

one has by the same formula

$$(f_2 0^+)(y) = \lim_{\lambda \downarrow 0} [\lambda^{-1}\langle y, Qy \rangle + \langle a, y \rangle + \lambda \alpha]$$

$$= \begin{cases} \langle a, y \rangle & \text{if } Qy = 0, \\ +\infty & \text{if } Qy \neq 0. \end{cases}$$

In particular, in the case where Q is positive definite (i.e. also non-singular) one has $f_2 0^+ = \delta(\cdot \mid 0)$. The latter formula would also hold, of course, for any proper convex function whose effective domain was bounded.

An especially interesting example is

$$f_3(x) = \log(e^{\xi_1} + \cdots + e^{\xi_n}), \qquad x = (\xi_1, \ldots, \xi_n), n > 1.$$

(The convexity of f_3 follows from Theorem 4.5 by a classical argument, but a separate derivation will also be given following Theorem 16.4.) The reader may calculate as an exercise that

$$(f_3 0^+)(y) = \max \{\eta_j \mid j = 1, \ldots, n\}, \qquad y = (\eta_1, \ldots, \eta_n).$$

Thus $f_3 0^+$ is not differentiable, even though $f_3 0^+$ is finite everywhere and f_3 itself is analytic.

The recession function of a closed proper convex function f will be characterized in Theorem 13.3 as the support function of the effective domain of the convex function conjugate to f.

THEOREM 8.6.   *Let f be a proper convex function, and let y be a vector. If one has*

$$\liminf_{\lambda \to +\infty} f(x + \lambda y) < +\infty,$$

for a given x, then x actually has the property that $f(x + \lambda y)$ is a non-increasing function of λ, $-\infty < \lambda < +\infty$. This property holds for every x if and only if $(f0^+)(y) \leq 0$. When f is closed, this property holds for every x if it holds for even one $x \in \text{dom } f$.

PROOF.   By definition, $(f0^+)(y) \leq 0$ if and only if the recession cone of epi f contains the vector $(y, 0)$, which means that $f(z + \lambda y) \leq f(z)$ for every z and every $\lambda \geq 0$. Thus $(f0^+)(y) \leq 0$ if and only if $f(x + \lambda y)$ is a non-increasing function of λ, $-\infty < \lambda < +\infty$, for every x. If f is closed, we have $(f0^+)(y) \leq 0$ by the last formula in Theorem 8.5 if there exists even one $x \in \text{dom } f$ such that $f(x + \lambda y)$ is non-increasing in λ. Suppose now that x is a point such that

$$\liminf_{\lambda \to +\infty} f(x + \lambda y) < \alpha,$$

where $\alpha \in R$, and let h be the proper convex function on R defined by

$h(\lambda) = f(x + \lambda y)$. The epigraph of h contains a sequence of points of the form (λ_k, α), $k = 1, 2, \ldots$, such that $\lambda_k \to +\infty$. The convex hull of this sequence is a half-line in the direction of the vector $(1, 0)$, and this half-line is contained in the closed convex set epi (cl h). Hence $(1, 0)$ belongs to the recession cone of epi (cl h), i.e. cl h is a non-increasing function on R. The effective domain of cl h must be an interval unbounded above. The closure operation at most lowers the value of h at the boundary of its effective domain (Theorem 7.4), so h itself must be a non-increasing function on R. Thus $f(x + \lambda y)$ is a non-increasing function of λ.   ‖

COROLLARY 8.6.1.   *Let f be a proper convex function and let y be a vector. In order that $f(x + \lambda y)$ be a constant function of λ, $-\infty < \lambda < \infty$, for every x, it is necessary and sufficient that $(f0^+)(y) \leq 0$ and $(f0^+)(-y) \leq 0$. When f is closed, this condition is satisfied if there exists one x such that, for some real number α,*

$$f(x + \lambda y) \leq \alpha, \qquad \forall \lambda \in R.$$

COROLLARY 8.6.2.   *A convex function f is constant on any affine set M where it is finite and bounded above.*

PROOF.   Redefining f to be $+\infty$ outside M if necessary, we can assume that $M = \operatorname{dom} f$. Then f is closed (Corollary 7.4.2). By the preceding corollary, f is constant along every line in M. Since M contains the line through any two of its (different) points, f must have the same value at all points of M.   ‖

The set of all vectors y such that $(f0^+)(y) \leq 0$ will be called the *recession cone* of f (not to be confused, of course, with the recession cone of epi f). This is a convex cone containing 0, closed if f is closed. (It corresponds to the intersection of $0^+(\text{epi } f)$ with the horizontal hyperplane $\{(y, 0) \mid y \in R^n\}$ in R^{n+1}.) As suggested by Theorem 8.6, the directions of the vectors in the recession cone of f will be called *directions in which f recedes*, or *directions of recession* of f.

The set of vectors y such that $(f0^+)(y) \leq 0$ and $(f0^+)(-y) \leq 0$ is the largest subspace contained in the recession cone of f (Theorem 2.7). We shall call it the *constancy space* of f, in view of Corollary 8.6.1. The directions of the vectors in the constancy space of f will be called *directions in which f is constant*.

In the examples preceding Theorem 8.6, the recession cone and constancy space of f_1 are both equal to $\{y \mid Qy = 0\}$, whereas the recession cone and constancy space of f_2 are

$$\{y \mid Qy = 0, \langle a, y \rangle \leq 0\} \quad \text{and} \quad \{y \mid Qy = 0, \langle a, y \rangle = 0\},$$

respectively. The recession cone of f_3 is the non-positive orthant of R^n, but the constancy space of f_3 consists of the zero vector alone.

THEOREM 8.7. *Let f be a closed proper convex function. Then all the non-empty level sets of the form $\{x \mid f(x) \leq \alpha\}$, $\alpha \in R$, have the same recession cone and the same lineality space, namely the recession cone and the constancy space of f, respectively.*

PROOF. This follows from Theorem 8.6: y belongs to the recession cone of $\{x \mid f(x) \leq \alpha\}$ if and only if $f(x + \lambda y) \leq \alpha$ whenever $f(x) \leq \alpha$ and $\lambda \geq 0$. ‖

COROLLARY 8.7.1. *Let f be a closed proper convex function. If the level set $\{x \mid f(x) \leq \alpha\}$ is non-empty and bounded for one α, it is bounded for every α.*

PROOF. Apply Theorem 8.4. ‖

THEOREM 8.8. *For any proper convex function f, the following conditions on a vector y and a real number v are equivalent:*
(a) $f(x + \lambda y) = f(x) + \lambda v$ *for every vector x and $\lambda \in R$;*
(b) (y, v) *belongs to the lineality space of* epi f;
(c) $-(f0^+)(-y) = (f0^+)(y) = v$.
When f is closed, y satisfies these conditions with $v = (f0^+)(y)$ if there is even one $x \in$ dom f such that $f(x + \lambda y)$ is an affine function of λ.

PROOF. Under (a), $f(x + y) - f(x) = v$ for every $x \in$ dom f, so that $v = (f0^+)(y)$ and $-v = (f0^+)(-y)$ by the first formula in Theorem 8.5. Thus (a) implies (c). Now (c) says that (y, v) and $(-y, -v)$ both belong to epi $(f0^+)$, i.e. (y, v) and $-(y, v)$ both belong to $0^+(\text{epi } f)$. This is the same as condition (b). Finally, (b) implies

$$(\text{epi } f) - \lambda(y, v) = \text{epi } f, \qquad \forall \lambda \in R.$$

For any λ, the set on the left is epi g, where g is the function defined by

$$g(x) = f(x + \lambda y) - \lambda v,$$

so (a) must hold. Thus (a), (b) and (c) are equivalent. The last assertion in the theorem follows from the last formula in Theorem 8.5. ‖

The set of vectors y such that $(f0^+)(-y) = -(f0^+)(y)$ will be called the *lineality space* of the proper convex function f. It is a subspace of R^n, the image of the lineality space of the convex set epi f under the projection $(y, v) \to y$, and on it $f0^+$ is linear (Theorem 4.8). The directions of the vectors in the lineality space of f will be called *directions in which f is affine.* The dimension of the lineality space is the *lineality* of f. The *rank* of f is defined to be the dimension of f minus the lineality of f.

The proper convex functions of rank 0 are the *partial affine functions,* i.e. the functions which agree with an affine function along a certain affine set and are $+\infty$ elsewhere. A closed proper convex function f has

$$\text{rank} f = \dim f$$

if and only if it is not affine along any line in dom f.

The rank of a convex set plainly coincides with the rank of its indicator function.

Some Closedness Criteria

There are many operations for convex sets which preserve relative interiors but have a more complicated behavior with respect to closures. For example, given a convex set C and a linear transformation A, one has ri $(AC) = A(\text{ri } C)$, but in general only cl $(AC) \supset A(\text{cl } C)$ (Theorem 6.6). When is cl (AC) actually equal to $A(\text{cl } C)$? When is the image of a closed convex set closed?

Such questions are worth careful attention. One reason is that they are connected with the preservation of lower semi-continuity. The epigraph of the image Ah of a proper convex function h under a linear transformation A is of the form $F \cup F_0$, where F is the image of epi h under the linear transformation $B : (x, \mu) \rightarrow (Ax, \mu)$ and F_0 is the "lower boundary" of F (in the sense of Theorem 5.3). If F is closed, one actually has $F = $ epi (Ah), so that Ah is lower semi-continuous (Theorem 7.1). One is thus led to study conditions under which the image of epi h under B is closed. The condition that epi h itself be closed, i.e. that h be lower semi-continuous, is generally not sufficient: if h is the closed proper convex function on R^2 given by

$$h(x) = \begin{cases} \exp\left[-(\xi_1\xi_2)^{1/2}\right] & \text{if } x = (\xi_1, \xi_2) \geq 0, \\ +\infty \text{ otherwise,} \end{cases}$$

and A is the projection $(\xi_1, \xi_2) \rightarrow \xi_1$, then the image of epi h is not closed and in fact

$$(Ah)(\xi_1) = \begin{cases} 0 & \text{if } \xi_1 > 0, \\ 1 & \text{if } \xi_1 = 0, \\ +\infty & \text{if } \xi_1 < 0, \end{cases}$$

so that Ah is not lower semi-continuous at 0.

A second reason for interest in closedness criteria is the bearing they have on the existence of solutions to extremum problems. For instance, $(Ah)(y)$ is the infimum of h on the affine set $\{x \mid Ax = y\}$. The infimum is attained if and only if the vertical line $\{(y, \mu) \mid \mu \in R\}$ intersects the above set F in a closed half-line (or the empty set), which would be true if F were closed and did not have the "downward" direction as a direction of recession. Again we need conditions under which the image F of epi h is sure to be closed.

Simple conditions for the preservation of closedness under various operations will be deduced below from the theory of recession cones. Several of these conditions will be dualized in §16 and sharpened in §19 and §20 to take advantage of polyhedral convexity.

The theorem we are about to prove will be at the root of all the other results in this section. For motivation, it is good to think about a case where the image of a closed convex set C under a projection A is not closed, as when

$$C = \{(\xi_1, \xi_2) \mid \xi_1 > 0, \xi_2 \geq \xi_1^{-1}\},$$

$$A : (\xi_1, \xi_2) \to \xi_1.$$

The source of difficulty here is that the hyperbolic convex set C is "asymptotic" to a line which A carries onto a point. It seems clear that, if C were instead some closed convex set in R^2 whose intersection with each of the lines parallel to the ξ_2-axis was bounded, then the image would be closed as desired. This condition could be expressed in terms of recession cones: 0^+C should not contain any vector in the direction of $(0, 1)$ or $(0, -1)$.

THEOREM 9.1. *Let C be a non-empty convex set in R^n, and let A be a linear transformation from R^n to R^m. Assume that every non-zero vector $z \in 0^+(\text{cl } C)$ satisfying $Az = 0$ actually belongs to the lineality space of cl C. Then cl $(AC) = A(\text{cl } C)$, and $0^+A(\text{cl } C) = A(0^+(\text{cl } C))$. In particular, if C is closed, and $z = 0$ is the only $z \in 0^+C$ such that $Az = 0$, then AC is closed.*

PROOF. We already know that cl $(AC) \supset A(\text{cl } C)$. Let y be any point of cl (AC). We shall show that $y = Ax$ for some $x \in$ cl C. Let L be the intersection of the lineality space of cl C and the null space of A, i.e.

$$L = (-0^+(\text{cl } C)) \cap 0^+(\text{cl } C) \cap \{z \mid Az = 0\}.$$

This L is a subspace of R^n, and by the hypothesis on $0^+(\text{cl } C)$ we also have

$$L = 0^+(\text{cl } C) \cap \{z \mid Az = 0\}.$$

The set $L^\perp \cap$ cl C has the same image under A as cl C, inasmuch as

$$\text{cl } C = (L^\perp \cap \text{cl } C) + L.$$

Furthermore, y is in the closure of this image. Hence, for every $\varepsilon > 0$, the intersection

$$C_\varepsilon = L^\perp \cap (\text{cl } C) \cap D_\varepsilon, \qquad D_\varepsilon = \{x \mid |y - Ax| \leq \varepsilon\},$$

is non-empty. Clearly C_ε is a closed convex set in R^n. Furthermore, C_ε is bounded. We prove this via Theorem 8.4 by showing that 0^+C_ε contains only the zero vector: by Corollary 8.3.3,

$$0^+C_\varepsilon = 0^+L^\perp \cap 0^+(\mathrm{cl}\ C) \cap 0^+D_\varepsilon$$
$$= L^\perp \cap 0^+(\mathrm{cl}\ C) \cap \{z \mid Az = 0\} = L^\perp \cap L = \{0\}.$$

Now, since the sets C_ε for $\varepsilon > 0$ form a nest of closed bounded subsets of R^n, the intersection of these sets is not empty. For any x in this intersection, we have $x \in \mathrm{cl}\ C$ and $y - Ax = 0$.

All that remains to be proved is that $A(0^+C) = 0^+(AC)$ if C is closed. Consider the convex cone

$$K = \{(\lambda, x) \mid \lambda > 0, x \in \lambda C\} \subset R^{n+1}$$

and the linear transformation

$$B:(\lambda, x) \to (\lambda, Ax).$$

Assuming that C is closed, we have

$$\mathrm{cl}\ K = 0^+(\mathrm{cl}\ K) = K \cup \{(0, z) \mid z \in 0^+C\}$$

(Theorem 8.2). The vectors (λ, z) whose image under B is the origin are those such that $\lambda = 0$ and $Az = 0$. Therefore the part of the theorem which has already been proved can be applied to K and B. Thus $\mathrm{cl}\ (BK) = B(\mathrm{cl}\ K)$, where

$$B(\mathrm{cl}\ K) = \{(\lambda, Ax) \mid \lambda > 0, x \in \lambda C\} \cup \{(0, Az) \mid z \in 0^+C\}.$$

Since AC is closed, we also have

$$\mathrm{cl}\ (BK) = \mathrm{cl}\ \{(\lambda, y) \mid \lambda > 0, y \in A(\lambda C) = \lambda AC\}$$
$$= \{(\lambda, y) \mid \lambda > 0, y \in \lambda AC\} \cup \{(0, y) \mid y \in 0^+(AC)\}$$

(Theorem 8.2). The equality of $\mathrm{cl}\ (BK)$ and $B(\mathrm{cl}\ K)$ implies that the set $\{Az \mid z \in 0^+C\}$ is the same as $0^+(AC)$. ‖

It should be noted that $0^+(AC)$ can differ from $A(0^+C)$ sometimes, even if C and AC are closed, for example if

$$C = \{(\xi_1, \xi_2) \mid \xi_2 \geq \xi_1^2\}, \qquad A:(\xi_1, \xi_2) \to \xi_1.$$

COROLLARY 9.1.1. *Let $C_1, \ldots, C_m$ be non-empty convex sets in R^n satisfying the following condition: if $z_1, \ldots, z_m$ are vectors such that $z_i \in 0^+(\mathrm{cl}\ C_i)$ and $z_1 + \cdots + z_m = 0$, then actually z_i belongs to the lineality space of $\mathrm{cl}\ C_i$ for $i = 1, \ldots, m$. Then*

$$\mathrm{cl}\ (C_1 + \cdots + C_m) = \mathrm{cl}\ C_1 + \cdots + \mathrm{cl}\ C_m,$$
$$0^+(\mathrm{cl}\ (C_1 + \cdots + C_m)) = 0^+(\mathrm{cl}\ C_1) + \cdots + 0^+(\mathrm{cl}\ C_m).$$

In particular, $C_1 + \cdots + C_m$ is closed under this hypothesis if the sets $C_1, \ldots, C_m$ are all closed.

PROOF. Let C be the direct sum $C_1 \oplus \cdots \oplus C_m$ in R^{mn}, and let A be the linear transformation

$$(x_1, \ldots, x_m) \to x_1 + \cdots + x_m, \qquad x_i \in R^n.$$

Then $AC = C_1 + \cdots + C_m$. Since

$$\text{cl } C = \text{cl } C_1 \oplus \cdots \oplus \text{cl } C_m,$$

we have (as an elementary consequence of the definition of "recession cone")

$$0^+(\text{cl } C) = 0^+(\text{cl } C_1) \oplus \cdots \oplus 0^+(\text{cl } C_m).$$

Apply the theorem. ‖

COROLLARY 9.1.2. *Let C_1 and C_2 be non-empty closed convex sets in R^n. Assume there is no direction of recession of C_1 whose opposite is a direction of recession of C_2. (This is true in particular if either C_1 or C_2 is bounded.) Then $C_1 + C_2$ is closed, and*

$$0^+(C_1 + C_2) = 0^+C_1 + 0^+C_2.$$

PROOF. Specialize the preceding corollary to $m = 2$. ‖

Refinements of Corollary 9.1.2 will be given in Corollary 19.3.2 and Theorem 20.3.

COROLLARY 9.1.3. *Let $K_1, \ldots, K_m$ be non-empty convex cones in R^n satisfying the following condition: if $z_i \in \text{cl } K_i$ for $i = 1, \ldots, m$ and $z_1 + \cdots + z_m = 0$, then z_i belongs to the lineality space of $\text{cl } K_i$ for $i = 1, \ldots, m$. Then*

$$\text{cl } (K_1 + \cdots + K_m) = \text{cl } K_1 + \cdots + \text{cl } K_m.$$

PROOF. Take $C_i = K_i$ in Corollary 9.1.1. ‖

These results will now be applied to convex functions.

THEOREM 9.2. *Let h be a closed proper convex function on R^n, and let A be a linear transformation from R^n to R^m. Assume that $Az \neq 0$ for every z such that $(h0^+)(z) \leq 0$ and $(h0^+)(-z) > 0$. Then the function Ah, where*

$$(Ah)(y) = \inf \{h(x) \mid Ax = y\},$$

is a closed proper convex function, and $(Ah)0^+ = A(h0^+)$. Moreover, for each y such that $(Ah)(y) \neq +\infty$, the infimum in the definition of $(Ah)(y)$ is attained for some x.

PROOF. Consider the non-empty closed convex set epi h and the linear transformation $B:(x, \mu) \to (Ax, \mu)$. The recession cone of epi h is epi $(h0^+)$, and the lineality space of epi h consists of the vectors (z, μ) such that $(h0^+)(z) \leq \mu$ and $(h0^+)(-z) \leq -\mu$. Thus epi h and B satisfy the hypothesis of Theorem 9.1, and we may conclude that $B(\text{epi } h)$ is a non-empty closed convex set whose recession cone is $B(\text{epi } (h0^+))$. Moreover

$$B(\text{epi } h) = \text{epi } (Ah),$$

$$B(\text{epi } (h0^+)) = \text{epi } (A(h0^+)).$$

The conclusions of the theorem will follow if we can establish that epi (Ah) contains no vertical lines. The presence of vertical lines would imply that the recession cone $B(\text{epi } (h0^+))$ contained a vector of the form $(0, \mu)$ with $\mu < 0$. Then epi $(h0^+)$ would contain some (z, μ) with $Az = 0$ and $\mu < 0$. For this z we have $(h0^+)(z) < 0$ and

$$(h0^+)(-z) \geq -(h0^+)(z) > 0$$

(Corollary 4.7.2), contrary to the hypothesis of the theorem.   ‖

The hypothesis of Theorem 9.2 concerning $h0^+$ is trivially satisfied, of course, if h has no directions of recession and in particular if dom h is bounded. Observe that this hypothesis is violated in the example given at the beginning of this section.

COROLLARY 9.2.1.   *Let $f_1, \ldots, f_m$ be closed proper convex functions on R^n. Assume that $z_1 + \cdots + z_m \neq 0$ for all choices of vectors $z_1, \ldots, z_m$ such that*

$$(f_1 0^+)(z_1) + \cdots + (f_m 0^+)(z_m) \leq 0,$$

$$(f_1 0^+)(-z_1) + \cdots + (f_m 0^+)(-z_m) > 0.$$

Then the infimal convolute $f_1 \square \cdots \square f_m$ is a closed proper convex function on R^n, and the infimum in the definition of $(f_1 \square \cdots \square f_m)(x)$ is attained for each x. Moreover,

$$(f_1 \square \cdots \square f_m)0^+ = f_1 0^+ \square \cdots \square f_m 0^+.$$

PROOF.   Let A be the "addition" linear transformation from R^{mn} to R^n:

$$A:(x_1, \ldots, x_m) \to x_1 + \cdots + x_m, \qquad x_i \in R^n,$$

and let h be the closed proper convex function on R^{mn} defined by

$$h(x_1, \ldots, x_m) = f_1(x_1) + \cdots + f_m(x_m), \qquad x_i \in R^n.$$

The result is obtained by applying Theorem 9.2 to this h and A. The details are left to the reader as an exercise.   ‖

Other forms of Corollary 9.2.1 will appear in Corollaries 19.3.4 and 20.1.1.

COROLLARY 9.2.2.   *Let f_1 and f_2 be closed proper convex functions on R^n such that*

$$(f_1 0^+)(z) + (f_2 0^+)(-z) > 0, \qquad \forall z \neq 0.$$

Then $f_1 \square f_2$ is a closed proper convex function, and the infimum in the formula

$$(f_1 \square f_2)(x) = \inf_y \{f_1(x - y) + f_2(y)\}$$

is attained for each x by some y.

PROOF. Take $m = 2$ in the preceding corollary. ‖

As an illustration of Corollary 9.2.2, let $f = f_2$ be an arbitrary closed proper convex function, and let f_1 be the indicator function of $-C$, where C is a non-empty closed convex set. Then

$$(f_1 \square f_2)(x) = \inf \{\delta(x - y \mid -C) + f(y) \mid y \in R^n\}$$
$$= \inf \{f(y) \mid y \in (C + x)\}.$$

The recession condition in the corollary is satisfied if f and C have no common direction of recession. In that case, the infimum of f over the translate $C + x$ is attained for each x, and it is a lower semi-continuous (convex) function of x.

Taking C to be the non-negative orthant of R^n, for instance, we have

$$C + x = \{y \mid y \geq x\}$$

for each x. If f is a closed proper convex function on R^n whose recession cone contains no non-negative non-zero vectors, we may conclude that the infimum in the formula

$$g(x) = \inf \{f(y) \mid y \geq x\}$$

is attained for each x, and that g is a closed proper convex function on R^n. Note that g is the greatest function such that $g \leq f$ and $g(\xi_1, \ldots, \xi_n)$ is a non-decreasing function of the real variable ξ_j for $j = 1, \ldots, n$.

The closure properties of other operations for convex sets and functions are as follows.

THEOREM 9.3. *Let $f_1, \ldots, f_m$ be proper convex functions on R^n. If every f_i is closed and $f_1 + \cdots + f_m$ is not identically $+\infty$, then $f_1 + \cdots + f_m$ is a closed proper convex function and*

$$(f_1 + \cdots + f_m)0^+ = f_1 0^+ + \cdots + f_m 0^+.$$

If the f_i are not all closed, but there exists a point common to every ri (dom f_i), *then*

$$\text{cl}\,(f_1 + \cdots + f_m) = \text{cl}\,f_1 + \cdots + \text{cl}\,f_m.$$

PROOF. Let $f = f_1 + \cdots f_m$ and let

$$x \in \text{ri}\,(\text{dom}\,f) = \text{ri}\,(\bigcap_{i=1}^m \text{dom}\,f_i).$$

For every y, we have

$$(\mathrm{cl}\, f)(y) = \lim_{\lambda \uparrow 1} f((1 - \lambda)x + \lambda y) = \sum_{i=1}^{m} \lim_{\lambda \uparrow 1} f_i((1 - \lambda)x + \lambda y)$$

(Theorem 7.5). If each f_i is closed, the latter sum is $f_1(y) + \cdots + f_m(y)$, so that $\mathrm{cl}\, f = f$. On the other hand, if the sets ri (dom f_i) have a point in common, then

$$\bigcap_{i=1}^{m} \mathrm{ri}\, (\mathrm{dom}\, f_i) = \mathrm{ri}\, (\mathrm{dom}\, f)$$

by Theorem 6.5. In this case $x \in \mathrm{ri}\, (\mathrm{dom}\, f_i)$ for $i = 1, \ldots, m$ and the f_i limit in the above sum is $(\mathrm{cl}\, f_i)(y)$; thus $\mathrm{cl}\, f = \mathrm{cl}\, f_1 + \cdots + \mathrm{cl}\, f_m$. The formula for $f0^+$ follows from the limit formula in Theorem 8.5. $\parallel$

THEOREM 9.4.   *Let f_i be a proper convex function on R^n for $i \in I$ (an arbitrary index set), and let*

$$f = \sup \{f_i \mid i \in I\}.$$

If f is finite somewhere and every f_i is closed, then f is closed and proper, and

$$f0^+ = \sup \{f_i 0^+ \mid i \in I\}.$$

If the f_i are not all closed, but there exists a point $\bar{x}$ common to every ri (dom f_i) such that $f(\bar{x})$ is finite, then

$$\mathrm{cl}\, f = \sup \{\mathrm{cl}\, f_i \mid i \in I\}.$$

PROOF.   Since epi f is the intersection of the sets epi f_i, it is closed when every f_i is closed. The formula for $f0^+$ follows from Corollary 8.3.3. The closure formula follows from Theorem 6.5 and Lemma 7.3: the intersection of the sets ri (epi f_i) will contain the point $(\bar{x}, f(\bar{x}) + 1)$. $\parallel$

THEOREM 9.5.   *Let A be a linear transformation from R^n to R^m, and let g be a proper convex function on R^m such that gA is not identically $+\infty$. If g is closed, then gA is closed and $(gA)0^+ = (g0^+)A$. If g is not closed, but $Ax \in \mathrm{ri}\, (\mathrm{dom}\, g)$ for some x, then $\mathrm{cl}\, (gA) = (\mathrm{cl}\, g)A$.*

PROOF.   We already know gA is a proper convex function (Theorem 5.7). The epigraph of gA is the inverse image of epi g under the (continuous) linear transformation $B:(x, \mu) \to (Ax, \mu)$, so gA is closed if g is closed. The formula for $(gA)0^+$ is then immediate from Corollary 8.3.4. The closure formula follows from Theorem 6.7 and Lemma 7.3. $\parallel$

THEOREM 9.6.   *Let C be a non-empty closed convex set not containing the origin, and let K be the convex cone generated by C. Then*

$$\mathrm{cl}\, K = K \cup 0^+C = \cup \{\lambda C \mid \lambda > 0 \quad \text{or} \quad \lambda = 0^+\}.$$

PROOF. Let K' be the convex cone in R^{n+1} generated by $\{(1, x) \mid x \in C\}$. Then

$$\text{cl } K' = \{(\lambda, x) \mid \lambda > 0, x \in \lambda C\} \cup \{(0, x) \mid x \in 0^+C\}$$

(Theorem 8.2). Under the linear transformation $A : (\lambda, x) \rightarrow x$, the image of cl K' is $K \cup 0^+C$. There is no non-zero (λ, x) in cl $K' = 0^+(\text{cl } K')$ having 0 as its image under A, so

$$A(\text{cl } K') = \text{cl } (AK') = \text{cl } K$$

by Theorem 9.1. ‖

COROLLARY 9.6.1. *If C is a non-empty closed bounded convex set not containing the origin, then the convex cone K generated by C is closed.*

PROOF. Here $0^+C = \{0\}$. ‖ (This result is also easy to prove by a direct argument using compactness.)

The need for the condition $0 \notin C$ in Theorem 9.6 and Corollary 9.6.1 is shown by the case where C is a closed ball with the origin on its boundary. The need for the boundedness assumption in Corollary 9.6.1 is shown by the case where C is a line not passing through the origin.

THEOREM 9.7. *Let f be a closed proper convex function on R^n with $f(0) > 0$, and let k be the positively homogeneous convex function generated by f. Then k is proper and* ·

$$(\text{cl } k)(x) = \inf \{(f\lambda)(x) \mid \lambda > 0 \quad \text{or} \quad \lambda = 0^+\},$$

the infimum being attained for each x. If $0 \in \text{dom } f$, k is itself closed, and $\lambda = 0^+$ can be omitted from the infimum (but the infimum then might not be attained).

PROOF. Here epi f is a non-empty closed convex set in R^{n+1} not containing the origin. The closed convex cone it generates, which is cl (epi k), is then the union of the sets $\lambda (\text{epi} f) = \text{epi } (f\lambda)$ for $\lambda \geq 0^+$ by the preceding theorem. This union does not contain any vectors $(0, \mu)$ with $\mu < 0$, so it is actually epi (cl k) and k is proper. The formula follows at once. If $0 \in \text{dom } f$, we have

$$(f0^+)(x) = \lim_{\alpha \uparrow + \infty} [f(\alpha x) - f(0)]/\alpha = \lim_{\lambda \downarrow 0} (f\lambda)(x)$$

by the last formula in Theorem 8.5, so it is enough to take the infimum of $(f\lambda)(x)$ over $\lambda > 0$. This infimum gives k itself, by definition. ‖

COROLLARY 9.7.1. *Let C be a closed convex set in R^n containing 0. The gauge function $\gamma(\cdot \mid C)$ of C is then closed. One has*

$$\{x \mid \gamma(x \mid C) \leq \lambda\} = \lambda C$$

for any $\lambda > 0$, and

$$\{x \mid \gamma(x \mid C) = 0\} = 0^+C.$$

PROOF. Apply the theorem to $f(x) = \delta(x \,|\, C) + 1$. One has $k = \gamma(\cdot \,|\, C)$ by definition,

$$(f\lambda)(x) = \delta(x \,|\, \lambda C) + \lambda, \qquad \forall \lambda > 0,$$

and $f0^+ = \delta(\cdot \,|\, 0^+ C)$. ‖

THEOREM 9.8. *Let* $C_1, \ldots, C_m$ *be non-empty closed convex sets in* R^n *satisfying the following condition: if* $z_1, \ldots, z_m$ *are vectors such that* $z_i \in 0^+ C_i$ *and* $z_1 + \cdots + z_m = 0$, *then* z_i *belongs to the lineality space of* C_i *for* $i = 1, \ldots, m$. *Let* $C = \mathrm{conv}\,(C_1 \cup \cdots \cup C_m)$. *Then*

$$\mathrm{cl}\,C = \cup \,\{\lambda_1 C_1 + \cdots + \lambda_m C_m \,|\, \lambda_i \geq 0^+, \lambda_1 + \cdots + \lambda_m = 1\}$$

(*where the notation* $\lambda_i \geq 0^+$ *means that* $\lambda_i C_i$ *is taken to be* $0^+ C_i$ *rather than* $\{0\}$ *when* $\lambda_i = 0$). *Moreover*

$$0^+(\mathrm{cl}\,C) = 0^+ C_1 + \cdots + 0^+ C_m.$$

PROOF. Let K_i be the convex cone in R^{n+1} generated by $\{(1, x_i) \,|\, x_i \in C_i\}$, $i = 1, \ldots, m$. We have

$$\mathrm{cl}\,K_i = \{(\lambda_i, x_i) \,|\, \lambda_i > 0, x_i \in \lambda_i C_i\} \cup \{(0, x_i) \,|\, x_i \in 0^+ C_i\}$$

(Theorem 8.2). Corollary 9.1.3 is applicable to the cones K_i by virtue of the condition on the cones $0^+ C_i$. Thus

$$\mathrm{cl}\,(K_1 + \cdots + K_m) = \mathrm{cl}\,K_1 + \cdots + \mathrm{cl}\,K_m.$$

The intersection of $\mathrm{cl}\,(K_1 + \cdots + K_m)$ with $H_1 = \{(1, x) \,|\, x \in R^n\}$ is the closure of the intersection of $K_1 + \cdots + K_m$ with H_1, and that consists of the vectors $(1, x)$ such that x belongs to some convex combination $\lambda_1 C_1 + \cdots + \lambda_m C_m$. The union of all such convex combinations is C (Theorem 3.3). Therefore

$$\mathrm{cl}\,(K_1 + \cdots + K_m) \cap H_1 = \{(1, x) \,|\, x \in \mathrm{cl}\,C\}.$$

This same set coincides on the other hand with the intersection of $\mathrm{cl}\,K_1 + \cdots + \mathrm{cl}\,K_m$ and H_1, and that consists of the pairs $(1, x)$ such that x belongs to the union described in the theorem. This establishes the formula for $\mathrm{cl}\,C$. From what we have shown, $\mathrm{cl}\,(K_1 + \cdots + K_m)$ must actually be the closure of the convex cone in R^{n+1} generated by $\{(1, x) \,|\, x \in \mathrm{cl}\,C\}$, so the vectors it contains of the form $(0, x)$ are those with $x \in 0^+(\mathrm{cl}\,C)$ (Theorem 8.2). The vectors of the form $(0, x)$ contained in $\mathrm{cl}\,K_1 + \cdots + \mathrm{cl}\,K_m$ are those with $x \in 0^+ C_1 + \cdots + 0^+ C_m$. Thus $0^+(\mathrm{cl}\,C)$ is the same as $0^+ C_1 + \cdots + 0^+ C_m$. ‖

COROLLARY 9.8.1. *If* $C_1, \ldots, C_m$ *are non-empty closed convex sets in* R^n *all having the same recession cone* K, *then the convex set* $C = \mathrm{conv}\,(C_1 \cup \cdots \cup C_m)$ *is closed and has* K *as its recession cone.*

PROOF. Suppose $z_1, \ldots, z_m$ are vectors such that $z_i \in K$ and $z_1 + \cdots + z_m = 0$. Then

$$-z_1 = z_2 + \cdots + z_m \in (-K) \cap K,$$

and similarly for $z_2, \ldots, z_m$. Thus z_i belongs to the lineality space of C_i for $i = 1, \ldots, m$, and the theorem is applicable. It is unnecessary to substitute 0^+C_i for $\{0\} = 0C_i$ in the union in the theorem, because

$$0^+C_i + \lambda_j C_j = \lambda_j(K + C_j) = \lambda_j C_j = 0C_i + \lambda_j C_j$$

for any index j with $\lambda_j > 0$. Thus cl $C = C$ (Theorem 3.3). ‖

COROLLARY 9.8.2. *If $C_1, \ldots, C_m$ are closed bounded convex sets in R^n, then* conv $(C_1 \cup \cdots \cup C_m)$ *is likewise closed and bounded.*

PROOF. Any C_i which is empty can be omitted without changing the convex hull, and every other C_i has $0^+C_i = \{0\}$. ‖

A stronger result than Corollary 9.8.2 will be given in Theorem 17.2.

A result analogous to Theorem 9.8 can obviously be stated for convex functions. We shall only treat the analogue of Corollary 9.8.1, however.

COROLLARY 9.8.3. *Let $f_1, \ldots, f_m$ be closed proper convex functions on R^n all having the same recession function k. Then $f = $ conv $\{f_1, \ldots, f_m\}$ is closed and proper and likewise has k as its recession function. In the formula for $f(x)$ in Theorem 5.6, the infimum is attained for each x by some convex combination.*

PROOF. Here we invoke Corollary 9.8.1 with $C_i = $ epi f_i, $K \to $ epi k. The convex hull C of the sets C_i is a non-empty closed convex set in R^{n+1}, and by the nature of its recession cone K it must be the epigraph of a closed proper convex function. This function can be none other than f, and $f0^+$ must therefore be k. The numbers μ expressible as one of the combinations over which the infimum is taken in Theorem 5.6 are just those such that (x, μ) belongs to C, as explained in the proof of Theorem 5.6. Here $C = $ epi f, so $\mu = f(x)$ itself is so expressible, i.e. the infimum is attained. ‖

Continuity of Convex Functions

The closure operation for convex functions alters a function "slightly" to make it lower semi-continuous. We shall now describe some common situations where a convex function f is automatically upper semi-continuous, so that $\text{cl} f$ (or f itself to the extent that it agrees with $\text{cl} f$) is actually continuous. Uniform continuity and equicontinuity will also be considered. In every case, a strong conclusion about continuity follows from an elementary hypothesis, because of convexity.

A function f on R^n is said to be *continuous relative to a subset S* of R^n if the restriction of f to S is a continuous function. Continuity relative to S means, in other words, that, for $x \in S$, $f(y)$ has to approach $f(x)$ as y approaches x along S, but not necessarily as y approaches x from the outside of S.

The following continuity theorem is the most important, although stronger results will be stated in Theorems 10.2 and 10.4.

THEOREM 10.1. *A convex function f on R^n is continuous relative to any relatively open convex set C in its effective domain, in particular relative to* ri $(\text{dom} f)$.

PROOF. The function g which agrees with f on C but is $+\infty$ everywhere else has C as its effective domain. Replacing f by g if necessary, we can reduce the theorem to the case where $C = \text{dom} f$. We can also assume without loss of generality that C is n-dimensional (and hence open, rather than merely relatively open). If f is improper, it is identically $-\infty$ on C (Theorem 7.2), and continuity is trivial. Assume therefore that f is proper, i.e. finite on C. We have $(\text{cl} f)(x) = f(x)$ for $x \in C$ (Theorem 7.4), so f is lower semi-continuous on C. To prove continuity, it suffices to prove that the level sets $\{x \mid f(x) \geq \alpha\}$ are all closed, since that will imply f is upper semi-continuous everywhere (Theorem 7.1). Since $C = \text{dom} f$ is open, we have by Lemma 7.3 that

$$\text{int } (\text{epi} f) = \{(x, \mu) \mid \mu > f(x)\}.$$

Therefore, for any $\alpha \in R$, $\{x \mid f(x) < \alpha\}$ is the projection on R^n of the (open convex) intersection of int $(\text{epi} f)$ and the half-space $\{(x, \mu) \mid \mu < \alpha\}$ in

R^{n+1}, implying that $\{x \mid f(x) < \alpha\}$ is open and its complement $\{x \mid f(x) \geq \alpha\}$ is closed. ‖

COROLLARY 10.1.1. *A convex function finite on all of R^n is necessarily continuous.*

One source of usefulness of this continuity result is the fact that convexity is preserved by certain operations that could not usually be expected to preserve continuity.

For example, let f be a real-valued function on $R^n \times T$ (where T is an arbitrary set), such that $f(x, t)$ is convex as a function of x for each t and bounded above as a function of t for each x. (This situation would arise, say, if one had a finite convex function on R^n continuously dependent on the time t over a certain closed interval T.) Then

$$h(x) = \sup \{f(x, t) \mid t \in T\}$$

depends continuously on x. To deduce this from Corollary 10.1.1, one only has to observe that h is finite everywhere by hypothesis and, being a pointwise supremum of a collection of convex functions, h is convex.

As another interesting example, consider any convex function f finite on all of R^n and any non-empty convex set C in R^n. For each $x \in R^n$, let $h(x)$ be the infimum of f over the translate $C + x$. We claim $h(x)$ depends continuously on x. In the first place,

$$h(x) = \inf_z \{f(x - z) + \delta(z \mid -C)\} = (f \,\square\, g)(x)$$

where g is the indicator function of $-C$. Thus h is a convex function on R^n. Since f is finite everywhere, dom $h = R^n$. Therefore, either h is identically $-\infty$ or it is finite everywhere (Theorem 7.2). At all events, h is continuous.

What can be said about continuity at relative boundary points of effective domains? Here is an instructive example of what can go wrong. On R^2, let

$$f(\xi_1, \xi_2) = \begin{cases} \xi_2^2/2\xi_1 & \text{if } \xi_1 > 0, \\ 0 & \text{if } \xi_1 = 0, \xi_2 = 0, \\ +\infty & \text{otherwise.} \end{cases}$$

As a matter of fact, f is the support function of the parabolic convex set

$$C = \{(\xi_1, \xi_2) \mid \xi_1 + (\xi_2^2/2) \leq 0\},$$

whence its convexity. Observe that f is continuous everywhere, except at $(0, 0)$, where it is only lower semi-continuous. The limit of $f(\xi_1, \xi_2)$ is α as (ξ_1, ξ_2) approaches $(0, 0)$ along the parabolic path with $\xi_1 = \xi_2^2/2\alpha$; here α can be any positive real number. However, the limit is 0 along any line segment joining the origin to a point x in the open right half-plane; this can be seen directly, but it also follows from Theorem 7.5. Trouble only arises, it seems, when the origin is approached along a path "tangent" to the boundary of dom f. When the path stays within a fixed simplex in dom f having the origin as one vertex, the limit is $0 = f(0, 0)$.

The example leads one to conjecture that a closed convex function is necessarily continuous on any simplex in its effective domain. The conjecture is valid in the case where the simplex is a line segment, by Corollary 7.5.1. We shall show that an even stronger conjecture is actually valid in general.

Let us agree to agree to call a subset S of R^n *locally simplicial* if for each $x \in S$ there exists a finite collection of simplices $S_1, \ldots, S_m$ contained in S such that, for some neighborhood U of x,

$$U \cap (S_1 \cup \cdots \cup S_m) = U \cap S.$$

A locally simplicial set need not be convex or closed. The class of locally simplicial sets includes, besides line segments and other simplices, all polytopes and polyhedral convex sets. This will be verified later, in Theorem 20.5. It also includes all relatively open convex sets.

In the proof below, we shall make use of the following intuitively obvious fact. Let C be a simplex with vertices $x_0, x_1, \ldots, x_m$, and let $x \in C$. Then C can be *triangulated* into simplices having x as a vertex, i.e. each $y \in C$ belongs to a simplex whose vertices are x and m of the $m + 1$ vertices of C. (The argument can obviously be reduced to the case where x cannot be expressed as a convex combination of fewer than $m + 1$ of the vertices of C, i.e. the case where $x \in \text{ri } C$. Each $y \in C$ lies on some line segment joining x with a relative boundary point z of C. This z can be expressed as a convex combination of m vertices of C, say $x_1, \ldots, x_m$. The points $x, x_1, \ldots, x_m$ are affinely independent, and the simplex they generate contains y.)

THEOREM 10.2. *Let f be a convex function on R^n, and let S be any locally simplicial subset of $\text{dom } f$. Then f is upper semi-continuous relative to S, so that if f is closed f is actually continuous relative to S.*

PROOF. Let $x \in S$, and let $S_1, \ldots, S_m$ be simplices in S such that some neighborhood of x has the same intersection with $S_1 \cup \cdots \cup S_m$ as it has with S. Each of the simplices S_i which contains x can be triangulated into finitely many other simplices, each having x as one vertex, as explained above. Let the simplices obtained this way be $T_1, \ldots, T_k$. Thus each T_j has x as one of its vertices, and some neighborhood of x has the same intersection with $T_1 \cup \cdots \cup T_k$ as it has with S. If we can show that f is upper semi-continuous at x relative to each of the sets T_j, it will follow that f is upper semi-continuous at x relative to $T_1 \cup \cdots \cup T_k$, and hence that f is upper semi-continuous at x relative to S. Thus the argument is reduced to showing that, if T is a simplex contained in dom f and x is a vertex of T, then f is upper semi-continuous at x relative to T. There is no loss of generality in supposing that T is n-dimensional. In fact, applying an affine transformation if necessary, we can suppose that $x = 0$ and that the vertices of T other than 0 are $e_1 = (1, 0, \ldots, 0)$, $\ldots, e_n = (0, \ldots, 0, 1)$. Then for any $z = (\zeta_1, \ldots, \zeta_n)$ in T we have

$$f(z) \leq (1 - \zeta_1 - \cdots - \zeta_n)f(0) + \zeta_1 f(e_1) + \cdots + \zeta_n f(e_n)$$

by the convexity of f. (This holds even though f might be improper; the expression $\infty - \infty$ cannot arise here because f nowhere has the value $+\infty$ on T.) The "lim sup" of the left side of this inequality as z goes to 0 in T cannot exceed the "lim sup" of the right side, which is $f(0)$. Thus f is upper semi-continuous at 0 relative to T. ‖

The uses of Theorem 10.2 are well demonstrated by the following application to the problem of extensions.

THEOREM 10.3. *Let C be a locally simplicial convex set, and let f be a finite convex function on* ri C *which is bounded above on every bounded subset of* ri C. *Then f can be extended in one and only one way to a continuous finite convex function on the whole of C.*

PROOF. Set $f(x) = +\infty$ for $x \notin$ ri C, and form cl f. The function cl f is convex, closed and proper, and it agrees with f on ri C (Theorem 7.4), moreover, cl f is finite on the relative boundary of C by the boundedness condition on f. By Theorem 10.2, cl f is continuous on C. Thus the restriction of cl f to C is a continuous finite convex extension of f. There can be only one such extension, since $C \subset$ cl (ri C). ‖

The extension in Theorem 10.3 can be effected, of course, by setting $f(x)$ (for a relative boundary point x of C) equal to the limit of $f(y)$ as y approaches x along any line segment joining x with a point of ri C.

As an example, consider the case where C is the non-negative orthant of R^n (which is locally simplicial according to Theorem 20.5). The interior of C is the positive orthant. Let f be any finite convex function on the positive orthant which is non-decreasing, in the sense that $f(\xi_1, \ldots, \xi_n)$ is a non-decreasing function of ξ_j for $j = 1, \ldots, n$. For each positive real number λ, we have

$$f(\xi_1, \ldots, \xi_n) \leq f(\lambda, \ldots, \lambda)$$

for all the vectors $x = (\xi_1, \ldots, \xi_n)$ such that $0 < \xi_j \leq \lambda$ for all j. Therefore f is bounded above on every bounded subset of the positive orthant. It follows from Theorem 10.3 that f can be extended uniquely to a finite continuous (non-decreasing) convex function on the whole non-negative orthant.

A real-valued function f on a set $S \subset R^n$ will be called *Lipschitzian relative to* S if there exists a real number $\alpha \geq 0$ such that

$$|f(y) - f(x)| \leq \alpha |y - x|, \qquad \forall y \in S, \qquad \forall x \in S.$$

This condition implies in particular that f is uniformly continuous relative to S.

The following theorem gives a significant refinement of Theorem 10.1.

THEOREM 10.4. *Let f be a proper convex function, and let S be any closed bounded subset of* ri $(\operatorname{dom} f)$. *Then f is Lipschitzian relative to* S.

PROOF. There is no loss of generality if we suppose that $\operatorname{dom} f$ is n-dimensional in R^n, so that S is actually in the interior of $\operatorname{dom} f$. Let B be the Euclidean unit ball. For each $\varepsilon > 0$, $S + \varepsilon B$ is a closed bounded set (the image of the compact set $S \times B$ under the continuous transformation $(x, u) \to x + \varepsilon u$). The nest of sets

$$(S + \varepsilon B) \cap (R^n \setminus \operatorname{int} (\operatorname{dom} f)), \qquad \varepsilon > 0,$$

has an empty intersection, and hence one of the sets in the nest is empty. Hence, for a certain $\varepsilon > 0$,

$$S + \varepsilon B \subset \operatorname{int} (\operatorname{dom} f).$$

By Theorem 10.1, f is continuous on $S + \varepsilon B$. Since $S + \varepsilon B$ is a closed bounded set, it follows that f is bounded on $S + \varepsilon B$. Let α_1 and α_2 be lower and upper bounds, respectively. Let x and y be any two distinct points in S, and let

$$z = y + (\varepsilon/|y - x|)(y - x).$$

Then $z \in S + \varepsilon B$ and

$$y = (1 - \lambda)x + \lambda z, \qquad \lambda = |y - x|/(\varepsilon + |y - x|).$$

From the convexity of f, we have

$$f(y) \leq (1 - \lambda)f(x) + \lambda f(z) = f(x) + \lambda(f(z) - f(x))$$

and consequently

$$f(y) - f(x) \leq \lambda(\alpha_2 - \alpha_1) \leq \alpha |y - x|, \qquad \alpha = (\alpha_2 - \alpha_1)/\varepsilon.$$

This inequality is valid for any x and y in S, so f is Lipschitzian relative to S. ‖

A finite convex function f on R^n is uniformly continuous, even Lipschitzian, relative to every bounded set by Theorem 10.4, but f need not be uniformly continuous or Lipschitzian relative to R^n as a whole. The circumstances under which f has these additional properties will now be described.

THEOREM 10.5. *Let f be a finite convex function on R^n. In order that $f be uniformly continuous relative to R^n, it is necessary and sufficient that the recession function $f0^+$ of f be finite everywhere. In this event, f is actually Lipschitzian relative to R^n.*

PROOF. Suppose that f is uniformly continuous. Choose any $\varepsilon > 0$. There exists a $\delta > 0$ such that $|z| \leq \delta$ implies

$$f(x + z) - f(x) \leq \varepsilon, \qquad \forall x.$$

For this δ, one has $(f0^+)(z) \leq \varepsilon$ when $|z| \leq \delta$, by the first formula in Theorem 8.5. Since $f0^+$ is a positively homogeneous proper convex function, this implies $f0^+$ is finite everywhere.

Conversely, suppose $f0^+$ is finite everywhere. Then $f0^+$ is continuous everywhere, according to Corollary 10.1.1, and hence

$$\infty > \alpha = \sup \{(f0^+)(z) \mid |z| = 1\}$$
$$= \sup \{|z|^{-1}(f0^+)(z) \mid z \neq 0\}.$$

It follows that

$$\alpha |y - x| \geq (f0^+)(y - x) \geq f(y) - f(x), \qquad \forall x, \qquad \forall y$$

(Corollary 8.5.1). Thus f is Lipschitzian and in particular uniformly continuous relative to R^n. ‖

COROLLARY 10.5.1. *A finite convex function f is Lipschitzian relative to R^n if*

$$\liminf_{\lambda \to \infty} f(\lambda y)/\lambda < \infty, \qquad \forall y.$$

PROOF. The limit equals $(f0^+)(y)$ by Theorem 8.5. ‖

COROLLARY 10.5.2. *Let g be any finite convex function Lipschitzian relative to R^n (for instance, $g(x) = \alpha |x| + \beta, \alpha > 0$). Then every finite convex function f such that $f \leq g$ is likewise Lipschitzian relative to R^n.*

PROOF. One has $f0^+ \leq g0^+$ when $f \leq g$. ‖

Theorem 10.5 will be dualized in Corollary 13.3.3.

We turn now to the continuity properties of collections of convex functions and closely related properties of convergence.

Let $\{f_i \mid i \in I\}$ be a collection of real-valued functions on a subset S of R^n. We shall say that $\{f_i \mid i \in I\}$ is *equi-Lipschitzian* relative to S if there exists a real number $\alpha \geq 0$ such that

$$|f_i(y) - f_i(x)| \leq \alpha |y - x|, \qquad \forall y \in S, \qquad \forall x \in S, \qquad \forall i \in I.$$

When this condition is satisfied, the collection is in particular *uniformly equicontinuous* relative to S, i.e. for every $\varepsilon > 0$ there exists a $\delta > 0$ such that

$$|f_i(y) - f_i(x)| \leq \varepsilon, \qquad \forall i \in I,$$

whenever $y \in S$, $x \in S$ and $|y - x| \leq \delta$. The collection $\{f_i \mid i \in I\}$ is said to be *pointwise bounded* on S, of course, if the set of real numbers $f_i(x)$, $i \in I$, is bounded for each $x \in S$. It is said to be *uniformly bounded* on S if there exist real numbers α_1 and α_2 such that

$$\alpha_1 \leq f_i(x) \leq \alpha_2, \qquad \forall x \in S, \qquad \forall i \in I.$$

THEOREM 10.6.   *Let C be a relatively open convex set, and let $\{f_i \mid i \in I\}$ be an arbitrary collection of convex functions finite and pointwise bounded on C. Let S be any closed bounded subset of C. Then $\{f_i \mid i \in I\}$ is uniformly bounded on S and equi-Lipschitzian relative to S.*

The conclusion remains valid if the pointwise boundedness assumption is weakened to the following pair of assumptions:

(a) *There exists a subset C' of C such that* conv $(\text{cl } C') \supset C$ *and* $\sup \{f_i(x) \mid i \in I\}$ *is finite for every $x \in C'$;*

(b) *There exists at least one $x \in C$ such that* $\inf \{f_i(x) \mid i \in I\}$ *is finite.*

PROOF.   There is no loss of generality if we suppose that C is actually open. Assuming (a) and (b), we shall show that $\{f_i \mid i \in I\}$ is uniformly bounded on every closed bounded subset of C. The equi-Lipschitzian property will then follow by the proof of Theorem 10.4, since the Lipschitz constant α constructed in that proof depended only on the given lower and upper bounds α_1 and α_2. Let

$$f(x) = \sup \{f_i(x) \mid i \in I\}.$$

This f is a convex function, and by (a) we have, since cl dom f contains cl C' and hence conv cl C' and C,

$$\text{dom } f \supset \text{int } (\text{cl } (\text{dom } f)) \supset \text{int } C = C.$$

(The first inclusion holds by Theorem 6.3, since dom f is convex.) Therefore f is continuous relative to C (Theorem 10.1). In particular, f is bounded above on every closed bounded subset of C, i.e. $\{f_i \mid i \in I\}$ is uniformly bounded from above on every closed bounded subset of C. To prove that $\{f_i \mid i \in I\}$ is also uniformly bounded from below on every closed bounded subset of C, it is enough to construct a continuous real-valued function g such that

$$f_i(x) \geq g(x), \qquad \forall x \in C, \qquad \forall i \in I.$$

Making use of (b), select any point $\bar{x} \in C$ such that

$$-\infty < \beta_1 = \inf\{f_i(\bar{x}) \mid i \in I\}.$$

Choose $\varepsilon > 0$ so small that $\bar{x} + \varepsilon B \subset C$, where B is the Euclidean unit ball, and let β_2 be a positive upper bound to the values of f on $\bar{x} + \varepsilon B$. Given any $x \in C$, $x \neq \bar{x}$, we have $\bar{x} = (1 - \lambda)z + \lambda x$ for

$$z = \bar{x} + (\varepsilon/|\bar{x} - x|)(\bar{x} - x),$$

$$\lambda = \varepsilon/(\varepsilon + |\bar{x} - x|).$$

Since $0 < \lambda < 1$ and $|z - \bar{x}| = \varepsilon$, we have (for any $i \in I$)

$$\beta_1 \leq f_i(\bar{x}) \leq (1 - \lambda)f_i(z) + \lambda f_i(x)$$

$$\leq (1 - \lambda)\beta_2 + \lambda f_i(x) \leq \beta_2 + \lambda f_i(x)$$

and consequently

$$f_i(x) \geq (\beta_1 - \beta_2)/\lambda = (\varepsilon + |\bar{x} - x|)(\beta_1 - \beta_2)/\varepsilon.$$

The quantity on the right depends continuously on x. The inequality is valid for every $x \in C$ and every $i \in I$, so the theorem is proved. ‖

THEOREM 10.7. *Let C be a relatively open convex set in R^n, and let T be any locally compact topological space (for instance, any open or closed subset of R^m). Let f be a real-valued function on $C \times T$ such that $f(x, t)$ is convex in x for each $t \in T$ and continuous in t for each $x \in C$. Then f is continuous on $C \times T$, i.e. $f(x, t)$ is jointly continuous in x and t.*

The conclusion remains valid if the assumption about continuity in t is weakened to the following: there exists a subset C' of C such that $\text{cl } C' \supset C$ and $f(x, \cdot)$ is a continuous function on T for each $x \in C'$.

PROOF. Let (x_0, t_0) be any point of $C \times T$. Let T_0 be any compact neighborhood of t_0 in T. For each $x \in C'$, the function $f(x, \cdot)$ is continuous on T_0 and hence bounded on T_0. Thus $\{f(\cdot, t) \mid t \in T_0\}$ is a collection of finite convex functions on C which is pointwise bounded on C'. It follows from Theorem 10.6 that the collection $\{f(\cdot, t) \mid t \in T_0\}$ is equi-Lipschitzian on closed bounded subsets of C and in particular equicontinuous at x_0. Given any $\varepsilon > 0$, we can therefore find a $\delta > 0$ such that

$$|f(x, t) - f(x_0, t)| \leq \varepsilon/4, \qquad \forall t \in T_0,$$

whenever $|x - x_0| \leq \delta$. Let x_1 be a point of C' such that $|x_1 - x_0| \leq \delta$. Since $f(x_1, \cdot)$ is continuous at t_0, we can find a neighborhood V of t_0 in T_0 such that

$$|f(x_1, t) - f(x_1, t_0)| \leq \varepsilon/4, \qquad \forall t \in V.$$

For any (x, t) such that $|x - x_0| \leq \delta$ and $t \in V$, we have

$$|f(x, t) - f(x_0, t_0)| \leq |f(x, t) - f(x_0, t)| + |f(x_0, t) - f(x_1, t)|$$
$$+ |f(x_1, t) - f(x_1, t_0)| + |f(x_1, t_0) - f(x_0, t_0)|$$
$$\leq (\varepsilon/4) + (\varepsilon/4) + (\varepsilon/4) + (\varepsilon/4) = \varepsilon.$$

This shows that f is continuous at (x_0, t_0).   ‖

THEOREM 10.8.   *Let C be a relatively open convex set, and let $f_1, f_2, \ldots$, be a sequence of finite convex functions on C. Suppose that the sequence converges pointwise on a dense subset of C, i.e. that there exists a subset C' of C such that* cl $C' \supset C$ *and, for each $x \in C'$, the limit of $f_1(x)$, $f_2(x), \ldots$, exists and is finite. The limit then exists for every $x \in C$, and the function f, where*

$$f(x) = \lim_{i \to \infty} f_i(x),$$

is finite and convex on C. Moreover the sequence $f_1, f_2, \ldots$, converges to f uniformly on each closed bounded subset of C.

PROOF.   There is no loss of generality if we assume C to be open. The collection $\{f_i \mid i = 1, 2, \ldots\}$ is pointwise bounded on C', and hence by Theorem 10.6 it is equi-Lipschitzian on each closed bounded subset of C. Let S be any closed bounded subset of C. Let S' be a closed bounded subset of C such that int $S' \supset S$. (The argument which establishes the existence of S' is given at the beginning of the proof of Theorem 10.4.) There exists a real number $\alpha > 0$ such that

$$|f_i(y) - f_i(x)| \leq \alpha \, |y - x|, \qquad \forall y \in S', \qquad \forall x \in S', \qquad \forall i.$$

Given any $\varepsilon > 0$, there exists a finite subset C_0' of $C' \cap S'$ such that each point of S lies within the distance $\varepsilon/3\alpha$ of at least one point of C_0'. Since C_0' is finite and the functions f_i converge pointwise on C_0', there exists an integer i_0 such that

$$|f_i(z) - f_j(z)| \leq \varepsilon/3, \qquad \forall i \geq i_0, \qquad \forall j \geq i_0, \qquad \forall z \in C_0'.$$

Given any $x \in S$, let z be one of the points of C_0' such that $|z - x| \leq \varepsilon/3\alpha$. Then, for every $i \geq i_0$ and $j \geq i_0$, we have

$$|f_i(x) - f_j(x)| \leq |f_i(x) - f_i(z)| + |f_i(z) - f_j(z)| + |f_j(z) - f_j(x)|$$
$$\leq \alpha \, |x - z| + (\varepsilon/3) + \alpha \, |z - x| \leq \varepsilon.$$

This proves that, given any $\varepsilon > 0$, there exists an integer i_0 such that

$$|f_i(x) - f_j(x)| \leq \varepsilon, \qquad \forall i \geq i_0, \qquad \forall j \geq j_0, \qquad \forall x \in S.$$

It follows that, for each $x \in S$, the real numbers $f_1(x), f_2(x), \ldots$, form a

Cauchy sequence, so that the limit $f(x)$ exists and is finite. Moreover, given any $\varepsilon > 0$, there exists an integer i_0 such that

$$|f_i(x) - f(x)| = \lim_{j \to \infty} |f_i(x) - f_j(x)| \leq \varepsilon, \qquad \forall x \in S, \qquad \forall i \geq i_0.$$

Thus the functions f_i converge to f uniformly on S. Since S was any closed bounded subset of C, we may conclude in particular that f exists throughout C. Of course, the convexity inequality

$$f_i((1 - \lambda)x + \lambda y) \leq (1 - \lambda)f_i(x) + \lambda f_i(y)$$

is preserved for each $x \in C$, $y \in C$ and $\lambda \in [0, 1]$ as $i \to \infty$, so f is convex. ‖

COROLLARY 10.8.1. *Let f be a finite convex function on a relatively open convex set C. Let $f_1, f_2, \ldots,$ be a sequence of finite convex functions on C such that*

$$\limsup_{i \to \infty} f_i(x) \leq f(x), \qquad \forall x \in C.$$

Then, for each closed bounded subset S of C and each $\varepsilon > 0$, there exists an index i_0 such that

$$f_i(x) \leq f(x) + \varepsilon, \qquad \forall i \geq i_0, \qquad \forall x \in S.$$

PROOF. Let $g_i(x) = \max\{f_i(x), f(x)\}$. The sequence of finite convex functions g_i converges pointwise to f on C, and hence it converges uniformly to f on S. ‖

THEOREM 10.9. *Let C be a relatively open convex set, and let $f_1, f_2, \ldots,$ be a sequence of finite convex functions on C. Suppose that the real number sequence $f_1(x), f_2(x), \ldots,$ is bounded for each $x \in C$ (or merely for each $x \in C'$, where C' is a dense subset of C). It is then possible to select a subsequence of $f_1, f_2, \ldots,$ which converges uniformly on closed bounded subsets of C to some finite convex function f.*

PROOF. A basic fact is needed: if C' is any subset of R^n, there exists a countable subset C'' of C' such that cl $C'' \supset C'$. (Outline of proof: let Q_1 be the collection of all closed (Euclidean) balls in R^n whose centers have rational coordinates and whose radii are rational. Let Q be the subcollection consisting of the balls in Q_1 which have a non-empty intersection with C'. Form C'' by selecting a point of $D \cap C'$ for each $D \in Q$.)

We apply this fact to a subset C' of C such that cl $C' \supset C$ and $\{f_i(x) \mid i = 1, 2, \ldots,\}$ is bounded for every $x \in C'$. The C'' we obtain has these same properties, and it is countable as well. In view of Theorem 10.8, all we need to show is that there is a subsequence of $f_1, f_2, \ldots,$ converging pointwise on C''. Let $x_1, x_2, \ldots,$ be the elements of C'' arranged in a sequence. The real number sequence $\{f_i(x_1) \mid i = 1, 2, \ldots\}$ is bounded, and consequently it has at least one convergent subsequence. Thus we can find a real number α_1 and an infinite subset I_1 of $\{1, 2, \ldots\}$, such that the values of the functions f_i in the subsequence corresponding to I_1 converge at x_1 to α_1. Next, since $\{f_i(x_2) \mid i \in I_1\}$ is bounded, we can find a real number α_2 and an infinite subset I_2 of I_1, not containing the first (i.e. least) integer in I_1 such that the values of the functions f_i in the subsequence corresponding to I_2 converge at x_2 to α_2 (as well as converge at x_1 to α_1). Then we can find a real number α_3 and an infinite subset I_3 of I_2, not containing the first integer in I_2, such that $f_i(x_3)$ for $i \in I_3$ converges to α_3, etc. Continuing in this way, we get an I_j and α_j for each x_j. Let I be the infinite set consisting of the first integer in I_1, the first integer in I_2, etc. The sequence of real numbers $f_i(x_j)$, $i \in I$, converges to α_j for each j. Thus the sequence of functions f_i, $i \in I$, converges pointwise on C''. ‖

Part III · Duality Correspondences

Separation Theorems

The notion of separation has proved to be one of the most fertile notions in convexity theory and its applications. It is based on the fact that a hyperplane in R^n divides R^n evenly in two, in the sense that the complement of the hyperplane is the union of two disjoint open convex sets, the open half-spaces associated with the hyperplane.

Let C_1 and C_2 be non-empty sets in R^n. A hyperplane H is said to *separate* C_1 and C_2 if C_1 is contained in one of the closed half-spaces associated with H and C_2 lies in the opposite closed half-space. It is said to separate C_1 and C_2 *properly* if C_1 and C_2 are not *both* actually contained in H itself. It is said to separate C_1 and C_2 *strongly* if there exists some $\varepsilon > 0$ such that $C_1 + \varepsilon B$ is contained in one of the open half-spaces associated with H and $C_2 + \varepsilon B$ is contained in the opposite open half-space, where B is the unit Euclidean ball $\{x \mid |x| \leq 1\}$. (Of course, $C_i + \varepsilon B$ consists of the points x such that $|x - y| \leq \varepsilon$ for at least one $y \in C_i$.)

Other kinds of separation are sometimes considered, for instance *strict* separation, where C_1 and C_2 must simply belong to opposing open half-spaces. Proper separation and strong separation seem the most useful by far, however, perhaps because they correspond in a natural way to extrema of linear functions.

THEOREM 11.1. *Let C_1 and C_2 be non-empty sets in R^n. There exists a hyperplane separating C_1 and C_2 properly if and only if there exists a vector b such that*

(a) $\inf \{\langle x, b \rangle \mid x \in C_1\} \geq \sup \{\langle x, b \rangle \mid x \in C_2\}$,

(b) $\sup \{\langle x, b \rangle \mid x \in C_1\} > \inf \{\langle x, b \rangle \mid x \in C_2\}$.

There exists a hyperplane separating C_1 and C_2 strongly if and only if there exists a vector b such that

(c) $\inf \{\langle x, b \rangle \mid x \in C_1\} > \sup \{\langle x, b \rangle \mid x \in C_2\}$.

PROOF. Suppose that b satisfies condition (a) and (b), and choose any β between the infimum over C_1 and the supremum over C_2. We have $b \neq 0$ and $\beta \in R$, so that $H = \{x \mid \langle x, b \rangle = \beta\}$ is a hyperplane (Theorem 1.3).

95

The half-space $\{x \mid \langle x, b \rangle \geq \beta\}$ contains C_1, while $\{x \mid \langle x, b \rangle \leq \beta\}$ contains C_2. Condition (b) implies C_1 and C_2 are not both contained in H. Thus H separates C_1 and C_2 properly.

Conversely, when C_1 and C_2 can be separated properly, the separating hyperplane and associated closed half-spaces containing C_1 and C_2 can be expressed in the manner just described for some b and β. One has $\langle x, b \rangle \geq \beta$ for every $x \in C_1$ and $\langle x, b \rangle \leq \beta$ for every $x \in C_2$, with strict inequality for at least one $x \in C_1$ or $x \in C_2$. Thus b satisfies conditions (a) and (b).

If b satisfies the stronger condition (c), we can actually choose $\beta \in R$ and $\delta > 0$ such that $\langle x, b \rangle \geq \beta + \delta$ for every $x \in C_1$, and $\langle x, b \rangle \leq \beta - \delta$ for every $x \in C_2$. Since the unit ball B is bounded, $\varepsilon > 0$ can be chosen so small that $|\langle y, b \rangle| < \delta$ for every y in εB. Then

$$C_1 + \varepsilon B \subset \{x \mid \langle x, b \rangle > \beta\},$$
$$C_2 + \varepsilon B \subset \{x \mid \langle x, b \rangle < \beta\},$$

so that $H = \{x \mid \langle x, b \rangle = \beta\}$ separates C_1 and C_2 strongly. Conversely, if C_1 and C_2 can be separated strongly, the inclusions just described hold for a certain b, β and $\varepsilon > 0$. Then

$$\beta \leq \inf \{\langle x, b \rangle + \varepsilon \langle y, b \rangle \mid x \in C_1, y \in B\} < \inf \{\langle x, b \rangle \mid x \in C_1\},$$
$$\beta \geq \sup \{\langle x, b \rangle + \varepsilon \langle y, b \rangle \mid x \in C_2, y \in B\} > \sup \{\langle x, b \rangle \mid x \in C_2\},$$

so that condition (c) holds. ‖

Whether or not two sets can be separated is an existence question, so it is not surprising that the most celebrated applications of separation theory occur in the proofs of various existence theorems. Typically, what happens is that one needs vectors b with certain properties, and one is able to construct a pair of convex sets C_1 and C_2 such that the vectors b in question correspond to the hyperplanes separating C_1 and C_2 (if any). One then invokes a theorem which says that C_1 and C_2 can indeed be separated in the required sense.

As it happens, the existence of separating hyperplanes in R^n is a relatively elementary matter, not involving the axiom of choice. The fundamental construction is given in the proof of the following theorem.

THEOREM 11.2. *Let C be a non-empty relatively open convex set in R^n, and let M be a non-empty affine set in R^n not meeting C. Then there exists a hyperplane H containing M, such that one of the open half-spaces associated with H contains C.*

PROOF. If M itself is a hyperplane, one of the associated open half-spaces must contain C, for otherwise M would meet C contrary to the hypothesis. (If C contained points x and y in the two opposing open half-spaces, some point of the line segment between x and y would lie in the mutual boundary M of the half-spaces.) Suppose therefore that M is not a hyperplane. We shall show how to construct an affine set M' of one higher dimension than M which again does not meet C. This construction will furnish a hyperplane H with the desired properties after n steps or less, and hence will prove the theorem.

Translating if necessary, we can suppose that $0 \in M$, so that M is a subspace. The convex set $C - M$ includes C but not 0. Since M is not a hyperplane, the subspace $M^\perp$ contains a two-dimensional subspace P. Let $C' = P \cap (C - M)$. This is a relatively open convex set in P (Corollary 6.5.1 and Corollary 6.6.2), and it does not contain 0. All we have to do is find a line L through 0 in P not meeting C', for then $M' = M + L$ will be a subspace of one higher dimension than M not meeting C. (Indeed, $(M + L) \cap C \neq \emptyset$ would imply $L \cap (C - M) \neq \emptyset$, contrary to $L \cap C' = \emptyset$.) For simplicity, we can identify the plane P with R^2. The existence of the line L is trivial if C' is empty or zero-dimensional. If aff C' is a line not containing 0, we can take L to be the parallel line through 0. If aff C' is a line containing 0, we can take L to be the perpendicular line through 0. In the only remaining case, C' is two-dimensional and hence open. The set $K = \bigcup \{\lambda C' \mid \lambda > 0\}$ is the smallest convex cone containing C' (Corollary 2.6.3), and it is open because it is a union of open sets. It does not contain 0. Therefore K is an open sector of R^2 corresponding to an angle no greater than π. We can take L to be the line extending one of the two boundary rays of the sector. ‖

The main separation theorem follows.

THEOREM 11.3. *Let C_1 and C_2 be non-empty convex sets in R^n. In order that there exist a hyperplane separating C_1 and C_2 properly, it is necessary and sufficient that ri C_1 and ri C_2 have no point in common.*

PROOF. Consider the convex set $C = C_1 - C_2$. Its relative interior is ri C_1 − ri C_2 by Corollary 6.6.2, so $0 \notin$ ri C if and only if ri C_1 and ri C_2 have no point in common. Now if $0 \notin$ ri C there exists by the preceding theorem a hyperplane containing $M = \{0\}$ such ri C is contained in one of the associated open half-spaces; the closure of that half-space then contains C, since $C \subset$ cl (ri C). Thus if $0 \notin$ ri C there exists some vector b such that

$$0 \leq \inf_{x \in C} \langle x, b \rangle = \inf_{x_1 \in C_1} \langle x_1, b \rangle - \sup_{x_2 \in C_2} \langle x_2, b \rangle,$$

$$0 < \sup_{x \in C} \langle x, b \rangle = \sup_{x_1 \in C_1} \langle x_1, b \rangle - \inf_{x_2 \in C_2} \langle x_2, b \rangle.$$

But this means C_1 and C_2 can be separated properly, according to Theorem 11.1. These conditions imply in turn that $0 \notin \text{ri } C$, for they assert the existence of a half-space $D = \{x \mid \langle x, b \rangle \geq 0\}$ containing C whose interior $\text{ri } D = \{x \mid \langle x, b \rangle > 0\}$ meets C. In that situation $\text{ri } C \subset \text{ri } D$ (Corollary 6.5.2). ‖

Proper separation allows that one (but not both) of the sets be contained in the separating hyperplane itself. That this provision is needed in Theorem 11.3 is shown by the sets

$$C_1 = \{(\xi_1, \xi_2) \mid \xi_1 > 0, \xi_2 \geq \xi_1^{-1}\},$$

$$C_2 = \{(\xi_1, 0) \mid \xi_1 \geq 0\}$$

in R^2. These convex sets are disjoint. The only separating hyperplane is the ξ_1-axis, which contains all of C_2. This example also shows that not every pair of disjoint closed convex sets can be separated strongly.

THEOREM 11.4.   Let C_1 and C_2 be non-empty convex sets in R^n. In order that there exist a hyperplane separating C_1 and C_2 strongly, it is necessary and sufficient that

$$\inf \{|x_1 - x_2| \mid x_1 \in C_1, x_2 \in C_2\} > 0,$$

in other words that $0 \notin \text{cl } (C_1 - C_2)$.

PROOF.   If C_1 and C_2 can be separated strongly, then, for some $\varepsilon > 0$, $C_1 + \varepsilon B$ does not meet $C_2 + \varepsilon B$. On the other hand, if the latter holds the convex sets $C_1 + \varepsilon B$ and $C_2 + \varepsilon B$ can separated properly, according to the preceding theorem. Since $\varepsilon B = \varepsilon' B + \varepsilon' B$ for $\varepsilon' = \varepsilon/2$, the sets $(C_1 + \varepsilon' B) + \varepsilon' B$ and $(C_2 + \varepsilon' B) + \varepsilon' B$ then belong to opposite closed half-spaces, so that $C_1 + \varepsilon' B$ and $C_2 + \varepsilon' B$ are in opposite open half-spaces. Thus C_1 and C_2 can be separated strongly if and only if, for some $\varepsilon > 0$, the origin does not belong to the set

$$(C_1 + \varepsilon B) - (C_2 + \varepsilon B) = C_1 - C_2 - 2\varepsilon B.$$

This condition means that

$$2\varepsilon B \cap (C_1 - C_2) = \emptyset$$

for some $\varepsilon > 0$, in other words $0 \notin \text{cl } (C_1 - C_2)$. ‖

COROLLARY 11.4.1.   Let C_1 and C_2 be non-empty disjoint closed convex sets in R^n having no common directions of recession. Then there exists a hyperplane separating C_1 and C_2 strongly.

PROOF.   We have $0 \notin (C_1 - C_2)$ since C_1 and C_2 are disjoint. But $\text{cl } (C_1 - C_2) = C_1 - C_2$ under the recession condition by Corollary 9.1.2. ‖

COROLLARY 11.4.2. *Let C_1 and C_2 be non-empty convex sets in R^n whose closures are disjoint. If either set is bounded, there exists a hyperplane separating C_1 and C_2 strongly.*

PROOF. Apply the first corollary to cl C_1 and cl C_2, one of which has no directions of recession at all. ‖

Special separation results which take advantage of polyhedral convexity will be presented in Corollary 19.3.3, Theorem 20.2, Corollary 20.3.1 and Theorem 22.6.

The set of solutions x to a system of weak linear inequalities $\langle x, b_i \rangle \le \beta_i$, $i \in I$, is a closed convex set, since it is an intersection of closed half-spaces. We shall now show that every closed convex set in R^n can be represented as some such solution set.

THEOREM 11.5. *A closed convex set C is the intersection of the closed half-spaces which contain it.*

PROOF. We can assume $\emptyset \ne C \ne R^n$, since the theorem is trivial otherwise. Given any $a \notin C$, the sets $C_1 = \{a\}$ and $C_2 = C$ satisfy the condition in Theorem 11.4. Hence there exists a hyperplane separating $\{a\}$ and C strongly. One of the closed half-spaces associated with this hyperplane contains C but does not contain a. Thus the intersection of the closed half-spaces containing C contains no points other than those in C. ‖

COROLLARY 11.5.1. *Let S be any subset of R^n. Then cl (conv S) is the intersection of all the closed half-spaces containing S.*

PROOF. A closed half-space contains $C = $ cl (conv S) if and only if it contains S. ‖

COROLLARY 11.5.2. *Let C be a convex subset of R^n other than R^n itself. Then there exists a closed half-space containing C. In other words, there exists some $b \in R^n$ such that the linear function $\langle \cdot, b \rangle$ is bounded above on C.*

PROOF. The hypothesis implies that cl $C \ne R^n$ (for otherwise $R^n = $ ri (cl C) $\subset C$). By the theorem, a point belongs to cl C if and only if it belongs to every closed half-space containing cl C, so the collection of closed half-spaces containing cl C cannot be empty. ‖

A sharper version of Theorem 11.5 will be given in Theorem 18.8.

The geometric concept of tangency is one of the most important tools in analysis. Tangent lines to curves and tangent planes to surfaces are defined classically in terms of differentiation. In convex analysis, the opposite approach is exploited. A generalized tangency is defined geometrically in terms of separation. This notion is subsequently used to develop a generalized theory of differentiation.

The generalized tangency is expressed by "supporting" hyperplanes and half-spaces. Let C be a convex set in R^n. A *supporting half-space* to C is a closed half-space which contains C and has a point of C in its boundary. *A supporting hyperplane* to C, is a hyperplane which is the boundary of a supporting half-space to C. The supporting hyperplanes to C, in other words, are the hyperplanes which can be represented in the form $H = \{x \mid \langle x, b \rangle = \beta\}$, $b \neq 0$, where $\langle x, b \rangle \leq \beta$ for every $x \in C$ and $\langle x, b \rangle = \beta$ for at least one $x \in C$. Thus a supporting hyperplane to C is associated with a linear function which achieves its maximum on C. The supporting hyperplanes passing through a given point $a \in C$ correspond to vectors b *normal* to C at a, as defined earlier.

If C is not n-dimensional, so that aff $C \neq R^n$, we can always extend aff C to a hyperplane containing *all* of C. Such supporting hyperplanes are hardly of interest, so we usually speak only of *non-trivial* supporting hyperplanes to C, i.e. ones which do not contain C itself.

THEOREM 11.6. *Let C be a convex set, and let D be a non-empty convex subset of C (for instance, a subset consisting of a single point). In order that there exist a non-trivial supporting hyperplane to C containing D, it is necessary and sufficient that D be disjoint from* ri C.

PROOF. Since $D \subset C$, the non-trivial supporting hyperplanes to C which contain D are the same as the hyperplanes which separate D and C properly. By Theorem 11.3, such a hyperplane exists if and only if ri D is disjoint from ri C. This condition is equivalent to D being disjoint from ri C (Corollary 6.5.2). ‖

COROLLARY 11.6.1. *A convex set has a non-zero normal at each of its boundary points.*

COROLLARY 11.6.2. *Let C be a convex set. An $x \in C$ is a relative boundary point of C if and only if there exists a linear function h not constant on C such that h achieves its maximum over C at x.*

The preceding results can be refined slightly in the case of convex cones.

THEOREM 11.7. *Let C_1 and C_2 be non-empty subsets of R^n, at least one of which is a cone. If there exists a hyperplane which separates C_1 and C_2 properly, then there exists a hyperplane which separates C_1 and C_2 properly and passes through the origin.*

PROOF. Assume that C_2, say, is a cone. If C_1 and C_2 can be separated properly, there exists a vector b satisfying the first two conditions in Theorem 11.1. Let

$$\beta = \sup \{\langle x, b \rangle \mid x \in C_2\}.$$

Then, as shown in the proof of Theorem 11.1, the set

$$H = \{x \mid \langle x, b \rangle = \beta\}$$

is a hyperplane which separates C_1 and C_2 properly. Since C_2 is a cone, we have

$$\lambda\langle x, b\rangle = \langle \lambda x, b\rangle \leq \beta < \infty, \qquad \forall x \in C_2, \qquad \forall \lambda > 0.$$

This implies that $\beta \geq 0$ and $\langle x, b\rangle \leq 0$ for every x in C_2. Hence $\beta = 0$ and $0 \in H$. ‖

COROLLARY 11.7.1. *A non-empty closed convex cone in R^n is the intersection of the homogeneous closed half-spaces which contain it (a homogeneous half-space being one with the origin on its boundary).*

PROOF. Use the theorem to refine the proof of Theorem 11.5. ‖

COROLLARY 11.7.2. *Let S be any subset of R^n, and let K be the closure of the convex cone generated by S. Then K is the intersection of all the homogeneous closed half-spaces containing S.*

PROOF. A homogeneous closed half-space is in particular a closed convex cone containing the origin, and such a cone includes S if and only if it includes K. Apply the preceding corollary. ‖

COROLLARY 11.7.3. *Let K be a convex cone in R^n other than R^n itself. Then K is contained in some homogeneous closed half-space of R^n. In other words, there exists some vector $b \neq 0$ such that $\langle x, b\rangle \leq 0$ for every $x \in K$.*

PROOF. Like Corollary 11.5.2. ‖

Conjugates of Convex Functions

There are two ways of viewing a classical curve or surface like a conic, either as a locus of points or as an envelope of tangents. This fundamental duality enters the theory of convexity in a slightly different form: a closed convex set in R^n is the intersection of the closed half-spaces which contain it (Theorem 11.5). Many intriguing duality correspondences exist as embodiments of this fact, among them conjugacy of convex functions, polarity of convex cones or of other classes of convex sets or functions, and the correspondence between convex sets and their support functions. The basic theory of conjugacy will be developed here. It will be used subsequently to deduce the theorems about the other correspondences.

The definition of the conjugate of a function grows naturally out of the fact that the epigraph of a closed proper convex function on R^n is the intersection of the closed half-spaces in R^{n+1} which contain it. The first step is to translate this geometric result into the language of functions.

The hyperplanes in R^{n+1} can be represented by means of the linear functions on R^{n+1}, and these can in turn be represented in the form

$$(x, \mu) \to \langle x, b \rangle + \mu \beta_0, \qquad b \in R^n, \qquad \beta_0 \in R.$$

Since non-zero linear functions which are scalar multiples of each other give the same hyperplanes, only the cases where $\beta_0 = 0$ or $\beta_0 = -1$ need concern us. The hyperplanes for $\beta_0 = 0$ are of the form

$$\{(x, \mu) \mid \langle x, b \rangle = \beta\}, \qquad 0 \neq b \in R^n, \qquad \beta \in R.$$

These we call *vertical*. The hyperplanes for $\beta_0 = -1$ are of the form

$$\{(x, \mu) \mid \langle x, b \rangle - \mu = \beta\}, \qquad b \in R^n, \qquad \beta \in R.$$

These are the graphs of the affine functions $h(x) = \langle x, b \rangle - \beta$ on R^n. Every closed half-space in R^{n+1} is thus of one of the following types:

1. $\{(x, \mu) \mid \langle x, b \rangle \leq \beta\} = \{(x, \mu) \mid h(x) \leq 0\}, b \neq 0,$
2. $\{(x, \mu) \mid \mu \geq \langle x, b \rangle - \beta\} = \text{epi } h,$
3. $\{(x, \mu) \mid \mu \leq \langle x, b \rangle - \beta\}.$

We shall refer to these types as *vertical, upper* and *lower*, respectively.

THEOREM 12.1. *A closed convex function f is the pointwise supremum of the collection of all affine functions h such that $h \leq f$.*

PROOF. We can take f to be proper, since the theorem is trivial otherwise (by the definition of the closure operation for improper convex functions). Inasmuch as epi f is a closed convex set, epi f is the intersection of the closed half-spaces in R^{n+1} containing it, as already pointed out. Of course, no lower half-space can contain a set like epi f, so only vertical and upper closed half-spaces are involved in the intersection. The half-spaces involved cannot all be vertical, for that would imply that epi f was a union of vertical lines in R^{n+1}, contrary to properness. The upper closed half-spaces containing epi f are just the epigraphs of the affine functions h such that $h \leq f$. Their intersection is the epigraph of the pointwise supremum of such functions h. Thus, to prove the theorem, we must show that the intersection of the vertical and upper closed half-spaces containing epi f is identical to the intersection of just the upper closed half-spaces containing epi f. Suppose that

$$V = \{(x, \mu) \mid 0 \geq \langle x, b_1 \rangle - \beta_1 = h_1(x)\}$$

is a vertical half-space containing epi f, and that (x_0, μ_0) is a point not in V. It is enough to demonstrate that there exists an affine function h such that $h \leq f$ and $\mu_0 < h(x_0)$. We already know there exists at least one affine function h_2 such that epi $h_2 \supset$ epi f, i.e. $h_2 \leq f$. For every $x \in \mathrm{dom}\, f$ we have $h_1(x) \leq 0$ and $h_2(x) \leq f(x)$, and hence

$$\lambda h_1(x) + h_2(x) \leq f(x), \qquad \forall \lambda \geq 0.$$

The same inequality holds trivially when $x \notin \mathrm{dom}\, f$, because then $f(x) = +\infty$. Thus if we fix any $\lambda \geq 0$ and define h by

$$h(x) = \lambda h_1(x) + h_2(x) = \langle x, \lambda b_1 + b_2 \rangle - (\lambda \beta_1 + \beta_2)$$

we will have $h \leq f$. Since $h_1(x_0) > 0$, a sufficiently large λ will ensure that $h(x_0) > \mu_0$ as desired. ‖

COROLLARY 12.1.1. *If f is any function from R^n to $[-\infty, \infty]$, then* cl (conv f) *is the pointwise supremum of the collection of all affine functions on R^n majorized by f.*

PROOF. Since cl (conv f) is the greatest closed convex function majorized by f, the affine functions h such that $h \leq$ cl (conv f) are the same as those such that $h \leq f$. ‖

COROLLARY 12.1.2. *Given any proper convex function f on R^n, there exists some $b \in R^n$ and $\beta \in R$ such that $f(x) \geq \langle x, b \rangle - \beta$ for every x.*

Notice, by the way, that Theorem 12.1 contains the corresponding theorem for convex sets, Theorem 11.5, as a special case. In fact, if f is the indicator function of a convex set C and $h(x) = \langle x, b \rangle - \beta$, we have $h \leq f$ if and only if $h(x) \leq 0$ for every $x \in C$, i.e. if and only if $C \subset \{x \mid \langle x, b \rangle \leq \beta\}$.

Let f be any closed convex function on R^n. According to Theorem 12.1, there is a dual way of describing f: one can describe the set F^* consisting of all pairs (x^*, μ^*) in R^{n+1} such that the affine function $h(x) = \langle x, x^* \rangle - \mu^*$ is majorized by f. We have $h(x) \leq f(x)$ for every x if and only if

$$\mu^* \geq \sup \{\langle x, x^* \rangle - f(x) \mid x \in R^n\}.$$

Thus F^* is actually the epigraph of the function f^* on R^n defined by

$$f^*(x^*) = \sup_x \{\langle x, x^* \rangle - f(x)\} = -\inf_x \{f(x) - \langle x, x^* \rangle\}.$$

This f^* is called the *conjugate* of f. It is actually the pointwise supremum of the affine functions $g(x^*) = \langle x, x^* \rangle - \mu$ such that (x, μ) belongs to the set $F = \text{epi } f$. Hence f^* is another convex function, in fact a closed convex function. Since f is the pointwise supremum of the affine functions $h(x) = \langle x, x^* \rangle - \mu^*$ such that $(x^*, \mu^*) \in F^* = \text{epi } f^*$, we have

$$f(x) = \sup_{x^*} \{\langle x, x^* \rangle - f^*(x^*)\} = -\inf_{x^*} \{f^*(x^*) - \langle x, x^* \rangle\}.$$

But this says that the conjugate f^{**} of f^* is f.

The constant functions $+\infty$ and $-\infty$ are plainly conjugate to each other. Since these are the only improper closed convex functions, all the other conjugate pairs must be proper.

The conjugate f^* of an arbitrary function f from R^n to $[-\infty, +\infty]$ can be defined by the same formula as above. Since f^* simply describes the affine functions majorized by f, f^* is then the same as the conjugate of $\text{cl} (\text{conv } f)$ (Corollary 12.1.1).

The main facts are summarized as follows.

THEOREM 12.2. *Let f be a convex function. The conjugate function f^* is then a closed convex function, proper if and only if f is proper. Moreover,* $(\text{cl} f)^* = f^*$ *and* $f^{**} = \text{cl} f$.

COROLLARY 12.2.1. *The conjugacy operation $f \rightarrow f^*$ induces a symmetric one-to-one correspondence in the class of all closed proper convex functions on R^n.*

COROLLARY 12.2.2. *For any convex function f on R^n, one actually has*

$$f^*(x^*) = \sup \{\langle x, x^* \rangle - f(x) \mid x \in \text{ri} (\text{dom} f)\}.$$

PROOF. The supremum gives $g^*(x^*)$, where g is the function which agrees with f on $\text{ri} (\text{dom} f)$ but is $+\infty$ elsewhere. We have $\text{cl} g = \text{cl} f$ (Corollary 7.3.4), and hence $g^* = f^*$ by the theorem. ‖

Taking conjugates clearly reverses functional inequalities: $f_1 \leq f_2$ implies $f_1^* \geq f_2^*$.

The theory of conjugacy can be regarded as the theory of the "best" inequalities of the type

$$\langle x, y \rangle \leq f(x) + g(y), \qquad \forall x, \qquad \forall y,$$

where f and g are functions from R^n to $(-\infty, +\infty]$. Let W denote the set of all function pairs (f, g) for which this inequality is valid. The "best" pairs (f, g) in W are those for which the inequality cannot be tightened, i.e. those such that, if $(f', g') \in W$, $f' \leq f$ and $g' \leq g$, then $f' = f$ and $g' = g$. Clearly, one has $(f, g) \in W$ if and only if

$$g(y) \geq \sup_x \{\langle x, y \rangle - f(x)\} = f^*(y), \qquad \forall y,$$

or equivalently

$$f(x) \geq \sup_y \{\langle x, y \rangle - g(y)\} = g^*(x), \qquad \forall x.$$

Therefore, the "best" pairs in W are precisely those such that $g = f^*$ and $f = g^*$. The "best" inequalities thus correspond to the pairs of mutually conjugate closed proper convex functions.

It is useful to remember, in particular, that the inequality

$$\langle x, x^* \rangle \leq f(x) + f^*(x^*), \qquad \forall x, \qquad \forall x^*,$$

holds for any proper convex function f and its conjugate f^*. We shall refer to this relation as *Fenchel's inequality*. The pairs (x, x^*) for which Fenchel's inequality is satisfied as an equation form the graph of a certain multivalued mapping ∂f called the subdifferential of f; see Theorem 23.5. Many properties of this mapping are described in §23, §24, and §25.

The conjugacy operation $f \to f^*$ is closely related to the classical Legendre transformation in the case of differentiable convex functions. This relationship is discussed in detail in §26.

Various examples of conjugate functions follow.

As a start, consider the closed proper convex function $f(x) = e^x$, $x \in R$. By definition

$$f^*(x^*) = \sup_x \{xx^* - e^x\}, \qquad \forall x^* \in R.$$

If $x^* < 0$, $xx^* - e^x$ can be made arbitrarily large by taking x very negative, so the supremum is $+\infty$. If $x^* > 0$, the elementary calculus can be used to determine the supremum, which turns out to be $x^* \log x^* - x^*$. If $x^* = 0$, the supremum is 0 trivially. Thus the function conjugate to the exponential function is

$$f^*(x^*) = \begin{cases} x^* \log x^* - x^* & \text{if } x^* > 0, \\ 0 & \text{if } x^* = 0, \\ +\infty & \text{if } x^* < 0. \end{cases}$$

Notice that the value of f^* at $x^* = 0$ could also have been determined as the limit of $x^* \log x^* - x^*$ as $x^* \downarrow 0$ (Corollary 7.5.1). The conjugate of f^* is in turn given by

$$\sup_{x^*} \{xx^* - f^*(x^*)\} = \sup \{xx^* - x^* \log x^* + x^* \mid x^* > 0\},$$

and this supremum is e^x by the calculus. Of course, the calculus is superfluous in reaching this conclusion, for we already know from Corollary 12.2.1 that $f^{**} = f$.

Notice from this example that a function which is finite everywhere need not have a conjugate which is finite everywhere. Properties of the effective domain of the conjugate of a convex function f will be correlated with properties of f in §13.

Here are some other conjugate pairs of closed proper convex functions on R (where $(1/p) + (1/q) = 1$):

1. $f(x) = (1/p)|x|^p, \ 1 < p < +\infty,$

   $f^*(x^*) = (1/q)|x^*|^q, \ 1 < q < +\infty.$

2. $f(x) = \begin{cases} -(1/p)x^p & \text{if } x \geq 0, \ 0 < p < 1, \\ +\infty & \text{if } x < 0, \end{cases}$

   $f^*(x^*) = \begin{cases} -(1/q)|x^*|^q & \text{if } x^* < 0, \ -\infty < q < 0, \\ +\infty & \text{if } x^* \geq 0. \end{cases}$

3. $f(x) = \begin{cases} -(a^2 - x^2)^{1/2} & \text{if } |x| \leq a, \ a \geq 0, \\ +\infty & \text{if } |x| > a, \end{cases}$

   $f^*(x^*) = a(1 + x^{*2})^{1/2}.$

4. $f(x) = \begin{cases} -\tfrac{1}{2} - \log x & \text{if } x > 0, \\ +\infty & \text{if } x \leq 0, \end{cases}$

   $f^*(x^*) = \begin{cases} -\tfrac{1}{2} - \log(-x^*) & \text{if } x^* < 0, \\ +\infty & \text{if } x^* \geq 0. \end{cases}$

In the last example, one has $f^*(x^*) = f(-x^*)$. There are actually many convex functions which satisfy this identity. The identity $f^* = f$ is much more restrictive, however: it has a unique solution on R^n, namely $f = w$, where

$$w(x) = (1/2)\langle x, x \rangle.$$

Indeed, one can see by direct calculation of w^* that $w^* = w$. On the other hand, if f is any convex function such that $f^* = f$, then f is proper, and by Fenchel's inequality

$$\langle x, x \rangle \leq f(x) + f^*(x) = 2f(x).$$

Thus $f \geq w$. This inequality implies that $f^* \leq w^*$. Since $f^* = f$ and $w^* = w$, it must be that $f = w$.

For quite a different example of conjugacy, consider the case where f is the indicator function of a subspace L of R^n. Then

$$f^*(x^*) = \sup_x \{\langle x, x^* \rangle - \delta(x \mid L)\} = \sup\{\langle x, x^* \rangle \mid x \in L\}.$$

The latter supremum is 0 if $\langle x, x^* \rangle = 0$ for every $x \in L$, but otherwise it is

$+\infty$. Thus f^* is the indicator function of the orthogonally complementary subspace $L^\perp$. The relation $f^{**} = f$ corresponds to the relation $L^{\perp\perp} = L$. In this sense, the orthogonality correspondence for subspaces can be regarded as a special case of the conjugacy correspondence for convex functions. This observation will be broadened at the beginning of §14.

We can generalize the orthogonality correspondence for subspaces slightly by taking the basic element to be, not a subspace, but a non-empty affine subset on which a certain affine function (perhaps identically zero) is given. Such elements can be identified, of course, with the *partial affine functions*, i.e. the proper convex functions f such that dom f is an affine set and f is affine on dom f. It turns out that *the conjugate of a partial affine function is another partial affine function*. Since a partial affine function is necessarily closed (Corollary 7.4.2), it is the conjugate of its conjugate. Thus partial affine functions, like subspaces, come in dual pairs. It is easy to derive a formula for this duality. Any partial affine function can be expressed (non-uniquely) in the form

$$f(x) = \delta(x \mid L + a) + \langle x, a^* \rangle + \alpha,$$

where L is a subspace, a and a^* are vectors, and α is a real number. The conjugate partial affine function is then

$$f^*(x^*) = \delta(x^* \mid L^\perp + a^*) + \langle x^*, a \rangle + \alpha^*$$

where $\alpha^* = -\alpha - \langle a, a^* \rangle$. This result is obtained by applying the following theorem to $h = \delta(\cdot \mid L)$, $A = I$.

THEOREM 12.3. *Let h be a convex function on R^n, and let*

$$f(x) = h(A(x - a)) + \langle x, a^* \rangle + \alpha,$$

where A is a one-to-one linear transformation from R^n onto R^n, a and a^* are vectors in R^n, and $\alpha \in R$. Then

$$f^*(x^*) = h^*(A^{*-1}(x^* - a^*)) + \langle x^*, a \rangle + \alpha^*,$$

where A^* is the adjoint of A and $\alpha^* = -\alpha - \langle a, a^* \rangle$.

PROOF. The substitution $y = A(x - a)$ enables us to calculate f^* as follows:

$$\begin{aligned}
f^*(x^*) &= \sup_x \{\langle x, x^* \rangle - h(A(x - a)) - \langle x, a^* \rangle - \alpha\} \\
&= \sup_y \{\langle A^{-1}y + a, x^* \rangle - h(y) - \langle A^{-1}y + a, a^* \rangle - \alpha\} \\
&= \sup_y \{\langle A^{-1}y, x^* - a^* \rangle - h(y)\} + \langle a, x^* - a^* \rangle - \alpha \\
&= \sup_y \{\langle y, A^{*-1}(x^* - a^*) \rangle - h(y)\} + \langle x^*, a \rangle + \alpha^*
\end{aligned}$$

The supremum in the last expression is $h^*(A^{*-1}(x^* - a^*))$ by definition. ‖

The conjugacy correspondence for partial affine functions can be expressed conveniently by means of Tucker representations of affine sets. Let f be any n-dimensional partial affine function on R^N, $0 < n < N$. Each Tucker representation of dom f (as described at the end of §1) yields an expression of f of the form

$$f(x) = \begin{cases} \alpha_{01}\xi_{\bar{1}} + \cdots + \alpha_{0n}\xi_{\bar{n}} - \alpha_{00} & \text{if} \\ \xi_{\overline{n+i}} = \alpha_{i1}\xi_{\bar{1}} + \cdots + \alpha_{in}\xi_{\bar{n}} - \alpha_{i0} & \text{for} \quad i = 1, \ldots, m, \\ +\infty & \text{otherwise.} \end{cases}$$

Here ξ_j is the jth component of x, $m = N - n$, and $\bar{1}, \ldots, \bar{N}$ is some permutation of the indices $1, \ldots, N$. (The coefficients α_{ij} are uniquely determined once the permutation has been given.) If we have such an expression of f, we can immediately write down a corresponding expression of f^*, namely

$$f^*(x^*) = \begin{cases} \beta_{01}\xi^*_{\overline{n+1}} + \cdots + \beta_{0m}\xi^*_{\overline{n+m}} - \beta_{00} & \text{if} \\ \xi^*_{\bar{j}} = \beta_{j1}\xi^*_{\overline{n+1}} + \cdots + \beta_{jm}\xi^*_{\overline{n+m}} - \beta_{j0} & \text{for} \quad i = 1, \ldots, n, \\ +\infty & \text{otherwise,} \end{cases}$$

where $\beta_{ji} = -\alpha_{ij}$ for $i = 0, 1, \ldots, m$ and $j = 0, 1, \ldots, n$. This is proved by direct calculation of f^* in terms of f.

The conjugates of all the *quadratic* convex functions on R^n can be obtained from the formula in Theorem 12.3 (with $A = I$) as soon as one knows the conjugates of the functions of the form

$$h(x) = (1/2)\langle x, Qx \rangle,$$

where Q is a symmetric positive semi-definite $n \times n$ matrix. If Q is non-singular, one can show by the calculus that the supremum of $\langle x, x^* \rangle - h(x)$ in x is attained uniquely at $x = Q^{-1}x^*$, so that

$$h^*(x^*) = (1/2)\langle x^*, Q^{-1}x^* \rangle.$$

If Q is singular, Q^{-1} does not exist, but there nevertheless exists a unique symmetric positive semi-definite $n \times n$ matrix Q' (easily calculated from Q) such that

$$QQ' = Q'Q = P,$$

where P is the matrix of the linear transformation which projects R^n orthogonally onto the orthogonal complement L of the subspace $\{x \mid Qx = 0\}$. For this Q' one has

$$h^*(x^*) = \begin{cases} (1/2)\langle x^*, Q'x^* \rangle & \text{if} \quad x^* \in L, \\ +\infty & \text{if} \quad x^* \notin L. \end{cases}$$

The verification of this is an exercise in linear algebra.

Let us call a proper convex function f a *partial quadratic convex function* if f can be expressed in the form

$$f(x) = q(x) + \delta(x \mid M),$$

where q is a finite quadratic convex function on R^n and M is an affine set in R^n. For example, the formula

$$h(z) = (1/2)(\lambda_1 \zeta_1^2 + \cdots + \lambda_n \zeta_n^2), \qquad 0 \le \lambda_j \le +\infty,$$

defines a partial quadratic convex function with

$$\operatorname{dom} h = \{z = (\zeta_1, \ldots, \zeta_n) \mid \zeta_j = 0, \forall j \text{ such that } \lambda_j = +\infty\}.$$

Such a function h may be called an *elementary* partial quadratic convex function. The conjugate of h is another function of the same type. Indeed, by direct calculation we have

$$h^*(z^*) = (1/2)(\lambda_1^* \zeta_1^{*2} + \cdots + \lambda_n^* \zeta_n^{*2}), \qquad 0 \le \lambda_j^* \le +\infty,$$

where $\lambda_j^* = 1/\lambda_j$ (with $1/\infty$ interpreted as 0 and $1/0$ interpreted as $+\infty$). It can be seen that, in general, f is a partial quadratic convex function on R^n if and only if f can be expressed in the form

$$f(x) = h(A(x - a)) + \langle x, a^* \rangle + \alpha,$$

where h is an elementary partial quadratic convex function on R^n, A is a one-to-one linear transformation from R^n onto itself, a and a^* are vectors in R^n, and α is a real number. Given such an expression of f, we have a similar expression for f^* by Theorem 12.3. It follows that *the conjugate of a partial quadratic convex function is a partial quadratic convex function.*

Let f be any closed proper convex function, so that $f^{**} = f$. By definition,

$$\inf_x f(x) = -\sup_x \{\langle x, 0 \rangle - f(x)\} = -f^*(0),$$

and dually

$$\inf_{x^*} f^*(x^*) = -f^{**}(0) = -f(0).$$

Therefore, the relation

$$\inf_x f(x) = 0 = f(0)$$

holds if and only if

$$\inf_{x^*} f^*(x^*) = 0 = f^*(0).$$

In other words, *the conjugacy correspondence preserves the class of non-negative closed convex functions which vanish at the origin.*

A closed convex function f is *symmetric*, i.e. satisfies

$$f(-x) = f(x), \qquad \forall x,$$

if and only if its conjugate is symmetric. The direct verification of this fact is simple enough, but it is also a special case of a more general symmetry result which can be deduced from Theorem 12.3. Let G be any set of orthogonal linear transformations of R^n onto itself. A function f is said to be *symmetric with respect to G* if

$$f(Ax) = f(x), \qquad \forall x, \qquad \forall A \in G.$$

Ordinary symmetry corresponds to the case where G consists of the single transformation $A: x \rightarrow -x$.

COROLLARY 12.3.1. *A closed convex function f is symmetric with respect to a given set G of orthogonal linear transformations if and only if f^* is symmetric with respect to G.*

PROOF. Specializing Theorem 12.3 to the case where $h = f$, $a = 0 = a^*$, $\alpha = 0$, we see that $fA = f$ implies $f^*A^{*-1} = f^*$. When A is orthogonal, $A^{*-1} = A$ by definition. Thus if $fA = f$ for every $A \in G$, then $f^*A = f^*$ for every $A \in G$. When f is closed, the converse implication is also valid, since $f^{**} = f$. ‖

The functions on R^n which are symmetric with respect to the set of *all* orthogonal transformations of R^n are, of course, those of the form

$$f(x) = g(|x|),$$

where $| \cdot |$ is the Euclidean norm and g is a function on $[0, +\infty)$. Such an f is a closed proper convex function if and only if g is a non-decreasing lower semi-continuous convex function with $g(0)$ finite (Theorem 5.1, Theorem 7.1). In the latter case, the conjugate function must be of the same type, i.e.

$$f^*(x^*) = g^+(|x^*|)$$

where g^+ is a non-decreasing lower semi-continuous convex function on $[0, +\infty)$ with $g^+(0)$ finite. As a matter of fact, we have

$$\begin{aligned}
f^*(x^*) &= \sup_x \{\langle x, x^* \rangle - f(x)\} \\
&= \sup_{\zeta \geq 0} \sup_{|x| = \zeta} \{\langle x, x^* \rangle - g(\zeta)\} \\
&= \sup_{\zeta \geq 0} \{\zeta |x^*| - g(\zeta)\},
\end{aligned}$$

so that g^+ must be given by the formula

$$g^+(\zeta^*) = \sup \{\zeta\zeta^* - g(\zeta) \mid \zeta \geq 0\}.$$

We shall call g^+ the *monotone conjugate* of g. Since $f^{**} = f$, we have $g^{++} = g$, i.e.

$$g(\zeta) = \sup \{\zeta\zeta^* - g^+(\zeta^*) \mid \zeta^* \geq 0\}.$$

Monotone conjugacy thus defines a symmetric one-to-one correspondence in the class of all non-decreasing lower semi-continuous convex functions on $[0, +\infty)$ which are finite at 0.

In the preceding paragraph, the Euclidean norm can be replaced by any closed gauge function in a sense to be described in Theorem 15.3.

Monotone conjugacy can be generalized to n dimensions. Consider the class of functions f on R^n which are *symmetric in each coordinate*, i.e.

which are symmetric with respect to $G = \{A_1, \ldots, A_n\}$, where A_j is the (orthogonal) linear transformation which reverses the sign of the jth component of every vector in R^n. Clearly f belongs to this class if and only if

$$f(x) = g(\text{abs } x),$$

where g is a function on the non-negative orthant of R^n and

$$\text{abs } (\xi_1, \ldots, \xi_n) = (|\xi_1|, \ldots, |\xi_n|).$$

In order that f be a closed proper convex function, it is necessary and sufficient that g be lower semi-continuous, convex, finite at the origin and non-decreasing (in the sense that $g(x) \leq g(x')$ when $0 \leq x \leq x'$, i.e. when $0 \leq \xi_j \leq \xi_j'$ for $j = 1, \ldots, n$). In this case, by Corollary 12.3.1,

$$f^*(x^*) = g^+(\text{abs } x^*),$$

where g^+ is a certain other non-decreasing lower semicontinuous convex function on the non-negative orthant of R^n such that $g^+(0)$ is finite. It is easily established that

$$g^+(z^*) = \sup \{\langle z, z^* \rangle - g(z) \mid z \geq 0\}, \qquad \forall z^* \geq 0.$$

In view of this formula, g^+ is called the *monotone conjugate* of g. We may draw the following conclusion.

THEOREM 12.4. *Let g be a non-decreasing lower semi-continuous convex function on the non-negative orthant of R^n such that $g(0)$ is finite. The monotone conjugate g^+ of g is then another such function, and the monotone conjugate of g^+ is in turn g.*

It can be shown that the formulas

$$g^-(z^*) = \inf \{\langle z, z^* \rangle - g(z) \mid z \geq 0\},$$
$$g(z) = \inf \{\langle z, z^* \rangle - g^-(z^*) \mid z^* \geq 0\},$$

similarly give a one-to-one symmetric correspondence in the class of all non-decreasing *upper* semi-continuous *concave* functions on the non-negative orthant of R^n which have values in $[-\infty, +\infty)$ and are not identically $-\infty$. (The proof is obtained by associating with each g in this class the closed proper convex function f which agrees with $-g$ on the non-negative orthant and has the value $+\infty$ everywhere else; the properties of f are dualized to properties of f^*, and f^* turns out to be expressible in terms of g^- in a certain way.) This correspondence is called *monotone conjugacy for concave functions*.

The general conjugacy correspondence for concave functions, which is closely related to the one for convex functions, will be considered in §30.

Other examples of conjugate convex functions will be given in the next three sections, especially §16.

Support Functions

A common sort of extremum problem is that of maximizing a linear function $\langle \cdot, x^* \rangle$ over a convex set C in R^n. One fruitful approach to such a problem is to study what happens as x^* varies. This leads to the consideration of the function which expresses the dependence of the supremum on x^*, namely the *support function* $\delta^*(\cdot \mid C)$ of C:

$$\delta^*(x^* \mid C) = \sup \{\langle x, x^* \rangle \mid x \in C\}.$$

The appropriateness of the δ^* notation for the support function will be clear below.

Minimization of linear functions over C, as well as maximization, can be studied in terms of $\delta^*(\cdot \mid C)$, because

$$\inf \{\langle x, x^* \rangle \mid x \in C\} = -\delta^*(-x^* \mid C).$$

The support function of C describes all the closed half-spaces which contain C. Indeed, one has

$$C \subset \{x \mid \langle x, x^* \rangle \leq \beta\}$$

if and only if

$$\beta \geq \delta^*(x^* \mid C).$$

The effective domain of $\delta^*(\cdot \mid C)$ is the barrier cone of C. Clearly, for any convex set C, one has

$$\delta^*(x^* \mid C) = \delta^*(x^* \mid \operatorname{cl} C) = \delta^*(x^* \mid \operatorname{ri} C), \forall x^*.$$

Separation theory yields the following result.

THEOREM 13.1.   *Let C be a convex set. Then $x \in \operatorname{cl} C$ if and only if*

$$\langle x, x^* \rangle \leq \delta^*(x^* \mid C)$$

for every vector x^*. On the other hand, $x \in \operatorname{ri} C$ if and only if the same condition holds, but with strict inequality for each x^* such that $-\delta^*(-x^* \mid C) \neq \delta^*(x^* \mid C)$. One has $x \in \operatorname{int} C$ if and only if

$$\langle x, x^* \rangle < \delta^*(x^* \mid C)$$

for every $x^* \neq 0$. Finally, assuming $C \neq \emptyset$, one has $x \in$ aff C if and only if

$$\langle x, x^* \rangle = \delta^*(x^* \mid C)$$

for every x^* such that $-\delta^*(-x^* \mid C) = \delta^*(x^* \mid C)$.

PROOF. The characterizations of cl C and ri C are immediate from Corollary 11.5.1 and Corollary 11.6.2, respectively. The case where ri C is actually int C is the case where C is not contained in any hyperplane, i.e. where $-\delta^*(-x^* \mid C) \neq \delta^*(x^* \mid C)$ for every $x^* \neq 0$. This yields the characterization of int C. The characterization of aff C expresses the fact that the smallest affine set containing C is the same as the intersection of all the hyperplanes containing C (Corollary 1.4.1). ‖

COROLLARY 13.1.1. *For convex sets C_1 and C_2 in R^n, one has cl $C_1 \subset$ cl C_2 if and only if $\delta^*(\cdot \mid C_1) \leq \delta^*(\cdot \mid C_2)$.*

It follows that a closed convex set C can be expressed as the set of solutions to a system of inequalities given by its support function:

$$C = \{x \mid \langle x, x^* \rangle \leq \delta^*(x^* \mid C), \forall x^*\}.$$

Thus C is completely determined by its support function. This fact is interesting, because it shows there is an important one-to-one correspondence between the closed convex sets in R^n and objects of quite a different sort, certain functions on R^n.

This correspondence has many remarkable properties. For example, the support function of the sum of two non-empty convex sets C_1 and C_2 is given by

$$\begin{aligned}
\delta^*(x^* \mid C_1 + C_2) &= \sup \{\langle x_1 + x_2, x^* \rangle \mid x_1 \in C_1, x_2 \in C_2\} \\
&= \sup \{\langle x_1, x^* \rangle \mid x_1 \in C_1\} + \sup \{\langle x_2, x^* \rangle \mid x_2 \in C_2\} \\
&= \delta^*(x^* \mid C_1) + \delta^*(x^* \mid C_2).
\end{aligned}$$

Addition of sets is therefore converted into addition of functions. Further properties of this sort will be encountered in §16.

Just what class of functions is involved? Given a function on R^n, how does one recognize whether it is the support function of some set C? This question will be answered in a moment.

It happens that the support function correspondence can be regarded as a special case of conjugacy. We need only keep in mind the trivial one-to-one correspondence between convex sets C and indicator functions $\delta(\cdot \mid C)$. The conjugate of $\delta(\cdot \mid C)$ is by definition given by

$$\sup_{x \in R^n} \{\langle x, x^* \rangle - \delta(x \mid C)\} = \sup_{x \in C} \langle x, x^* \rangle = \delta^*(x^* \mid C).$$

The conjugate of $\delta^*(x^* \mid C)$ then satisfies

$$(\delta^*(\cdot \mid C))^* = \text{cl } \delta(\cdot \mid C) = \delta(\cdot \mid \text{cl } C),$$

according to the nature of the conjugacy correspondence (Theorem 12.2).

THEOREM 13.2. *The indicator function and the support function of a closed convex set are conjugate to each other. The functions which are the support functions of non-empty convex sets are the closed proper convex functions which are positively homogeneous.*

PROOF. Practically everything is obvious from Theorem 12.2 and the remarks just made. We only have to show that a closed proper convex function f has no values other than 0 and $+\infty$ if and only if its conjugate is positively homogeneous. The first property of f is equivalent to having $f(x) = \lambda f(x)$ for every x and $\lambda > 0$. The second property is equivalent to having

$$f^*(x^*) = \lambda f^*(\lambda^{-1}x^*) = (f^*\lambda)(x^*)$$

for every x^* and $\lambda > 0$. But

$$(\lambda f)^*(x^*) = \sup_x \{\langle x, x^* \rangle - \lambda f(x)\}$$
$$= \sup_x \{\lambda(\langle x, \lambda^{-1}x^* \rangle - f(x))\} = \lambda f^*(\lambda^{-1}x^*).$$

Thus $f = \lambda f$ for every $\lambda > 0$ if and only $f^* = f^*\lambda$ for every $\lambda > 0$, when f is a closed convex function. ‖

In particular, Theorem 13.2 says that $\delta^*(x^* \mid C)$ is a lower semi-continuous function of x^*, and

$$\delta^*(x_1^* + x_2^* \mid C) \leq \delta^*(x_1^* \mid C) + \delta^*(x_2^* \mid C), \qquad \forall x_1^*, \forall x_2^*.$$

COROLLARY 13.2.1. *Let f be any positively homogeneous convex function which is not identically $+\infty$. Then cl f is the support function of a certain closed convex set C, namely*

$$C = \{x^* \mid \forall x, \langle x, x^* \rangle \leq f(x)\}.$$

PROOF. Either cl f is a closed proper positively homogeneous convex function, or cl f is the constant function $-\infty$ (the support function of $\emptyset$). Thus cl $f = \delta^*(\cdot \mid C)$ for a certain closed convex set C. It follows that, by definition, $f^* = (\text{cl } f)^* = \delta(\cdot \mid C)$, and $C = \{x^* \mid f^*(x^*) \leq 0\}$. But $f^*(x^*) \leq 0$ if and only if $\langle x, x^* \rangle - f(x) \leq 0$ for every x. ‖

COROLLARY 13.2.2. *The support functions of the non-empty bounded convex sets are the finite positively homogeneous convex functions.*

PROOF. A finite convex function is necessarily closed (Corollary 7.4.2). In view of the characterization of support functions in the theorem, we need only observe that a convex set C is bounded if and only if

$\delta^*(x^* \mid C) < +\infty$ for every x^*. Indeed, a subset C of R^n is bounded if and only if it is contained in some cube, and that is true if and only if every linear function is bounded above on C. ‖

The Euclidean norm, for instance, must be the support function of some set, because it is a finite positively homogeneous convex function. What is the set? The Cauchy-Schwarz inequality

$$|\langle x, y \rangle| \leq |x| \cdot |y|$$

implies that $\langle x, y \rangle \leq |x|$ when $|y| \leq 1$. Of course, $\langle x, y \rangle = |x|$ if $x = 0$ or if $y = |x|^{-1}x$. Thus

$$|x| = \sup \{\langle x, y \rangle \mid |y| \leq 1\} = \delta^*(x \mid B),$$

where B is the unit Euclidean ball. More generally, the support function of the ball $a + \lambda B$, $\lambda \geq 0$, is

$$f(x) = \langle x, a \rangle + \lambda |x|.$$

As further examples, the support functions of the sets

$$C_1 = \{x = (\xi_1, \ldots, \xi_n) \mid \xi_j \geq 0, \xi_1 + \cdots + \xi_n = 1\},$$
$$C_2 = \{x = (\xi_1, \ldots, \xi_n) \mid |\xi_1| + \cdots + |\xi_n| \leq 1\},$$
$$C_3 = \{x = (\xi_1, \xi_2) \mid \xi_1 < 0, \xi_2 \leq \xi_1^{-1}\},$$
$$C_4 = \{x = (\xi_1, \xi_2) \mid 2\xi_1 + \xi_2^2 \leq 0\},$$

are readily calculated to be

$$\delta^*(x^* \mid C_1) = \max \{\xi_j^* \mid j = 1, \ldots, n\},$$
$$\delta^*(x^* \mid C_2) = \max \{|\xi_j^*| \mid j = 1, \ldots, n\},$$
$$\delta^*(x^* \mid C_3) = \begin{cases} -2(\xi_1^* \xi_2^*)^{1/2} & \text{if } x^* = (\xi_1^*, \xi_2^*) \geq 0, \\ +\infty & \text{otherwise,} \end{cases}$$
$$\delta^*(x^* \mid C_4) = \begin{cases} \xi_2^{*2}/2\xi_1^* & \text{if } \xi_1^* > 0, \\ 0 & \text{if } \xi_1^* = 0 = \xi_2^*, \\ +\infty & \text{otherwise.} \end{cases}$$

The support functions of convex sets are positively homogeneous convex functions according to Theorem 13.2, but so are the gauge functions of convex sets. Relationships between support functions and gauge functions will be explored in §14.

A convex function f is accompanied by various convex sets, such as its effective domain, epigraph, and level sets. We shall show how the support functions of these sets may be derived from the conjugate convex function f^*.

THEOREM 13.3. *Let f be a proper convex function. The support function of* dom *f is then the recession function f*$^{*}0^{+}$ *of f**. *If f is closed, the support function of* dom *f** *is the recession function f*0^{+} *of f.*

PROOF. By definition, f^{*} is the pointwise supremum of the affine functions $h(x^{*}) = \langle x, x^{*} \rangle - \mu$, $(x, \mu) \in$ epi f. Therefore epi f^{*} is the (non-empty) intersection of the corresponding closed half-spaces epi h. The recession cone $0^{+}(\text{epi } f^{*})$ is then the intersection of the sets $0^{+}(\text{epi } h)$ (Corollary 8.3.3). This means that $f^{*}0^{+}$ is the pointwise supremum of the functions $h0^{+}$. Trivially, $(h0^{+})(x^{*}) = \langle x, x^{*} \rangle$ when $h(x^{*}) = \langle x, x^{*} \rangle - \mu$. Thus $f^{*}0^{+}$ is the pointwise supremum of the linear functions $\langle x, \cdot \rangle$ such that $(x, \mu) \in$ epi f for some μ, i.e.

$$(f^{*}0^{+})(x^{*}) = \sup \{\langle x, x^{*} \rangle \mid x \in \text{dom} f\} = \delta^{*}(x^{*} \mid \text{dom} f).$$

The second assertion of the theorem follows by duality, because $f^{**} = f$ when f is closed. ∥

A convex function f will be called *co-finite* if f is closed and proper and epi f contains no non-vertical half-lines, i.e.

$$(f0^{+})(y) = +\infty, \qquad \forall y \neq 0.$$

The latter condition is in particular satisfied, of course, if dom f is bounded.

COROLLARY 13.3.1. *Let f be a closed convex function on* R^{n}. *In order that f** *be finite everywhere, so that* dom $f^{*} = R^{n}$, *it is necessary and sufficient that f be co-finite.*

PROOF. We have dom $f^{*} = R^{n}$ if and only if dom f^{*} is not contained in any closed half-space of R^{n} (Corollary 11.5.2). This is equivalent to the condition that $\delta^{*}(x \mid \text{dom} f^{*}) < +\infty$ only for $x = 0$. ∥

COROLLARY 13.3.2. *Let f be a closed proper convex function. In order that* dom f^{*} *be an affine set, it is necessary and sufficient that* $(f0^{+})(y) = +\infty$ *for every y which is not actually in the lineality space of f.*

PROOF. As an exercise in separation theory, it can be shown that a convex set C is affine if and only if every linear function which is bounded above on C is constant on C. This condition means that $-\delta^{*}(-y \mid C) = \delta^{*}(y \mid C)$ whenever $\delta^{*}(y \mid C) < +\infty$. For $C = \text{dom} f^{*}$, we have $\delta^{*}(y \mid C) = (f0^{+})(y)$, and the vectors y such that $-(f0^{+})(-y) = (f0^{+})(y)$ are by definition the vectors in the lineality space of f. ∥

COROLLARY 13.3.3. *Let f be a proper convex function. In order that* dom f^{*} *be bounded, it is necessary and sufficient that f be finite everywhere and that there exist a real number* $\alpha \geq 0$ *such that*

$$|f(z) - f(x)| \leq \alpha |z - x|, \qquad \forall z, \qquad \forall x.$$

The smallest α *for which this Lipschitz condition holds is then*

$$\alpha = \sup \{|x^{*}| \mid x^{*} \in \text{dom} f^{*}\}.$$

PROOF. We can assume that f is closed, because f and $\mathrm{cl}\, f$ have the same conjugate, and the Lipschitz condition is satisfied by f if and only if it is satisfied by $\mathrm{cl}\, f$. The first assertion then follows from Theorem 10.5, since $\mathrm{dom}\, f^*$ is bounded if and only if its support function, which is $f0^+$ by Theorem 13.3, is finite everywhere. Now the Lipschitz condition on f is equivalent to having

$$f(x + y) \leq f(x) + \alpha |y|, \qquad \forall x, \qquad \forall y,$$

and that is in turn equivalent to

$$(f0^+)(y) \leq \alpha |y|, \qquad \forall y$$

(Corollary 8.5.1). But $g(y) = \alpha |y|$ is the support function of αB, where B is the unit Euclidean ball. Hence $f0^+ \leq g$ means $\mathrm{cl}\,(\mathrm{dom}\, f^*) \subset \alpha B$ (Corollary 13.1.1). This shows that the Lipschitz condition holds for a given α if and only if $|x^*| \leq \alpha$ for every $x^* \in \mathrm{dom}\, f^*$. ‖

COROLLARY 13.3.4. *Let f be a closed proper convex function. Let x^* be a fixed vector and let $g(x) = f(x) - \langle x, x^* \rangle$. Then*

(a) $x^* \in \mathrm{cl}\,(\mathrm{dom}\, f^*)$ *if and only if* $(g0^+)(y) \geq 0$ *for every y;*

(b) $x^* \in \mathrm{ri}\,(\mathrm{dom}\, f^*)$ *if and only if* $(g0^+)(y) > 0$ *for all vectors y except those satisfying* $-(g0^+)(-y) = (g0^+)(y) = 0$;

(c) $x^* \in \mathrm{int}\,(\mathrm{dom}\, f^*)$ *if and only if* $(g0^+)(y) > 0$ *for every $y \neq 0$;*

(d) $x^* \in \mathrm{aff}\,(\mathrm{dom}\, f^*)$ *if and only if* $(g0^+)(y) = 0$ *for every vector y such that* $-(g0^+)(-y) = (g0^+)(y)$.

PROOF. Let $C = (\mathrm{dom}\, f^*) - x^*$. Clearly $x^* \in \mathrm{cl}\,(\mathrm{dom}\, f^*)$ if and only if $0 \in \mathrm{cl}\, C$, and so forth. We have $g^*(y^*) = f^*(y^* + x^*)$ (Theorem 12.3), and hence $\mathrm{dom}\, g^* = C$. The support function of C is therefore $g0^+$ by Theorem 13.3, and conditions (a), (b), (c) and (d) follow immediately from the corresponding support function conditions in Theorem 13.1. ‖

THEOREM 13.4. *Let f be a proper convex function on R^n. The lineality space of f^* is then the orthogonal complement of the subspace parallel to $\mathrm{aff}\,(\mathrm{dom}\, f)$. Dually, if f is closed the subspace parallel to $\mathrm{aff}\,(\mathrm{dom}\, f^*)$ is the orthogonal complement of the lineality space of f, and one has*

$$\text{lineality}\, f^* = n - \text{dimension}\, f,$$
$$\text{dimension}\, f^* = n - \text{lineality}\, f.$$

PROOF. The lineality space L of f^* consists of the vectors x^* such that $-(f^*0^+)(-x^*) = (f^*0^+)(x^*)$. By Theorem 13.3, $(f^*0^+)(x^*)$ and $-(f^*0^+)(-x^*)$ are the supremum and infimum of the linear function $\langle \cdot, x^* \rangle$ on $\mathrm{dom}\, f$, respectively. Thus $x^* \in L$ if and only if $\langle \cdot, x^* \rangle$ is constant on $\mathrm{dom}\, f$, or equivalently constant on $\mathrm{aff}\,(\mathrm{dom}\, f)$ (since the hyperplanes containing $\mathrm{aff}\,(\mathrm{dom}\, f)$ and $\mathrm{dom}\, f$ are the same). A linear function $\langle \cdot, x^* \rangle$

is constant on a non-empty affine set M if and only if

$$0 = \langle x_1, x^* \rangle - \langle x_2, x^* \rangle = \langle x_1 - x_2, x^* \rangle, \qquad \forall x_1 \in M, \qquad \forall x_2 \in M$$

This condition means that $x^* \in (M - M)^\perp$. Thus $L = (M - M)^\perp$, where $M = \text{aff}\,(\text{dom} f)$. But $M - M$ is the subspace parallel to M (Theorem 1.2). This establishes the first assertion of the theorem. Since the dimensions of orthogonally complementary subspaces in R^n add up to n, and affine sets parallel to each other have the same dimension, it follows that

$$\dim M + \dim L = n.$$

By definition, however, $\dim M$ is the dimension of f and $\dim L$ is the lineality of f^*. The second assertion of the theorem and the second dimensionality formula must be true, because $f^{**} = f$ when f is closed.   ‖

COROLLARY 13.4.1.   *Closed proper convex functions conjugate to each other have the same rank.*

PROOF.   This is immediate from the formulas in the theorem and the definition of rank.   ‖

COROLLARY 13.4.2.   *Let f be a closed proper convex function. Then dom f^* has a non-empty interior if and only if there are no lines along which f is (finite and) affine.*

PROOF.   The dimension of f^* is n if and only if the lineality of f is 0.   ‖

Given a convex function h, a level set of the form

$$C = \{ x \mid h(x) \leq \beta + \langle x, b^* \rangle \}$$

can always be expressed as $\{ x \mid f(x) \leq 0 \}$, where

$$f(x) = h(x) - \langle x, b^* \rangle - \beta.$$

The conjugate of f is

$$f^*(x^*) = h^*(x^* + b^*) + \beta.$$

The following theorem then gives the support function of C.

THEOREM 13.5.   *Let f be a closed proper convex function. The support function of $\{ x \mid f(x) \leq 0 \}$ is then cl g, where g is the positively homogeneous convex function generated by f^*. Dually, the closure of the positively homogeneous convex function k generated by f is the support function of $\{ x^* \mid f^*(x^*) \leq 0 \}$.*

PROOF.   Only the second assertion needs to be proved, by virtue of the duality between f and f^*. By Corollary 13.2.1, cl k is the support function of D, where D is the set of vectors x^* such that $\langle \cdot, x^* \rangle \leq k$. The linear functions majorized by k correspond to the upper closed half-spaces in R^{n+1} which are convex cones containing epi k. But, by the definition of k, the closed convex cones containing epi k are the same as those containing epi f. Thus D consists of the vectors x^* such that $\langle x, x^* \rangle \leq f(x)$ for every x, in other words $f^*(x^*) \leq 0$. ‖

The support function of epi f may be obtained by dualizing the following result (and reversing signs).

COROLLARY 13.5.1. *Let f be a closed proper convex function on R^n. The function k on R^{n+1} defined by*

$$k(\lambda, x) = \begin{cases} (f\lambda)(x) & \text{if } \lambda > 0, \\ (f0^+)(x) & \text{if } \lambda = 0, \\ +\infty & \text{if } \lambda < 0, \end{cases}$$

is then the support function of

$$C = \{(\lambda^*, x^*) \mid \lambda^* \leq -f^*(x^*)\} \subset R^{n+1}.$$

PROOF. Let $h(\lambda, x) = f(x) + \delta(\lambda \mid 1)$ on R^{n+1}. The closure of the positively homogeneous convex function generated by h is k, as pointed out after Theorem 8.5. Hence k is the support function of

$$\{(\lambda^*, x^*) \mid h^*(\lambda^*, x^*) \leq 0\}$$

by the present theorem. But

$$h^*(\lambda^*, x^*) = \sup \{\lambda\lambda^* + \langle x, x^* \rangle - f(x) - \delta(\lambda \mid 1) \mid \lambda \in R, x \in R^n\}$$

$$= \sup_x \{\lambda^* + \langle x, x^* \rangle - f(x)\} = \lambda^* + f^*(x^*).$$

Thus $h^*(\lambda^*, x^*) \leq 0$ means that $\lambda^* \leq -f^*(x^*)$. ‖

More explicit formulas for the support functions in Theorem 13.5 can be obtained from the formulas in Theorem 9.7 for the positively homogeneous function generated by a given function.

As an example, let us calculate the support function of an "elliptic" convex set

$$C = \{x \mid (1/2)\langle x, Qx \rangle + \langle a, x \rangle + \alpha \leq 0\}.$$

where Q is a positive definite $n \times n$ symmetric matrix. We have $C = \{x \mid f(x) \leq 0\}$ for a certain finite convex function f on R^n. By Theorem 13.5, $\delta^*(\cdot \mid C)$ is the closure of the positively homogeneous convex function g generated by f^*. As seen in the preceding section,

$$f^*(x^*) = (1/2)\langle x^* - a, Q^{-1}(x^* - a) \rangle - \alpha$$

$$= (1/2)\langle x^*, Q^{-1}x^* \rangle + \langle b, x^* \rangle + \beta,$$

where $b = -Q^{-1}a$ and $\beta = (1/2)\langle a, Q^{-1}a \rangle - \alpha$. For any $x^* \neq 0$, $g(x^*)$ is by definition the infimum of $(f^*\lambda)(x^*) = \lambda f^*(\lambda^{-1}x^*)$ in $\lambda > 0$. Since dom $f^* = R^n$, we have dom $g = R^n$. Hence g is itself closed, and

$$\delta^*(x^* \mid C) = g(x^*) = \inf_{\lambda > 0} \{(1/2\lambda)\langle x^*, Q^{-1}x^* \rangle + \langle b, x^* \rangle + \lambda\beta\}.$$

This infimum is readily calculated. Assuming $C \neq \emptyset$, we have

$$0 \leq \sup_x \{-f(x)\} = f^*(0) = \beta.$$

If $\beta = 0$, the infimum is plainly $\langle b, x^* \rangle$. If $\beta > 0$, we can get the infimum by taking the derivative with respect to λ and setting that equal to 0. The general formula so obtained is

$$\delta^*(x^* \mid C) = \langle b, x^* \rangle + [2\beta\langle x^*, Q^{-1}x^* \rangle]^{1/2}.$$

Polars of Convex Sets

The correspondence between convex sets and their support functions reflects a certain duality between positive homogeneity and the property of being an indicator function. Namely, suppose f is a proper convex function on R^n. If f is an indicator function, its conjugate f^* is positively homogeneous (Theorem 13.2). If f is positively homogeneous, f^* is an indicator function (Corollary 13.2.1). It follows that, if f is a positively homogeneous indicator function, then f^* is a positively homogeneous indicator function. Of course, the positively homogeneous indicator functions are simply the indicator functions of cones. Thus, if $f(x) = \delta(x \mid K)$ for a non-empty convex cone K, then $f^*(x^*) = \delta(x^* \mid K^\circ)$ for a certain other non-empty convex cone K°, which must be closed since f^* is closed. This K° is called the *polar* of K. By Corollary 13.2.1, we have

$$K^\circ = \{x^* \mid \forall x, \langle x, x^* \rangle \leq \delta(x \mid K)\}$$
$$= \{x^* \mid \forall x \in K, \langle x, x^* \rangle \leq 0\}.$$

The polar $K^{\circ\circ}$ of K° is cl K, since the conjugate of $f^* = \delta(\cdot \mid K^\circ)$ is in turn cl $f = \delta(\cdot \mid$ cl $K)$. Also, (cl $K)^\circ = K^\circ$ (inasmuch as (cl $f)^* = f^*$). The conjugacy correspondence among convex functions thus includes a special symmetric one-to-one correspondence among convex cones, as follows.

THEOREM 14.1. *Let K be a non-empty closed convex cone. The polar K° of K is then another non-empty closed convex cone, and $K^{\circ\circ} = K$. The indicator functions of K and K° are conjugate to each other.*

The first assertion of Theorem 14.1 could also be derived directly from the fact that a non-empty closed convex cone is the intersection of the homogeneous closed half-spaces which contain it (Corollary 11.7.1).

The second assertion of Theorem 14.1 is noteworthy because the indicators of convex cones appear frequently in extremum problems, and their conjugates are needed in determining the corresponding dual problems.

Observe that, if K is a subspace of R^n, then K° is the orthogonally complementary subspace. In general, for any non-empty closed convex cone K, K° consists of all the vectors normal to K at 0, while K consists of all the vectors normal to K° at 0.

If K is the non-negative orthant of R^n, then $K^\circ = -K$ (the non-positive orthant). If K is the convex cone generated by a non-empty vector collection $\{a_i \mid i \in I\}$, then K consists of all non-negative linear combinations x of the a_i's, and it follows that

$$K^\circ = \{x^* \mid \forall x \in K, \langle x, x^* \rangle \leq 0\}$$
$$= \{x^* \mid \forall i \in I, \langle a_i, x^* \rangle \leq 0\}.$$

The polar of K° is in turn cl K by the above. *Thus the polar of a convex cone of the form*

$$\{y \mid \forall i \in I, \langle a_i, y \rangle \leq 0\}$$

is the closure of the convex cone generated by the a_i's. If the latter cone is closed (as is always the case for example when the collection $\{a_i \mid i \in I\}$ is *finite*, as will be seen in Theorem 19.1), the polar consists of all non-negative linear combinations of the a_i's.

An extension of the polarity correspondence to a more general class of convex sets will be discussed below, but first we shall describe some further connections between polars of convex cones and conjugates of convex functions.

THEOREM 14.2.   *Let f be a proper convex function. The polar of the convex cone generated by dom f is then the recession cone of f^*. Dually, if f is closed, the polar of the recession cone of f is the closure of the convex cone generated by* dom f^*.

PROOF.   For any $\alpha > \inf f^*$, the recession cone of f^* is by Theorem 8.7 the same as the recession cone 0^+C of the (non-empty closed) convex set

$$C = \{x^* \mid f^*(x^*) \leq \alpha\}$$
$$= \{x^* \mid \langle x, x^* \rangle - f(x) \leq \alpha, \forall x\}$$
$$= \{x^* \mid \langle x, x^* \rangle \leq \alpha + f(x), \forall x \in \operatorname{dom} f\}.$$

It is clear from the latter expression that a vector y^* has the property that

$$x^* + \lambda y^* \in C, \quad \forall x^* \in C, \quad \forall \lambda \geq 0,$$

if and only if

$$\langle x, y^* \rangle \leq 0, \quad \forall x \in \operatorname{dom} f.$$

Therefore

$$0^+C = \{y^* \mid \langle x, y^* \rangle \leq 0, \forall x \in \operatorname{dom} f\}$$
$$= \{y^* \mid \langle y, y^* \rangle \leq 0, \forall y \in K\},$$

where

$$K = \{y \mid \exists x \in \operatorname{dom} f, \exists \lambda \geq 0, y = \lambda x\}.$$

Thus $0^+C = K°$, where K is the convex cone generated by dom f. The dual part of the theorem follows by the fact that $f^{**} = f$ when f is closed.   ‖

COROLLARY 14.2.1.   *The polar of the barrier cone of a non-empty closed convex set C is the recession cone of C.*

PROOF.   Take f to be the support function of C (so that f^* is the indicator function of C by Theorem 13.2).   ‖

COROLLARY 14.2.2.   *Let f be a closed proper convex function. In order that $\{x \mid f(x) \leq \alpha\}$ be a bounded set for every $\alpha \in R$, it is necessary and sufficient that $0 \in$ int (dom f^*).*

PROOF.   We have $0 \in$ int (dom f^*) if and only if the convex cone K generated by dom f^* is all of R^n (Corollary 6.4.1). On the other hand, the level sets $\{x \mid f(x) \leq \alpha\}$ are all bounded if and only if the recession cone of f, which is $K°$, consists of the zero vector alone (Theorem 8.7 and Theorem 8.4). We have $K° = \{0\}$ if and only if cl $K = \{0\}° = R^n$, and cl $K = R^n$ implies that actually $K = R^n$.   ‖

THEOREM 14.3.   *Let f be a closed proper convex function such that $f(0) > 0 > \inf f$. The closed convex cones generated by $\{x \mid f(x) \leq 0\}$ and by $\{x^* \mid f^*(x^*) \leq 0\}$ are then polar to each other.*

PROOF.   Since $f^*(0) = -\inf f$ and $f(0) = -\inf f^*$, the hypothesis implies that $f^*(0) > 0 > \inf f^*$. Thus $\{x \mid f(x) \leq 0\}$ and $\{x^* \mid f^*(x^*) \leq 0\}$ are non-empty closed convex sets not containing the origin. Let k be the positively homogeneous convex function generated by f. Since cl k and the indicator function of $\{x^* \mid f^*(x^*) \leq 0\}$ are conjugate to each other (Theorem 13.5, Theorem 13.2), the recession cone K of cl k and the closure of the convex cone generated by $\{x^* \mid f^*(x^*) \leq 0\}$ are polar to each other (Theorem 14.2). We must show that K is the closure of the convex cone generated by $\{x \mid f(x) \leq 0\}$. We have (cl $k)0^+ = $ cl k by positive homogeneity, so

$$K = \{x \mid (\text{cl } k)(x) \leq 0\}$$

by definition. Therefore

$$K = \text{cl } \{x \mid k(x) \leq 0\} = \text{cl } \{x \mid k(x) < 0\}$$

by Theorem 7.6, provided the last set is not empty. Now $k(x)$ is the infimum of $(f\lambda)(x)$ in $\lambda > 0$ for each $x \neq 0$. Moreover, $(f\lambda)(x) \leq 0$ for a positive λ if and only if $\lambda^{-1}x \in \{y \mid f(y) \leq 0\}$; likewise with $\leq$ replaced by $<$. Since $\inf f < 0$, the set $\{x \mid k(x) < 0\}$ is not empty. The convex cone generated by $\{x \mid f(x) \leq 0\}$ lies between $\{x \mid k(x) < 0\}$ and $\{x \mid k(x) \leq 0\}$, so its closure must be K.   ‖

The polarity correspondence for convex cones has been derived from the conjugacy correspondence for convex functions, but the converse derivation is also possible. Recall that each closed proper convex function

f on R^n corresponds to a certain non-empty closed convex cone in R^{n+2}, namely cl K where K is the convex cone generated by the triples $(1, x, \mu)$ such that $(x, \mu) \in$ epi f. This cone completely determines f, of course. Actually, from the discussion of recession cones and functions, cl K is just the set of $(\lambda, x, \mu) \in R^{n+2}$ such that $\lambda > 0$ and $\mu \geq (f\lambda)(x)$, or $\lambda = 0$ and $\mu \geq (f0^+)(x)$. We shall now show that the conjugate of f can be obtained from the polar of K with minor changes.

THEOREM 14.4.   *Let f be a closed proper convex function on R^n, and let K be the convex cone generated by the vectors $(1, x, \mu) \in R^{n+2}$ such that $\mu \geq f(x)$. Let K^* be the convex cone generated by the $(1, x^*, \mu^*) \in R^{n+2}$ such that $\mu^* \geq f^*(x^*)$.   Then*

$$\text{cl } K^* = \{(\lambda^*, x^*, \mu^*) \mid (-\mu^*, x^*, -\lambda^*) \in K^\circ\}.$$

PROOF.   Since f is proper, cl K contains the vector $(0, 0, 1)$ but not $(0, 0, -1)$. It follows that the polar cone $(\text{cl } K)^\circ = K^\circ$ is contained in the half-space

$$H = \{(-\mu^*, x^*, -\lambda^*) \mid \lambda^* \geq 0\},$$

but not in the boundary hyperplane of H. Thus K° is the closure of its intersection with the interior of H (Corollary 6.5.2). It follows that K° is the closure of the convex cone generated by the intersection of K° with the hyperplane

$$\{(-\mu^*, x^*, -\lambda^*) \mid \lambda^* = 1\}.$$

A vector belongs to K° if and only if it has a non-positive inner product with every vector of the form $\lambda(1, x, \mu)$ such that $\lambda \geq 0$ and $\mu \geq f(x)$. Thus $(-\mu^*, x^*, -1)$ belongs to K° if and only if

$$-\mu^* + \langle x, x^* \rangle - \mu \leq 0$$

whenever $\mu \geq f(x)$, i.e. if and only if

$$\mu^* \geq \sup_x \{\langle x, x^* \rangle - f(x)\} = f^*(x^*).$$

This shows that the image of K° under the mapping

$$(\lambda^*, x^*, \mu^*) \to (-\mu^*, x^*, -\lambda^*)$$

is the closure of the convex cone generated by the vectors $(1, x^*, \mu^*)$ with $\mu^* \geq f^*(x^*)$, i.e. the closure of K^*.   ‖

The polarity correspondence for convex cones can be generalized to a polarity correspondence for the class of all closed convex sets containing the origin. This can be seen by taking conjugates of the gauge functions of convex sets instead of the indicator functions of convex cones. The gauge and indicator functions of a non-empty convex set coincide, of course, when the set is a cone.

Let C be a non-empty convex set. By definition, the gauge function $\gamma(\cdot \mid C)$ is the positively homogeneous convex function generated by $f = \delta(\cdot \mid C) + 1$. The closure of $\gamma(\cdot \mid C)$ is the support function of $\{x^* \mid f^*(x^*) \leq 0\}$ (Theorem 13.5). But $f^* = \delta^*(\cdot \mid C) - 1$. Thus

$$\mathrm{cl}\, \gamma(\cdot \mid C) = \delta^*(\cdot \mid C^\circ),$$

where C° is the closed convex set given by

$$\begin{aligned} C^\circ &= \{x^* \mid \delta^*(x^* \mid C) - 1 \leq 0\} \\ &= \{x^* \mid \forall x \in C, \langle x, x^* \rangle \leq 1\}. \end{aligned}$$

The set C° is called the *polar* of C. Note that C° contains the origin. The polar of C° is

$$\begin{aligned} C^{\circ\circ} &= \{x \mid \forall x^* \in C^\circ, \langle x, x^* \rangle \leq 1\} \\ &= \{x \mid \delta^*(x \mid C^\circ) \leq 1\} = \{x \mid \mathrm{cl}\, \gamma(x \mid C) \leq 1\}. \end{aligned}$$

If C itself contains the origin and is closed, the latter set is just C according to Corollary 9.7.1. In general $C^\circ = D^\circ$, where

$$D = \mathrm{cl}\, (\mathrm{conv}\, (C \cup \{0\})),$$

because a set of the form $\{x \mid \langle x, x^* \rangle \leq 1\}$ contains C if and only if it contains D. Since $D^{\circ\circ} = D$, it follows that

$$C^{\circ\circ} = \mathrm{cl}\, (\mathrm{conv}\, (C \cup \{0\})).$$

In particular, we have another symmetric one-to-one correspondence.

THEOREM 14.5. *Let C be a closed convex set containing the origin. The polar C° is then another closed convex set containing the origin, and $C^{\circ\circ} = C$. The gauge function of C is the support function of C°. Dually, the gauge function of C° is the support function of C.*

COROLLARY 14.5.1. *Let C be a closed convex set containing the origin. Then C° is bounded if and only if $0 \in \mathrm{int}\, C$. Dually, C is bounded if and only if $0 \in \mathrm{int}\, C^\circ$.*

PROOF. We have C° bounded if and only if the support function $\gamma(\cdot \mid C)$ of C° is finite everywhere (Corollary 13.2.2.). On the other hand, $\gamma(\cdot \mid C)$ is finite everywhere if and only if $0 \in \mathrm{int}\, C$ (Corollary 6.4.1). $\parallel$

The polar of a convex cone K, as previously defined, coincides of course with the polar of K as a convex set, since the half-space $\{x \mid \langle x, x^* \rangle \leq 1\}$ contains K if and only if $\{x \mid \langle x, x^* \rangle \leq 0\}$ contains K.

Observe that polarity is order-inverting, i.e. $C_1 \subset C_2$ implies $C_1^\circ \supset C_2^\circ$.

As examples, the polars of the closed convex sets

$$C_1 = \{x = (\xi_1, \ldots, \xi_n) \mid \xi_j \geq 0, \xi_1 + \cdots + \xi_n \leq 1\},$$

$$C_2 = \{x = (\xi_1, \ldots, \xi_n) \mid |\xi_1| + \cdots + |\xi_n| \leq 1\},$$

$$C_3 = \{x = (\xi_1, \xi_2) \mid (\xi_1 - 1)^2 + \xi_2^2 \leq 1\},$$

$$C_4 = \{x = (\xi_1, \xi_2) \mid \xi_1 \leq 1 - (1 + \xi_2^2)^{1/2}\},$$

may be determined to be

$$C_1^\circ = \{x^* = (\xi_1^*, \ldots, \xi_n^*) \mid \xi_j^* \leq 1 \quad \text{for} \quad j = 1, \ldots, n\},$$

$$C_2^\circ = \{x^* = (\xi_1^*, \ldots, \xi_n^*) \mid |\xi_j^*| \leq 1 \quad \text{for} \quad j = 1, \ldots, n\},$$

$$C_3^\circ = \{x^* = (\xi_1^*, \xi_2^*) \mid \xi_1^* \leq (1 - \xi_2^{*2})/2\},$$

$$C_4^\circ = \text{conv}\,(P \cup \{0\}), \quad \text{where}$$

$$P = \{x^* = (\xi_1^*, \xi_2^*) \mid \xi_1^* \geq (1 + \xi_2^{*2})/2\}.$$

Other examples will be given following Corollary 15.3.2.

THEOREM 14.6. *Let C and C° be a polar pair of closed convex sets containing the origin. Then the recession cone of C and the closure of the convex cone generated by C° are polar to each other. The lineality space of C and the subspace generated by C° are orthogonally complementary to each other. Dually, also, with C and C° interchanged.*

PROOF. The recession cone of C is a closed convex cone, and since $0 \in C$ it is the largest such cone contained in C (Corollary 8.3.2). Its polar must be the smallest closed convex cone containing C°, and that is the closure of the convex cone generated by C°. Similarly, the lineality space of C is the largest subspace contained in C, inasmuch as $0 \in C$, so its orthogonal complement (which is the same as its polar) must be the smallest subspace containing C°. ‖

COROLLARY 14.6.1. *Let C be a closed convex set in R^n containing the origin. Then*

$$\text{dimension } C^\circ = n - \text{lineality } C,$$

$$\text{lineality } C^\circ = n - \text{dimension } C,$$

$$\text{rank } C^\circ = \text{rank } C.$$

PROOF. When a convex set contains 0, the subspace it generates coincides with the affine set it generates (Theorem 1.1). The dimensionality relations between C and C° follow therefore from the orthogonality relations in the theorem. ‖

Ordinarily, there is no simple polarity relation between the level sets of a convex function and the level sets of its conjugate. A useful inequality does hold, nevertheless, for an important class of functions.

THEOREM 14.7. *Let f be a non-negative closed convex function which vanishes at the origin. Then f* likewise is non-negative and vanishes at the origin, and for $0 < \alpha < \infty$ one has*

$$\{x \mid f(x) \le \alpha\}^\circ \subset \alpha^{-1}\{x^* \mid f^*(x^*) \le \alpha\} \subset 2\{x \mid f(x) \le \alpha\}^\circ.$$

PROOF. By hypothesis, $\inf f = f(0) = 0$. Since $\inf f = -f^*(0)$ and $\inf f^* = -f^{**}(0) = -f(0)$, we have $\inf f^* = f^*(0) = 0$ too, as already noted in §12. Let $C = \{x \mid f(x) \le \alpha\}$, $0 < \alpha < \infty$. This C is a closed convex set containing the origin. We can write $C = \{x \mid h(x) \le 0\}$, where $h(x) = f(x) - \alpha$. Then $h^*(x^*) = f^*(x^*) + \alpha$, and the closure of the positively homogeneous convex function generated by h^* is the support function $\delta^*(\cdot \mid C)$ of C (Theorem 13.5). But $\delta^*(x^* \mid C) = \gamma(x^* \mid C^\circ)$ by Theorem 14.5. Since $0 < h^*(0) < \infty$, the positively homogeneous convex function generated by h^* is itself closed (Theorem 9.7), and we have the formula

$$\gamma(x^* \mid C^\circ) = \inf \{(h^*\lambda)(x^*) \mid \lambda > 0\}.$$

In particular, $\gamma(x^* \mid C^\circ) \le h^*(x^*)$, so that

$$\{x^* \mid f^*(x^*) \le \alpha\} = \{x^* \mid h^*(x^*) \le 2\alpha\}$$
$$\subset \{x^* \mid \gamma(x^* \mid C^\circ) \le 2\alpha\} = 2\alpha C^\circ.$$

This establishes the second inclusion in the theorem. To establish the first inclusion, it is enough to show that

$$\{x^* \mid \gamma(x^* \mid C^\circ) < \alpha\} \subset \{x^* \mid f^*(x^*) \le \alpha\},$$

since the first set has αC° as its closure and the second set is closed (f^* being closed). Given any vector x^* such that $\gamma(x^* \mid C^\circ) < \alpha$, there exists (by the formula above) some $\lambda > 0$ such that

$$\alpha > (h^*\lambda)(x^*) = \lambda f^*(\lambda^{-1}x^*) + \lambda\alpha.$$

Since $f^* \ge 0$, λ has to be less than 1. We have

$$f^*(x^*) = f^*((1 - \lambda)0 + \lambda(\lambda^{-1}x^*)) \le (1 - \lambda)f^*(0) + \lambda f^*(\lambda^{-1}x^*)$$
$$= \lambda f^*(\lambda^{-1}x^*) < (1 - \lambda)\alpha.$$

Thus $f^*(x^*) < \alpha$. ‖

Polars of Convex Functions

A function k on R^n will be called a *gauge* if k is a non-negative positively homogeneous convex function such that $k(0) = 0$, i.e. if epi k is a convex cone in R^{n+1} containing the origin but not containing any vectors (x, μ) such that $\mu < 0$. Gauges are thus the functions k such that

$$k(x) = \gamma(x \mid C) = \inf \{\mu \geq 0 \mid x \in \mu C\}$$

for some non-empty convex set C. Of course, C is not uniquely determined by k in general, although one always has $\gamma(\cdot \mid C) = k$ for

$$C = \{x \mid k(x) \leq 1\}.$$

If k is closed, the latter C is the *unique closed* convex set containing the origin such that $\gamma(\cdot \mid C) = k$.

The *polar* of a gauge k is the function k° defined by

$$k^\circ(x^*) = \inf \{\mu^* \geq 0 \mid \langle x, x^* \rangle \leq \mu^* k(x), \forall x\}.$$

If k is finite everywhere and positive except at the origin, this formula can be written as

$$k^\circ(x^*) = \sup_{x \neq 0} \frac{\langle x, x^* \rangle}{k(x)}.$$

Note that, if k is the indicator function of a convex cone K, k° is the same as the conjugate of k, the indicator function of the polar convex cone K°.

Polars of convex functions more general than gauge functions will be defined by a modified formula later in this section.

THEOREM 15.1. *If k is a gauge function, then the polar k° of k is a closed gauge function, and $k^{\circ\circ} = \operatorname{cl} k$. In fact, if $k = \gamma(\cdot \mid C)$, where C is a non-empty convex set, then $k^\circ = \gamma(\cdot \mid C^\circ)$, where C° is the polar of C.*

PROOF. Let C be a non-empty convex set such that $k = \gamma(\cdot \mid C)$. For $\mu^* > 0$, the condition

$$\langle x, x^* \rangle \leq \mu^* \gamma(x \mid C), \qquad \forall x,$$

in the definition of k° can be expressed as

$$\langle \mu y, \mu^{*-1} x^* \rangle \leq \mu, \qquad \forall y \in C, \qquad \forall \mu \geq 0,$$

and this is equivalent to

$$\langle y, \mu^{*-1} x^* \rangle \leq 1, \qquad \forall y \in C,$$

i.e. $\mu^{*-1} x^* \in C^\circ$. For $\mu^* = 0$, on the other hand, the same condition implies $x^* = 0$. Thus

$$k^\circ(x^*) = \inf \{ \mu^* \geq 0 \mid x^* \in \mu^* C^\circ \} = \gamma(x^* \mid C^\circ).$$

In particular, k° is closed (Corollary 9.7.1). Now let $D = \{ x \mid k(x) \leq 1 \}$. This D is a convex set containing the origin, and $\gamma(\cdot \mid D) = k$. Hence it follows that $k^\circ = \gamma(\cdot \mid D^\circ)$ and $k^{\circ\circ} = \gamma(\cdot \mid D^{\circ\circ})$. Of course, $D^{\circ\circ} = (\text{cl } D)^{\circ\circ} = \text{cl } D$ (Theorem 14.5). Since

$$\{ x \mid (\text{cl } k)(x) \leq 1 \} = \text{cl } \{ x \mid k(x) \leq 1 \}$$

(Theorem 7.6), we have cl $k = \gamma(\cdot \mid \text{cl } D)$. Therefore $k^{\circ\circ} = \text{cl } k$. ‖

COROLLARY 15.1.1. *The polarity operation* $k \to k^\circ$ *induces a one-to-one symmetric correspondence in the class of all closed gauges on* R^n. *Two closed convex sets containing the origin are polar to each other if and only if their gauge functions are polar to each other.*

COROLLARY 15.1.2. *If* C *is a closed convex set containing the origin, the gauge function of* C *and the support function of* C *are gauges polar to each other.*

PROOF. This is immediate from Theorem 14.5. ‖

General norms, to be discussed below, are in particular closed gauges; some examples of polar gauges of this type will be given following Theorem 15.2 and Corollary 15.3.2. An example of a polar pair of closed gauges which are not norms is

$$k(x) = (\xi_1^2 + \xi_2^2)^{1/2} + \xi_1, \qquad x = (\xi_1, \xi_2) \in R^2,$$

$$k^\circ(x^*) = \begin{cases} [(\xi_2^{*2}/\xi_1^*) + \xi_1^*]/2 & \text{if } \xi_1^* > 0, \\ 0 & \text{if } \xi_1^* = 0 = \xi_2^*, \\ +\infty & \text{otherwise, where } x^* = (\xi_1^*, \xi_2^*). \end{cases}$$

Observe that gauges polar to each other have the property that

$$\langle x, x^* \rangle \leq k(x) k^\circ(x^*), \qquad \forall x \in \text{dom } k, \qquad \forall x^* \in \text{dom } k^\circ.$$

The theory of such inequalities is, in fact, one of the classical reasons for studying polar convex sets. Just as conjugate pairs of convex functions correspond to the "best" inequalities of the type

$$\langle x, y \rangle \leq f(x) + g(y), \qquad \forall x, \qquad \forall y,$$

as explained in §12, polar pairs of gauges correspond to the "best" inequalities of the type

$$\langle x, y \rangle \leq h(x)j(y), \qquad \forall x \in H, \qquad \forall y \in J,$$

where H and J are subsets of R^n and h and j are non-negative real-valued functions on H and J, respectively. Namely, given any inequality of the latter type, one can get a "better" inequality as follows. Let

$$k(x) = \inf \{\mu \geq 0 \mid \langle x, y \rangle \leq \mu j(y), \forall y \in J\}.$$

This formula expresses the epigraph of k as the intersection of a certain collection of closed half-spaces in R^{n+1} whose boundary hyperplanes pass through the origin, so k is a closed gauge. We have

$$\langle x, y \rangle \leq k(x)j(y), \qquad \forall x \in \text{dom } k, \qquad \forall y \in J,$$

and this inequality is "better" than the given one in the sense that dom $k \supset H$ and

$$k(x) \leq h(x), \qquad \forall x \in H.$$

The new inequality implies that dom $k° \supset J$ and

$$k°(y) \leq j(y), \qquad \forall y \in J.$$

Hence there is an even "better" inequality, namely

$$\langle x, y \rangle \leq k(x)k°(y), \qquad \forall x \in \text{dom } k, \qquad \forall y \in \text{dom } k°.$$

It follows that the "best" inequalities, i.e. the ones which cannot be tightened by replacing h or j by lesser functions on larger domains, are precisely those such that, if one sets $h(x) = +\infty$ for $x \notin H$ and $j(y) = +\infty$ for $y \notin J$, h and j are closed gauges polar to each other.

In the case where k is the Euclidean norm, k is both the gauge function and the support function of the Euclidean unit ball, so that $k° = k$. The corresponding inequality is then just the Schwarz inequality:

$$\langle x, y \rangle \leq |x| \cdot |y|.$$

In general, a gauge k is called a *norm* if it is finite everywhere, symmetric, and positive except at the origin. Norms are thus characterized (in view of

Theorem 4.7) as the real-valued functions k such that

(a) $k(x) > 0$, $\forall x \neq 0$,
(b) $k(x_1 + x_2) \leq k(x_1) + k(x_2)$, $\forall x_1$, $\forall x_2$,
(c) $k(\lambda x) = \lambda k(x)$, $\forall x$, $\forall \lambda > 0$,
(d) $k(-x) = k(x)$, $\forall x$.

Properties (c) and (d) can be combined as

$$k(\lambda x) = |\lambda|\, k(x), \qquad \forall x, \qquad \forall \lambda.$$

THEOREM 15.2. *The relations*

$$k(x) = \gamma(x \mid C), \qquad C = \{x \mid k(x) \leq 1\},$$

define a one-to-one correspondence between the norms k and the symmetric closed bounded convex sets C such that $0 \in \text{int } C$. The polar of a norm is a norm.

PROOF. Norms, being finite convex functions, are continuous (Theorem 10.1) and hence closed. We already know that the relations in the theorem define a one-to-one correspondence between the closed gauge functions k and the closed convex sets C containing the origin. Symmetry of k is obviously equivalent to symmetry of C. The condition that k be finite everywhere is equivalent to the condition that C contain a positive multiple of every vector, and this is satisfied if and only if $0 \in \text{int } C$ (Corollary 6.4.1). The condition that $k(x) > 0$ for $x \neq 0$ is equivalent to the condition that C contain no half-line of the form $\{\lambda x \mid \lambda \geq 0\}$, and this is satisfied if and only if C is bounded (Theorem 8.4). If C is a symmetric closed bounded convex set such that $0 \in \text{int } C$, the support function of C is finite everywhere, symmetric, and positive except at the origin. The support function of C is the gauge function of C°, the polar of $\gamma(\cdot \mid C)$, so in this case the polar of $\gamma(\cdot \mid C)$ is a norm. $\|$

An example of non-Euclidean norms polar to each other is

$$k(x) = \max\{|\xi_1|, \ldots, |\xi_n|\}, \qquad x = (\xi_1, \ldots, \xi_n),$$
$$k^\circ(x^*) = |\xi_1^*| + \cdots + |\xi_n^*|, \qquad x^* = (\xi_1^*, \ldots, \xi_n^*).$$

More examples will be given below.

If k is a norm, the inequality associated with k can be expressed as

$$|\langle x, x^* \rangle| \leq k(x) k^\circ(x^*), \qquad \forall x, \qquad \forall x^*,$$

by virtue of the finiteness and symmetry of k and k°.

The concept of a norm is natural to the study of certain metric structures and corresponding approximation problems. By definition, a *metric* on

R^n is a real-valued function ρ on $R^n \times R^n$ such that

(a) $\rho(x, y) > 0$ if $x \neq y$, $\rho(x, y) = 0$ if $x = y$,
(b) $\rho(x, y) = \rho(y, x)$, $\forall x$, $\forall y$,
(c) $\rho(x, z) \leq \rho(x, y) + \rho(y, z)$, $\forall x$, $\forall y$, $\forall z$.

The quantity $\rho(x, y)$ is interpreted as the *distance* between x and y with respect to ρ. Generally speaking, a metric on R^n need not have any relation with the algebraic structure of R^n: an extreme example is the metric defined by

$$\rho(x, y) = \begin{cases} 0 & \text{if } x = y, \\ 1 & \text{if } x \neq y. \end{cases}$$

Two properties which may naturally be demanded of a metric ρ, in order that it be compatible with vector addition and scalar multiplication, are

(d) $\rho(x + z, y + z) = \rho(x, y)$, $\forall x, y, z$,
(e) $\rho(x, (1 - \lambda)x + \lambda y) = \lambda\rho(x, y)$, $\forall x, y$, $\forall \lambda \in [0, 1]$.

Property (d) says that distances remain invariant under translation, and (e) says that distances behave linearly along line segments. A metric which has these two extra properties is called a *Minkowski metric* on R^n.

There is a one-to-one correspondence between Minkowski metrics and norms. If k is a norm, then

$$\rho(x, y) = k(x - y)$$

defines a Minkowski metric; moreover, each Minkowski metric is defined in this way by a uniquely determined norm. These facts are easy to prove, and we leave them as an exercise for the reader.

It follows by Theorem 15.2 that there is a one-to-one correspondence between Minkowski metrics and symmetric closed bounded convex sets C such that $0 \in \text{int } C$. Given any such C, there is a unique Minkowski metric ρ such that

$$\{y \mid \rho(x, y) \leq \varepsilon\} = x + \varepsilon C, \qquad \forall x, \qquad \forall \varepsilon > 0.$$

Note that, since C is bounded and $0 \in \text{int } C$, there exist positive scalars α and β such that

$$\alpha B \subset C \subset \beta B,$$

where B is the unit Euclidean ball. For such scalars one has

$$\alpha^{-1}d(x, y) \geq \rho(x, y) \geq \beta^{-1}d(x, y), \qquad \forall x, \qquad \forall y,$$

where $d(x, y)$ is the Euclidean distance. This implies that all the Minkowski metrics on R^n are "equivalent" to the Euclidean metric, i.e. they all define the same open and closed sets and Cauchy sequences in the sense of metric theory.

Some important examples of convex functions conjugate to each other can be constructed from gauges polar to each other, namely certain gauge-like convex functions. An extended-real-valued function f on R^n is said to be *gauge-like* if $f(0) = \inf f$ and the various level sets

$$\{x \mid f(x) \leq \alpha\}, \qquad f(0) < \alpha < +\infty,$$

are all proportional, i.e. can all be expressed as positive scalar multiples of a single set.

THEOREM 15.3. *A function f is a gauge-like closed proper convex function if and only if it can be expressed in the form*

$$f(x) = g(k(x)),$$

where k is a closed gauge and g is a non-constant non-decreasing lower semi-continuous convex function on $[0, +\infty]$ such $g(\zeta)$ is finite for some $\zeta > 0$. ($g(+\infty)$ is to be interpreted as $+\infty$ in the formula for f.) If f is of this type, then f^* is gauge-like too. In fact

$$f^*(x^*) = g^+(k^\circ(x^*)),$$

where g^+, the monotone conjugate of g, satisfies the same conditions as g.

PROOF. Suppose first that f is a function given by $f(x) = g(k(x))$, where g and k have the properties described. Let I be the interval where g is finite, and let $C = \{x \mid k(x) \leq 1\}$. The conditions on g imply that $g(\zeta) \to +\infty$ as $\zeta \to +\infty$ (Theorem 8.6). For any real $\alpha > f(0) = g(0)$, the number

$$\lambda = \sup \{\zeta \geq 0 \mid g(\zeta) \leq \alpha\}$$

is finite and positive, and one has

$$\{x \mid f(x) \leq \alpha\} = \{x \mid k(x) \leq \lambda\} = \lambda C.$$

This shows that f is gauge-like. The conjugate of f may be calculated as

$$f^*(x^*) = \sup_x \{\langle x, x^* \rangle - g(k(x))\} = \sup_{\zeta \in I} \sup_{x \in \zeta C} \{\langle x, x^* \rangle - g(\zeta)\}$$

$$= \sup_{\zeta \in I} \{\zeta(\sup_{y \in C} \langle y, x^* \rangle) - g(\zeta)\}.$$

The inner supremum is $\delta^*(x^* \mid C)$ by definition, and that is the same as $\gamma(x^* \mid C^\circ)$ (Theorem 14.5). In fact it is the same as $k^\circ(x^*)$, since $k = \gamma(\cdot \mid C)$. On the other hand, for $\zeta^* \geq 0$,

$$\sup_{\zeta \in I} \{\zeta \zeta^* - g(\zeta)\} = \sup_{\zeta \geq 0} \{\zeta \zeta^* - g(\zeta)\} = g^+(\zeta^*).$$

It follows that $f^*(x^*) = g^+(k°(x^*))$. Our discussion of monotone conjugacy towards the end of §12 makes it clear that g^+ satisfies the same conditions that we have imposed on g. Therefore f^* is gauge-like, and by the same calculation used for f^* we have

$$f^{**}(x) = g^{++}(k^{°°}(x)) = g(k(x)) = f(x).$$

Inasmuch as $f^{**} = f$, and $f(0)$ is finite by the conditions on g, this shows that f is a closed proper convex function (Theorem 12.2).

It remains only to show that, given any gauge-like closed proper convex function f, we have $f(x) = g(k(x))$ for some g and k as described. The conditions on f imply that the level sets

$$C_\alpha = \{x \mid f(x) \le \alpha\}, \qquad \alpha > \alpha_0 = f(0) = \inf f,$$

are closed convex sets containing the origin, and they are all positive multiples of a certain C. If they are all actually the same multiple λC, then trivially

$$f(x) = f(0) + \delta(x \mid \lambda C) = g(k(x)),$$

where k is the gauge of C and

$$g(\zeta) = \begin{cases} \alpha_0 & \text{if } 0 \le \zeta \le \lambda, \\ +\infty & \text{if } \zeta > \lambda. \end{cases}$$

We can suppose therefore that C is not a cone, and that f is not merely constant on dom f. In this case we define g instead by

$$g(\zeta) = \inf \{\alpha \mid C_\alpha \supset \zeta C\}, \qquad \zeta \ge 0.$$

Clearly g is non-decreasing, non-constant and

$$\alpha_0 = g(0) = \inf \{g(\zeta) \mid \zeta > 0\} < \infty.$$

For every vector x, we have

$$\begin{aligned} f(x) &= \inf \{\alpha \mid \alpha > \alpha_0, x \in C_\alpha\} \\ &= \inf \{\alpha \mid \zeta > 0, x \in \zeta C \subset C_\alpha\} \\ &= \inf \{g(\zeta) \mid \zeta > 0, x \in \zeta C\} \\ &= \inf \{g(\zeta) \mid \zeta \ge \gamma(x \mid C) = k(x)\} = g(k(x)). \end{aligned}$$

Since C is not a cone, there exist vectors x such that $k(x) = 1$, and for such a vector we have

$$g(\zeta) = g(\zeta k(x)) = g(k(\zeta x)) = f(\zeta x), \qquad \forall \zeta \ge 0.$$

The convexity and lower semi-continuity of f are therefore inherited by g, and it follows that g has all the required properties. ‖

The main application of Theorem 15.3 is to functions f with the property that, for a certain exponent p, $1 < p < \infty$,

$$f(\lambda x) = \lambda^p f(x), \qquad \forall \lambda > 0, \qquad \forall x.$$

Such a function is said to be *positively homogeneous of degree p*.

COROLLARY 15.3.1. *A closed proper convex function f is positively homogeneous of degree p, where $1 < p < \infty$, if and only if it is of the form*

$$f(x) = (1/p)k(x)^p$$

for a certain closed gauge k. For such an f, the conjugate of f is positively homogeneous of degree q, where $1 < q < \infty$ and $(1/p) + (1/q) = 1$; in fact

$$f^*(x^*) = (1/q)k^\circ(x^*)^q.$$

PROOF. If f is positively homogeneous of degree p, then f is gauge-like. The corollary follows from the fact that the function $g(\zeta) = (1/p)\zeta^p$, $\zeta \geq 0$, satisfies the conditions of the theorem and has $g^+(\zeta^*) = (1/q)\zeta^{*q}$. ‖

Of course, if $f = (1/p)k^p$ as in the corollary, then $(pf)^{1/p} = k$. Thus:

COROLLARY 15.3.2. *Let f be a closed proper convex function positively homogeneous of degree p, where $1 < p < \infty$. Then $(pf)^{1/p}$ is a closed gauge whose polar is $(qf^*)^{1/q}$, where $1 < q < \infty$ and $(1/p) + (1/q) = 1$. Thus one has*

$$\langle x, x^* \rangle \leq [pf(x)]^{1/p}[qf^*(x^*)]^{1/q}, \qquad \forall x \in \operatorname{dom} f, \qquad \forall x^* \in \operatorname{dom} f^*,$$

and the closed convex sets

$$C = \{x \mid [pf(x)]^{1/p} \leq 1\} = \{x \mid f(x) \leq 1/p\},$$
$$C^* = \{x^* \mid [qf^*(x^*)]^{1/q} \leq 1\} = \{x^* \mid f^*(x^*) \leq 1/q\},$$

are polar to each other.

PROOF. This is immediate from the preceding corollary and the general properties that $k = (pf)^{1/p}$ has by virtue of its being a closed gauge. ‖

For example, for any p, $1 < p < \infty$, define

$$f(\xi_1, \ldots, \xi_n) = (1/p)(|\xi_1|^p + \cdots + |\xi_n|^p).$$

Then f is a closed proper convex function on R^n positively homogeneous of degree p, and the conjugate of f is given by

$$f^*(\xi_1^*, \ldots, \xi_n^*) = (1/q)(|\xi_1^*|^q + \cdots + |\xi_n^*|^q),$$

where $1 < q < \infty$ and $(1/p) + (1/q) = 1$, as is readily calculated. By Corollary 15.3.2, the function

$$k(\xi_1, \ldots, \xi_n) = (|\xi_1|^p + \cdots + |\xi_n|^p)^{1/p}$$

is a closed gauge whose polar is given by

$$k°(\xi_1^*, \ldots, \xi_n^*) = (|\xi_1^*|^q + \cdots + |\xi_n^*|^q)^{1/q},$$

and the closed convex sets

$$C = \{x = (\xi_1, \ldots, \xi_n) \mid |\xi_1|^p + \cdots + |\xi_n|^p \leq 1\},$$
$$C^* = \{x^* = (\xi_1^*, \ldots, \xi_n^*) \mid |\xi_1^*|^q + \cdots + |\xi_n^*|^q \leq 1\},$$

are polar to each other. As a matter of fact, k and $k°$ are in this case *norms* polar to each other.

For another example, let Q be any symmetric positive definite $n \times n$ matrix, and let

$$f(x) = (1/2)\langle x, Qx \rangle.$$

As pointed out in §12, f is a (closed proper) convex function on R^n whose conjugate is given by

$$f^*(x^*) = (1/2)\langle x^*, Q^{-1}x^* \rangle.$$

Since f is positively homogeneous of degree 2, we have by Corollary 15.3.2 that

$$k(x) = \langle x, Qx \rangle^{1/2}$$

is a gauge—in fact a norm—with polar

$$k°(x^*) = \langle x^*, Q^{-1}x^* \rangle^{1/2}.$$

Moreover, the polar of the convex set

$$C = \{x \mid \langle x, Qx \rangle \leq 1\}$$

is given by

$$C° = \{x^* \mid \langle x^*, Q^{-1}x^* \rangle \leq 1\}.$$

Thus, for instance, the polar of the elliptic disk

$$C = \{(\xi_1, \xi_2) \mid (\xi_1^2/\alpha_1^2) + (\xi_2^2/\alpha_2^2) \leq 1\},$$

is the elliptic disk

$$C° = \{(\xi_1^*, \xi_2^*) \mid \alpha_1^2 \xi_1^{*2} + \alpha_2^2 \xi_2^{*2} \leq 1\}.$$

It follows further that, for any g satisfying the hypothesis of Theorem 15.3, a pair of closed proper convex functions conjugate to each other is given by

$$f(x) = g(\langle x, Qx \rangle^{1/2}), \qquad f^*(x^*) = g^+(\langle x^*, Q^{-1}x^* \rangle^{1/2}).$$

Gauges belong in particular to the class of non-negative convex functions f which vanish at the origin. The polarity correspondence for gauges can actually be extended to this larger class by defining the *polar $f°$* of f by

$$f°(x^*) = \inf \{\mu^* \geq 0 \mid \langle x, x^* \rangle \leq 1 + \mu^* f(x), \forall x\}.$$

If f is a gauge, this definition reduces to the definition already given, because of the positive homogeneity of f. If f is the indicator function of a convex set C containing the origin, then f° is the indicator function of C°.

THEOREM 15.4. *Let f be a non-negative convex function which vanishes at the origin. The polar f° of f is then a non-negative closed convex function which vanishes at the origin, and $f^{\circ\circ} = \mathrm{cl}\, f$.*

PROOF. Certainly f° is non-negative and $f^\circ(0) = 0$. The epigraph of f° consists of the vectors (x^*, μ^*) in R^{n+1} such that

$$\langle x, x^* \rangle - \mu\mu^* \le 1, \qquad \forall (x, \mu) \in \mathrm{epi}\, f,$$

and consequently one has

$$\mathrm{epi}\, f^\circ = (A(\mathrm{epi}\, f))^\circ = A((\mathrm{epi}\, f)^\circ),$$

where A is the vertical reflection in R^{n+1}, i.e. the linear transformation $(x^*, \mu^*) \to (x^*, -\mu^*)$. Thus $\mathrm{epi}\, f^\circ$ is a closed convex set (implying that f° is a closed convex function). Moreover,

$$\mathrm{epi}\, (f^{\circ\circ}) = (A\, (\mathrm{epi}\, f^\circ))^\circ = (AA((\mathrm{epi}\, f)^\circ))^\circ$$

$$= (\mathrm{epi}\, f)^{\circ\circ} = \mathrm{cl}\, (\mathrm{epi}\, f) = \mathrm{epi}\, (\mathrm{cl}\, f)$$

(Theorem 14.5), so that $f^{\circ\circ} = \mathrm{cl}\, f$. ‖

COROLLARY 15.4.1. *The polarity operation $f \to f^\circ$ induces a symmetric one-to-one correspondence in the class of all non-negative closed convex functions which vanish at the origin.*

Note that functions polar to each other in the extended sense always satisfy

$$\langle x, x^* \rangle \le 1 + f(x)f^\circ(x^*), \qquad \forall x \in \mathrm{dom}\, f, \qquad \forall x^* \in \mathrm{dom}\, f^\circ.$$

They yield the "best" inequalities of a certain type. The details of this may be developed as a simple exercise.

Let f be a non-negative closed convex function which vanishes at the origin. Then f° has these same properties, as we have just seen. But so does the conjugate function f^*, as is apparent from its definition. What is the relationship between f° and f^*? The answer to this question can be reached through a geometric analysis of the epigraph of the function $g = f^{*\circ}$ in comparison with the epigraph of f.

First we calculate g from f^*. By definition, if $g(x) < \lambda < \infty$ we have $\lambda > 0$ and

$$1 \ge \sup_{x^*} \{ \langle x, x^* \rangle - \lambda f^*(x^*) \} = \lambda \sup_{x^*} \{ \langle \lambda^{-1}x, x^* \rangle - f^*(x^*) \}$$

$$= \lambda f^{**}(\lambda^{-1}x) = \lambda f(\lambda^{-1}x) = (f\lambda)(x).$$

On the other hand, by the same computation, if $0 < \lambda < \infty$ and $(f\lambda)(x) \leq 1$ we have $\lambda \geq g(x)$. Therefore

$$g(x) = \inf \{\lambda > 0 \mid (f\lambda)(x) \leq 1\}.$$

We shall call this function g the *obverse* of f.

Notice that, if f is the indicator function of a closed convex set C containing the origin, then g is the gauge of C. On the other hand, if f is the gauge of C, then g is the indicator function of C. The indicator and gauge functions of C are thus the obverses of each other.

In general, there is a simple geometric relationship between epi f and epi g. Since $(f\lambda)(x)$ approaches $(f0^+)(x)$ in the above formula as $\lambda \downarrow 0$ (Corollary 8.5.2), we have

$$\text{epi } g = \{(x, \lambda) \mid h(\lambda, x) \leq 1\},$$

where

$$h(\lambda, x) = \begin{cases} (f\lambda)(x) & \text{if } \lambda > 0, \\ (f0^+)(x) & \text{if } \lambda = 0, \\ +\infty & \text{if } \lambda < 0. \end{cases}$$

As we have observed in §8, $P = \text{epi } h$ is a closed convex cone in R^{n+2}, and it is the smallest such cone containing $\{(1, x, \mu) \mid \mu \geq f(x)\}$. The intersection of P with the hyperplane $\{(\lambda, x, \mu) \mid \lambda = 1\}$ thus corresponds to epi f. The calculation above shows that the intersection of P with the hyperplane $\{(\lambda, x, \mu) \mid \mu = 1\}$ corresponds to epi g. What is more, P must be the smallest closed convex cone containing $\{(\lambda, z, 1) \mid \lambda \geq g(x)\}$, since P is the closure of its intersection with the open half-space $\{(\lambda, x, \mu) \mid \mu \geq 0\}$ (inasmuch as $f \geq 0$). Thus f and g lead to the same closed convex cone P in R^{n+2}, except that in passing between f and g the roles of λ and μ are reversed.

THEOREM 15.5. *Let f be a non-negative closed convex function which vanishes at the origin, and let g be the obverse of f. Then g is a non-negative closed convex function which vanishes at the origin, and f is the obverse of g. One has $f^\circ = g^*$ and $f^* = g^\circ$. Moreover f° and f^* are the obverses of each other.*

PROOF. The fact that f is the obverse of g is clear from the symmetry just explained. Thus $f = g^{*\circ}$. This implies $f^\circ = g^{*\circ\circ} = g^*$. On the other hand, $g = f^{*\circ}$ implies $g^\circ = f^{*\circ\circ} = f^*$. The obverse of f^* is $f^{**\circ} = f^\circ$. ‖

COROLLARY 15.5.1. *If f is any non-negative closed convex function vanishing at the origin, one has $f^{*\circ} = f^{\circ*}$.*

PROOF. $f^{\circ*} = g^{**} = g = f^{*\circ}$. ‖

Ordinarily, the level sets of f° are *not* simply the polars of the level sets of f, and the gauge functions k and $k°$ in Theorem 15.3 cannot be replaced by arbitrary polar pairs of functions. For the obverse g of f, we have $(f\lambda)(x) \leq \mu$ if and only if $(g\mu)(x) \leq \lambda$ (assuming $\lambda > 0$ and $\mu > 0$). Consequently, for $0 < \alpha < \infty$, one does have

$$\{x \mid g(x) \leq \alpha\} = \{x \mid (f\alpha)(x) \leq 1\} = \alpha\{x \mid f(x) \leq \alpha^{-1}\}.$$

Since $f°$ is the obverse of f^*, we may conclude that

$$\{x^* \mid f°(x^*) \leq \alpha^{-1}\} = \alpha^{-1}\{x^* \mid f^*(x^*) \leq \alpha\}, \qquad \forall \alpha > 0.$$

Note that this set is the middle set in the inequality in Theorem 14.7.

Dual Operations

Suppose we perform some operation on given convex functions $f_1, \ldots, f_m$, such as adding them. How is the conjugate of the resulting function related to the conjugate functions $f_1^*, \ldots, f_m^*$? Similar questions can be asked about the behavior of set or functional operations under the polarity correspondences. In most cases, it turns out that the duality correspondence converts a familiar operation into another familiar operation (modulo some details about closures). The operations thus arrange themselves in dual pairs.

We begin with some simple cases already covered by Theorem 12.3. Let h be any convex function on R^n. If we translate h by a, that is if we replace h by $f(x) = h(x - a)$, we get $f^*(x^*) = h^*(x^*) + \langle a, x^* \rangle$. On the other hand, if we add a linear function to h to form $f(x) = h(x) + \langle x, a^* \rangle$, the conjugate of f is given by $f^*(x^*) = h^*(x^* - a^*)$, a translate of h^*.

For a real constant α, the conjugate of $h + \alpha$ is $h^* - \alpha$.

For a convex set C, the support function of a translate $C + a$ is given by $\delta^*(x^* \mid C) + \langle a, x^* \rangle$. This is easy enough to demonstrate directly, but one should note that it is also a special case of what we have just pointed out. The conjugate of the indicator function $h = \delta(\cdot \mid C)$ is the support function $\delta^*(\cdot \mid C)$, and translating C is the same as translating its indicator function.

The operations of left and right non-negative scalar multiplication are dual to each other:

THEOREM 16.1. *For any proper convex function f, one has $(\lambda f)^* = f^*\lambda$ and $(f\lambda)^* = \lambda f^*$, $0 \leq \lambda < \infty$.*

PROOF. When $\lambda > 0$, this is elementary to verify from the definition of the conjugate. When $\lambda = 0$, it simply expresses the fact that the constant function 0 is conjugate to the indicator function $\delta(\cdot \mid 0)$. ‖

COROLLARY 16.1.1. *For any non-empty convex set C, one has $\delta^*(x^* \mid \lambda C) = \lambda \delta^*(x^* \mid C)$, $0 \leq \lambda < \infty$.*

PROOF. Take $f(x) = \delta(x \mid C)$. ‖

The polar of a convex set C is a level set of the support function of C namely

$$C° = \{x^* \mid \delta^*(x^* \mid C) \leq 1\}.$$

Any support function result like Corollary 16.1.1 can therefore be translated immediately into a polarity result.

COROLLARY 16.1.2. *For any non-empty convex set C one has $(\lambda C)° = \lambda^{-1}C°$ for $0 < \lambda < \infty$.*

In dealing with the duality of various other operations for convex sets and functions, we need to invoke the conditions of §9 to settle questions about closures. These conditions will first be dualized.

LEMMA 16.2. *Let L be a subspace of R^n and let f be a proper convex function. Then L meets ri $(\text{dom} f)$ if and only if there exists no vector $x^* \in L^\perp$ such that $(f^*0^+)(x^*) \leq 0$ and $(f^*0^+)(-x^*) > 0$.*

PROOF. Since L is relatively open, we have $L \cap \text{ri} (\text{dom} f)$ empty if and only if there exists a hyperplane separating L and $\text{dom} f$ properly (Theorem 11.3). Proper separation corresponds to the existence of some $x^* \in R^n$ such that

$$\inf \{\langle x, x^* \rangle \mid x \in L\} \geq \sup \{\langle x, x^* \rangle \mid x \in \text{dom} f\},$$

$$\sup \{\langle x, x^* \rangle \mid x \in L\} > \inf \{\langle x, x^* \rangle \mid x \in \text{dom} f\}.$$

(See Theorem 11.1.) The supremum and infimum over $\text{dom} f$ are

$$(f^*0^+)(x^*) \quad \text{and} \quad -(f^*0^+)(-x^*),$$

respectively, since f^*0^+ is the support function of $\text{dom} f$ (Theorem 13.3). The infimum over L is 0 if $x^* \in L^\perp$ and $-\infty$ if $x^* \notin L^\perp$. The two extremal conditions on x^* are therefore equivalent to the conditions that $x^* \in L^\perp$, $0 \geq (f^*0^+)(x^*)$ and $0 > -(f^*0^+)(-x^*)$. $\|$

COROLLARY 16.2.1. *Let A be a linear transformation from R^n to R^m. Let g be a proper convex function on R^m. In order that there exist no vector $y^* \in R^m$ such that*

$$A^*y^* = 0, \quad (g^*0^+)(y^*) \leq 0, \quad (g^*0^+)(-y^*) > 0,$$

it is necessary and sufficient that $Ax \in \text{ri} (\text{dom} g)$ for at least one $x \in R^n$.

PROOF. For the subspace $L = \{Ax \mid x \in R^n\}$ one has

$$L^\perp = \{y^* \mid A^*y^* = 0\}.$$

Apply the lemma to L and g. $\|$

COROLLARY 16.2.2. *Let $f_1, \ldots, f_m$ be proper convex functions on R^n.*

In order that there do not exist vectors $x_1^*, \ldots, x_m^*$ *such that*

$$x_1^* + \cdots + x_m^* = 0,$$

$$(f_1^*0^+)(x_1^*) + \cdots + (f_m^*0^+)(x_m^*) \leq 0,$$

$$(f_1^*0^+)(-x_1^*) + \cdots + (f_m^*0^+)(-x_m^*) > 0,$$

it is necessary and sufficient that

$$\text{ri}\,(\text{dom}\,f_1) \cap \cdots \cap \text{ri}\,(\text{dom}\,f_m) \neq \emptyset.$$

PROOF. Regard R^{mn} as the space of m-tuples $x = (x_1, \ldots, x_m)$, $x_i \in R^n$, so that the inner product is expressed by

$$\langle x, x^* \rangle = \langle x_1, x_1^* \rangle + \cdots + \langle x_m, x_m^* \rangle.$$

The convex function f on R^{mn} defined by

$$f(x_1, \ldots, x_m) = f_1(x_1) + \cdots + f_m(x_m)$$

then has as its conjugate

$$f^*(x_1^*, \ldots, x_m^*) = f_1^*(x_1^*) + \cdots + f_m^*(x_m^*),$$

and the recession function of f^* is given by

$$(f^*0^+)(x_1^*, \ldots, x_m^*) = (f_1^*0^+)(x_1^*) + \cdots + (f_m^*0^+)(x_m^*).$$

The subspace

$$L = \{x \mid x_1 = x_2 = \cdots = x_m\}$$

has as its orthogonal complement

$$L^\perp = \{x^* \mid x_1^* + \cdots + x_m^* = 0\}.$$

Apply the lemma to f and L. ‖

We shall now show that the two functional operations in §5 involving linear transformations are dual to each other.

THEOREM 16.3. *Let A be a linear transformation from R^n to R^m. For any convex function f on R^n, one has*

$$(Af)^* = f^*A^*.$$

For any convex function g on R^m, one has

$$((\text{cl}\,g)A)^* = \text{cl}\,(A^*g^*).$$

If there exists an x such that $Ax \in \text{ri}\,(\text{dom}\,g)$, the closure operation can be omitted from the second formula; then

$$(gA)^*(x^*) = \inf\{g^*(y^*) \mid A^*y^* = x^*\},$$

where for each x^* the infimum is attained (or is $+\infty$ vacuously).

Proof. Direct calculation proves the first relation:

$$(Af)^*(y^*) = \sup_y \{\langle y, y^* \rangle - \inf_{Ax=y} f(x)\} = \sup_y \sup_{Ax=y} \{\langle y, y^* \rangle - f(x)\}$$

$$= \sup_x \{\langle Ax, y^* \rangle - f(x)\} = \sup_x \{\langle x, A^*y^* \rangle - f(x)\} = f^*(A^*y^*).$$

Applying this relation to A^* and g^*, we get

$$(A^*g^*)^* = g^{**}A^{**} = (\text{cl } g)A,$$

and consequently

$$((\text{cl } g)A)^* = (A^*g^*)^{**} = \text{cl } (A^*g^*).$$

The rest of the theorem is trivial if g takes on $-\infty$ somewhere (since then $g(y) = -\infty$ throughout ri (dom g) according to Theorem 7.2, so that g^* and $(gA)^*$ are identically $+\infty$). Assume, therefore, that $g(y) > -\infty$ for every y and that, for some x, Ax belongs to ri (dom g). Theorem 9.5 asserts that in this case (cl g)A = cl (gA). Hence $((\text{cl } g)A)^* = (gA)^*$. On the other hand, Corollary 16.2.1 says that g^* and A^* satisfy the condition in Theorem 9.2. This condition guarantees that cl $(A^*g^*) = A^*g^*$ and that the infimum in the definition of A^*g^* be attained. ‖

COROLLARY 16.3.1. *Let A be a linear transformation from R^n to R^m. For any convex set C in R^n, one has*

$$\delta^*(y^* \mid AC) = \delta^*(A^*y^* \mid C), \qquad \forall y^* \in R^m.$$

For any convex set D in R^m, one has

$$\delta^*(\cdot \mid A^{-1}(\text{cl } D)) = \text{cl } (A^* \delta^*(\cdot \mid D)).$$

If there exists some x such that $Ax \in$ ri D, the closure operation can be omitted in the second formula, and

$$\delta^*(x^* \mid A^{-1}D) = \inf \{\delta^*(y^* \mid D) \mid A^*y^* = x^*\},$$

where for each x^* the infimum is attained (or is $+\infty$ vacuously).

Proof. Take $f(x) = \delta(x \mid C)$, $g(y) = \delta(y \mid D)$. ‖

COROLLARY 16.3.2. *Let A be a linear transformation from R^n to R^m. For any convex set C in R^n, one has*

$$(AC)^\circ = A^{*-1}(C^\circ).$$

For any convex set D in R^m, one has

$$(A^{-1}(\text{cl } D))^\circ = \text{cl } (A^*(D^\circ)).$$

If there exists some x such that $Ax \in$ ri D, the closure operation can be omitted from the second formula.

Proof. Immediate from the preceding corollary. ‖

It will follow from Corollary 19.3.1 that, when g is "polyhedral" in the sense that the epigraph of g is a polyhedral convex set, the condition $Ax \in \mathrm{ri}\,(\mathrm{dom}\,g)$ in Theorem 16.3 can be weakened to $Ax \in \mathrm{dom}\,g$. Of course, the need for the relative interior condition in the general case can be seen via Corollary 16.2.1 from the need for the corresponding recession function condition in Theorem 9.2, shown by the example at the beginning of §9.

As an illustration of Theorem 16.3, suppose that

$$h(\xi_1) = \inf_{\xi_2} f(\xi_1, \xi_2), \qquad \xi_1 \in R,$$

where f is a convex function on R^2. Then $h = Af$, where A is the projection $(\xi_1, \xi_2) \to \xi_1$. The adjoint A^* is the transformation $\xi_1^* \to (\xi_1^*, 0)$, so we have

$$h^*(\xi_1^*) = f^*(\xi_1^*, 0).$$

For another example, consider a convex function h on R^n of the form

$$h(x) = g_1(\langle a_1, x \rangle) + \cdots + g_m(\langle a_m, x \rangle),$$

where $a_1, \ldots, a_m$ are elements of R^n and $g_1, \ldots, g_m$ are closed proper convex functions of a single real variable. To determine the conjugate of h, we observe that $h = gA$, where A is the linear transformation

$$x \to (\langle a_1, x \rangle, \ldots, \langle a_m, x \rangle)$$

and g is the closed proper convex function on R^m given by

$$g(y) = g_1(\eta_1) + \cdots + g_m(\eta_m) \quad \text{for} \quad y = (\eta_1, \ldots, \eta_m).$$

The adjoint A^* is the linear transformation

$$y^* = (\eta_1^*, \ldots, \eta_m^*) \to \eta_1^* a_1 + \cdots + \eta_m^* a_m,$$

while obviously

$$g^*(y^*) = g_1^*(\eta_1^*) + \cdots + g_m^*(\eta_m^*).$$

Therefore $(A^*g^*)(x^*)$ is for each $x^* \in R^n$ the infimum of

$$g_1^*(\eta_1^*) + \cdots + g_m^*(\eta_m^*)$$

over all the choices of the real numbers $\eta_1^*, \ldots, \eta_m^*$ such that

$$\eta_1^* a_1 + \cdots + \eta_m^* a_m = x^*.$$

The conjugate of h is the closure of this convex function A^*g^* by Theorem 16.3. If there exists an $x \in R^n$ such that

$$\langle a_i, x \rangle \in \mathrm{ri}\,(\mathrm{dom}\,g_i) \quad \text{for} \quad i = 1, \ldots, m,$$

then the infimum in the definition of $(A^*g^*)(x^*)$ is attained for each x^* by some choice of $\eta_1^*, \ldots, \eta_m^*$, and we have simply $h^* = A^*g^*$.

Observe in Theorem 16.3 that, in the case where the closure operation can be omitted, the formula $(gA)^* = A^*g^*$ says that (for any $x^* \in R^n$)

$$\sup \{\langle x, x^* \rangle - g(Ax) \mid x \in R^n\} = \inf \{g^*(y^*) \mid A^*y^* = x^*\}.$$

Thus Theorem 16.3 yields a non-trivial fact about the relationship between two different extremum problems. Similar results are embodied in Theorems 16.4 and 16.5 below. The derivation and analysis of such "inf = sup" formulas is the subject matter of the general theory of dual extremum problems to be developed in §30 and §31.

We proceed now to show that the operations of addition and infimal convolution of convex functions are dual to each other. This is the most important case of dual operations as far as applications to extremum problems are concerned.

THEOREM 16.4. *Let $f_1, \ldots, f_m$ be proper convex functions on R^n. Then*

$$(f_1 \,\square\, \cdots \,\square\, f_m)^* = f_1^* + \cdots + f_m^*,$$
$$(\text{cl}\, f_1 + \cdots + \text{cl}\, f_m)^* = \text{cl}\, (f_1^* \,\square\, \cdots \,\square\, f_m^*).$$

If the sets $\text{ri}\,(\text{dom}\, f_i)$, $i = 1, \ldots, m$, have a point in common, the closure operation can be omitted from the second formula, and

$$(f_1 + \cdots + f_m)^*(x^*)$$
$$= \inf \{f_1^*(x_1^*) + \cdots + f_m^*(x_m^*) \mid x_1^* + \cdots + x_m^* = x^*\},$$

where for each x^* the infimum is attained.

PROOF. By definition,

$$(f_1 \,\square\, \cdots \,\square\, f_m)^*(x^*) = \sup_x \{\langle x, x^* \rangle - \inf_{x_1 + \cdots + x_m = x} \{f_1(x_1) + \cdots + f_m(x_m)\}\}$$

$$= \sup_x \sup_{x_1 + \cdots + x_m = x} \{\langle x, x^* \rangle - f_1(x_1) - \cdots - f_m(x_m)\}$$

$$= \sup_{x_1, \ldots, x_m} \{\langle x_1, x^* \rangle + \cdots + \langle x_m, x^* \rangle$$
$$- f_1(x_1) - \cdots - f_m(x_m)\}$$

$$= f_1^*(x^*) + \cdots + f_m^*(x^*).$$

This implies that

$$(f_1^* \,\square\, \cdots \,\square\, f_m^*)^* = f_1^{**} + \cdots + f_m^{**} = \text{cl}\, f_1 + \cdots + \text{cl}\, f_m,$$

and hence that

$$(\text{cl}\, f_1 + \cdots + \text{cl}\, f_m)^* = (f_1^* \,\square\, \cdots \,\square\, f_m^*)^{**} = \text{cl}\, (f_1^* \,\square\, \cdots \,\square\, f_m^*).$$

If the sets ri (dom f_i) have a point in common, cl $f_1 + \cdots + $ cl f_m is the same as cl $(f_1 + \cdots + f_m)$ according to Theorem 9.3. The conjugate of the latter function is $(f_1 + \cdots + f_m)^*$. On the other hand, Corollary 16.2.2 says that, under the same intersection condition, $f_1^*, \ldots, f_m^*$ satisfy the hypothesis of Corollary 9.2.1, which ensures that $f_1^* \square \cdots \square f_m^*$ be closed and that the infimum in the definition of $f_1^* \square \cdots \square f_m^*$ always be attained.   ‖

COROLLARY 16.4.1.   *Let* $C_1, \ldots, C_m$ *be non-empty convex sets in* R^n. *Then*

$$\delta^*(\cdot \mid C_1 + \cdots + C_m) = \delta^*(\cdot \mid C_1) + \cdots + \delta^*(\cdot \mid C_m),$$

$$\delta^*(\cdot \mid \text{cl } C_1 \cap \cdots \cap \text{cl } C_m) = \text{cl } (\delta^*(\cdot \mid C_1) \square \cdots \square \delta^*(\cdot \mid C_m)).$$

If the sets ri C_i, $i = 1, \ldots, m$, *have a point in common, the closure operation can be omitted from the second formula, and one has*

$$\delta^*(x^* \mid C_1 \cap \cdots \cap C_m)$$
$$= \inf \{\delta^*(x_1^* \mid C_1) + \cdots + \delta^*(x_m^* \mid C_m) \mid x_1^* + \cdots + x_m^* = x^*\},$$

where for each x^* *the infimum is attained.*

PROOF.   Take $f_i = \delta(\cdot \mid C_i)$.   ‖

COROLLARY 16.4.2.   *Let* $K_1, \ldots, K_m$ *be non-empty convex cones in* R^n. *Then*

$$(K_1 + \cdots + K_m)^\circ = K_1^\circ \cap \cdots \cap K_m^\circ,$$

$$(\text{cl } K_1 \cap \cdots \cap \text{cl } K_m)^\circ = \text{cl } (K_1^\circ + \cdots + K_m^\circ).$$

If the cones ri K, $i = 1, \ldots, m$, *have a point in common, the closure operation can be omitted from the second formula.*

PROOF.   Apply the theorem to $f_i = \delta(\cdot \mid K_i)$. One has $f_i^* = \delta(\cdot \mid K_i^\circ)$, as explained at the beginning of §14.   ‖

An important refinement of the last part of Theorem 16.4 in the case where some of the functions f_i are "polyhedral" will be given in Theorem 20.1.

An example of the use of Theorem 16.4 is the calculation of the conjugate of the distance function

$$f(x) = d(x, C) = \inf \{|x - y| \mid y \in C\},$$

where C is a given non-empty convex set. As remarked after Theorem 5.4, we have $f = f_1 \square f_2$, where

$$f_1(x) = |x|, \qquad f_2(x) = \delta(x \mid C).$$

Therefore

$$f^*(x^*) = f_1^*(x^*) + f_2^*(x^*) = \begin{cases} \delta^*(x^* \mid C) & \text{if } |x^*| \leq 1, \\ +\infty & \text{otherwise.} \end{cases}$$

For a similar example of interest in approximation theory, consider the function

$$f(x) = \inf \{\|x - \zeta_1 a_1 - \cdots - \zeta_m a_m\|_\infty \mid \zeta_i \in R\},$$

where $a_1, \ldots, a_m$ are given elements of R^n and

$$\|x\|_\infty = \max \{|\xi_j| \mid j = 1, \ldots, n\} \quad \text{for} \quad x = (\xi_1, \ldots, \xi_n).$$

Here $f = f_1 \,\square\, f_2$, where

$$f_1(x) = \|x\|_\infty, \qquad f_2(x) = \delta(x \mid L),$$

L being the subspace of R^n generated by $a_1, \ldots, a_m$. Since f_1 is the support function of the set

$$D = \{x^* = (\xi_1^*, \ldots, \xi_n^*) \mid |\xi_1^*| + \cdots + |\xi_n^*| \leq 1\},$$

f_1^* is the indicator of D (Theorem 13.2). On the other hand, f_2^* is the indicator of the orthogonally complementary subspace

$$L^\perp = \{x^* \mid \langle x^*, a_i \rangle = 0, \quad i = 1, \ldots, m\}.$$

Therefore f^*, which is $f_1^* + f_2^*$ by Theorem 16.4, is the indicator function of $D \cap L^\perp$. It follows that f itself is the support function of the (polyhedral) convex set $D \cap L^\perp$.

The second part of Theorem 16.4 is illustrated by the calculation of the conjugate of

$$f(x) = \begin{cases} h(x) & \text{if } x \geq 0, \\ +\infty & \text{if } x \not\geq 0, \end{cases}$$

where h is a given closed proper convex function on R^n. We have $f = h + \delta(\cdot \mid K)$, where K is the non-negative orthant of R^n. The conjugate of $\delta(\cdot \mid K)$ is by Theorem 14.1 the indicator of the polar cone K°, which happens to be $-K$, the non-positive orthant. By Theorem 16.4, f^* is the closure of the convex function

$$g = h^* \,\square\, \delta(\cdot \mid -K),$$

and this g is given by

$$g(x^*) = \inf \{h^*(z^*) \mid z^* \geq x^*\}.$$

If ri (dom h) meets ri K (the positive orthant), we have the formula

$$f^*(x^*) = \min \{h^*(z^*) \mid z^* \geq x^*\}.$$

We shall see in Theorem 20.1 that (since K is polyhedral) the latter is valid even if ri (dom h) merely meets the non-negative orthant K itself, rather than ri K.

As a final example of the way in which Theorem 16.4 can be used to determine conjugate functions, we calculate the conjugate of the important function

$$f(\xi_1, \ldots, \xi_n) = \begin{cases} \xi_1 \log \xi_1 + \cdots + \xi_n \log \xi_n & \text{if } \xi_j \geq 0 \text{ for} \\ \qquad j = 1, \ldots, n \text{ and } \xi_1 + \cdots + \xi_n = 1, \\ +\infty & \text{otherwise} \end{cases}$$

(where $0 \log 0 = 0$). Note that f is a closed proper convex function on R^n, since

$$f(x) = g(x) + \delta(x \mid C)$$

where

$$C = \{x = (\xi_1, \ldots, \xi_n) \mid \xi_1 + \cdots + \xi_n = 1\},$$
$$g(x) = k(\xi_1) + \cdots + k(\xi_n),$$
$$k(\xi) = \begin{cases} \xi \log \xi & \text{for } \xi > 0, \\ 0 & \text{for } \xi = 0, \\ +\infty & \text{for } \xi < 0. \end{cases}$$

The relative interiors of the effective domains of g and $\delta(\cdot \mid C)$ have a non-empty intersection, so by the last part of Theorem 16.4 we have

$$f^* = [g + \delta(\cdot \mid C)]^* = g^* \;\square\; \delta(\cdot \mid C)^* = g^* \;\square\; \delta^*(\cdot \mid C),$$

in other words

$$f^*(x^*) = \inf_{y^*} \{g^*(x^* - y^*) + \delta^*(y^* \mid C)\},$$

where for each x^* the infimum is attained by some y^*. Obviously

$$g^*(x^*) = k^*(\xi_1^*) + \cdots + k^*(\xi_n^*),$$

and by elementary calculation

$$k^*(\xi^*) = e^{\xi^* - 1}.$$

On the other hand,

$$\delta^*(x^* \mid C) = \begin{cases} \lambda & \text{if } x^* = \lambda(1, \ldots, 1) \text{ for a certain } \lambda \in R, \\ +\infty & \text{otherwise.} \end{cases}$$

Therefore

$$f^*(x^*) = \min_{\lambda \in R} \{\lambda + \sum_{j=1}^n e^{\xi_j^* - \lambda - 1}\}.$$

This minimum can be calculated by taking the derivative with respect to λ and setting it equal to zero. The result is the formula

$$f^*(x^*) = \log(e^{\xi_1^*} + \cdots + e^{\xi_n^*}).$$

The fact that the conjugacy correspondence is order-inverting leads to the duality of the pointwise supremum operation and the convex hull operation for convex functions.

THEOREM 16.5. *Let f_i be a proper convex function on R^n for each $i \in I$ (an arbitrary index set). Then*

$$(\text{conv } \{f_i \mid i \in I\})^* = \sup \{f_i^* \mid i \in I\},$$

$$(\sup \{\text{cl } f_i \mid i \in I\})^* = \text{cl } (\text{conv } \{f_i^* \mid i \in I\}).$$

If I is finite and the sets $\text{cl } (\text{dom } f_i)$ are all the same set C (as is the case of course when every f_i is finite throughout R^n), then the closure operation can be omitted from the second formula. Moreover, in this case

$$(\sup \{f_i \mid i \in I\})^* = \inf \{\textstyle\sum_{i \in I} \lambda_i f_i^*(x_i^*)\},$$

where for each x^* the infimum (taken over all representations of x^* as a convex combination $\sum_{i \in I} \lambda_i x_i^*$) is attained.

PROOF. Let $f = \text{conv } \{f_i \mid i \in I\}$. The elements (x^*, μ^*) of $\text{epi } f^*$ correspond to the affine functions $h = \langle \cdot, x^* \rangle - \mu^*$ such that $h \leq f$. These functions h are the same as the ones such that $h \leq f_i$ for every i. Thus $\mu^* \geq f^*(x^*)$ if and only if $\mu^* \geq f_i^*(x^*)$ for every i, which proves the first formula. Applying this formula to the f_i^*, we get

$$(\text{conv } \{f_i^* \mid i \in I\})^* = \sup \{f_i^{**} \mid i \in I\} = \sup \{\text{cl } f_i \mid i \in I\},$$

and consequently

$$(\sup \{\text{cl } f_i \mid i \in I\})^* = (\text{conv } \{f_i^* \mid i \in I\})^{**} = \text{cl } (\text{conv } \{f_i^* \mid i \in I\}).$$

If there exists a point common to the $\text{ri } (\text{dom } f_i)$ at which the supremum of the f_i is finite, we have by Theorem 9.4

$$(\sup \{\text{cl } f_i \mid i \in I\})^* = (\text{cl } (\sup \{f_i \mid i \in I\}))^* = (\sup \{f_i \mid i \in I\})^*.$$

This is valid in particular when I is finite and $\text{cl } (\text{dom } f_i) = C$ for every i. In the latter case, the support functions of the sets $\text{dom } f_i$, which are the recession functions $f_i^* 0^+$ (Theorem 13.3), are all equal to $\delta^*(\cdot \mid C)$, so that by Corollary 9.8.3 $\text{conv } \{f_i^* \mid i \in I\}$ is closed and is given by the infimum formula as described. ‖

COROLLARY 16.5.1. *Let C_i be a non-empty convex set in R^n for each $i \in I$ (an index set). Then the support function of the convex hull D of the* *union of the sets C_i is given by*

$$\delta^*(\cdot \mid D) = \sup \{\delta^*(\cdot \mid C_i) \mid i \in I\},$$

while the support function of the intersection C of the sets cl C_i *is given by*

$$\delta^*(\cdot \mid C) = \text{cl} \, (\text{conv} \, \{\delta^*(\cdot \mid C_i) \mid i \in I\}).$$

PROOF. Take $f_i = \delta(\cdot \mid C_i)$ in the theorem. ‖

COROLLARY 16.5.2. *Let C_i be a convex set in R^n for each $i \in I$ (an index set). Then*

$$(\text{conv} \, \{C_i \mid i \in I\})^\circ = \bigcap \, \{C_i^\circ \mid i \in I\},$$

$$(\bigcap \, \{\text{cl} \, C_i \mid i \in I\})^\circ = \text{cl} \, (\text{conv} \, \{C_i^\circ \mid i \in I\}).$$

PROOF. This is obvious from the preceding corollary. It also follows directly from the fact that the polarity correspondence is order-inverting. ‖

An illustration of Theorem 16.5 is the calculation of the conjugate of

$$f(x) = \max \, \{|x - a_i| \mid i = 1, \ldots, m\},$$

where $a_1, \ldots, a_m$ are given elements of R^n. Here f is the pointwise maximum of the convex functions

$$f_i(x) = |x - a_i|, \qquad i = 1, \ldots, m,$$

whose conjugates are given by

$$f_i^*(x^*) = \delta(x^* \mid B) + \langle a_i, x^* \rangle,$$

where B is the Euclidean unit ball. Since the functions f_i have the same effective domain, namely all of R^n, the last part of Theorem 16.5 is applicable and we may conclude that, for each x^*, $f^*(x^*)$ is the minimum of

$$\lambda_1 \langle a_1, x_1^* \rangle + \cdots + \lambda_m \langle a_m, x_m^* \rangle$$

over all x_i^* and λ_i satisfying

$$\lambda_1 x_1^* + \cdots + \lambda_m x_m^* = x^*,$$

$$|x_i^*| \leq 1, \, \lambda_i \geq 0, \, \lambda_1 + \cdots + \lambda_m = 1.$$

Part IV · Representation and Inequalities

Carathéodory's Theorem

If S is a subset of R^n, the convex hull of S can be obtained by forming all convex combinations of elements of S. According to the classical theorem of Carathéodory, it is not really necessary to form combinations involving more than $n + 1$ elements at a time. One can limit attention to convex combinations $\lambda_1 x_1 + \cdots + \lambda_m x_m$ such that $m \leq n + 1$ (or even to combinations such that $m = n + 1$, if one does not insist on the vectors x_i being distinct).

Carathéodory's Theorem is the fundamental dimensionality result in convexity theory, and it is the source of many other results in which dimensionality is prominent. We shall use it in §21 to prove Helly's Theorem, concerning intersections of convex sets, as well as various results about infinite systems of linear inequalities.

In order to formulate a comprehensive version of Carathéodory's Theorem which covers the generation of convex cones and other unbounded convex sets as well as the generation of ordinary convex hulls, we consider the convex hulls of sets S which consist of both points and directions (points at infinity).

Let S_0 be a set of points of R^n, and let S_1 be a set of directions of R^n as defined in §8. We define the *convex hull* conv S of $S = S_0 \cup S_1$ to be the smallest convex set C in R^n such that $C \supset S_0$ and C recedes in all the directions in S_1. Obviously, this smallest C exists. In fact

$$C = \operatorname{conv}(S_0 + \operatorname{ray} S_1) = \operatorname{conv} S_0 + \operatorname{cone} S_1,$$

where ray S_1 consists of the origin and all the vectors whose directions belong to S_1, and

$$\operatorname{cone} S_1 = \operatorname{conv}(\operatorname{ray} S_1),$$

i.e. cone S_1 is the convex cone generated by all the vectors whose directions belong to S_1. Algebraically, a vector x belongs to conv S if and only if it can be expressed in the form

$$x = \lambda_1 x_1 + \cdots + \lambda_k x_k + \lambda_{k+1} x_{k+1} + \cdots + \lambda_m x_m,$$

where $x_1, \ldots, x_k$ are vectors from S_0 and $x_{k+1}, \ldots, x_m$ are arbitrary vectors having directions in $S_1(1 \leq k \leq m)$, all the coefficients λ_i are non-negative, and $\lambda_1 + \cdots + \lambda_k = 1$. Let us call such an x a *convex combination of m points and directions in S*. Such convex combinations correspond to the non-negative linear combinations

$$\lambda_1(1, x_1) + \cdots + \lambda_k(1, x_k) + \lambda_{k+1}(0, x_{k+1}) + \cdots + \lambda_m(0, x_m)$$

in R^{n+1} which lie in the hyperplane $H = \{(1, x) \mid x \in R^n\}$. Thus another way to obtain conv S is to intersect the hyperplane H with the convex cone in R^{n+1} generated by S', where S' consists of all vectors in R^{n+1} of the form $(1, x)$ with $x \in S_0$ or $(0, x)$ with $x \in S_1'$, S_1' being any subset of R^n such that the set of directions of the vectors in S_1' is S_1.

The convex cone generated by a set $T \subseteq R^n$ can be thought of equally well as the convex hull of the set S consisting of the origin and all the directions of vectors in T. A convex combination x of m elements of this S is necessarily a convex combination of 0 and $m - 1$ directions in S, and hence it is simply a non-negative linear combination of $m - 1$ vectors in T.

The *affine hull* aff S of a mixed set of points and directions in R^n is defined of course to be aff (conv S), the smallest affine set which contains all the points of S and recedes in all the directions of S. Trivially, aff $S =$ conv $S = \emptyset$ if S contains directions only. We say S is *affinely independent* if (aff S) $= m - 1$, where m is the total number of points and directions in S. For S non-empty, this condition means that S contains at least one point and that the vectors

$$(1, x_1), \ldots, (1, x_k), \qquad (0, x_{k+1}), \ldots, (0, x_m)$$

are linearly independent in R^{n+1}, where $x_1, \ldots, x_k$ are the points in S and $x_{k+1}, \ldots, x_m$ are any vectors whose directions are the different directions in S.

By a *generalized m-dimensional simplex*, we shall mean a set which is the convex hull of $m + 1$ affinely independent points and directions, the points being called the *ordinary vertices* of the simplex and the directions the *vertices at infinity*. Thus the one-dimensional generalized simplices are the line segments and the closed half-lines. The two-dimensional generalized simplices are the triangles, the closed strips (the convex hulls of pairs of distinct parallel closed half-lines) and the closed quadrants (the convex hulls of pairs of distinct closed half-lines with the same end-point).

A generalized m-dimensional simplex with one ordinary vertex and $m - 1$ vertices at infinity will be called an m-dimensional (skew) *orthant*. The m-dimensional orthants in R^n are just the images of the non-negative orthant of R^m under one-to-one affine transformations from R^m into R^n. These orthants are all closed sets, since the non-negative orthant of R^m is closed.

More generally, every generalized m-dimensional simplex in R^n is closed, since such a set can be identified with the intersection of an $(m + 1)$-dimensional orthant in R^{n+1} and the hyperplane $\{(1, x) \mid x \in R^n\}$ as indicated above.

THEOREM 17.1 (Carathéodory's Theorem). *Let S be any set of points and directions in R^n, and let $C = \text{conv } S$. Then $x \in C$ if and only if x can be expressed as a convex combination of $n + 1$ of the points and directions in S (not necessarily distinct). In fact C is the union of all the generalized $\bar{d}$-dimensional simplices whose vertices belong to S, where $d = \dim C$.*

PROOF. Let S_0 be the set of points in S and S_1 the set of directions in S. Let S_1' be a set of vectors in R^n such that the set of directions of the vectors in S_1' is S_1. Let S' be the subset of R^{n+1} consisting of all the vectors of the form $(1, x)$ with $x \in S_0$ or of the form $(0, x)$ with $x \in S_1'$. Let K be the convex cone generated by S'. As pointed out above, conv S can be identified with the intersection of K and the hyperplane $\{(1, x) \mid x \in R^n\}$. Translating the statement of the theorem into this context in R^{n+1}, we see that it is only necessary to show that any non-zero vector $y \in K$, which is in any case a non-negative linear combination of elements of S', can actually be expressed as a non-negative linear combination of $d + 1$ linearly independent vectors of S', where $d + 1$ is the dimension of K ($=$ the dimension of the subspace of R^{n+1} generated by S'). The argument is algebraic, and it does not depend on the relationship between S' and S. Given $y \in K$, let $y_1, \ldots, y_m$ be vectors in S' such that $y = \lambda_1 y_1 + \cdots + \lambda_m y_m$, where the coefficients λ_i are all non-negative. Assuming the vectors y_i are not themselves linearly independent, we can find scalars $\mu_1, \ldots, \mu_m$, at least one of which is positive, such that $\mu_1 y_1 + \cdots + \mu_m y_m = 0$. Let λ be the greatest scalar such that $\lambda \mu_i \leq \lambda_i$ for $i = 1, \ldots, m$, and let $\lambda_i' = \lambda_i - \lambda \mu_i$. Then

$$\lambda_1' y_1 + \cdots + \lambda_m' y_m = \lambda_1 y_1 + \cdots + \lambda_m y_m - \lambda(\mu_1 y_1 + \cdots + \mu_m y_m) = y.$$

By the choice of λ, the new coefficients λ_i' are non-negative, and at least one of them is 0. We therefore have an expression of y as a non-negative linear combination of fewer than m elements of S'. If these remaining elements are not linearly independent, we can repeat the argument and eliminate another of them. After a finite number of steps we get an expression of y as a non-negative linear combination of linearly independent vectors $z_1, \ldots, z_r$ of S'. Then $r \leq d + 1$ by definition of $d + 1$. Choosing additional vectors $z_{r+1}, \ldots, z_{d+1}$ from S' if necessary to make a basis for the subspace generated by S', we get the desired expression of y by adding the term $0z_{r+1} + \cdots + 0z_{d+1}$ to the expression in terms of $z_1, \ldots, z_r$. ‖

COROLLARY 17.1.1. *Let $\{C_i \mid i \in I\}$ be an arbitrary collection of convex sets in R^n, and let C be the convex hull of the union of the collection. Then* *every point of C can be expressed as a convex combination of $n + 1$ or fewer affinely independent points, each belonging to a different C_i.*

PROOF. By the theorem, each $x \in C$ can be expressed as a convex combination $\lambda_0 x_0 + \cdots + \lambda_d x_d$, where $x_0, x_1, \ldots, x_d$ are affinely independent points in the union of the collection and $d = \dim C \leq n$. Points with zero coefficients can be dropped from this expression. If two of the points with non-zero coefficients belong to the same C_i, say x_0 and x_1, the corresponding term $\lambda_0 x_0 + \lambda_1 x_1$ can be coalesced to μy, where $\mu = \lambda_0 + \lambda_1$ and

$$y = (\lambda_0/\mu)x_0 + (\lambda_1/\mu)x_1 \in C_i.$$

This y is affinely independent of $x_2, \ldots, x_d$. This shows that the expression of x can be reduced to one involving points belonging to different sets in the collection. ‖

COROLLARY 17.1.2. *Let $\{C_i \mid i \in I\}$ be an arbitrary collection of non-empty convex sets in R^n, and let K be the convex cone generated by the union of the collection. Then every non-zero vector of K can be expressed as a non-negative linear combination of n or fewer linearly independent vectors, each belonging to a different C_i.*

PROOF. Take S in the theorem to consist of the origin and all the directions of vectors in the sets C_i. By the theorem, each $x \in K$ belongs to a d-dimensional orthant with the origin as vertex, where $d = \dim K$. Thus each non-zero $x \in K$ can be expressed as a non-negative linear combination of d linearly independent vectors in the union of the sets C_i. By the argument given in the proof of the preceding corollary, this expression can be reduced to one in which no two vectors belong to the same C_i. ‖

COROLLARY 17.1.3. *Let $\{f_i \mid i \in I\}$ be an arbitrary collection of proper convex functions on R^n, and let f be the convex hull of the collection. Then, for every vector x,*

$$f(x) = \inf \{\textstyle\sum_{i \in I} \lambda_i f_i(x_i) \mid \sum_{i \in I} \lambda_i x_i = x\},$$

where the infimum is taken over all expressions of x as a convex combination in which at most $n + 1$ of the coefficients λ_i are non-zero and the vectors x_i with non-zero coefficients are affinely independent.

PROOF. This is proved by applying Corollary 17.1.1 to the sets $C_i = \operatorname{epi} f_i$. The argument is the same as in Theorem 5.6, except for one feature. From Corollary 17.1.1, we have the fact that $\mu > f(x)$ if and only if there is some $\alpha < \mu$ such that (x, α) belongs to a simplex with vertices in sets $\operatorname{epi} f_i$ with different indices i. Now when (x, α) belongs to a simplex in R^{n+1}, there is a minimal $\alpha' \leq \alpha$ such that (x, α') belongs to the same simplex. The vertices of the simplex needed to express (x, α') as a convex combination generate a "subsimplex" which contains no "vertical" line segment. These vertices $(y_1, \alpha_1), \ldots, (y_m, \alpha_m)$ thus have the property that $y_1, \ldots, y_m$ are themselves affinely independent. Therefore $f(x)$ is the infimum of the values of α such that (x, α) can be expressed as a convex combination of points $(y_1, \alpha_1), \ldots, (y_m, \alpha_m)$ (belonging to sets epi f_i with different indices i) such that $y_1, \ldots, y_m$ are affinely independent. The affine independence of $y_1, \ldots, y_m$ implies of course that $m \leq n + 1$, and the desired formula follows at once. ‖

We would like to point out that Corollary 17.1.3 contains Corollary 17.1.1 as a special case. (Take f_i to be the indicator function of C_i.)

COROLLARY 17.1.4.   *Let* $\{f_i \mid i \in I\}$ *be an arbitrary collection of proper convex functions on* R^n. *Let f be the greatest positively homogeneous convex function such that* $f \leq f_i$ *for every* $i \in I$, *i.e. the positively homogeneous convex function generated by* conv $\{f_i \mid i \in I\}$. *Then, for every vector* $x \neq 0$,

$$f(x) = \inf \{\textstyle\sum_{i \in I} \lambda_i f_i(x_i) \mid \sum_{i \in I} \lambda_i x_i = x\},$$

where the infimum is taken over all expressions of x as a non-negative linear combination in which at most n of the coefficients λ_i *are non-zero and the vectors* x_i *with non-zero coefficients are linearly independent.*

PROOF.   This is proved just like the last corollary, except that one applies Corollary 17.1.2 (instead of Corollary 17.1.1) to the sets $C_i = $ epi f_i. The convex cone K generated by the sets C_i yields epi f, of course, when its "lower boundary" is adjoined (in the sense of Theorem 5.3). ‖

COROLLARY 17.1.5.   *Let f be an arbitrary function from* R^n *to* $(-\infty, +\infty]$. *Then*

$$(\operatorname{conv} f)(x) = \inf \{\textstyle\sum_{i=1}^{n+1} \lambda_i f(x_i) \mid \sum_{i=1}^{n+1} \lambda_i x_i = x\},$$

where the infimum is taken over all expressions of x as a convex combination of $n + 1$ *points. (The formula is also valid if one takes only the combinations in which the* $n + 1$ *points are affinely independent.)*

PROOF.   Apply Theorem 17.1 to $S = $ epi f in R^{n+1}, and use the argument in the proof of Corollary 17.1.3 again to reduce the number of points needed from $n + 2$ to $n + 1$. ‖

COROLLARY 17.1.6.   *Let f be an arbitrary function from* R^n *to* $(-\infty, +\infty]$, *and let k be the positively homogeneous convex function generated by* f *(i.e. by* conv f). *Then, for each vector* $x \neq 0$,

$$k(x) = \inf \{\textstyle\sum_{i=1}^{n} \lambda_i f(x_i) \mid \sum_{i=1}^{n} \lambda_i x_i = x\},$$

where the infimum is taken over all expressions of x as a non-negative linear combination of n vectors. (The formula is also valid if one takes only the convex combinations in which the n vectors are linearly independent.)

PROOF.   Apply Theorem 17.1 to the set S in R^{n+1} consisting of the origin and the directions of the vectors in epi f, and use the argument in the proof of Corollary 17.1.3 to reduce the number of vectors needed from $n + 1$ to n. ‖

One of the most important consequences of Carathéodory's Theorem concerns the closedness of convex hulls. In general, of course, the convex hull of a closed set of points need not be closed. For example, conv S is not closed when S is the union of a line in R^2 and a single point not on the line.

THEOREM 17.2.   *If S is a bounded set of points in R^n, then* cl (conv S) = conv (cl S). *In particular, if S is closed and bounded, then* conv S *is closed and bounded.*

PROOF.   Let $m = (n + 1)^2$, and let Q be the set of all vectors of the form

$$(\lambda_0, \ldots, \lambda_n, x_0, \ldots, x_n) \in R^m$$

such that the components $\lambda_i \in R$ and $x_i \in R^n$ satisfy

$$\lambda_i \geq 0, \qquad \lambda_0 + \cdots + \lambda_n = 1, \qquad x_i \in \text{cl } S.$$

The image of Q under the continuous mapping

$$\theta : (\lambda_0, \ldots, \lambda_n, x_0, \ldots, x_n) \to \lambda_0 x_0 + \cdots + \lambda_n x_n$$

from R^m to R^n is conv (cl S) by Carathéodory's Theorem. If S is bounded in R^n, Q is closed and bounded in R^m, and hence the image of Q under θ is closed and bounded too. Then

$$\text{conv (cl } S) = \text{cl (conv (cl } S)) \supset \text{cl (conv } S).$$

Of course, in general

$$\text{cl (conv } S) = \text{conv (cl (conv } S)) \supset \text{conv (cl } S),$$

so the commutativity of "conv" and "cl" follows. ‖

COROLLARY 17.2.1.   *Let S be a non-empty closed bounded set in R^n. Let f be a continuous real-valued function on S, and let $f(x) = +\infty$ for $x \notin S$. Then* conv f *is a closed proper convex function.*

PROOF.   Let F be the graph of f over S, i.e. the subset of R^{n+1} consisting of the points of the form $(x, f(x))$, $x \in S$. Since S is closed and bounded and f is continuous, F is closed and bounded. It follows from the theorem that conv F is closed and bounded. Let K be the vertical ray $\{(0, \mu) \mid \mu \geq 0\}$ in R^{n+1}. The non-empty convex set $K + \text{conv } F$ is closed (as can be seen by an elementary argument on the closedness of K and compactness of conv F), and it contains no "vertical" lines. It is therefore the epigraph of a certain closed proper convex function. This function must in fact be conv f. ‖

Note that, under the hypothesis of Corollary 17.2.1, the convex function h defined by

$$h(z) = \sup\{\langle z, x \rangle - f(x) \mid x \in S\}$$

is finite everywhere (and hence continuous everywhere). Corollary 17.2.1 implies that the conjugate of h is the function conv f (Theorem 12.2).

Carathéodory's Theorem concerns the convex hull of a given set S of points and directions. Results dual to Carathéodory's Theorem concern instead the intersection of a given set of half-spaces.

Any closed half-space H in R^n can, of course, be represented by a vector (x^*, μ^*) in R^{n+1} with $x^* \neq 0$:

$$H = \{x \in R^n \mid \langle x, x^* \rangle \leq \mu^*\}.$$

Suppose that S^* is a given non-empty set of vectors (x^*, μ^*) in R^{n+1}, and consider the closed convex set C which is the intersection of the closed half-spaces corresponding to these vectors, i.e.

$$C = \{x \mid \forall (x^*, \mu^*) \in S^*, \langle x, x^* \rangle \leq \mu^*\}.$$

In general, there will be other closed half-spaces containing C besides the ones corresponding to vectors of S^*. How may the vectors (x^*, μ^*) representing these other half-spaces be expressed in terms of the vectors in S^*?

The vectors representing closed half-spaces containing C are of course the vectors (x^*, μ^*), $x^* \neq 0$, in the epigraph of the support function of C, since the inequality $\langle x, x^* \rangle \leq \mu^*$ holds for every $x \in C$ if and only if

$$\mu^* \geq \sup\{\langle x, x^* \rangle \mid x \in C\} = \delta^*(x^* \mid C).$$

In order that a function k be the support function of a convex set D such that D is contained in all the half-spaces corresponding to vectors of S^*, it is necessary and sufficient that k be a positively homogeneous closed convex function on R^n such that $S^* \subset$ epi k (Theorem 13.2). Since C is the largest of such sets D, its support function must be the greatest of such functions. It follows that $\delta^*(\cdot \mid C) = \text{cl} f$, where f is the positively homogeneous convex function generated by S^*, i.e. the function defined by

$$f(x^*) = \inf\{\mu^* \mid (x^*, \mu^*) \in K\},$$

where K is the convex cone in R^{n+1} generated by S^* and the "vertical" vector $(0, 1)$ in R^{n+1}. Assuming C is not empty, we have

$$\text{epi } \delta^*(\cdot \mid C) = \text{epi (cl} f) = \text{cl (epi} f) = \text{cl } K.$$

The most general closed half-space containing C therefore corresponds to a vector (x^*, μ^*), $x^* \neq 0$, which is a limit of vectors in K. The vectors in

K itself, on the other hand, can be represented in terms of the vectors in S^*. One has $(x^*, \mu^*) \in K$ if and only if there exist vectors $(x_i^*, \mu_i^*) \in S^*$, $i = 1, \ldots, m$, such that

$$(x^*, \mu^*) = \lambda_0(0, 1) + \lambda_1(x_1^*, \mu_1^*) + \cdots + \lambda_m(x_m^*, \mu_m^*)$$

for certain non-negative scalars $\lambda_0, \lambda_1, \ldots, \lambda_m$. This condition says that $x^* = \lambda_1 x_1^* + \cdots + \lambda_m x_m^*$ and $\mu^* \geq \lambda_1 \mu_1^* + \cdots + \lambda_m \mu_m^*$. Applying Cara-théodory's Theorem to the set S consisting of the origin, the "upward" direction in R^{n+1} and the directions of the vectors in S^* (this S has conv $S = K$), we see that m can always be taken $\leq n + 1$. Actually, here only the "bottoms" of the simplices in R^{n+1} are really needed, so (as in the proof of Corollary 17.1.3) m can always be taken $\leq n$. It follows that, when cl $K = K$, we can represent every closed half-space containing C in terms of n or fewer of the given half-spaces corresponding to S^*.

Here is one example of such a representation.

THEOREM 17.3. *Let S^* be a non-empty closed bounded set of vectors (x^*, μ^*) in R^{n+1}, and let*

$$C = \{x \mid \forall(x^*, \mu^*) \in S^*, \langle x, x^* \rangle \leq \mu^*\}.$$

Suppose the convex set C is n-dimensional. Then, for a given vector (x^*, μ^*), $x^* \neq 0$, the half-space

$$H = \{x \mid \langle x, x^* \rangle \leq \mu^*\}$$

contains C if and only if there exist vectors $(x_i^*, \mu_i^*) \in S^*$ and coefficients $\lambda_i \geq 0$, $i = 1, \ldots, m$, where $m \leq n$, such that

$$\lambda_1 x_1^* + \cdots + \lambda_m x_m^* = x^*, \qquad \lambda_1 \mu_1^* + \cdots + \lambda_m \mu_m^* \leq \mu^*.$$

PROOF. Let D be the union of S^* and $(0, 1)$ in R^{n+1}. Let K be the convex cone generated by D. By the preceding remarks, all we have to do is to show that cl $K = K$. Since D is closed and bounded, conv D is also closed and bounded by Theorem 17.2. Furthermore, K is the same as the convex cone generated by conv D. If the origin of R^{n+1} does not belong to conv D, then cl $K = K$ as desired (Corollary 9.6.1). To show that the origin is not in conv D, we make use of the n-dimensionality of C. The n-dimensionality implies the existence of a point $\bar{x}$ in int C. For such an $\bar{x}$, we have $\langle \bar{x}, x^* \rangle < \mu^*$ for every $(x^*, \mu^*) \in S^*$. Thus the open upper half-space

$$\{(x^*, \mu^*) \mid \langle \bar{x}, x^* \rangle - \mu^* < 0\}$$

in R^{n+1} contains D (and hence conv D), but it does not contain the origin. ‖

The condition in Theorem 17.3 is, of course, equivalent to the existence of n half-spaces of the form

$$H_i = \{x \mid \langle x, x_i^* \rangle \le \mu_i^* \}, \qquad (x_i^*, \mu_i^*) \in S^*$$

(not necessarily distinct) with the property that

$$H_1 \cap \cdots \cap H_n \subset H.$$

Extreme Points and Faces of Convex Sets

Given a convex set C, there exist various point sets S such that $C = $ conv S. For any such S, the points of C can be expressed as convex combinations of the points S as in Carathéodory's Theorem. One may call this an "internal representation" of C, in distinction to an "external representation" of C as the intersection of some collection of half-spaces. Representations of the form $C = $ conv S or $C = $ cl (conv S) can also be considered in which S contains both points and directions, as in the preceding section. Of course, the smaller or more special S is, the more significant the internal representation of C. A smallest S actually exists in the most important cases. We shall demonstrate this below from the general theory of facial structure.

A *face* of a convex set C is a convex subset C' of C such that every (closed) line segment in C with a relative interior point in C' has both endpoints in C'. The empty set and C itself are faces of C. The zero-dimensional faces of C are called the *extreme points* of C. Thus a point $x \in C$ is an extreme point of C if and only if there is no way to express x as a convex combination $(1 - \lambda)y + \lambda z$ such that $y \in C$, $z \in C$ and $0 < \lambda < 1$, except by taking $y = z = x$.

For convex cones, the concept of an extreme point is not of much use, since the origin would be the only candidate for an extreme point. One studies extreme rays of the cone instead, an *extreme ray* being a face which is a half-line emanating from the origin. In general, if C' is a half-line face of a convex set C, we shall call the direction of C' an *extreme direction* of C (extreme point of C at infinity). The extreme rays of a convex cone are thus in one-to-one correspondence with the extreme directions of the cone.

If C' is the set of points where a certain linear function h achieves its maximum over C, then C' is a face of C. (Namely, C' is convex because it is the intersection of C and $\{x \mid h(x) = \alpha\}$, where α is the maximum. If the maximum is achieved on the relative interior of a line segment $L \subset C$, then h must be constant on L, so that $L \subset C'$.) A face of this type is called an *exposed face*. The exposed faces of C (aside from C itself and possibly the empty set) are thus precisely the sets of the form $C \cap H$, where H is a non-trivial supporting hyperplane to C. An *exposed point* of C is an exposed face which is a point, i.e. a point through which there is a supporting hyperplane which contains no other point of C. We define the *exposed directions* (exposed points at infinity) of C to be the directions of the exposed half-line faces of C. An *exposed ray* of a convex cone is an exposed face which is a half-line emanating from the origin. Notice that an exposed point is an extreme point, an exposed direction is an extreme direction, and an exposed ray is an extreme ray.

Faces are not always exposed. For example, let C be the convex hull of a torus, and let D be one of the two closed disks forming the sides of C. The relative boundary points of D are extreme points of C but not exposed points of C. (They are exposed points of D, however, and D is an exposed face of C.)

If C'' is a face of C' and C' is a face of C, then C'' is a face of C. This is immediate from the definition of "face." In particular, *an extreme point or extreme direction of a face of C is an extreme point or extreme direction of C itself*. The parallel statement for exposed faces is not true, as the torus example shows.

If C' is a face of C and D is a convex set such that $C' \subset D \subset C$, then C' is *a fortiori* a face of D. If C' is exposed in C, it is also exposed in D.

For example, let C be a closed convex set, let C' be a half-line face of C with endpoint x, and let $D = x + 0^+C$. Then $C' \subset D \subset C$ (Theorem 8.3), so C' is a half-line face of D and $C' - x$ is an extreme ray of the cone 0^+C. It follows that every extreme direction of C is also an extreme direction of 0^+C. Similarly, every exposed direction of C is an exposed direction of 0^+C. The converses do not hold: if C is a parabolic convex set in R^2, say, 0^+C is the ray in the direction of the axis of C; in this case 0^+C has one extreme (actually exposed) direction, while C itself has no half-line faces and hence no extreme or exposed directions at all.

The definition of "face" implies a stronger property involving arbitrary convex subsets, not just line segments:

THEOREM 18.1. *Let C be a convex set, and let C' be a face of C. If D is a convex set in C such that* ri D *meets* C', *then* $D \subset C'$.

PROOF. Let $z \in C' \cap$ ri D. If x is any point of D other than z, there exists a $y \in D$ such that z is in the relative interior of the line segment between x and y. Since C' is a face, x and y must be in C'. Thus $D \subset C'$. ‖

COROLLARY 18.1.1. *If C' is a face of a convex set C, then $C' = C \cap$ cl C'. In particular, C' is closed if C is closed.*

PROOF. Take $D = C \cap$ cl C'. ‖

COROLLARY 18.1.2. *If C' and C'' are faces of a convex set C such that* ri C' *and* ri C'' *have a point in common, then actually* $C' = C''$.

PROOF. $C'' \subset C'$ because ri C'' meets C', and likewise $C' \subset C''$ because ri C' meets C''.   ‖

COROLLARY 18.1.3.   *Let C be a convex set, and let C' be a face of C other than C itself. Then C' is entirely contained in the relative boundary of C, so that dim $C' <$ dim C.*

PROOF. If ri C met C', we would have $C \subset C'$. The assertion about dimensions stems from Corollary 6.3.3.   ‖

Let $F(C)$ be the collection of all faces of a given convex set C. Regarded as a partially ordered set under inclusion, $F(C)$ has a greatest element and a least element (C and $\emptyset$). The intersection of an arbitrary set of faces is obviously another face, so every set of elements of $F(C)$ has a greatest lower bound in the partial ordering. Every set of elements then has a least upper bound too (since the set of all its upper bounds has a greatest lower bound). Thus $F(C)$ is a complete lattice. Any strictly decreasing sequence of faces must be finite in length, because the dimensions of the faces must be strictly decreasing by Corollary 18.1.3.

THEOREM 18.2.   *Let C be a non-empty convex set, and let U be the collection of all relative interiors of non-empty faces of C. Then U is a partition of C, i.e. the sets in U are disjoint and their union is C. Every relatively open convex subset of C is contained in one of the sets in U, and these are the maximal relatively open convex subsets of C.*

PROOF. The relative interiors of different faces of C are disjoint by Corollary 18.1.2. Given any non-empty relatively open convex subset D of C (for instance D may consist of a single point), let C' be the smallest face of C containing D (the intersection of the collection of faces which contain D). If D were contained in the relative boundary of C', there would be a supporting hyperplane H to C' containing D but not all of C' (Theorem 11.6). Then D would be in the exposed face $C'' = C' \cap H$ of C' which would be a face of C properly smaller the C'. Thus D cannot be entirely contained in the relative boundary of C' and must meet ri C'. This implies that ri $D \subset$ ri C' (Corollary 6.5.2). But ri $D = D$. Thus D is contained in one of the sets in U. Since none of the sets in U is contained in any other, we may conclude that the sets in U are the maximal relatively open convex subsets of C and that their union is C.   ‖

Note that, given two different points x and y in C, there exists a relatively open convex subset D of C containing both x and y if and only if there is a line segment in C having both x and y in its relative interior. If we define $x \sim y$ to mean that either x and y satisfy this line segment condition or $x = y$, it follows from the theorem that $\sim$ is an equivalence relation on C whose equivalence classes are the relative interiors of the non-empty faces of C.

If $C = \text{conv } S$, there is a one-to-one correspondence between the faces of C and certain subsets of S, according to the following theorem.

THEOREM 18.3. *Let $C = \text{conv } S$, where S is a set of points and directions, and let C' be a non-empty face of C. Then $C' = \text{conv } S'$, where S' consists of the points in S which belong to C' and the directions in S which are directions of recession of C'.*

PROOF. We have $C' \supset \text{conv } S'$ by definition. On the other hand, let x be any point of C'. We shall prove that $x \in \text{conv } S'$. Since $x \in \text{conv } S$, there exist points $x_1, \ldots, x_k$ in S and non-zero vectors $x_{k+1}, \ldots, x_m$ whose directions belong to S ($1 \leq k \leq m$), such that

$$x = \lambda_1 x_1 + \cdots + \lambda_k x_k + \lambda_{k+1} x_{k+1} + \cdots + \lambda_m x_m,$$

where $\lambda_i > 0$ for $i = 1, \ldots, m$ and $\lambda_1 + \cdots + \lambda_k = 1$. (See §17.) Let $D = \text{conv } S''$, where S'' consists of the points $x_1, \ldots, x_k$ and the directions of $x_{k+1}, \ldots, x_m$. Then $x \in \text{ri } D$, inasmuch as the coefficients λ_i in the above expression are all positive (Theorem 6.4). Hence ri D meets C'. By Theorem 18.1, $D \subset C'$. Thus $x_1, \ldots, x_k$ belong to C', and (assuming $k < m$) C' contains certain half-lines whose directions are those of x_{k+1}, $\ldots, x_m$. These directions are therefore directions in which cl C' is receding. They are also directions of recession of C (because they belong to S and $C = \text{conv } S$). Since $C' = C \cap \text{cl } C'$ by Corollary 18.1.1, they are in fact directions in which C' is receding. Therefore $S'' \subset S'$ and $x \in \text{conv } S'$. $\|$

COROLLARY 18.3.1. *Suppose $C = \text{conv } S$, where S is a set of points and directions. Then every extreme point of C is a point of S. If no half-line in C contains an unbounded set of points of S (which is true in particular if the set of all points in S is bounded), then every extreme direction of C is a direction in S.*

PROOF. Take C' to be a face of C which is a single point or a half-line. $\|$

A convex set C has no extreme points or extreme directions whatsoever, of course, if its lineality is non-zero. In this case, however, we have $C = C_0 + L$, where L is the lineality subspace of C and $C_0 = C \cap L^\perp$ is a convex set which has lineality zero. The faces C' of C are evidently in one-to-one correspondence with the faces C_0' of C_0 by the formulas $C' = C_0' + L$, $C_0' = C' \cap L^\perp$. Thus in studying faces it really suffices, for the most part, to consider convex sets with lineality zero.

We turn now to the question of internal representations. In the first place, when is a closed convex set C the convex hull of its relative boundary? This is obviously not true when C is an affine set or a *closed half* of an affine set (the intersection of an affine set and a closed half-space which meets it does not contain it). But it is true in all other cases by the following theorem.

THEOREM 18.4.   *Let C be a closed convex set which is not an affine set or a closed half of an affine set. Then each relative interior point of C lies on some line segment joining two relative boundary points of C.*

PROOF.   Let D be the relative boundary of C. Since C is not affine, D is not empty. We shall show first that D cannot be convex. If D were convex, there would be a non-trivial supporting hyperplane H to C with $H \supset D$ (Theorem 11.6). Let A be the corresponding open half-space containing ri C but disjoint from D. Since C is not a closed half of aff C, there must exist a point x in $A \cap$ aff C such that $x \notin$ ri C. Any line segment joining x with some point of ri C must intersect C in a line segment having one of its endpoints in D. This is incompatible with A being disjoint from D. Thus D is not convex, and there must exist distinct points x_1 and x_2 in D whose connecting line segment contains a point of ri C. Let M be the line through x_1 and x_2. The intersection of M with C must be the line segment connecting x_1 and x_2, for if it were any larger x_1 or x_2 would have to be in ri C by Theorem 6.1. Every line parallel to M must likewise have a (closed) bounded intersection with C by Corollary 8.4.1. Thus, given any $y \in$ ri C, the line through y parallel to M intersects C in a segment whose two endpoints are in D.   ‖

Here is the fundamental representation theorem. It is stated for closed convex sets with lineality zero, but there is an obvious extension to closed convex sets of arbitrary lineality, as seen from the remarks above.

THEOREM 18.5.   *Let C be a closed convex set containing no lines, and let S be the set of all extreme points and extreme directions of C. Then C = conv S.*

PROOF.   The theorem is trivial if dim $C \leq 1$ (in which case C is $\emptyset$, or a single point, or a closed line segment, or a closed half-line). Let us make the induction hypothesis that the theorem is true for all closed convex sets of dimension smaller than a given $m > 1$, and that C is itself m-dimensional. We have $C \supset$ conv S by definition, because the points in S belong to C and the directions in S are directions of recession of C. Since C contains no lines and is not itself a half-line, it is the convex hull of its relative boundary by Theorem 18.4. Therefore, to show $C \subset$ conv S, we need only show that every relative boundary point of C belongs to conv S. By Theorem 18.2, a relative boundary point x is contained in the relative interior of some face C' other than C itself. This C' is closed by Corollary 18.1.1, and it has a smaller dimension than C by Corollary 18.1.3. The theorem is valid for C' by induction, so $x \in$ conv S', where S' is the set of extreme points and extreme directions of C'. Since $S' \subset S$, we have $x \in$ conv S.   ‖

COROLLARY 18.5.1. *A closed bounded convex set is the convex hull of its extreme points.*

COROLLARY 18.5.2. *Let K be a closed convex cone containing more than just the origin but containing no lines. Let T be any set of vectors in K such that each extreme ray of K is generated by some $x \in T$. Then K is the convex cone generated by T.*

PROOF. To say that K is the convex cone generated by T is to say that $K = \text{conv } S$, where S consists of the origin and the directions of the vectors in T. Here the origin is the unique extreme point of K, and the directions of the vectors in T are the extreme directions of K. ∥

COROLLARY 18.5.3. *A non-empty closed convex set containing no lines has at least one extreme point.*

PROOF. If the S in the theorem contained only directions, conv S would be empty by definition. ∥

Observe that the set S in Theorem 18.5 is *minimal* in the sense that (by Corollary 18.3.1) if S' is any set of points and directions such that $C = \text{conv } S'$ and no half-line contains an unbounded set of points of S', then $S' \supset S$.

The set of extreme points of a closed bounded convex set C need not be closed. For example, let C_1 be a closed circular disk in R^3, and let C_2 be a line segment perpendicular to C_1 whose midpoint is a relative boundary point of C_1. The convex hull C of $C_1 \cup C_2$ is closed. But the set of extreme points of C consists of the two endpoints of C_2 and all the relative boundary points of C_1 other than the midpoint of C_2, and this set is not closed.

By means of the following theorem, one obtains internal representations of the type $C = \text{cl (conv } S)$. Extreme points are replaced by exposed points.

THEOREM 18.6 (Straszewicz's Theorem). *For any closed convex set C, the set of exposed points of C is a dense subset of the set of extreme points of C. Thus every extreme point is the limit of some sequence of exposed points.*

PROOF. Let B be the unit Euclidean ball. For any $\alpha > 0$, the points x with $|x| < \alpha$ which are extreme or exposed points of C are the same as those which are extreme or exposed points of $C \cap \alpha B$. It suffices therefore to prove the theorem in the case where C is bounded (and non-empty). Let S be the set of exposed points of C. Of course, S is contained in the set of extreme points of C, and cl $S \subset C$. We must show that every extreme point belongs to cl S. Assume x is an extreme point of C not in cl S. Then x cannot be in $C_0 = \text{conv (cl } S)$ (Corollary 18.3.1). Since C_0 is closed (Theorem 17.2), there exists a closed half-space H containing C_0

but not x (Theorem 11.5). We shall construct an exposed point p of C not in H, and this contradiction will establish the theorem. Let e be an "outward" normal to H with $|e| = 1$. Let ε be the smallest positive scalar such that $(x - \varepsilon e) \in H$, and let $y = x - \lambda e$ for some $\lambda > \varepsilon$. Consider the Euclidean ball B_0 of radius λ with center y. The boundary of B_0 contains x. However, the points of H which are not interior to B_0 are at a distance of at least $(2\varepsilon\lambda)^{1/2}$ from x, as one can calculate from the Pythagorean Theorem. Assume now that λ was chosen so large that $(2\varepsilon\lambda)^{1/2} > r$, where r is the supremum of $|z - x|$ for $z \in C \cap H$. Then C contains points whose distance from y is at least λ (namely x), but no point of $C \cap H$ meets this description. Choose a $p \in C$ which maximizes $|p - y|$ (i.e. a farthest point of C from y). Then $p \notin H$. Let B_1 be the Euclidean ball with center y having p on its boundary. The supporting hyperplane to B_1 at p contains no point of B_1 other than p. Since $p \in C \subset B_1$, it follows that p is an exposed point of C.   ‖

THEOREM 18.7.   *Let C be a closed convex set containing no lines, and let S be the set of all exposed points and exposed directions of C. Then $C = \mathrm{cl}\,(\mathrm{conv}\,S)$.*

PROOF.   We can assume for simplicity that C is n-dimensional in R^n, and that $n \geq 2$ (the theorem being true trivially when C is less than two-dimensional). Since the S specified here is contained in the one in Theorem 18.5, we have $C \supset \mathrm{cl}\,(\mathrm{conv}\,S)$. Also, $\mathrm{cl}\,(\mathrm{conv}\,S)$ is a closed convex set which contains all the extreme points of C (Theorem 18.6) and hence is non-empty (Corollary 18.5.3). Suppose $\mathrm{cl}\,(\mathrm{conv}\,S)$ is not all of C; we shall argue from this to a contradiction. By Theorem 11.5, there exists a hyperplane H which meets C but does not meet $\mathrm{cl}\,(\mathrm{conv}\,S)$. The convex set $C \cap H$ must have at least one extreme point (Corollary 18.5.3) and hence at least one exposed point x (Theorem 18.6). According to the definition of "exposed point," there exists in H an $(n - 2)$-dimensional affine set M which meets $C \cap H$ only at x. In particular, this M does not meet the (non-empty) interior of C, so by Theorem 11.2 we can extend M to a hyperplane H' which does not meet $\mathrm{int}\,C$. This H' is a supporting hyperplane to C, and $C' = C \cap H'$ is a (closed) exposed face of C. The extreme points or exposed points of C' are extreme points of C too, and consequently they belong to $\mathrm{cl}\,(\mathrm{conv}\,S)$ and not to H. The hyperplane H meets C' at x alone. Inasmuch as x cannot be an exposed point of C', we must have $\{x\} = H \cap \mathrm{ri}\,C'$ and consequently $\dim C' = 1$. By the hypothesis, C' cannot be a line. Nor can C' be the line segment between two points, for these points (being extreme points of C') would belong to S contrary to $x \notin \mathrm{conv}\,S$. The only other possibility is that C' is a closed half-line with its endpoint in S. The direction of C' then belongs to S, since it is an exposed direction of C by definition. That implies $C' \subset$ conv S contrary to $x \notin$ conv S. ‖

COROLLARY 18.7.1. *Let K be a closed convex cone containing more than just the origin but containing no lines. Let T be any set of vectors in K such that each exposed ray of K is generated by some $x \in T$. Then K is the closure of the convex cone generated by T.*

The exposed points of a closed convex set C will be characterized in Corollary 25.1.3 as the gradients of the support function of C.

The concept dual to that of "exposed point" is that of "tangent hyperplane." A hyperplane H is said to be *tangent* to a closed convex set C at a point x if H is the unique supporting hyperplane to C at x. A *tangent half-space* to C is a supporting half-space whose boundary is tangent to C at some point. It will be seen later, from the discussion of the differentiability of convex functions, that tangent hyperplanes can also be defined by differential limits as in classical analysis.

The following "external" representation theorem may be viewed as the dual of Theorem 18.7. It is a stronger version of Theorem 11.5.

THEOREM 18.8. *An n-dimensional closed convex set C in R^n is the intersection of the closed half-spaces tangent to it.*

PROOF. Let G be the epigraph of the support function $\delta^*(\cdot \mid C)$. This G is a closed convex cone in R^{n+1} containing more than just the origin. Since C is n-dimensional, C has a non-empty interior, so that

$$-\delta^*(-x^* \mid C) < \delta^*(x^* \mid C)$$

for every $x^* \neq 0$. Hence G can contain no lines through the origin. By Corollary 18.7.1, we have $G = \text{cl (conv } S)$, where S is the union of all exposed rays of G. It follows that the linear functions $\langle x, \cdot \rangle$ majorized by $\delta^*(\cdot \mid C)$, which correspond of course to the points x of C (Theorem 13.1), are the same as the linear functions whose epigraphs contain every "non-vertical" exposed ray of G. Put another way, C is the intersection of all the half-spaces $\{x \mid \langle x, x^* \rangle \leq \alpha\}$ such that the set of non-negative multiples of (x^*, α) is a "non-vertical" exposed ray of G. The latter condition means that there is some non-vertical supporting hyperplane to G (the graph of a certain linear function $\langle y, \cdot \rangle$) which intersects G only in the ray generated by (x^*, α). In other words, there is some $y \in C$ such that $\langle y, x^* \rangle = \delta^*(x^* \mid C) = \alpha$ but $\langle y, y^* \rangle < \delta^*(y^* \mid C)$ for every y^* which is not a non-negative multiple of x^*. This says that the half-space $\{x \mid \langle x, x^* \rangle \leq \alpha\}$ is tangent to C at y. Thus C is the intersection of all such half-spaces. ‖

Polyhedral Convex Sets and Functions

A *polyhedral* convex set in R^n is by definition a set which can be expressed as the intersection of some finite collection of closed half-spaces, i.e. as the set of solutions to some finite system of inequalities of the form

$$\langle x, b_i \rangle \leq \beta_i, \qquad i = 1, \ldots, m.$$

Actually, of course, the set of solutions to any finite mixed system of linear equations and weak linear inequalities is a polyhedral convex set, since an equation $\langle x, b \rangle = \beta$ can always be expressed as two inequalities: $\langle x, b \rangle \leq \beta$ and $\langle x, -b \rangle \leq -\beta$. Every affine set (including the empty set and R^n) is polyhedral (Corollary 1.4.1).

It is clear that a polyhedral convex set is a cone if and only if it can be expressed as the intersection of a finite collection of closed half-spaces whose boundary hyperplanes pass through the origin. A polyhedral convex cone is thus the set of solutions to some finite system of *homogeneous* ($\beta_i = 0$) weak linear inequalities.

The property of being "polyhedral" is a finiteness condition on the "external" representations of a convex set. There is a dual property of importance which is a finiteness condition on the "internal" representations of a convex set. A *finitely generated* convex set is defined to be a set which is the convex hull of a finite set of points and directions (in the sense of §17). Thus C is a finitely generated convex set if and only if there exist vectors $a_1, \ldots, a_m$ such that, for a fixed integer k, $0 \leq k \leq m$, C consists of all the vectors of the form

$$x = \lambda_1 a_1 + \cdots + \lambda_k a_k + \lambda_{k+1} a_{k+1} + \cdots + \lambda_m a_m,$$

with

$$\lambda_1 + \cdots + \lambda_k = 1, \qquad \lambda_i \geq 0 \quad \text{for} \quad i = 1, \ldots, m.$$

The finitely generated convex sets which are cones are the sets which can be expressed this way with $k = 0$, i.e. with no requirement about certain coefficients adding up to 1; in such an expression, $\{a_1, \ldots, a_m\}$ is called a set of *generators* for the cone. A finitely generated convex cone is thus the convex hull of the origin and finitely many directions.

The finitely generated convex sets which are bounded are the *polytopes*, including the simplices. The unbounded finitely generated convex sets may be regarded as *generalized polytopes* having certain vertices at infinity, like the generalized simplices in §17.

It turns out that the polyhedral convex sets are the same as the finitely generated ones. This classical result is an outstanding example of a fact which is completely obvious to geometric intuition, but which wields important algebraic content and is not trivial to prove. The proof we shall give is based on the theory of faces of convex sets. This approach stresses the intuitive reasons why the theorem is true. Self-contained algebraic proofs which require less elaborate machinery are also possible, however.

THEOREM 19.1. *The following properties of a convex set C are equivalent:*
(a) *C is polyhedral;*
(b) *C is closed and has only finitely many faces;*
(c) *C is finitely generated.*

PROOF. (a) implies (b): Let $H_1, \ldots, H_m$ be closed half-spaces whose intersection is C. Let C' be a non-empty face of C. For each i, ri C' must be contained in int H_i or in the boundary hyperplane M_i of H_i. Let D be the intersection of the finite collection consisting of the relatively open convex sets int H_i or M_i containing ri C'. This D is a convex subset of C, and it is relatively open (Theorem 6.5). Since ri C' is a maximal relatively open convex subset of C (Theorem 18.2), we must really have ri $C' = D$. There are only finitely many intersections of the form D, and different faces of C have disjoint relative interiors (Corollary 18.1.2), so it follows that C can have only finitely many faces.

(b) implies (c): First consider the case where C contains no lines. According to Theorem 18.5, C is the convex hull of its extreme points and extreme directions. Since C has only finitely many faces, it has only finitely many extreme points and extreme directions. Hence C is finitely generated. Now suppose C does contain lines. Then $C = C_0 + L$, where L is the lineality space of C and C_0 is a closed convex set containing no lines, namely $C_0 = C \cap L^\perp$. The faces of C_0 are of the form $C_0' = C' \cap L^\perp$ where C' is a face of C, so C_0 has only finitely many faces. Hence C_0 is finitely generated. Let $b_1, \ldots, b_m$ be a basis for L. Any $x \in C$ can be expressed in the form

$$x = x_0 + \mu_1 b_1 + \cdots + \mu_m b_m + \mu_1'(-b_1) + \cdots + \mu_m'(-b_m),$$

where $x_0 \in C_0$, $\mu_i \geq 0$ and $\mu_i' \geq 0$ for $i = 1, \ldots, m$. Hence C itself is finitely generated.

(c) implies (b). Assuming that $C = \operatorname{conv} S$, where S is a finite set of points and directions, we can express C as the union of finitely many generalized simplices by Carathéodory's Theorem (Theorem 17.1). Each generalized simplex is a closed set, so C is closed too. There is a one-to-one correspondence between the faces of C and certain subsets of S by Theorem 18.3, so C can have only finitely many faces.

(b) implies (a). It suffices to treat the case where C is n-dimensional in R^n. In that case C is the intersection of its tangent closed half-spaces (Theorem 18.8). If H is the boundary hyperplane of a tangent closed half-space there exists by definition some $x \in C$ such that H is the unique supporting hyperplane to C through x. Thus H is the unique supporting hyperplane to C through the exposed face $C \cap H$. Since C has only finitely many faces, it follows that it can have only finitely many tangent closed half-spaces. Hence C is polyhedral. ‖

The proof of Theorem 19.1 shows, incidentally, that every face of a polyhedral convex set is itself polyhedral.

COROLLARY 19.1.1. *A polyhedral convex set has at most a finite number of extreme points and extreme directions.*

PROOF. This is immediate from the fact that extreme points and extreme directions correspond to faces which are points and half-lines, respectively. ‖

A *polyhedral convex function* is a convex function whose epigraph is polyhedral. Common examples of such functions are the affine (or partial affine) functions and the indicator functions of polyhedral convex sets (especially the non-negative orthant of R^n).

In general, for f to be a polyhedral convex function on R^n, epi f must be the intersection of finitely many closed half-spaces in R^{n+1} which are either "vertical" or the epigraphs of affine functions. In other words, f is a polyhedral convex function if and only if f can be expressed in the form

$$f(x) = h(x) + \delta(x \mid C),$$

where

$$h(x) = \max \{\langle x, b_1 \rangle - \beta_1, \ldots, \langle x, b_k \rangle - \beta_k\},$$

$$C = \{x \mid \langle x, b_{k+1} \rangle \leq \beta_{k+1}, \ldots, \langle x, b_m \rangle \leq \beta_m\}.$$

A convex function f is said to be *finitely generated* if there exist vectors $a_1, \ldots, a_k, a_{k+1}, \ldots, a_m$ and corresponding scalars α_i such that

$$f(x) = \inf \{\lambda_1 \alpha_1 + \cdots + \lambda_k \alpha_k + \lambda_{k+1} \alpha_{k+1} + \cdots + \lambda_m \alpha_m\},$$

where the infimum is taken over all choices of the coefficients λ_i such that

$$\lambda_1 a_1 + \cdots + \lambda_k a_k + \lambda_{k+1} a_{k+1} + \cdots + \lambda_m a_m = x$$

$$\lambda_1 + \cdots + \lambda_k = 1, \qquad \lambda_i \geq 0 \quad \text{for} \quad i = 1, \ldots, m.$$

This condition on f means that

$$f(x) = \inf \{\mu \mid (x, \mu) \in F\},$$

where F is the convex hull of a certain finite set of points and directions in R^{n+1}, namely the points (a_i, α_i), $i = 1, \ldots, k$, and the directions of (a_i, α_i), $i = k + 1, \ldots, m$, along with the direction of $(0, 1)$ ("up"). According to Theorem 19.1, such an F is a closed set and hence coincides entirely with epi f. This implies in particular that, for any x such that $f(x)$ is finite, the point $(x, f(x))$ belongs to F, and hence the infimum defining $f(x)$ is actually attained. We may draw the following conclusions.

COROLLARY 19.1.2. *A convex function is polyhedral if and only if it is finitely generated. Such a function, if proper, is necessarily closed. The infimum for a given x in the definition of "finitely generated convex function," if finite, is attained by some choice of the coefficients λ_i.*

The absolute value function is a polyhedral convex function on R. More generally, the function f defined by

$$f(x) = |\xi_1| + \cdots + |\xi_n|, \qquad x = (\xi_1, \ldots, \xi_n)$$

is polyhedral convex on R^n, since it is the pointwise supremum of the 2^n linear functions of the form

$$x \rightarrow \varepsilon_1 \xi_1 + \cdots + \varepsilon_n \xi_n, \qquad \varepsilon_j = +1 \quad \text{or} \quad -1.$$

Note that f is actually a norm. Another commonly encountered polyhedral convex norm is the Tchebycheff norm f defined by

$$f(x) = \max \{|\xi_1|, \ldots, |\xi_n|\}.$$

This f is the pointwise supremum of the $2n$ linear functions of the form

$$x \rightarrow \varepsilon_j \xi_j, \qquad \varepsilon_j = +1 \quad \text{or} \quad -1, \qquad j = 1, \ldots, n.$$

We shall demonstrate now that the property of being "polyhedral" is preserved under many important operations. We begin with duality.

THEOREM 19.2. *The conjugate of a polyhedral convex function is polyhedral.*

PROOF. If f is polyhedral, it is finitely generated and can be expressed as above for certain vectors $a_1, \ldots, a_k, a_{k+1}, \ldots, a_m$ and corresponding scalars α_i. Substituting this formula for f into the formula which defines the conjugate function f^*, we get

$$f^*(x^*) = \sup \{\langle \sum_{i=1}^m \lambda_i a_i, x^* \rangle - \sum_{i=1}^m \lambda_i \alpha_i\},$$

where the supremum is taken over

$$\lambda_1 \geq 0, \ldots, \lambda_m \geq 0, \qquad \lambda_1 + \cdots + \lambda_k = 1.$$

It is easy to see that when

$$\langle a_i, x^* \rangle - \alpha_i \le 0 \quad \text{for} \quad i = k + 1, \ldots, m$$

one has

$$f^*(x^*) = \max \{\langle a_i, x^* \rangle - \alpha_i \mid i = 1, \ldots, k\},$$

but otherwise $f^*(x^*) = +\infty$. Thus f^* is polyhedral.   ‖

COROLLARY 19.2.1.   *A closed convex set C is polyhedral if and only if its support function $\delta^*(\cdot \mid C)$ is polyhedral.*

PROOF.   The indicator function and support function of C are conjugate to each other, so by the theorem one is polyhedral if and only if the other is.   ‖

As an example, consider the problem of maximizing a linear function $\langle a, \cdot \rangle$ over the set C which consists of all the solutions to a certain finite system of weak linear inequalities. The supremum is $\delta^*(a \mid C)$. Since C is polyhedral, it follows from Corollary 19.2.1 that the supremum is a polyhedral convex function of a.

If f is any polyhedral convex function, the level sets $\{x \mid f(x) \le \alpha\}$ are obviously polyhedral convex sets. Since the polar C° of a convex set C is the level set of the support function $\delta^*(\cdot \mid C)$ corresponding to $\alpha = 1$, we have:

COROLLARY 19.2.2.   *The polar of a polyhedral convex set is polyhedral.*

The intersection of finitely many polyhedral convex sets is polyhedral. Likewise, the pointwise supremum of finitely many polyhedral convex functions is polyhedral.

THEOREM 19.3.   *Let A be a linear transformation from R^n to R^m. Then AC is a polyhedral convex set in R^m for each polyhedral convex set C in R^n, and $A^{-1}D$ is a polyhedral convex set in R^n for each polyhedral convex set D in R^m.*

PROOF.   Let C be polyhedral in R^n. By Theorem 19.1, C is finitely generated, so there exist vectors $a_1, \ldots, a_k, a_{k+1}, \ldots, a_r$ such that

$$C = \{\textstyle\sum_{i=1}^r \lambda_i a_i \mid \lambda_1 + \cdots + \lambda_k = 1, \lambda_i \ge 0 \quad \text{for} \quad i = 1, \ldots, r\}.$$

Let b_i be the image of a_i under A. Then

$$AC = \{\textstyle\sum_{i=1}^r \lambda_i b_i \mid \lambda_1 + \cdots + \lambda_k = 1, \lambda_i \ge 0 \quad \text{for} \quad i = 1, \ldots, r\}.$$

Thus AC is finitely generated, and hence polyhedral by Theorem 19.1. Now let D be a polyhedral convex set in R^m. Express D as the set of solutions y to a certain system

$$\langle y, a_i^* \rangle \le \alpha_i, \quad i = 1, \ldots, s.$$

Then $A^{-1}D$ is the set of solutions x to

$$\langle Ax, a_i^* \rangle \leq \alpha_i, \qquad i = 1, \ldots, s.$$

This is a finite system of weak linear inequalities on x, so $A^{-1}D$ is polyhedral. ‖

COROLLARY 19.3.1. *Let A be a linear transformation from R^n to R^m. For each polyhedral convex function f on R^n, the convex function Af is polyhedral on R^m, and the infimum in its definition, if finite, is attained. For each polyhedral convex function g on R^m, gA is polyhedral on R^n.*

PROOF. The image of epi f under the linear transformation $(x, \mu) \to (Ax, \mu)$ is a polyhedral convex set, and it equals epi (Af). The inverse image of epi g under this same transformation is a polyhedral convex set, and it equals epi (gA). ‖

COROLLARY 19.3.2. *If C_1 and C_2 are polyhedral convex sets in R^n, then $C_1 + C_2$ is polyhedral.*

PROOF. Let $C = \{(x_1, x_2) \mid x_1 \in C_1, x_2 \in C_2\}$. It is clear that C can be expressed as the intersection of finitely many closed half-spaces in R^{2n}. Hence C is polyhedral. The image of C under the linear transformation $A : (x_1, x_2) \to x_1 + x_2$ is polyhedral too, by the theorem, and this image is $C_1 + C_2$. ‖

COROLLARY 19.3.3. *If C_1 and C_2 are non-empty disjoint polyhedral convex sets, there exists a hyperplane separating C_1 and C_2 strongly.*

PROOF. We have $0 \notin C_1 - C_2$, and $C_1 - C_2$ is closed because it is polyhedral by the preceding corollary. Strong separation is then possible according to Theorem 11.4. ‖

COROLLARY 19.3.4. *If f_1 and f_2 are proper polyhedral convex functions on R^n, then $f_1 \square f_2$ is a polyhedral convex function too. Moreover, if $f_1 \square f_2$ is proper, the infimum in the definition of $(f_1 \square f_2)(x)$ is attained for each x.*

PROOF. epi f_1 + epi f_2 is a polyhedral convex set, and it equals epi $(f_1 \square f_2)$. ‖

Theorem 19.3 implies in particular that the orthogonal projection of a polyhedral convex set $C \subset R^n$ on a subspace L is another polyhedral convex set.

To illustrate Theorem 19.3 further, along with Corollary 19.3.2, let A be a linear transformation from R^n to R^m, and let

$$C = \{z \in R^n \mid \exists x \geq z, Ax \in \text{conv } \{b_1, \ldots, b_r\}\},$$

where $b_1, \ldots, b_r$ are fixed elements of R^m. We have

$$C = A^{-1}D - K,$$

where K is the non-negative orthant of R^n (a polyhedral convex cone) and

$$D = \text{conv } \{b_1, \ldots, b_r\}$$

(a finitely generated convex set), and therefore C is a polyhedral convex set. A good illustration of Corollary 19.3.4 is the case where

$$f_1(x) = \max \{|\xi_j| \,|\, j = 1, \ldots, n\} = \|x\|_\infty,$$
$$f_2(x) = \delta(x \,|\, C),$$
$$C = \{x \,|\, \langle a_i, x \rangle \leq \alpha_i, i = 1, \ldots, m\}.$$

Here

$$(f_1 \,\square\, f_2)(y) = \inf_x \{f_1(y - x) + f_2(x)\}$$
$$= \inf \{\|y - x\|_\infty \,|\, x \in C\},$$

and this quantity is of interest when y is to be approximated as closely as possible with respect to the Tchebycheff norm $\|\cdot\|_\infty$ by some solution x to the system

$$\langle a_i, x \rangle \leq \alpha_i, \qquad i = 1, \ldots, m.$$

Since f_1 and f_2 are polyhedral convex functions, $(f_1 \,\square\, f_2)(y)$ is a polyhedral convex function of y.

THEOREM 19.4.   *If f_1 and f_2 are proper polyhedral convex functions, then $f_1 + f_2$ is polyhedral.*

PROOF.   We have $f_i(x) = h_i(x) + \delta(x \,|\, C_i)$ for $i = 1, 2$, where C_1 and C_2 are polyhedral convex sets and

$$h_1(x) = \max \{\langle x, a_i \rangle - \alpha_i \,|\, i = 1, \ldots, k\},$$
$$h_2(x) = \max \{\langle x, b_j \rangle - \beta_j \,|\, j = 1, \ldots, r\}.$$

Let $C = C_1 \cap C_2$, $d_{ij} = a_i + b_j$ and $\mu_{ij} = \alpha_i + \beta_j$. Then C is a polyhedral convex set, and

$$(f_1 + f_2)(x) = h(x) + \delta(x \,|\, C),$$

where

$$h(x) = \max \{\langle x, d_{ij} \rangle - \mu_{ij} \,|\, i = 1, \ldots, k \text{ and } j = 1, \ldots, r\}.$$

Thus $f_1 + f_2$ is polyhedral. $\|$

Obviously λf is polyhedral for $\lambda \geq 0$ if f is a polyhedral convex function.

THEOREM 19.5.   *Let C be a non-empty polyhedral convex set. Then λC is polyhedral for every scalar λ. The recession cone 0^+C is also polyhedral. In fact, if C is represented as conv S, where S is a finite set of points and directions, then $0^+C = \text{conv } S_0$, where S_0 consists of the origin and the directions in S.*

PROOF.   Express C as the set of solutions to a finite system of inequalities: $\langle x, b_i \rangle \leq \beta_i, i = 1, \ldots, m$. Then λC, for $\lambda > 0$, is the set of solutions to: $\langle x, b_i \rangle \leq \lambda \beta_i, i = 1, \ldots, m$. Furthermore, 0^+C is the set of solutions to $\langle x, b_i \rangle \leq 0$, $i = 1, \ldots, m$. Thus λC for $\lambda > 0$, and 0^+C, are polyhedral. Trivially $0C$ is polyhedral, since by definition $0C = \{0\}$. Also $-C$ is polyhedral, since $-C$ is the image of C under the linear transformation $x \to -x$, and it follows that λC is polyhedral for $\lambda < 0$. Now suppose $C = \operatorname{conv} S$, where S consists of $a_1, \ldots, a_k$ and the directions of $a_{k+1}, \ldots, a_m$. Let K be the polyhedral convex cone in R^{n+1} generated by the vectors $(1, a_1), \ldots, (1, a_k), (0, a_{k+1}), \ldots, (0, a_m)$. The intersection of K with the hyperplane $\{(1, x) \in R^{n+1} \mid x \in R^n\}$ can be identified with C, and (since K is closed) the intersection of K with the hyperplane $\{(0, x) \in R^{n+1} \mid x \in R^n\}$ can be identified with 0^+C. Thus 0^+C is generated by $a_{k+1}, \ldots, a_m$. In other words, $0^+C = \operatorname{conv} S_0$. ‖

COROLLARY 19.5.1. *If f is a proper polyhedral convex function, then $f\lambda$ is polyhedral for $\lambda \geq 0$ and $\lambda = 0^+$.*

PROOF. Apply Theorem 19.5 to $C = \operatorname{epi} f$. ‖

The convex hull of the union of two polyhedral convex sets need not be polyhedral, as is seen for instance in the case of a line and a point not on the line. The difficulty is that the ordinary convex hull operation does not adequately take account of directions of recession. A pair of non-empty polyhedral convex sets C_1 and C_2 in R^n can be expressed as $C_1 = \operatorname{conv} S_1$ and $C_2 = \operatorname{conv} S_2$, where S_1 and S_2 are finite sets of points and directions, and one then has

$$\operatorname{conv} (C_1 \cup C_2) \subset \operatorname{conv} (S_1 \cup S_2),$$

but equality need not hold. In general, by Theorem 19.5 one has

$$\operatorname{conv} (S_1 \cup S_2) = (C_1 + 0^+C_2) \cup (0^+C_1 + C_2) \cup \operatorname{conv} (C_1 \cup C_2).$$

However, cl $(\operatorname{conv} (C_1 \cup C_2))$ must recede in all the directions in which C_1 and C_2 recede, since it is a closed convex set containing C_1 and C_2 (Theorem 8.3). Thus cl $(\operatorname{conv} (C_1 \cup C_2))$ contains $C_1 + 0^+C_2$ and $0^+C_1 + C_2$, and hence $\operatorname{conv} (S_1 \cup S_2)$. Since $\operatorname{conv} (S_1 \cup S_2)$, being finitely generated, is polyhedral and hence closed, this implies

$$\operatorname{conv} (S_1 \cup S_2) = \operatorname{cl} (\operatorname{conv} (C_1 \cup C_2)).$$

The following conclusion may be drawn.

THEOREM 19.6. *Let $C_1, \ldots, C_m$ be non-empty polyhedral convex sets in R^n, and let $C = \operatorname{cl} (\operatorname{conv} (C_1 \cup \cdots \cup C_m))$. Then C is a polyhedral convex set, and*

$$C = \cup \{\lambda_1 C_1 + \cdots + \lambda_m C_m\},$$

where the union is taken over all choices of $\lambda_i \geq 0$, $\lambda_1 + \cdots + \lambda_m = 1$, with 0^+C_i substituted for $0C_i$ when $\lambda_i = 0$.

The situation is quite similar when convex cones are generated.

THEOREM 19.7. *Let C be a non-empty polyhedral convex set, and let K be the closure of the convex cone generated by C. Then K is a polyhedral convex cone, and*

$$K = \cup \{\lambda C \mid \lambda > 0 \quad \text{or} \quad \lambda = 0^+\}.$$

PROOF. Let the latter union be denoted by K'. The convex cone generated by C is contained in K', and its closure K in turn contains K'. (Since K is a closed convex set containing C and 0, K must contain the recession cone 0^+C by Theorem 8.3.) Thus cl $K' = K$. It will be enough therefore to show that K' is polyhedral. Express C as conv S, where S consists of $a_1, \ldots, a_k$ and the directions of $a_{k+1}, \ldots, a_m$. For $\lambda > 0$, λC is the convex hull of $\lambda a_1, \ldots, \lambda a_k$ and the directions of $a_{k+1}, \ldots, a_m$, while 0^+C is by Theorem 19.5 the convex hull of the origin and the directions of $a_{k+1}, \ldots, a_m$. Thus K' is simply the set of all non-negative linear combinations of $a_1, \ldots, a_k, a_{k+1}, \ldots, a_m$. This shows that K' is finitely generated and hence polyhedral. ‖

COROLLARY 19.7.1. *If C is a polyhedral convex set containing the origin, the convex cone generated by C is polyhedral.*

PROOF. If $0 \in C$, 0^+C is contained in the sets λC for $\lambda > 0$ and consequently may be omitted from the union in the theorem. The union is then just the convex cone generated by C, and the theorem says that this union is polyhedral. ‖

As a miscellaneous exercise in polyhedral convexity, it may be shown that, if C is a convex polytope in R^n and S is an arbitrary non-empty subset of C, then

$$D = \{y \mid S + y \subset C\}$$

is a convex polytope. Also, under what circumstances are the "umbra" and "penumbra" defined at the end of §3 polyhedral convex sets?

Some Applications of Polyhedral Convexity

In this section, we shall show how certain separation theorems, closure conditions and other results which were proved earlier for general convex sets and functions may be refined when some of the convexity is polyhedral.

We begin with the general formula for the conjugate of a sum of proper convex functions (Theorem 16.4):

$$(\mathrm{cl}\, f_1 + \cdots + \mathrm{cl}\, f_m)^* = \mathrm{cl}\, (f_1^* \,\square\, \cdots \,\square\, f_m^*).$$

Suppose that every f_i is polyhedral, and that

$$\mathrm{dom}\, f_1 \cap \cdots \cap \mathrm{dom}\, f_m \neq \emptyset.$$

Then $\mathrm{cl}\, f_i = f_i$, and $f_1 + \cdots + f_m$ is a proper polyhedral convex function (Theorem 19.4). The conjugate of $f_1 + \cdots + f_m$ must be proper too, so $f_1^* \,\square\, \cdots \,\square\, f_m^*$ must be proper. Every f_i^* is polyhedral (Theorem 19.2). Hence $f_1^* \,\square\, \cdots \,\square\, f_m^*$ is polyhedral (in particular, closed) by Corollary 19.3.4, and it follows that

$$(f_1 + \cdots + f_m)^* = f_1^* \,\square\, \cdots \,\square\, f_m^*.$$

Moreover, the infimum in the definition of $(f_1^* \,\square\, \cdots \,\square\, f_m^*)(x^*)$ is attained for each x^* by Corollary 19.3.4. This result is a refinement of the second half of Theorem 16.4.

We shall now show that, in the general mixed case, where some of the functions f_i may be polyhedral and some not, the conjugation formula in Theorem 16.4 remains valid if ri $(\mathrm{dom}\, f_i)$ is replaced by $\mathrm{dom}\, f_i$ for each i such that f_i is polyhedral.

THEOREM 20.1. *Let $f_1, \ldots, f_m$ be proper convex functions on R^n such that $f_1, \ldots, f_k$ are polyhedral. Assume that the intersection*

$$\mathrm{dom}\, f_1 \cap \cdots \cap \mathrm{dom}\, f_k \cap \mathrm{ri}\, (\mathrm{dom}\, f_{k+1}) \cap \cdots \cap \mathrm{ri}\, (\mathrm{dom}\, f_m)$$

is not empty. Then

$$(f_1 + \cdots + f_m)^*(x^*) = (f_1^* \,\square\, \cdots \,\square\, f_m^*)(x^*)$$
$$= \inf \{ f_1^*(x_1^*) + \cdots + f_m^*(x_m^*) \mid x_1^* + \cdots + x_m^* = x^* \},$$

where for each x^* the infimum is attained.

PROOF.　We already know the validity of this formula in the case where all the sets ri $(\mathrm{dom} f_i)$, $i = 1, \ldots, m$, actually have a point in common (Theorem 16.4) and in the case where $k = m$ (as just described). Assume therefore that $1 \leq k < m$, and set

$$g_1 = f_1 + \cdots + f_k, \qquad g_2 = f_{k+1} + \cdots + f_m.$$

The formula is valid for calculating the conjugates of g_1 and g_2, so we have

$$g_1^*(y_1^*) = \inf \{ f_1^*(x_1^*) + \cdots + f_k^*(x_k^*) \mid x_1^* + \cdots + x_k^* = y_1^* \},$$
$$g_2^*(y_2^*) = \inf \{ f_{k+1}^*(x_{k+1}^*) + \cdots + f_m^*(x_m^*) \mid x_{k+1}^* + \cdots + x_m^* = y_2^* \},$$

where for each y_1^* and y_2^* the infima are attained. Hence it is enough to show that

$$(g_1 + g_2)^*(x^*) = \inf \{ g_1^*(y_1^*) + g_2^*(y_2^*) \mid y_1^* + y_2^* = x^* \},$$

where for each x^* the infimum is attained by some y_1^* and y_2^*. The convex functions g_1 and g_2 are proper, and g_1 is polyhedral (Theorem 19.4). Since

$$\mathrm{dom}\, g_1 = \mathrm{dom}\, f_1 \cap \cdots \cap \mathrm{dom}\, f_k,$$
$$\mathrm{dom}\, g_2 = \mathrm{dom}\, f_{k+1} \cap \cdots \cap \mathrm{dom}\, f_m,$$

we have

$$\mathrm{ri}\,(\mathrm{dom}\, g_2) = \mathrm{ri}\,(\mathrm{dom}\, f_{k+1} \cap \cdots \cap \mathrm{dom}\, f_m)$$
$$= \mathrm{ri}\,(\mathrm{dom}\, f_{k+1}) \cap \cdots \cap \mathrm{ri}\,(\mathrm{dom}\, f_m)$$

(Theorem 6.5), and hence

$$\mathrm{dom}\, g_1 \cap \mathrm{ri}\,(\mathrm{dom}\, g_2) \neq \emptyset.$$

This implies that, for the affine hull M of $\mathrm{dom}\, g_2$,

$$\mathrm{ri}\,(M \cap \mathrm{dom}\, g_1) \cap \mathrm{ri}\,(\mathrm{dom}\, g_2) \neq \emptyset.$$

The proper convex function $h = \delta(\cdot \mid M) + g_1$ has $M \cap \mathrm{dom}\, g_1$ as its effective domain, so

$$\mathrm{ri}\,(\mathrm{dom}\, h) \cap \mathrm{ri}\,(\mathrm{dom}\, g_2) \neq \emptyset$$

and the formula in the theorem is valid for calculating $(h + g_2)^*$. Furthermore, $h + g_2 = g_1 + g_2$. Thus

$$(g_1 + g_2)^*(x^*) = (h^* \,\square\, g_2^*)(x^*)$$
$$= \inf \{ h^*(z^*) + g_2^*(y^*) \mid z^* + y^* = x^* \},$$

where for each x^* the infimum is attained. On the other hand, since $\delta(\cdot \mid M)$ and g_1 are polyhedral, the formula in the theorem is also valid for calculating h^*:

$$h^*(z^*) = \inf \{\delta^*(u^* \mid M) + g_1^*(y_1^*) \mid u^* + y_1^* = z^*\}$$

with the infimum always attained. Therefore

$$(g_1 + g_2)^*(x^*)$$
$$= \inf \{\delta^*(u^* \mid M) + g_1^*(y_1^*) + g_2^*(y^*) \mid u^* + y_1^* + y^* = x^*\},$$

where for each x^* the infimum is attained. Since the relative interiors of the effective domains of $\delta(\cdot \mid M)$ and g_2 trivially have a point in common we can apply the established formula yet again to see that

$$\inf \{\delta^*(u^* \mid M) + g_2^*(y^*) \mid u^* + y^* = y_2^*\}$$
$$= (\delta(\cdot \mid M) + g_2)^*(y_2^*) = g_2^*(y_2^*)$$

with the infimum always attained. Thus we have simply

$$(g_1 + g_2)^*(x^*) = \inf \{g_1^*(y_1^*) + g_2^*(y_2^*) \mid y_1^* + y_2^* = x^*\},$$

where for each x^* the infimum is attained. This proves the theorem. $\quad \|$

COROLLARY 20.1.1. *Let* $f_1, \ldots, f_m$ *be closed proper convex functions on* R^n *such that* $f_1, \ldots, f_k$ *are polyhedral. Assume that the intersection*

$$\operatorname{dom} f_1^* \cap \cdots \cap \operatorname{dom} f_k^* \cap \operatorname{ri}(\operatorname{dom} f_{k+1}^*) \cap \cdots \cap \operatorname{ri}(\operatorname{dom} f_m^*)$$

is not empty. Then $f_1 \,\square\, \cdots \,\square\, f_m$ *is a closed proper convex function, and the infimum in its definition is always attained.*

PROOF. Apply the theorem to the conjugate functions $f_1^*, \ldots, f_m^*$. $\quad \|$

The following special separation theorem for polyhedral convex sets may be used to analyze the intersection condition in Theorem 20.1 and Corollary 20.1.1.

THEOREM 20.2. *Let* C_1 *and* C_2 *be non-empty convex sets in* R^n *such that* C_1 *is polyhedral. In order that there exist a hyperplane separating* C_1 *and* C_2 *properly and not containing* C_2, *it is necessary and sufficient that* $C_1 \cap \operatorname{ri} C_2 = \emptyset$.

PROOF. If H is a hyperplane separating C_1 and C_2 properly and not containing C_2, then $\operatorname{ri} C_2$ lies entirely in one of the open half-spaces associated with H and hence does not meet C_1. This shows the necessity of the condition.

To prove the sufficiency, we assume that $C_1 \cap \operatorname{ri} C_2 = \emptyset$. Let $D = C_1 \cap \operatorname{aff} C_2$. If $D = \emptyset$, we can separate the polyhedral convex sets C_1 and $\operatorname{aff} C_2$ strongly by Corollary 19.3.3, and any strongly separating hyperplane will in particular separate C_1 and C_2 properly without containing C_2.

We assume therefore that $D \neq \emptyset$. Since ri $D \cap$ ri $C_2 = \emptyset$, there exists by Theorem 11.3 a hyperplane H which separates D and C_2 properly. This H cannot contain C_2, for that would imply

$$H \supset \text{aff } C_2 \supset C_2 \cup D$$

contrary to what is meant by "proper" separation. Let C_2' be the intersection of aff C_2 with the closed half-space containing C_2 and having boundary H. Then C_2' is a closed half of aff C_2 such that $C_2' \supset C_2$ and ri $C_2' \supset$ ri C_2. Moreover, C_2' is polyhedral and

$$C_1 \cap \text{ri } C_2' = D \cap \text{ri } C_2' = \emptyset.$$

If actually $C_1 \cap C_2' = \emptyset$, we can separate C_1 and C_2' strongly (by Corollary 19.3.3 again), and the strongly separating hyperplane will in particular separate C_1 and C_2 as required. Hence we can suppose that $C_1 \cap C_2' \neq \emptyset$. In this case $C_1 \cap M \neq \emptyset$, where M is the affine set which is the relative boundary of C_2', i.e. $M = H \cap \text{aff } C_2$. Translating all the sets if necessary, we can suppose that the origin belongs to $C_1 \cap M$, so that M is a subspace and C_2' is a cone. The convex cone K generated by C_1 is polyhedral by Corollary 19.7.1, and $K \cap \text{ri } C_2' = \emptyset$. Let $C_1' = K + M$. Then C_1' is a polyhedral convex cone (Corollary 19.3.2), $C_1' \supset C_1$ and $C_1' \cap C_2' = M$. Express C_1' as the intersection of a finite collection of closed half-spaces $H_1, \ldots, H_p$, where each H_i has the origin on its boundary. Each H_i must contain M. If a given H_i contains a point of ri C_2', it must contain all of C_2' (because C_2' is a cone which is a closed half of an affine set). Since ri C_2' is not contained in C_1', it follows that one of the half-spaces H_i does not contain any point of ri C_2'. The boundary hyperplane of this H_i separates C_1' and C_2' properly and does not meet ri C_2'. Since $C_1 \subset C_1'$ and ri $C_2 \subset$ ri C_2', this hyperplane separates C_1 and C_2 properly and does not contain C_2.   ‖

The separation condition in Theorem 20.2 can be translated into a support function condition:

COROLLARY 20.2.1.   *Let C_1 and C_2 be non-empty convex sets in R^n such that C_1 is polyhedral. In order that $C_1 \cap \text{ri } C_2$ be non-empty, it is necessary and sufficient that every vector x^* which satisfies*

$$\delta^*(x^* \mid C_1) \leq -\delta^*(-x^* \mid C_2)$$

also satisfies

$$\delta^*(x^* \mid C_1) = \delta^*(x^* \mid C_2).$$

PROOF.   Suppose $x^* \neq 0$. By definition, $\delta^*(x^* \mid C_1)$ is the supremum of the linear function $\langle \cdot, x^* \rangle$ on C_1, while $-\delta^*(-x^* \mid C_2)$ is the infimum of $\langle \cdot, x^* \rangle$ on C_2. Thus the numbers α between $\delta^*(x^* \mid C_1)$ and

$-\delta^*(-x^* \mid C_2)$ correspond to the hyperplanes $\{x \mid \langle x, x^* \rangle = \alpha\}$ which separate C_1 and C_2. Such a hyperplane contains all of C_2 if and only if $\alpha = \delta^*(x^* \mid C_2)$. Thus the support function condition in the corollary asserts there is no hyperplane which separates C_1 and C_2 properly without containing C_2. By the theorem, this is equivalent to $C_1 \cap \text{ri } C_2$ being non-empty. ‖

Here is another closure condition which makes use of polyhedral convexity.

THEOREM 20.3. *Let C_1 and C_2 be non-empty convex sets in R^n such that C_1 is polyhedral and C_2 is closed. Suppose that every direction of recession of C_1 whose opposite is a direction of recession of C_2 is actually a direction in which C_2 is linear. Then $C_1 + C_2$ is closed.*

PROOF. The idea is to apply Corollary 20.1.1 to the functions $f_1 = \delta(\cdot \mid C_1)$ and $f_2 = \delta(\cdot \mid C_2)$. If $\text{dom} f_1^* \cap \text{ri} (\text{dom} f_2^*)$ is not empty, Corollary 20.1.1 implies that $f_1 \square f_2$ is closed, and this is the same as $C_1 + C_2$ being closed. Now the sets $K_1 = \text{dom} f_1^*$ and $K_2 = \text{dom} f_2^*$ are just the barrier cones of C_1 and C_2, and K_1 is polyhedral since f_1^* is polyhedral (Theorem 19.2). According to Corollary 20.2.1, $K_1 \cap \text{ri } K_2$ is non-empty if every vector x^* which satisfies

$$\delta^*(x^* \mid K_1) \leq -\delta^*(-x^* \mid K_2)$$

also satisfies

$$\delta^*(x^* \mid K_1) = \delta^*(x^* \mid K_2).$$

The support functions of the barrier cones K_1 and K_2 are simply the indicator functions of the polars of these cones, which are the recession cones 0^+C_1 and 0^+C_2 (Corollary 14.2.1). Thus the support function condition is just the condition that every x^* in $0^+C_1 \cap (-0^+C_2)$ be in 0^+C_2 (and therefore in the lineality space $0^+C_2 \cap (-0^+C_2)$ of C_2). This condition is the same as the direction condition in the theorem, and hence it is satisfied by hypothesis. ‖

COROLLARY 20.3.1. *Let C_1 and C_2 be non-empty convex sets in R^n such that C_1 is polyhedral, C_2 is closed and $C_1 \cap C_2 = \emptyset$. Suppose that C_1 and C_2 have no common directions of recession, except for directions in which C_2 is linear. Then there exists a hyperplane separating C_1 and C_2 strongly.*

PROOF. According to Theorem 11.4, strong separation is possible if $0 \notin \text{cl } (C_1 - C_2)$. We have $0 \notin C_1 - C_2$, of course, since C_1 and C_2 are disjoint. The direction hypothesis implies by the present theorem that $C_1 + (-C_2)$ is closed, i.e. that $C_1 - C_2 = \text{cl } (C_1 - C_2)$. ‖

A useful fact in some applications of polyhedral convexity is that any closed bounded convex set in R^n can be "approximated" as closely as one pleases by a polyhedral convex set:

THEOREM 20.4. *Let C be a non-empty closed bounded convex set, and let D be any convex set such that $C \subset$ int D. Then there exists a polyhedral convex set P such that $P \subset$ int D and $C \subset$ int P.*

PROOF. For each $x \in C$ it is possible to choose a simplex S_x such that $x \in$ int S_x and $S_x \subset$ int D. Since C is closed and bounded, we must have

$$C \subset \bigcup \{\text{int } S_x \mid x \in C_0\}$$

for a certain finite subset C_0 of C. Let

$$P = \text{conv} \bigcup \{S_x \mid x \in C_0\}.$$

Then int $D \supset P$ and int $P \supset C$. Moreover P is a polytope and hence is polyhedral by Theorem 19.1. ∥

The following result has already been cited in connection with the continuity of lower semi-continuous convex functions on locally simplicial sets (Theorem 10.2).

THEOREM 20.5. *Every polyhedral convex set is locally simplicial. In particular, every polytope is locally simplicial.*

PROOF. Let C be a polyhedral convex set, and let $x \in C$. Let U be a simplex with x in its interior. Then $U \cap C$ is a polyhedral convex set. Since $U \cap C$ is also bounded, it can be expressed as the convex hull of a finite set of points (Theorem 19.1). Then by Caratheodory's Theorem (Theorem 17.1)

$$U \cap C = S_1 \cup \cdots \cup S_m,$$

where $S_1, \ldots, S_m$ are certain simplices, and it follows by definition that C is locally simplicial. ∥

Helly's Theorem and Systems of Inequalities

By a system of *convex* inequalities in R^n, we shall mean a system which can be expressed in the form

$$f_i(x) \leq \alpha_i, \qquad \forall i \in I_1,$$

$$f_i(x) < \alpha_i, \qquad \forall i \in I_2,$$

where $I = I_1 \cup I_2$ is an arbitrary index set, each f_i is a convex function on R^n, and $-\infty \leq \alpha_i \leq +\infty$. The set of solutions x to such a system is, of course, a certain convex set in R^n, the intersection of the convex level sets

$$\{x \mid f_i(x) \leq \alpha_i\}, \qquad i \in I_1,$$

$$\{x \mid f_i(x) < \alpha_i\}, \qquad i \in I_2.$$

If every f_i is closed and there are no strict inequalities (i.e. $I_2 = \emptyset$), the set of solutions is closed. The system is said to be *inconsistent* if the set of solutions is empty; otherwise it is *consistent*.

If α_i is finite and g_i is the convex function $f_i - \alpha_i$, the inequality $f_i(x) \leq \alpha_i$ is the same as $g_i(x) \leq 0$, and $f_i(x) < \alpha_i$ is the same as $g_i(x) < 0$. For this reason, one simply considers, for the most part, systems of inequalities in which all the right-hand sides are 0.

Linear equations may be incorporated into a system of convex inequalities by the device of writing $\langle x, b \rangle = \beta$ as a pair of inequalities: $\langle x, b \rangle \leq \beta$ and $\langle x, -b \rangle \leq -\beta$.

The theorems proved below mostly concern the existence of solutions to certain finite and infinite systems of convex inequalities. The systems are generally nonlinear. In the case of finite systems of purely linear inequalities (weak or strict), there is a special, more refined existence and representation theory involving so-called elementary vectors. This will be treated in the next section.

The first result which we establish is a fundamental existence theorem expressed in the form of two mutually exclusive alternatives.

THEOREM 21.1.   *Let C be a convex set, and let $f_1, \ldots, f_m$ be proper convex functions such that $\operatorname{dom} f_i \supset \operatorname{ri} C$. Then one and only one of the following alternatives holds:*

(a) *There exists some $x \in C$ such that*

$$f_1(x) < 0, \ldots, f_m(x) < 0;$$

(b) *There exist non-negative real numbers $\lambda_1, \ldots, \lambda_m$, not all zero, such that*

$$\lambda_1 f_1(x) + \cdots + \lambda_m f_m(x) \geq 0, \qquad \forall x \in C.$$

PROOF.   Assume that (a) holds. Given any x satisfying (a) and any multipliers $\lambda_1 \geq 0, \ldots, \lambda_m \geq 0$, each term of the expression

$$\lambda_1 f_1(x) + \cdots + \lambda_m f_m(x)$$

is non-positive. Terms for which λ_i is not zero are actually negative, so that the whole expression must be negative if the multipliers λ_i are not all zero. Therefore (b) cannot hold.

Assume now that (a) does not hold. We must show that in this case (b) holds. We can suppose C to be non-empty since otherwise (b) holds trivially. Let

$$C_1 = \{z = (\zeta_1, \ldots, \zeta_m) \in R^m \mid \exists x \in C, f_i(x) < \zeta_i \quad \text{for} \quad i = 1, \ldots, m\}.$$

It is easy to see that C_1 is a non-empty convex set in R^m. Since (a) does not hold, C_1 does not contain any z with $\zeta_i \leq 0$ for $i = 1, \ldots, m$. The non-positive orthant

$$C_2 = \{z = (\zeta_1, \ldots, \zeta_m) \mid \zeta_i \leq 0 \quad \text{for} \quad i = 1, \ldots, m\}$$

(which is a convex set) is thus disjoint from C_1, so that C_1 and C_2 can be separated properly by some hyperplane (Theorem 11.3). Thus, for a certain non-zero vector $z^* = (\lambda_1, \ldots, \lambda_m)$ and a certain real number α, we have

$$\alpha \leq \langle z^*, z \rangle = \lambda_1 \zeta_1 + \cdots + \lambda_m \zeta_m, \qquad \forall z \in C_1,$$

$$\alpha \geq \langle z^*, z \rangle = \lambda_1 \zeta_1 + \cdots + \lambda_m \zeta_m, \qquad \forall z \in C_2.$$

Since C_2 is the non-positive orthant, the second of these two conditions implies that $\alpha \geq 0$ and $\lambda_i \geq 0$ for $i = 1, \ldots, m$. (If λ_1, say, were negative, the inequality $\alpha \geq \langle z^*, z \rangle$ would be violated by any z of the form $(\zeta_1, 0, \ldots, 0)$ with ζ_1 sufficiently negative.) From the first condition, we then have that

$$0 \leq \lambda_1 \zeta_1 + \cdots + \lambda_m \zeta_m$$

whenever there exists an $x \in C$ such that $\zeta_i > f_i(x)$ for $i = 1, \ldots, m$.

Therefore, for each x in the set

$$D = C \cap \operatorname{dom} f_1 \cap \cdots \cap \operatorname{dom} f_m,$$

we have

$$0 \leq \lambda_1[f_1(x) + \varepsilon] + \cdots + \lambda_m[f_m(x) + \varepsilon]$$

for arbitrary $\varepsilon > 0$, implying

$$0 \leq \lambda_1 f_1(x) + \cdots + \lambda_m f_m(x).$$

The convex function $f = \lambda_1 f_1 + \cdots + \lambda_m f_m$ is thus non-negative and finite on D. Then f is also non-negative on cl D (Corollary 7.3.3). Since ri $C \subset D$ by hypothesis, we have

$$C \subset \operatorname{cl}(\operatorname{ri} C) \subset \operatorname{cl} D,$$

and consequently $f(x) \geq 0$ for every $x \in C$. Thus (b) holds. ‖

Some condition like the condition ri $C \subset \operatorname{dom} f_i$ in Theorem 21.1 is necessary, as is shown by the following example. Let f_1 be the convex function on R defined by $f_1(x) = -x^{1/2}$ if $x \geq 0$, $f_1(x) = +\infty$ if $x < 0$. Let $f_2(x) = x$ and $C = R$. Then there is no $x \in C$ such that $f_1(x) < 0$ and $f_2(x) < 0$ (i.e. (a) does not hold), and yet the only non-negative multipliers λ_1 and λ_2 such that $\lambda_1 f_1(x) + \lambda_2 f_2(x) \geq 0$ for every $x \in C$ are $\lambda_1 = 0$, $\lambda_2 = 0$ (i.e. (b) does not hold either). The condition ri $C \subset \operatorname{dom} f_1$ is violated in this example.

The next result is a modification of Theorem 21.1 to take special account of affine functions. (Note that Theorem 21.1 can be regarded as the case where $k = m$.)

THEOREM 21.2. *Let C be a convex set, and let $f_1, \ldots, f_k$ be proper convex functions such that $\operatorname{dom} f_i \supset$ ri C. Let $f_{k+1}, \ldots, f_m$ be affine functions such that the system*

$$f_{k+1}(x) \leq 0, \ldots, f_m(x) \leq 0,$$

has at least one solution x in ri C. Then one and only one of the following alternatives holds:

(a) *There exists some $x \in C$ such that*

$$f_1(x) < 0, \ldots, f_k(x) < 0, \qquad f_{k+1}(x) \leq 0, \ldots, f_m(x) \leq 0;$$

(b) *There exist non-negative real numbers $\lambda_1, \ldots, \lambda_m$ such that at least one of the numbers $\lambda_1, \ldots, \lambda_k$ is not zero, and*

$$\lambda_1 f_1(x) + \cdots + \lambda_m f_m(x) \geq 0, \qquad \forall x \in C.$$

PROOF. The proof is like that of Theorem 21.1, except that a more careful separation argument is needed. It is evident, as in the previous proof, that if (a) holds (b) cannot hold. Assume now that (a) does not hold. We shall show that (b) holds. Let C_1 be the set of vectors $z = (\zeta_1, \ldots, \zeta_m)$ in R^m for which there exists an $x \in C$ satisfying

$$f_i(x) < \zeta_i \quad \text{for} \quad i = 1, \ldots, k,$$

and

$$f_i(x) = \zeta_i \quad \text{for} \quad i = k + 1, \ldots, m.$$

(Here $C_1 \neq \emptyset$, provided that $C \neq \emptyset$, as can be supposed without loss of generality.) Let C_2 be the non-positive orthant,

$$C_2 = \{z = (\zeta_1, \ldots, \zeta_m) \mid \zeta_i \leq 0, i = 1, \ldots, m\}.$$

Evidently C_1 and C_2 are convex sets, and C_2 is polyhedral. We have $C_1 \cap C_2 = \emptyset$, because (a) does not hold. According to a special separation theorem which we have proved for polyhedral convex sets, Theorem 20.2, there exists a hyperplane which separates C_1 and C_2 properly and does not contain C_1. Thus there exists a real number α and a vector $z^* = (\lambda_1, \ldots, \lambda_m)$ such that

$$\alpha \leq \lambda_1 \zeta_1 + \cdots + \lambda_m \zeta_m, \qquad \forall (\zeta_1, \ldots, \zeta_m) \in C_1,$$

$$\alpha \geq \lambda_1 \zeta_1 + \cdots + \lambda_m \zeta_m, \qquad \forall (\zeta_1, \ldots, \zeta_m) \in C_2,$$

and such that the first inequality holds as a strict inequality for at least one element of C_1. The second inequality implies that

$$\alpha \geq 0, \qquad \lambda_1 \geq 0, \ldots, \lambda_m \geq 0.$$

The first inequality implies then that

$$\lambda_1 \zeta_1 + \cdots + \lambda_k \zeta_k + \lambda_{k+1} f_{k+1}(x) + \cdots + \lambda_m f_m(x) \geq \alpha$$

whenever $x \in C$ and $\zeta_i > f_i(x)$ for $i = 1, \ldots, k$. Hence

$$\lambda_1 f_1(x) + \cdots + \lambda_k f_k(x) + \lambda_{k+1} f_{k+1}(x) + \cdots + \lambda_m f_m(x) \geq \alpha$$

for every x in the convex set

$$D = C \cap \operatorname{dom} f_1 \cap \cdots \cap \operatorname{dom} f_k.$$

Since the convex function $f = \lambda_1 f_1 + \cdots + \lambda_m f_m$ satisfies $f(x) \geq \alpha$ for every $x \in D$, it also satisfies $f(x) \geq \alpha$ for every $x \in \operatorname{cl} D$ (Corollary 7.3.3). By hypothesis ri $C \subset D$, so for every $x \in C$ we have $f(x) \geq \alpha$, and consequently

$$\lambda_1 f_1(x) + \cdots + \lambda_m f_m(x) \geq 0,$$

as desired. To complete the proof that (b) holds, we need only show that the multipliers $\lambda_1, \ldots, \lambda_k$ are not all zero. Suppose $\lambda_1 = \cdots = \lambda_k = 0$; we shall argue to a contradiction. We have $f = \lambda_{k+1} f_{k+1} + \cdots + \lambda_m f_m$, so f is affine. By the hypothesis of the theorem, there exists at least one $x \in \text{ri } C$ such that $f_i(x) \leq 0$ for $i = k + 1, \ldots, m$, and for such an x, we have $f(x) \leq 0$. But $f(x) \geq \alpha \geq 0$ for every $x \in C$, so this implies that $\alpha = 0$ and that the infimum of f over C is attained on ri C. Then f, being an affine function, must actually be constant on C, i.e. $f(x) = 0$ for every $x \in C$. On the other hand, according to the choice of the hyperplane separating C_1 and C_2, there is some $(\zeta_1, \ldots, \zeta_m) \in C_1$ such that

$$\alpha < \lambda_1 \zeta_1 + \cdots + \lambda_m \zeta_m.$$

Thus there is some $x \in C$ such that

$$\alpha < \lambda_{k+1} f_{k+1}(x) + \cdots + \lambda_m f_m(x) = f(x).$$

For this x we have $f(x) \neq 0$, and the constancy of f is contradicted. ‖

Our main existence theorem for solutions to systems of weak (rather than strict) convex inequalities is the following.

THEOREM 21.3. *Let $\{f_i \mid i \in I\}$ be a collection of closed proper convex functions on R^n, where I is an arbitrary index set. Let C be any non-empty closed convex set in R^n. Assume the functions f_i have no common direction of recession which is also a direction of recession of C. Then one and only one of the following alternatives holds:*

(a) *There exists a vector $x \in C$ such that*

$$f_i(x) \leq 0, \qquad \forall i \in I;$$

(b) *There exist non-negative real numbers λ_i, only finitely many non-zero, such that, for some $\varepsilon > 0$, one has*

$$\sum_{i \in I} \lambda_i f_i(x) \geq \varepsilon, \qquad \forall x \in C.$$

If alternative (b) *holds, the multipliers λ_i can actually be chosen so that at most $n + 1$ of them are non-zero.*

PROOF. Adding the indicator function of C to the given collection if necessary, we can reduce the theorem to the case where $C = R^n$. Obviously (a) and (b) cannot hold simultaneously. Assume that (a) does not hold. We shall prove that (b) holds, and that will establish the theorem.

Let k be the positively homogeneous convex function generated by

$$h = \text{conv } \{f_i^* \mid i \in I\}.$$

The conjugate of k is the indicator function of the convex set $\{x \mid h^*(x) \leq 0\}$ (Theorem 13.5). Since every f_i is closed by hypothesis, we have

$$h^* = \sup \{f_i^{**} \mid i \in I\} = \sup \{f_i \mid i \in I\}$$

by Theorem 16.5. Therefore k^* is the indicator function of

$$D = \{x \mid f_i(x) \leq 0, \forall i \in I\}.$$

But D is empty, because (a) does not hold. Thus k^* is the constant function $+\infty$, and cl $k = k^{**}$ must be the constant function $-\infty$. In particular, $(\text{cl } k)(0) = -\infty$.

The rest of the proof is in two parts: we show that alternative (b) holds if $k(0) = (\text{cl } k)(0)$, and then we prove that $k(0) = (\text{cl } k)(0)$ under our hypothesis about directions of recession.

Suppose $k(0) = -\infty$. Then $h(0) < 0$. Applying Carathéodory's Theorem in the form of Corollary 17.1.3, we get the existence of vectors x_i^* and non-negative scalars λ_i, at most $n + 1$ of which are non-zero, such that

$$\sum_{i \in I} \lambda_i x_i^* = 0, \quad \sum_{i \in I} \lambda_i f_i^*(x_i^*) < 0.$$

For notational simplicity, let us suppose that the indices i corresponding to non-zero scalars λ_i are just the integers $1, \ldots, m$ ($m \leq n + 1$). Setting $y_i^* = \lambda_i x_i^*$, we then have

$$y_1^* + \cdots + y_m^* = 0,$$

$$(f_1^* \lambda_1)(y_1^*) + \cdots + (f_m^* \lambda_m)(y_m^*) < 0.$$

Therefore

$$(f_1^* \lambda_1 \,\square \cdots \square\, f_m^* \lambda_m)(0) < 0.$$

The latter inequality implies a certain property of the function $f = \lambda_1 f_1 + \cdots + \lambda_m f_m$. Namely, we have

$$f^* = \text{cl} \,((\lambda_1 f_1)^* \,\square \cdots \square\, (\lambda_m f_m)^*)$$
$$= \text{cl} \,(f_1^* \lambda_1 \,\square \cdots \square\, f_m^* \lambda_m)$$

by Theorems 16.4 and 16.1, and consequently $f^*(0) < 0$. But, by definition,

$$f^*(0) = \sup_x \{\langle x, 0 \rangle - f(x)\} = -\inf_x f(x).$$

Therefore $\inf f > 0$, i.e. there exists some $\varepsilon > 0$ such that

$$\lambda_1 f_1(x) + \cdots + \lambda_m f_m(x) \geq \varepsilon, \qquad \forall x \in R^n.$$

The multipliers λ_i thus satisfy alternative (b).

We must prove now that $k(0) = (\text{cl } k)(0)$. The effective domain of k is the convex cone generated by the union of the sets $\text{dom} f_i^*$, $i \in I$. If the relative interior of this set contains 0, then certainly $k(0) = (\text{cl } k)(0)$ as desired. Now if $0 \notin \text{ri} (\text{dom } k)$, we can separate 0 from dom k (Theorem 11.3). In this case, therefore, there exists a non-zero vector y such that $\langle y, x^* \rangle \leq 0$ for every $x^* \in \text{dom } k$. We have

$$\langle y, x^* \rangle \leq 0, \forall x^* \in \text{dom} f_i^*, \forall i \in I,$$

so the direction of y is a direction of recession of f_i for every $i \in I$ (Theorem 13.3). But the existence of such directions has been excluded by hypothesis. Thus it is impossible that $0 \notin \text{ri (dom } k)$, and this finishes the proof. ∥

Theorem 21.3 applies both to infinite systems and finite systems. One of its main consequences, as far as infinite systems are concerned, is that existence questions for such systems can be reduced in the following sense to existence questions for finite systems.

COROLLARY 21.3.1. *Let $\{f_i \mid i \in I\}$ be a collection of closed proper convex functions on R^n, where I is an arbitrary index set. Let C be any non-empty closed convex set in R^n. Assume that the functions f_i have no common direction of recession which is also a direction of recession of C. Assume also that, for every $\varepsilon > 0$ and every set of m indices $i_1, \ldots, i_m$ in I with $m \leq n + 1$, the system*

$$f_{i_1}(x) < \varepsilon, \ldots, f_{i_m}(x) < \varepsilon,$$

is satisfied by at least one $x \in C$. Then there exists an $x \in C$ such that

$$f_i(x) \leq 0, \qquad \forall x \in I.$$

PROOF. It is enough to show that alternative (b) of the theorem is incompatible with the assumption here about subsystems having solutions. Under (b), there would exist a non-empty subset I' of I, containing $n + 1$ indices or less such that, for certain positive real numbers λ_i (where $i \in I'$) and a certain $\delta > 0$,

$$\sum_{i \in I'} \lambda_i f_i(x) \geq \delta, \qquad \forall x \in C.$$

Define $\lambda = \sum_{i \in I'} \lambda_i$ and $\varepsilon = \delta/\lambda$. Then

$$\sum_{i \in I'} (\lambda_i/\lambda) f_i(x) \geq \varepsilon, \qquad \forall x \in C,$$

and consequently

$$\sum_{i \in I'} (\lambda_i/\lambda)(f_i(x) - \varepsilon) \geq 0, \qquad \forall x \in C.$$

This is impossible since, by our hypothesis, there exists an $x \in C$ such that $f_i(x) < \varepsilon$ for every $i \in I'$. ∥

Corollary 21.3.1 includes an important classical result, known as Helly's Theorem.

COROLLARY 21.3.2 (Helly's Theorem). *Let $\{C_i \mid i \in I\}$ be a collection of non-empty closed convex sets in R^n, where I is an arbitrary index set. Assume that the sets C_i have no common direction of recession. If every subcollection consisting of $n + 1$ or fewer sets has a non-empty intersection, then the entire collection has a non-empty intersection.*

PROOF. Apply the preceding corollary to the functions $f_i = \delta(\cdot \mid C_i)$, with $C = R^n$. ∥

The recession hypothesis in Helly's Theorem is satisfied of course if one or more of the sets C_i is bounded. As a matter of fact, it is satisfied if and only if some finite subcollection of the C_i's has a bounded intersection, assuming that every finite subcollection has a non-empty intersection. The proof of this fact is left as an exercise.

A counterexample which shows the need for the hypothesis about directions of recession in Theorem 21.3 may be obtained by taking $C = R^2$, $I = \{1, 2\}$ and

$$f_1(x) = (\xi_1^2 + 1)^{1/2} - \xi_2,$$
$$f_2(x) = (\xi_2^2 + 1)^{1/2} - \xi_1,$$

for every $x = (\xi_1, \xi_2)$. The "hyperbolic" convex sets

$$\{x \mid f_1(x) \leq 0\} = \{(\xi_1, \xi_2) \mid \xi_2 \geq (\xi_1^2 + 1)^{1/2}\},$$
$$\{x \mid f_2(x) \leq 0\} = \{(\xi_1, \xi_2) \mid \xi_1 \geq (\xi_2^2 + 1)^{1/2}\},$$

have no point in common, so the first alternative of Theorem 21.3 does not hold. But the second alternative does not hold either, because for every choice of coefficients $\lambda_1 \geq 0$ and $\lambda_2 \geq 0$ the infimum of

$$\lambda_1 f_1(x) + \lambda_2 f_2(x)$$

is 0 along the ray emanating from the origin in the direction of the vector $(1, 1)$. The latter direction happens to be a direction of recession common to f_1 and f_2.

A similar counterexample shows the need for the hypothesis about directions of recession in Helly's Theorem. With f_1 and f_2 as above, consider the collection consisting of all the (non-empty closed convex) subsets of R^2 of the form

$$C_{k,\varepsilon} = \{x \mid f_k(x) \leq \varepsilon\}, \qquad \varepsilon > 0, \qquad k = 1, 2.$$

Every subcollection consisting of three ($= n + 1$) or fewer sets has a non-empty intersection, since each $C_{k,\varepsilon}$ contains the half-line

$$\{\lambda(1, 1) \mid \lambda \geq (1 - \varepsilon^2)/2\varepsilon\},$$

but the intersection of the entire collection is empty.

The two counterexamples just given depend on the fact that the convex sets involved have non-trivial asymptotes. It may be hoped, therefore, that stronger results can be obtained in cases where there is enough linearity or polyhedral convexity present to prevent unsuitable asymptotic behavior. Refinements of this sort will be described in the next two theorems.

THEOREM 21.4. *The hypothesis about directions of recession in Theorem 21.3 and Corollary 21.3.1 may be replaced by the following weaker hypothesis if $C = R^n$. There exists a finite subset I_0 of the index set I such that f_i is* *affine for each $i \in I_0$, and such that each direction which is a direction of recession of f_i for every $i \in I$ is actually a direction in which f_i is constant for every $i \in I \setminus I_0$.*

PROOF. We have to show how to modify the proof of Theorem 21.3 to fit the weaker hypothesis. Only the last part of the proof, concerned with showing that $(\mathrm{cl}\, k)(0) = k(0)$, is affected. It is assumed that alternative (a) does not hold.

Let $I_1 = I \setminus I_0$. It suffices to show that $(\mathrm{cl}\, k)(0) = k(0)$ in the case where I_0 and I_1 are both non-empty. (We could always add new indices to I_0 and I_1 with the corresponding new functions f_i identically zero. The augmented system would still satisfy the hypothesis, and alternative (a) or (b) would hold if and only if the same alternative held for the original system.) For $j = 0, 1$, let k_j be the positively homogeneous convex function generated by

$$h_j = \mathrm{conv}\, \{f_i^* \mid i \in I_j\}.$$

Let k be as in the proof of Theorem 21.3. It is clear that

$$k = \mathrm{conv}\, \{k_0, k_1\}.$$

Since the epigraphs of k_0 and k_1 are convex cones containing the origin, and the convex hull of two such cones is the same as their sum (Theorem 3.8), we have

$$k(x^*) = \inf \{\mu \mid (x^*, \mu) \in K\},$$

where $K = \mathrm{epi}\, k_0 + \mathrm{epi}\, k_1$. Therefore

$$k(x^*) = \inf \{k_0(x_0^*) + k_1(x_1^*) \mid x_1^* + x_2^* = x^*, x_j^* \in \mathrm{dom}\, k_j\}.$$

In particular, setting $x^* = 0$, we have

$$k(0) = \inf \{k_0(-z) + k_1(z) \mid z \in (-\mathrm{dom}\, k_0) \cap \mathrm{dom}\, k_1\}.$$

Holding this fact in reserve for the moment, let us consider the nature of k_0 and $\mathrm{dom}\, k_0$ more closely.

For each $i \in I_0$, the function f_i is affine by hypothesis, say

$$f_i(x) = \langle a_i, x \rangle - \alpha_i.$$

The conjugate function is then of the form

$$f_i^*(x^*) = \delta(x^* \mid a_i) + \alpha_i,$$

i.e. $\mathrm{epi}\, f_i^*$ is a vertical half-line in R^{n+1} emanating upward from the point (a_i, α_i). Thus $\mathrm{epi}\, k_0$ is the convex cone in R^{n+1} generated by the points (a_i, α_i) together with $(0, 1)$. Since

$$k_0(x^*) = \inf \{u \mid (x^*, \mu) \in \mathrm{epi}\, k_0\},$$

it follows that

$$k_0(x^*) = \inf \{\textstyle\sum_{i \in I_0} \lambda_i \alpha_i \mid \lambda_i \geq 0, \textstyle\sum_{i \in I_0} \lambda_i a_i = x^*\},$$

so that k_0 is a finitely generated convex function. Therefore k_0 and dom k_0 are polyhedral (Corollary 19.1.2).

As the next step, we claim that

$$(-\text{dom } k_0) \cap \text{ri (dom } k_1) \neq \emptyset.$$

This will be proved from the recession hypothesis by a separation argument. Suppose the polyhedral convex set $-\text{dom } k_0$ does not meet ri (dom k_1). By Theorem 20.2, there exists a hyperplane which separates $-\text{dom } k_0$ and dom k_1 properly without containing dom k_1. This hyperplane necessarily passes through the origin, since the origin belongs to both dom k_0 and dom k_1. Thus there exists some vector $y \neq 0$ such that

$$\langle y, x^* \rangle \geq 0, \qquad \forall x^* \in (-\text{dom } k_0)$$
$$\langle y, x^* \rangle \leq 0, \qquad \forall x^* \in \text{dom } k_1,$$

where $\langle y, x^* \rangle < 0$ for at least one $x^* \in \text{dom } k_1$. Then

$$\langle y, x^* \rangle \leq 0, \forall x^* \in \text{dom } f_i^*, \forall i \in I,$$

so the direction of y is a direction of recession of f_i for every $i \in I$ (Theorem 13.3). By hypothesis, such a direction is a direction in which f_i is constant for every $i \in I_1$. Thus f_i, for every $i \in I_1$, also has the direction of $-y$ as a direction of recession, so that

$$\langle -y, x^* \rangle \leq 0, \forall x^* \in \text{dom } f_i^*, \forall i \in I_1.$$

But dom k_1 is the convex cone generated by the sets dom f_i^*, $i \in I_1$. Hence we have

$$\langle -y, x^* \rangle \leq 0, \qquad \forall x^* \in \text{dom } k_1.$$

This contradicts the fact that $\langle y, x^* \rangle < 0$ for at least one $x^* \in \text{dom } k_1$. The contradiction shows that the intersection of $-\text{dom } k_0$ and ri (dom k_1) is non-empty as claimed.

If k_0 is improper, it must be identically $-\infty$ on dom k_0 (since it is a polyhedral convex function). If k_1 is improper, it is identically $-\infty$ on ri (dom k_1). In either case, for

$$z \in (-\text{dom } k_0) \cap \text{ri (dom } k_1)$$

we have

$$k_0(-z) + k_1(z) = -\infty,$$

and hence, by the formula for k which was established at the beginning of the proof, $k(0) = -\infty$. Then $k(0) = (\text{cl } k)(0)$, and there is nothing more to prove.

Assume therefore that k_0 and k_1 are both proper. Let g be the polyhedral convex function defined by $g(z) = k_0(-z)$, so that $\operatorname{dom} g = -\operatorname{dom} k_0$. In terms of g, we have

$$k(0) = \inf_z \{g(z) + k_1(z)\}$$
$$= -\sup_z \{\langle 0, z \rangle - (g + k_1)(z)\} = -(g + k_1)^*(0).$$

The conjugate of $(g + k_1)^*$ is given by the formula in Theorem 20.1, since g is polyhedral and $\operatorname{dom} g$ meets ri $(\operatorname{dom} k_1)$. Therefore

$$-k(0) = (g^* \,\square\, k_1^*)(0).$$

Now, by the same argument given for k at the beginning of the proof of Theorem 21.3, k_j^* is the indicator function of the convex set

$$C_j = \{x \mid f_i(x) \leq 0, \forall i \in I_j\}, \qquad j = 0, 1.$$

Consequently g^* is the indicator function of $-C_0$, and $g^* \,\square\, k_1^*$ is the indicator function of $-C_0 + C_1$. The set

$$D = C_0 \cap C_1$$

is empty, because alternative (a) is assumed not to hold. The origin thus does not belong to $-C_0 + C_1$, and we have

$$-k(0) = \delta(0 \mid -C_0 + C_1) = +\infty.$$

This implies again that $(\operatorname{cl} k)(0) = k(0) = -\infty$, and the proof of the theorem is complete. ‖

THEOREM 21.5. *The hypothesis in Helly's Theorem (Corollary 21.3.2) about directions of recession may be replaced by the following weaker hypothesis. There exists a finite subset I_0 of the index set I such that C_i is polyhedral for every $i \in I_0$, and such that each direction which is a direction of recession of C_i for every $i \in I$ is actually a direction in which C_i is linear for every $i \in I \setminus I_0$.*

PROOF. Let $\{C_i \mid i \in I\}$ be a collection satisfying this modified hypothesis for Helly's Theorem. First consider the case where, for every $i \in I_0$, C_i is a closed half-space. For $i \in I_0$, let f_i be an affine function such that

$$C_i = \{x \mid f_i(x) \leq 0\}.$$

For $i \in I \setminus I_0$, let f_i be the indicator function of C_i. Corollary 21.3.1 can be applied to $C = R^n$ and the collection $\{f_i \mid i \in I\}$ under the weaker hypothesis just established in Theorem 21.4, and from this we may conclude that the intersection of $\{C_i \mid i \in I\}$ is not empty. Now in the general polyhedral case, each C_i for $i \in I_0$ can be expressed as the intersection of a certain finite collection of closed half-spaces. Put all these half spaces together into one collection denoted by $\{C_i' \mid i \in I_0'\}$. Let $C_i' = C_i$ for $i \in I \setminus I_0$, and form the collection

$$\{C_i' \mid i \in I'\}, \quad I' = (I \setminus I_0) \cup I_0'.$$

This collection again has the $n + 1$ intersection property that $\{C_i \mid i \in I\}$ has. Any direction in which C_i' is receding for every $i \in I'$ is a direction in which C_i is receding for every $i \in I$, and hence is a direction in which C_i' is linear for every $i \in I' \setminus I_0'$. Thus the collection $\{C_i' \mid i \in I'\}$ satisfies the modified hypothesis for Helly's Theorem, with C_i' a closed half-space for $i \in I_0'$. The result has already been verified in the half-space case, so the intersection of $\{C_i' \mid i \in I'\}$ (which is the same as the intersection of $\{C_i \mid i \in I\}$) must be non-empty. ∥

In the case of a finite collection of convex sets, Helly's Theorem is true without any hypothesis about directions of recession, as we now demonstrate.

THEOREM 21.6.   *Let* $\{C_i \mid i \in I\}$ *be a finite collection of convex sets in* R^n *(not necessarily closed). If every subcollection consisting of* $n + 1$ *or fewer sets has a non-empty intersection, then the entire collection has a non-empty intersection.*

PROOF.   For each subcollection consisting of $n + 1$ or fewer sets, select one of the vectors in the intersection of the subcollection. The selected vectors then make up a certain finite subset S of R^n. For each $i \in I$, let C_i' be the convex hull of the non-empty finite set $S \cap C_i$. Each C_i' is a closed bounded convex set contained in C_i. If J is any set of $n + 1$ or fewer indices in I, the intersection of the sets C_i for $i \in J$ contains one of the vectors in S, and this vector then belongs to the intersection of the sets C_i' for $i \in J$. The collection $\{C_i' \mid i \in I\}$ thus satifies the hypothesis of Corollary 21.3.2 and has a non-empty intersection. This intersection is contained in the intersection of the original collection, so that too is non-empty. ∥

This version of Helly's Theorem is applicable to finite systems of convex inequalities:

COROLLARY 21.6.1.   *Let there be given a system of the form*

$$f_1(x) < 0, \ldots, f_k(x) < 0, \qquad f_{k+1}(x) \leq 0, \ldots, f_m(x) \leq 0,$$

where $f_1, \ldots, f_m$ *are convex functions on* R^n. *(The inequalities may be all strict or all weak.) If every subsystem consisting of* $n + 1$ *or fewer inequalities has a solution in a given convex set* C, *then the system as a whole has a solution in* C.

PROOF. Let $C_0 = C$,

$$C_i = \{x \mid f_i(x) < 0\} \quad \text{for} \quad i = 1, \ldots, k$$

and

$$C_i = \{x \mid f_i(x) \leq 0\} \quad \text{for} \quad i = k + 1, \ldots, m.$$

Apply the theorem to the collection $\{C_i \mid i = 0, \ldots, m\}$. ‖

COROLLARY 21.6.2. *If alternative* (b) *holds in Theorem 21.1 or Theorem 21.2, the numbers λ_i can actually be chosen so that no more than $n + 1$ of them differ from* 0.

PROOF. If alternative (a) fails in Theorem 21.1 or Theorem 21.2, it already fails for a subsystem consisting of $n + 1$ or fewer inequalities, according to Corollary 21.6.1. Alternative (b) holds for this subsystem. ‖

Linear Inequalities

This section treats the theory of finite systems of (weak or strict) linear inequalities. First we shall state various existence results as special cases of relatively difficult theorems that have been established in §21 for more general systems of inequalities. An alternate method of development will then be presented which yields the same results in an elementary way independent of the general theory of convexity.

THEOREM 22.1. *Let $a_i \in R^n$ and $\alpha_i \in R$ for $i = 1, \ldots, m$. Then one and only one of the following alternatives holds:*

(a) *There exists a vector $x \in R^n$ such that*

$$\langle a_i, x \rangle \leq \alpha_i, \qquad i = 1, \ldots, m,$$

(b) *There exist non-negative real numbers $\lambda_1, \ldots, \lambda_m$, such that*

$$\sum_{i=1}^m \lambda_i a_i = 0 \quad \text{and} \quad \sum_{i=1}^m \lambda_i \alpha_i < 0.$$

PROOF. Let $f_i(x) = \langle a_i, x \rangle - \alpha_i$ for $i = 1, \ldots, m$. The hypothesis of Theorem 21.4 is then satisfied with $I_0 = I = \{1, \ldots, m\}$. Hence one and only one of the alternatives in Theorem 21.3 is satisfied (with $C = R^n$). Alternative (a) is the same as the present alternative (a). Alternative (b) in Theorem 21.3 says that, for certain non-negative numbers $\lambda_1, \ldots, \lambda_m$, the function

$$f(x) = \sum_{i=1}^m \lambda_i f_i(x) = \left\langle \sum_{i=1}^m \lambda_i a_i, x \right\rangle - \sum_{i=1}^m \lambda_i \alpha_i$$

has a positive lower bound on R^n. Since f is an affine function, this can only happen if f is a positive constant function, and this is the meaning of alternative (b) in the present theorem. ‖

In situations where strict inequalities are involved, the following result can be used.

THEOREM 22.2. *Let $a_i \in R^n$ and $\alpha_i \in R$ for $i = 1, \ldots, m$, and let k be an integer, $1 \leq k \leq m$. Assume that the system*

$$\langle a_i, x \rangle \leq \alpha_i \quad \text{for} \quad i = k + 1, \ldots, m$$

is consistent. Then one and only one of the following alternatives holds:

(a) *There exists a vector x such that*

$$\langle a_i, x \rangle < \alpha_i \quad for \quad i = 1, \ldots, k,$$
$$\langle a_i, x \rangle \leq \alpha_i \quad for \quad i = k + 1, \ldots, m;$$

(b) *There exist non-negative real numbers* $\lambda_1, \ldots, \lambda_m$, *such that at least one of the numbers* $\lambda_1, \ldots, \lambda_k$ *is not zero, and*

$$\sum_{i=1}^{m} \lambda_i a_i = 0 \quad and \quad \sum_{i=1}^{m} \lambda_i \alpha_i \leq 0.$$

PROOF. Let $f_i(x) = \langle a_i, x \rangle - \alpha_i$, $i = 1, \ldots, m$. The hypothesis of Theorem 21.2 is satisfied with $C = R^n$. Alternatives (a) and (b) of Theroem 21.2 correspond to the present (a) and (b), just as in the preceding proof. ‖

Theorem 22.1 is, of course, applicable to the system in the hypothesis of Theorem 22.2. Thus the hypothesis of Theorem 22.2 fails to be satisfied if and only if there exist non-negative real numbers $\lambda_{k+1}, \ldots, \lambda_m$ such that

$$\sum_{i=k+1}^{m} \lambda_i a_i = 0 \quad and \quad \sum_{i=k+1}^{m} \lambda_i \alpha_i < 0$$

Altogether then, one has a necessary and sufficient condition for the existence of solutions to any finite system of (weak and/or strict) linear inequalities.

An inequality $\langle a_0, x \rangle \leq \alpha_0$ is said to be a *consequence* of the system

$$\langle a_i, x \rangle \leq \alpha_i, \quad i = 1, \ldots, m$$

if it is satisfied by every x which satisfies the system. For example, the inequality $\xi_1 + \xi_2 \geq 0$ is a consequence of the system

$$\xi_i \geq 0, \quad i = 1, 2;$$

this is the case where $(\xi_1, \xi_2) = x$, $a_0 = (-1, -1)$, $a_1 = (-1, 0)$, $a_2 = (0, -1)$ and $\alpha_0 = \alpha_1 = \alpha_2 = 0$.

THEOREM 22.3. *Assume that the system*

$$\langle a_i, x \rangle \leq \alpha_i, \quad i = 1, \ldots, m,$$

is consistent. An inequality $\langle a_0, x \rangle \leq \alpha_0$ *is then a consequence of this system if and only if there exist non-negative real numbers* $\lambda_1, \ldots, \lambda_m$ *such that*

$$\sum_{i=1}^{m} \lambda_i a_i = a_0 \quad and \quad \sum_{i=1}^{m} \lambda_i \alpha_i \leq \alpha_0.$$

PROOF. The inequality $\langle a_0, x \rangle \leq \alpha_0$ is a consequence if and only if the system

$$\langle -a_0, x \rangle < -\alpha_0, \quad \langle a_i, x \rangle \leq \alpha_i \quad for \quad i = 1, \ldots, m,$$

is inconsistent, i.e. has no solution x. By Theorem 22.2, such inconsistency is equivalent to the existence of non-negative real numbers $\lambda_0', \lambda_1', \ldots, \lambda_m'$

such that $\lambda'_0 \neq 0$,

$$\lambda'_0(-a_0) + \lambda'_1 a_1 + \cdots + \lambda'_m a_m = 0.$$

$$\lambda'_0(-\alpha_0) + \lambda'_1 \alpha_1 + \cdots + \lambda'_m \alpha_m \leq 0.$$

This condition is equivalent to the one in the theorem under the relations $\lambda_i = \lambda'_i/\lambda'_0$, $i = 1, \ldots, m$. ‖

COROLLARY 22.3.1 (Farkas' Lemma).   *An inequality* $\langle a_0, x \rangle \leq 0$ *is a consequence of the system*

$$\langle a_i, x \rangle \leq 0, \qquad i = 1, \ldots, m$$

if and only if there exist non-negative real numbers $\lambda_1, \ldots, \lambda_m$ *such that*

$$\sum_{i=1}^{m} \lambda_i a_i = a_0.$$

PROOF.   The hypothesis of Theorem 22.3 is satisfied, because the zero vector satisfies $\langle a_i, x \rangle \leq 0$ for $i = 1, \ldots, m$. ‖

Farkas' Lemma has a simple meaning in terms of polar convex cones. The set of all non-negative linear combinations of $a_1, \ldots, a_m$ is the convex cone K generated by $a_1, \ldots, a_m$, and the solutions x to the system $\langle a_i, x \rangle \leq 0$, $i = 1, \ldots, m$, form the cone K° polar to K. An inequality $\langle a_0, x \rangle \leq 0$ is a consequence of the system if and only if $\langle a_0, x \rangle \leq 0$ for every $x \in K^\circ$, in other words $a_0 \in K^{\circ\circ}$. Farkas' Lemma says that $K^{\circ\circ} = K$. This result could also have been reached in another way. For any convex cone K, one has $K^{\circ\circ} = \operatorname{cl} K$, as shown in §14. Here K is finitely generated, and hence closed (Theorem 19.1), so that $K^{\circ\circ} = K$.

Theorem 22.3 and Farkas' Lemma are also valid for certain *infinite* systems

$$\langle a_i, x \rangle \leq \alpha_i \qquad i \in I,$$

according to Theorem 17.3. The condition for validity is that the set of solutions to the system have a non-empty interior and that the set of points

$$\{(a_i, \alpha_i) \mid i \in I\}$$

be closed and bounded in R^{n+1}.

If one of the inequalities in alternative (a) of Theorem 22.1 is changed to an equality condition, the effect on alternative (b) is to remove the non-negativity requirement from the multiplier λ_i corresponding to this condition. The reader can prove this, as an exercise, by applying Theorem 22.1 to the modified system in which each equation is expressed by a pair of inequalities.

Theorem 22.3 can easily be generalized to mixed systems of weak and strict inequalities (using the same proof), but the statement of the result becomes somewhat complicated in this case.

It is often convenient to express systems of inequalities in matrix notation. In the case of Theorem 22.1, for example, let A be the $m \times n$ matrix whose rows are $a_1, \ldots, a_m$, and let $a = (\alpha_1, \ldots, \alpha_m)$. The system in alternative (a) can be expressed as

$$Ax \leq a,$$

using the convention that a vector inequality is to be interpreted component by component. Setting $w = (\lambda_1, \ldots, \lambda_m)$, we can express the conditions in alternative (b) by

$$w \geq 0, \qquad A^*w = 0, \qquad \langle w, a \rangle < 0,$$

where A^* is the transpose matrix. This formulation makes it clear that (b), like (a), simply concerns the existence of a solution to a certain system which can be expressed by a finite number of linear inequalities. The system in (b) may be called the *alternative* to the system in (a). The two systems are *dual* to each other, in the sense that, no matter what coefficients are chosen, one of the systems has a solution and the other does not.

Other dual pairs of inequality systems can be constructed. For example, an alternative to the system

$$x \geq 0, \qquad Ax = a,$$

may be found from Farkas' Lemma. Let $a_1, \ldots, a_n$ denote the columns of A. The given system concerns non-negative real numbers $\xi_1, \ldots, \xi_n$ (the components of x) such that $\xi_1 a_1 + \cdots + \xi_n a_n = a$. According to Farkas' Lemma, such numbers fail to exist if and only if there is a vector $w \in R^m$ such that $\langle a_j, w \rangle \leq 0$ for $j = 1, \ldots, n$ and $\langle a, w \rangle > 0$. The system

$$A^*w \leq 0, \qquad \langle a, w \rangle > 0,$$

is thus dual to the given system.

For the sake of an exercise, it can be shown that the system

$$x \geq 0, \qquad Ax \leq a,$$

and the system

$$w \geq 0, \qquad A^*w \geq 0, \qquad \langle a, w \rangle < 0,$$

are dual to each other.

There are many dual pairs of systems that can be obtained by considering various mixtures of equations and weak and strict inequalities, and one cannot hope to write them all down. Nevertheless, there is a unified way of approaching the subject which provides one with a good "formula" for the alternative to a given system.

We may say that, in general, we are interested in finding an alternative to a system which can be expressed in the form

$$\zeta_j \in I_j \quad \text{for} \quad j = 1, \ldots, N,$$

$$\zeta_{n+i} = \sum_{j=1}^{n} \alpha_{ij}\zeta_j \quad \text{for} \quad i = 1, \ldots, m,$$

where $N = m + n$, $A = (\alpha_{ij})$ is a given real coefficient matrix, and each I_j is a certain real interval. (By a *real interval*, we mean merely a convex subset of R; thus I_j may be open or closed or neither, and it may consist of just a single number.) For instance, the system $Ax \leq a$ corresponds to the case where $I_j = (-\infty, +\infty)$ for $j = 1, \ldots, n$ and $I_{n+i} = (-\infty, \alpha_i]$ for $i = 1, \ldots, m$. The system $x \geq 0$, $Ax = a$, corresponds to $I_j = [0, +\infty)$ for $j = 1, \ldots, n$ and $I_{n+i} = \{\alpha_i\}$ for $i = 1, \ldots, m$.

The alternative system in each of the cases we have already mentioned concerns a condition on the numbers $\zeta_1^*, \ldots, \zeta_N^*$ which satisfy

$$-\zeta_j^* = \sum_{i=1}^{m} \zeta_{n+i}^* \alpha_{ij} \quad \text{for} \quad j = 1, \ldots, n.$$

In the alternative to $Ax \leq a$, the condition is that

$$\zeta_j^* = 0 \quad \text{for} \quad j = 1, \ldots, n,$$

$$\zeta_{n+i}^* \geq 0 \quad \text{for} \quad i = 1, \ldots, m,$$

$$\zeta_{n+1}^* \alpha_1 + \cdots + \zeta_{n+m}^* \alpha_m < 0.$$

Now notice that this condition is equivalent to the condition that

$$\zeta_1^* \zeta_1 + \cdots + \zeta_N^* \zeta_N < 0$$

for every choice of numbers $\zeta_1, \ldots, \zeta_N$ such that $\zeta_{n+i} \leq \alpha_i$ for $i = 1, \ldots, m$, in other words for every choice of $\zeta_1, \ldots, \zeta_N$ such that $\zeta_j \in I_j$ for $j = 1, \ldots, N$. The condition in the alternative system can thus be expressed simply by

$$\zeta_1^* I_1 + \cdots + \zeta_N^* I_N < 0.$$

(Here "<0" really means $\subset (-\infty, 0)$.) Similarly, in the alternative system to $x \geq 0$, $Ax = a$, the condition is that

$$\zeta_j^* \geq 0 \quad \text{for} \quad j = 1, \ldots, n,$$

$$\zeta_{n+1}^* \alpha_1 + \cdots + \zeta_{n+m}^* \alpha_m > 0,$$

and this can be expressed in terms of the corresponding intervals I_j as

$$\zeta_1^* I_1 + \cdots + \zeta_N^* I_N > 0.$$

One may conjecture that in the general case, no matter what intervals $I_1, \ldots, I_N$ are specified, there will be an alternative system which can be expressed in the form

$$\zeta_1^* I_1 + \cdots + \zeta_N^* I_N > 0,$$
$$-\zeta_j^* = \sum_{i=1}^m \zeta_{n+i}^* \alpha_{ij} \quad \text{for} \quad j = 1, \ldots, n.$$

(Note that there is no loss of generality in taking only the case ">0," because a solution exists for this case if and only if a solution exists for the case "<0.") We shall prove below that this conjecture is true.

The vectors $z = (\zeta_1, \ldots, \zeta_N)$ in R^N which satisfy

$$\zeta_{n+i} = \sum_{j=1}^n \alpha_{ij} \zeta_j \quad \text{for} \quad i = 1, \ldots, m$$

form a certain n-dimensional subspace L of R^N. As pointed out at the end of §1, the orthogonally complementary subspace $L^\perp$ consists of the vectors $z^* = (\zeta_1^*, \ldots, \zeta_N^*)$ such that

$$-\zeta_j^* = \sum_{i=1}^m \zeta_{n+i}^* \alpha_{ij} \quad \text{for} \quad j = 1, \ldots, n.$$

We can simplify matters therefore by speaking of L and $L^\perp$ rather than of linear relationships given by a coefficient matrix A. (Any subspace and its orthogonal complement can, of course, be expressed in terms of a coefficient matrix by taking a Tucker representation, as in §1.)

Thus we may suppose we are given simply a subspace L of R^N and certain real intervals $I_1, \ldots, I_N$. The question is whether there exists a vector $(\zeta_1, \ldots, \zeta_N) \in L$ such that $\zeta_j \in I_j$ for $j = 1, \ldots, N$. The conjecture is that such a vector fails to exist if and only if there exists a vector $(\zeta_1^*, \ldots, \zeta_N^*) \in L^\perp$ such that $\sum_{j=1}^N \zeta_j^* I_j > 0$. The set $\sum_{j=1}^N \zeta_j^* I_j$ is a real interval, by the way, since a linear combination of convex sets is convex.

The conjecture is really in the form of a separation theorem: either the subspace L meets the generalized rectangle

$$C = \{(\zeta_1, \ldots, \zeta_N) \mid \zeta_j \in I_j, j = 1, \ldots, N\},$$

or there exists a hyperplane $\{z \mid \langle z, z^* \rangle = 0\}$ containing L and not meeting C. This furnishes some geometric motivation for the conjecture. The proof given below makes no use of the geometry, however, and it does not invoke any general theorems about convexity. It is a completely independent proof of a combinatorial nature, and it provides a good elementary way of deriving results like Farkas' Lemma directly.

Everything depends on the concept of an "elementary vector" of a subspace L of R^N. Thinking of a vector $z = (\zeta_1, \ldots, \zeta_N)$ as a real-valued function on the set $\{1, \ldots, N\}$ (the value of the function being ζ_j at the point j), one is led to define the *support* of z to be the set of indices j such that $\zeta_j \neq 0$. Each vector in L then has a certain subset of $\{1, \ldots, N\}$ assigned to it, namely its support. An *elementary vector* of L is a non-zero vector z in L whose support is minimal with respect to L, i.e. whose support does not properly include the support of any other non-zero vector in L. If z is an elementary vector of L, then obviously so is λz for any $\lambda \neq 0$.

The concept of an elementary vector is derived from an important example in the theory of graphs, which we shall now sketch for the sake of motivation.

A directed graph G may be defined formally, for purposes of this example, as a triplet $\{E, V, C\}$ where $E = \{e_1, \ldots, e_N\}$ is an abstract set of elements called *edges* (branches, lines, arcs or links), $V = \{v_1, \ldots, v_M\}$ is an abstract set of elements called *vertices* (nodes or points), and $C = (c_{ij})$ is an $M \times N$ matrix, called the *incidence matrix*, whose entries are all $+1$, -1 or 0, with exactly one $+1$ and one -1 in each column. The interpretation of the incidence matrix C is that, for each edge e_j, the vertex at the "initial end" of e_j is the v_i with $c_{ij} = +1$, while the vertex at the "terminal end" of e_j is the v_i with $c_{ij} = -1$.

Given a directed graph G, consider the subspace L of R^N consisting of all the vectors $z = (\zeta_1, \ldots, \zeta_N)$ such that

$$\sum_{j=1}^{N} c_{ij}\zeta_j = 0 \quad \text{for} \quad i = 1, \ldots, M.$$

If we think of G as a representation of a network of pipes, say, and interpret ζ_j as the amount of water per second flowing through pipe e_j (a positive ζ_j being regarded as a flow from the initial vertex of e_j to the terminal vertex and a negative ζ_j as a flow in the oppsoite direction), then the vectors in L can be interpreted as the *circulations* in G, i.e. flows which are conservative at every vertex. The support of such a vector z gives the set of edges e_j in which the associated amount of flow ζ_j is non-zero. An elementary vector of L therefore corresponds to a non-zero circulation in G which is non-zero in a minimal set of edges. Without going into the details, it can be said that the minimal sets of edges in question comprise the *elementary circuits* of G (closed "paths" which do not intersect themselves), and that each elementary vector of L is in fact of the form

$$z = \lambda(\varepsilon_1, \ldots, \varepsilon_N), \qquad \lambda \neq 0,$$

where $(\varepsilon_1, \ldots, \varepsilon_N)$ is the incidence vector for some elementary circuit ($\varepsilon_j = +1$ if the circuit passes through the edge e_j from the initial vertex to the terminal vertex, $\varepsilon_j = -1$ if the circuit passes through e_j in the opposite direction, $\varepsilon_j = 0$ if the circuit does not use e_j at all).

A further important example of elementary vectors can be obtained by considering the orthogonal complement $L^\perp$ of the circulation space L. Of course, $L^\perp$ is the subspace of R^N generated by the rows of the incidence matrix C, or in other words $L^\perp$ consists of the vectors $z^* = (\zeta_1^*, \ldots, \zeta_N^*)$ such that, for some vector $p = (\pi_1, \ldots, \pi_M)$, one has

$$\zeta_j^* = -\sum_{i=1}^{M} \pi_i c_{ij} \quad \text{for} \quad j = 1, \ldots, N.$$

If π_i is interpreted as the "potential" at the vertex v_i, this formula says that ζ_j^* is obtained simply by subtracting the potential at the initial vertex of e_j from the potential at the terminal vertex of e_j. Thus the vectors z^* in $L^\perp$ can be interpreted as the *tensions* in G, with ζ_j^* the amount of tension or potential difference across e_j. The support of such a vector gives the set of edges in which the associated amount of tension is non-zero. An elementary vector of $L^\perp$ corresponds to a tension in G which is non-zero in a minimal set of edges. It can be shown that such sets of edges comprise the so-called *elementary cocircuits* of G, and that the elementary vectors of $L^\perp$ are of the form

$$z^* = \lambda(\varepsilon_1, \ldots, \varepsilon_N), \qquad \lambda \neq 0,$$

where $(\varepsilon_1, \ldots, \varepsilon_N)$ is the incidence vector of some elementary cocircuit. (The elementary cocircuits of G, which correspond to "minimal cuts" of G, can be obtained as follows, assuming for simplicity that G is connected. Take any subset W of the vertex set V such that deletion from G of all the edges with one vertex in W and one vertex not in W would leave a directed graph with exactly two connected components. The elementary cocircuit associated with W consists of the edges just described, with $\varepsilon_j = +1$ if e_j has its initial vertex in W and its terminal vertex not in W, $\varepsilon_j = -1$ if e_j has its terminal vertex in W and its initial vertex not in W, and $\varepsilon_j = 0$ if e_j has neither or both of its vertices in W.)

In the case of a general subspace of R^N which is not the space of all circulations or tensions in some directed graph, it is not necessarily true, of course, that every elementary vector is a multiple of a vector whose components are all $+1$, -1 or 0. Nevertheless the elementary vectors do have certain special properties, as shown in the following lemmas which will be needed in the proof of the main result, Theorem 22.6.

LEMMA 22.4. *Let L be a subspace of R^N. If z and z' are elementary vectors of L having the same support, then z and z' are proportional, i.e. $z' = \lambda z$ for some $\lambda \neq 0$.*

PROOF. Let j be any index in the common support of z and z', and let $\lambda = \zeta_j'/\zeta_j$ (where ζ_j and ζ_j' are the jth components of z and z'). The vector $y = z' - \lambda z$ belongs to L. The support of y is contained in the support of z, and it is properly smaller than the support of z because it does not contain j. Since z is an elementary vector of L, it follows by definition that y has to be the zero vector. Thus $z' = \lambda z$. ∥

COROLLARY 22.4.1. *A subspace L of R^N has only finitely many elementary vectors, up to scalar multiples.*

PROOF. There are only finitely many subsets of $\{1, \ldots, N\}$ appearing as the supports of elementary vectors of L. The correspondence between these sets and the elementary vectors is one-to-one up to scalar multiples by the lemma. ∥

LEMMA 22.5.   *Every vector in a given subspace L can be expressed as a linear combination of elementary vectors of L.*

PROOF.   Let z be any non-zero vector of L. There must exist an elementary vector z_1 of L whose support is contained in that of z. Let j be one of the indices in the support of z_1, and let λ_1 be the quotient of the jth component of z by the jth component of z_1. The vector $z' = z - \lambda_1 z_1$ belongs to L and has its support properly contained in that of z. If z' is elementary (or if $z' = 0$), the expression $z = z' + \lambda_1 z_1$ meets the requirement. Otherwise, we can apply the same argument to z' to get a further decomposition

$$z = (z'' + \lambda_2 z_2) + \lambda_1 z_1,$$

where z_2 is an elementary vector of L and z'' is a vector of L whose support is properly contained in that of z'. (The support of z'' thus contains at least two indices fewer than that of z.) After a finite number of decomposition steps, the required expression of z must result.   ‖

In the proof below, we shall need one other intuitively obvious fact: if $J_1, \ldots, J_m$ are real intervals such that no two are disjoint, then there is a point common to all m intervals. This is a special case of Helly's Theorem (Theorem 21.6 with $n = 1$), but it is such a simple case that we want to point out an easy independent proof. Form a symmetric $m \times m$ matrix A by choosing an element α_{ij} from each intersection $J_i \cap J_j$. Let

$$\beta_1 = \max_i \, (\min_j \alpha_{ij}),$$

$$\beta_2 = \min_j \, (\max_i \alpha_{ij}) = \min_i \, (\max_j \alpha_{ij}).$$

Then $\beta_1 \leq \beta_2$. Let β be any number between β_1 and β_2. For $i = 1, \ldots, m$, we have

$$\min_j \alpha_{ij} \leq \beta \leq \max_j \alpha_{ij},$$

so that β lies between two numbers in J_i. Thus $\beta \in J_i$ for $i = 1, \ldots, m$.

THEOREM 22.6.   *Let L be a subspace of R^N, and let $I_1, \ldots, I_N$ be real intervals. Then one and only one of the following alternatives holds:*

(a) *There exists a vector $z = (\zeta_1, \ldots, \zeta_N) \in L$ such that*

$$\zeta_1 \in I_1, \ldots, \zeta_N \in I_N;$$

(b) *There exists a vector $z^* = (\zeta_1^*, \ldots, \zeta_N^*) \in L^\perp$ such that*

$$\zeta_1^* I_1 + \cdots + \zeta_N^* I_N > 0.$$

If alternative (b) *holds, z^* can actually be chosen to be an elementary vector of $L^\perp$.*

PROOF. Alternatives (a) and (b) cannot both hold, for that would entail the existence of two vectors z and z^* simultaneously satisfying $z^* \perp z$ and $\langle z^*, z \rangle > 0$. Assume that (b), in the stronger form involving elementary vectors, does not hold. In other words, assume that

$$0 \in (\zeta_1^* I_1 + \cdots + \zeta_N^* I_N)$$

for every elementary vector of $L^\perp$. We shall demonstrate that (a) holds. Let p be the number of *non-trivial* intervals among $I_1, \ldots, I_N$, i.e. intervals which contain more than one point but are not simply the whole of $(-\infty, +\infty)$. The proof is by induction on p.

In the case where $p = 0$, we can suppose (for simplicity of notation) that I_j consists of a single number α_j for $j = 1, \ldots, k$, and $I_j = (-\infty, +\infty)$ for $j = k + 1, \ldots, N$. Let L_0 be the subspace of R^N consisting of the vectors $z' = (\zeta_1', \ldots, \zeta_N')$ for which there exists a $z \in L$ with $\zeta_j' = \zeta_j$ for $j = 1, \ldots, k$. The subspace $L_0^\perp$ then consists of the vectors $z^* \in L^\perp$ such that $\zeta_j^* = 0$ for $j = k + 1, \ldots, N$. The elementary vectors of $L_0^\perp$ are just the elementary vectors of $L^\perp$ which belong to $L_0^\perp$. Since by assumption

$$0 \in [\zeta_1^* \alpha_1 + \cdots + \zeta_k^* \alpha_k + \zeta_{k+1}^* (-\infty, \infty) + \cdots + \zeta_N^*(-\infty, \infty)]$$

for every elementary vector z^* of $L^\perp$, we have

$$0 = \zeta_1^* \alpha_1 + \cdots + \zeta_k^* \alpha_k + \zeta_{k+1}^* \cdot 0 + \cdots + \zeta_N^* \cdot 0$$

for every elementary vector z^* of $L_0^\perp$. The vector $(\alpha_1, \ldots, \alpha_k, 0, \ldots, 0)$ is thus orthogonal to all the elementary vectors of $L_0^\perp$. Since $L_0^\perp$ is generated algebraically by its elementary vectors according to Lemma 22.5, we have

$$(\alpha_1, \ldots, \alpha_k, 0, \ldots, 0) \in L_0^{\perp\perp} = L_0.$$

This means there exists a vector $z \in L$ such that $\zeta_j = \alpha_j$ for $j = 1, \ldots, k$. This z satisfies alternative (a).

Now consider a case where at least one of the given intervals is non-trivial, say I_1. Make the induction hypothesis that (a) holds in all the cases where there are fewer non-trivial intervals than in the given case. We shall show that there exists a number $\alpha_1 \in I_1$ such that

$$0 \in (\zeta_1^* \alpha_1 + \zeta_2^* I_2 + \cdots + \zeta_N^* I_N)$$

for every elementary vector z^* of $L^\perp$. This will mean that I_1 can be replaced by a trivial subinterval, so that, by induction, (a) is satisfied.

The $\alpha_1 \in I_1$ which we need merely has to satisfy

$$\alpha_1 \in [\zeta_2^* I_2 + \cdots + \zeta_N^* I_N]$$

for every elementary vector z^* of $L^\perp$ such that $\zeta_1^* = -1$. By Lemma 22.4 there are only finitely many elementary vectors of this type. Denote the set of them by E, and for each $z^* \in E$ let J_{z^*} denote the interval $\zeta_2^* I_2 + \cdots + \zeta_N^* I_N$. To prove the existence of the desired α_1, we must show that the finite collection of intervals consisting of I_1 and the J_{z^*}, $z^* \in E$, has a nonempty intersection. It suffices to show that no two of the intervals in this collection are disjoint. For $z^* \in E$ we have $I_1 \cap J_{z^*} \neq \emptyset$, because

$$0 \in [(-1)I_1 + \zeta_2^* I_2 + \cdots + \zeta_N^* I_N] = -I_1 + J_{z^*}$$

by our assumption that $0 \in \sum_{j=1}^n \zeta_j^* I_j$ for every elementary vector z^* of $L^\perp$. Observe that the latter condition is still satisfied if I_1 is replaced by $(-\infty, +\infty)$. This replacement yields a system with fewer non-trivial intervals than the given system, and for this other system (a) holds by induction. Thus there exists a vector $z \in L$ such that $\zeta_2 \in I_2, \ldots, \zeta_N \in I_N$. For each $z^* \in E$, this z satisfies

$$0 = \langle z^*, z \rangle = (-1)\zeta_1 + \zeta_2^* \zeta_2 + \cdots + \zeta_N^* \zeta_N.$$

Thus $\zeta_1 \in J_{z^*}$ for every $z^* \in E$, and no two of the intervals J_{z^*} can be disjoint. The theorem now follows. ‖

In the case of Theorem 22.6 where L is the space of all circulations in some directed graph G, as described above, alternative (a) asserts the existence of a circulation z such that the amount of flow ζ_j in the edge e_j lies in a certain prescribed interval I_j for every j. Alternative (b), on the other hand, asserts something about the elementary vectors of $L^\perp$, which is the space of all tensions in G. In fact, bearing in mind the relationship between the elementary vectors of $L^\perp$ and the elementary cocircuits of G, we can express (b) as follows: there exists an elementary cocircuit of G whose incidence vector $(\varepsilon_1, \ldots, \varepsilon_N)$ has the property that

$$\varepsilon_1 I_1 + \cdots + \varepsilon_N I_N > 0.$$

Similarly, in the case of Theorem 22.6 where L is the space of all tensions in some directed graph G, so that $L^\perp$ is the space of all circulations in G, alternative (a) asserts the existence of a tension whose amounts lie in prescribed intervals, while alternative (b) asserts the existence of an elementary circuit of G whose incidence vector $(\varepsilon_1, \ldots, \varepsilon_N)$ has the property that

$$\varepsilon_1 I_1 + \cdots + \varepsilon_N I_N > 0.$$

As an application of Theorem 22.6, we shall prove:

THEOREM 22.7 (Tucker's Complementarity Theorem). *Given any subspace L of R^N, there exist a non-negative vector $z = (\zeta_1, \ldots, \zeta_N) \in L$ and a non-negative vector $z^* = (\zeta_1^*, \ldots, \zeta_N^*) \in L^\perp$ such that the supports of z*

and z^* are complementary (i.e. for each index i either $\zeta_i > 0$ and $\zeta_i^* = 0$, or $\zeta_i = 0$ and $\zeta_i^* > 0$). The supports of z and z^* (but not z and z^* themselves) are uniquely determined by L.

PROOF. We note first that, for each index k, one and only one of the following alternatives holds:

(a) There exists a non-negative $z \in L$ such that $\zeta_k > 0$;

(b) There exists a non-negative $z^* \in L^\perp$ such that $\zeta_k^* > 0$.

This is seen by applying Theorem 22.6 to the case where $I_i = [0, +\infty)$ for $i \neq k$ and $I_k = (0, +\infty)$. Now let S be the set of indices k such that (a) holds, and for each $k \in S$ let z_k be one of the non-negative vectors in L whose kth component is positive. Let S^* be the set of indices k such that (b) holds, and for each $k \in S^*$ let z_k^* be one of the non-negative vectors in $L^\perp$ whose kth component is positive. Then S and S^* are complementary subsets of $\{1, \ldots, N\}$, and the non-negative vectors

$$z = \sum_{k \in S} z_k \in L, \qquad z^* = \sum_{k \in S^*} z_k^* \in L^\perp$$

have S and S^* as their supports, respectively.   ‖

Part V · Differential Theory

Directional Derivatives and Subgradients

Convex functions have many useful differential properties, and one of these is the fact that one-sided directional derivatives exist universally. Just as the ordinary two-sided directional derivatives of a differentiable function f can be described in terms of gradient vectors, which correspond to tangent hyperplanes to the graph of f, the one-sided directional derivatives of any proper convex function f, not necessarily differentiable, can be described in terms of "subgradient" vectors, which correspond to supporting hyperplanes to the epigraph of f.

Let f be any function from R^n to $[-\infty, +\infty]$, and let x be a point where f is finite. The *one-sided directional derivative* of f at x with respect to a vector y is defined to be the limit

$$f'(x; y) = \lim_{\lambda \downarrow 0} \frac{f(x + \lambda y) - f(x)}{\lambda},$$

if it exists ($+\infty$ and $-\infty$ being allowed as limits). Note that

$$-f'(x; -y) = \lim_{\lambda \uparrow 0} \frac{f(x + \lambda y) - f(x)}{\lambda},$$

so that the one-sided directional derivative $f'(x; y)$ is two-sided if and only if $f'(x; -y)$ exists and

$$f'(x; -y) = -f'(x; y).$$

Of course, if f is actually differentiable at x, the directional derivatives $f'(x; y)$ are all finite and two-sided, and one has

$$f'(x; y) = \langle \nabla f(x), y \rangle, \qquad \forall y,$$

where $\nabla f(x)$ is the gradient of f at x. (See §25.)

THEOREM 23.1. *Let f be a convex function, and let x be a point where f is finite. For each y, the difference quotient in the definition of $f'(x; y)$ is a* *non-decreasing function of $\lambda > 0$, so that $f'(x; y)$ exists and*

$$f'(x; y) = \inf_{\lambda > 0} \frac{f(x + \lambda y) - f(x)}{\lambda} .$$

Moreover, $f'(x; y)$ is a positively homogeneous convex function of y, with $f'(x; 0) = 0$ and

$$-f'(x; -y) \leq f'(x; y), \qquad \forall y.$$

PROOF. The difference quotient for $\lambda > 0$ can be expressed as $\lambda^{-1}h(\lambda y)$, where $h(y) = f(x + y) - f(x)$. The convex set epi h is obtained by translating epi f so that the point $(x, f(x))$ is moved to $(0, 0)$. On the other hand $\lambda^{-1}h(\lambda y) = (h\lambda^{-1})(y)$, where by definition $h\lambda^{-1}$ is the convex function whose epigraph is λ^{-1} epi h. Since epi h contains the origin, the latter set increases, if anything, as λ^{-1} increases. In other words, for each y, the difference quotient $(h\lambda^{-1})(y)$ decreases if anything as λ decreases. It follows that

$$\inf_{\lambda > 0} (h\lambda^{-1})(y) = f'(x; y), \qquad \forall y.$$

Thus the directional derivative function $f'(x; \cdot)$ exists, and it is the positively homogeneous convex function generated by h. One has $f'(x; 0) = 0$ by definition. Furthermore, given any $\mu_1 > f'(x; -y)$ and any $\mu_2 > f'(x; y)$, one has

$$(1/2)\mu_1 + (1/2)\mu_2 \geq f'(x; (1/2)(-y) + (1/2)y) = 0$$

by convexity. Therefore $-f'(x; -y) \leq f'(x; y)$ for every y.   ‖

Observe that the effective domain of $f'(x; y)$ as a convex function of y is the convex cone generated by the translate $(\operatorname{dom} f) - x$ (which contains the origin).

In the case where f is a convex function on the real line R, the directional derivatives of f at x are completely described by the *right derivative*

$$f'_+(x) = f'(x; 1)$$

and the *left derivative*

$$f'_-(x) = -f'(x; -1).$$

According to Theorem 23.1, f'_+ and f'_- are well-defined throughout dom f, if f is proper, and $f'_-(x) \leq f'_+(x)$. This one-dimensional case will be treated in detail in §24.

A vector x^* is said to be a *subgradient* of a convex function f at a point x if

$$f(z) \geq f(x) + \langle x^*, z - x \rangle, \qquad \forall z.$$

This condition, which we refer to as the *subgradient inequality*, has a simple geometric meaning when f is finite at x: it says that the graph of the affine function $h(z) = f(x) + \langle x^*, z - x \rangle$ is a non-vertical supporting hyperplane to the convex set epi f at the point $(x, f(x))$.

The set of all subgradients of f at x is called the *subdifferential of f at x* and is denoted by $\partial f(x)$. The multivalued mapping $\partial f: x \to \partial f(x)$ is called the *subdifferential* of f. Obviously $\partial f(x)$ is a closed convex set, since by definition $x^* \in \partial f(x)$ if and only if x^* satisfies a certain infinite system of weak linear inequalities (one for each z). In general, $\partial f(x)$ may be empty, or it may consist of just one vector. If $\partial f(x)$ is not empty, f is said to be *subdifferentiable* at x.

For example, the Euclidean norm $f(x) = |x|$ is subdifferentiable at every $x \in R^n$, although it is differentiable only at every $x \neq 0$. The set $\partial f(0)$ consists of all the vectors x^* such that

$$|z| \geq \langle x^*, z \rangle, \qquad \forall z;$$

in other words, it is the Euclidean unit ball. For $x \neq 0$, $\partial f(x)$ consists of the single vector $|x|^{-1}x$. If f is the Tchebycheff norm instead of the Euclidean norm, i.e.

$$f(x) = \max \{|\xi_j| \mid j = 1, \ldots, n\} \quad \text{for} \quad x = (\xi_1, \ldots, \xi_n),$$

it can be seen that

$$\partial f(0) = \text{conv} \{\pm e_1, \ldots, \pm e_n\}$$

(where e_j is the vector forming the jth row of the $n \times n$ identity matrix), while for $x \neq 0$

$$\partial f(x) = \text{conv} \{(\text{sign } \xi_j)e_j \mid j \in J_x\},$$

with

$$J_x = \{j \mid |\xi_j| = f(x)\}.$$

An example of a convex function which is not subdifferentiable everywhere is

$$f(x) = \begin{cases} -(1 - |x|^2)^{1/2} & \text{if } |x| \leq 1, \\ +\infty & \text{otherwise.} \end{cases}$$

This f is subdifferentiable (in fact differentiable) at x when $|x| < 1$, but $\partial f(x) = \emptyset$ when $|x| \geq 1$, even though $x \in \text{dom} f$ for $|x| = 1$.

An important special case in the theory of subgradients is the case where f is the indicator of a non-empty convex set C. By definition, $x^* \in \partial \delta(x \mid C)$ if and only if

$$\delta(z \mid C) \geq \delta(x \mid C) + \langle x^*, z - x \rangle, \qquad \forall z.$$

This condition means that $x \in C$ and $0 \geq \langle x^*, z - x \rangle$ for every $z \in C$, i.e. that x^* is normal to C at x. Thus $\partial \delta(x \mid C)$ *is the normal cone to C at x*

(*empty if* $x \notin C$). The case where C is the non-negative orthant of R^n will be considered at the end of this section.

It will be shown in Theorem 25.1 that $\partial f(x)$ consists of a single vector x^* if and only if the convex function f is finite in a neighborhood of x, differentiable (in the ordinary sense) at x and has x^* as its gradient at x. In this event, of course, $\partial f(x)$ completely describes the directional derivatives of f at x. It turns out, however, that there is a close relationship between $\partial f(x)$ and the directional derivatives of f at x even when $\partial f(x)$ does not consist of just a single vector. This will be demonstrated in the next three theorems.

THEOREM 23.2.   *Let f be a convex function, and let x be a point where f is finite. Then x^* is a subgradient of f at x if and only if*

$$f'(x;y) \geq \langle x^*, y \rangle, \qquad \forall y.$$

In fact, the closure of $f'(x;y)$ as a convex function of y is the support function of the closed convex set $\partial f(x)$.

PROOF.   Setting $z = x + \lambda y$, we can turn the subgradient inequality into the condition that

$$[f(x + \lambda y) - f(x)]/\lambda \geq \langle x^*, y \rangle$$

for every y and $\lambda > 0$. Since the difference quotient decreases to $f'(x;y)$ as $\lambda \downarrow 0$, this inequality is equivalent to the one in the theorem. The theorem now follows from applying Corollary 13.2.1 to the positively homogeneous convex function $f'(x; \cdot)$. ‖

In the one-dimensional case of Theorem 23.2, the subgradients are the slopes x^* of the non-vertical lines in R^2 which pass through $(x, f(x))$ without meeting ri (epi f). These form the closed interval of real numbers between $f'_-(x)$ and $f'_+(x)$.

The consequences of Theorem 23.2 are many. First we give the main results on the existence of subgradients.

THEOREM 23.3.   *Let f be a convex function, and let x be a point where f is finite. If f is subdifferentiable at x, then f is proper. If f is not subdifferentiable at x, there must be some infinite two-sided directional derivative at x, i.e. there must exist some y such that*

$$f'(x;y) = -f'(x, -y) = -\infty;$$

in fact the latter must hold for every y of the form $z - x$ with $z \in$ ri (dom f).

PROOF.   Subdifferentiability at x implies that f majorizes a certain affine function, and hence that f is proper. The set $\partial f(x)$ is empty if and only if its support function is the constant function $-\infty$. This support function is cl $(f'(x; \cdot))$ by the preceding theorem. The closure of a convex function is identically $-\infty$ if and only if the function itself has the value $-\infty$ somewhere. Thus if f is not subdifferentiable at x there must exist some y such that $f'(x; y) = -\infty$ (in which case $-f'(x; -y) = -\infty$ too, since by Theorem 23.1 one always has $-f'(x; -y) \leq f'(x; y)$). In this case, $f'(x; \cdot)$ must have the value $-\infty$ throughout the relative interior of its effective domain D (Theorem 7.2). But D is the union of the convex sets λC over all $\lambda \geq 0$, where C is the translate $(\text{dom} f) - x$, and since $0 \in C$ this implies that

$$C \subset D \subset \text{aff } C.$$

Thus ri $C \subset$ ri D, both relative interiors being interiors relative to the same affine set. This shows that $f'(x; \cdot)$ must have the value $-\infty$ throughout $(\text{dom} f) - x$, and the proof is complete.   ‖

THEOREM 23.4.   *Let f be a proper convex function. For $x \notin \text{dom} f$, $\partial f(x)$ is empty. For $x \in \text{ri}(\text{dom} f)$, $\partial f(x)$ is non-empty, $f'(x; y)$ is closed and proper as a function of y, and*

$$f'(x; y) = \sup \{\langle x^*, y\rangle \mid x^* \in \partial f(x)\} = \delta^*(y \mid \partial f(x)).$$

Finally, $\partial f(x)$ is a non-empty bounded set if and only if $x \in \text{int}(\text{dom} f)$, in which case $f'(x; y)$ is finite for every y.

PROOF.   Taking $z \in \text{dom} f$ in the subgradient inequality, we see that the inequality cannot be satisfied by any x^* when $f(x) = +\infty$. If $x \in \text{ri}(\text{dom} f)$, the effective domain of $f'(x; \cdot)$ is an affine set, the subspace parallel to the affine hull of dom f. Since $f'(x; \cdot)$ vanishes at the origin, it cannot be identically $-\infty$ on this affine set. Hence $f'(x; \cdot)$ is proper (Theorem 7.2) and closed (Corollary 7.4.2). But then $f'(x; \cdot)$ is itself the support function of $\partial f(x)$ by Theorem 23.2, whence the supremum formula and the non-emptiness of $\partial f(x)$. If actually ri $(\text{dom} f) = \text{int}(\text{dom} f)$, the effective domain of $f'(x; \cdot)$ is the whole space, so that the support function $\delta^*(\cdot \mid \partial f(x))$ is finite everywhere. On the other hand, since $\delta^*(\cdot \mid \partial f(x))$ is the closure of $f'(x; \cdot)$, if $\delta^*(\cdot \mid \partial f(x))$ is finite everywhere $f'(x; \cdot)$ must be finite everywhere, implying by Corollary 6.4.1 that $x \in \text{int}(\text{dom} f)$. The last statement of theorem now follows from the fact that a non-empty convex set is bounded if and only if its support function is finite everywhere (Corollary 13.2.2).   ‖

There is also a more geometric way to prove that a proper convex function f is always subdifferentiable on ri $(\text{dom} f)$. For any $x \in \text{ri}(\text{dom} f)$, one has (x, μ) in ri $(\text{epi} f)$ for $f(x) < \mu < \infty$ (Lemma 7.3), whereas $(x, f(x))$ is itself a relative boundary point of epi f. By Theorem 11.6, there exists a non-trivial supporting hyperplane to epi f containing

$(x, f(x))$. This hyperplane cannot be vertical, so it is the graph of some affine function corresponding to a subgradient x^* at x.

An important case to keep in mind is where f is a finite convex function on R^n. Then, at each point x, the subdifferential $\partial f(x)$ is a non-empty closed bounded convex set, $f'(x; \cdot)$ is a finite positively homogeneous convex function, and for each vector y the directional derivative $f'(x; y)$ is the maximum of the various inner products $\langle x^*, y \rangle$ as x^* ranges over $\partial f(x)$.

A generalization of the assertion in Theorem 23.4 that $\partial f(x)$ is bounded when $x \in \text{int}\ (\text{dom}\ f)$ is this: for any $x \in \text{dom}\ f$ such that $\partial f(x) \neq \emptyset$, the recession cone of $\partial f(x)$ is the normal cone to $\text{dom}\ f$ at x. This may be proved as an exercise; the verification will be given later as part of the proof of Theorem 25.6, which explains how $\partial f(x)$ may be constructed from limits of sequences of ordinary gradients when $\text{int}\ (\text{dom}\ f)$ is not empty.

The set of points where a proper convex function is subdifferentiable lies between $\text{dom}\ f$ and $\text{ri}\ (\text{dom}\ f)$ according to Theorem 23.4, but it need not actually be convex. For example, on R^2 let

$$f(\xi_1, \xi_2) = \max \{g(\xi_1), |\xi_2|\},$$

where $g(\xi_1) = 1 - \xi_1^{1/2}$ if $\xi_1 \geq 0$, $g(\xi_1) = +\infty$ if $\xi_1 < 0$. The effective domain of f is the right closed half-plane, and f is subdifferentiable everywhere on this half-plane except in the relative interior of the line segment joining $(0, 1)$ and $(0, -1)$.

Duality is prevalent in the theory of subgradients, due to the following fact.

THEOREM 23.5. *For any proper convex function f and any vector x, the following four conditions on a vector x^* are equivalent to each other:*

(a) $x^* \in \partial f(x)$;
(b) $\langle z, x^* \rangle - f(z)$ *achieves its supremum in z at $z = x$;*
(c) $f(x) + f^*(x^*) \leq \langle x, x^* \rangle$;
(d) $f(x) + f^*(x^*) = \langle x, x^* \rangle$.

If $(\text{cl}\ f)(x) = f(x)$, three more conditions can be added to this list;

(a*) $x \in \partial f^*(x^*)$;
(b*) $\langle x, z^* \rangle - f^*(z^*)$ *achieves its supremum in z^* at $z^* = x^*$;*
(a**) $x^* \in \partial(\text{cl}\ f)(x)$.

PROOF. The subgradient inequality defining (a) can be rewritten as:

$$\langle x, x^* \rangle - f(x) \geq \langle z, x^* \rangle - f(z), \qquad \forall z.$$

This is (b). Since the supremum in (b) is $f^*(x^*)$ by definition, (b) is the same as (c) or (d). Dually, (a*), (b*) and (a**) are equivalent to

$$f^{**}(x) + f^*(x^*) = \langle x, x^* \rangle,$$

and this coincides with (d) when $f(x) = (\mathrm{cl}\, f)(x) = f^{**}(x)$. ‖

COROLLARY 23.5.1. *If f is a closed proper convex function, ∂f^* is the inverse of ∂f in the sense of multivalued mappings, i.e. $x \in \partial f^*(x^*)$ if and only if $x^* \in \partial f(x)$.*

COROLLARY 23.5.2. *If f is a proper convex function and x is a point where f is subdifferentiable, then $(\mathrm{cl}\, f)(x) = f(x)$ and $\partial(\mathrm{cl}\, f)(x) = \partial f(x)$.*

PROOF. In general,

$$f(x) \geq (\mathrm{cl}\, f)(x) = f^{**}(x) \geq \langle x, x^* \rangle - f^*(x^*).$$

If f is subdifferentiable at x, there exists at least one x^* such that (d) holds, implying $f(x) = (\mathrm{cl}\, f)(x)$. Then $\partial(\mathrm{cl}\, f)(x) = \partial f(x)$ by the equivalence of (a) and (a**) in the theorem. ‖

COROLLARY 23.5.3. *Let C be a non-empty closed convex set. Then, for each vector x^*, $\partial\delta^*(x^* \mid C)$ consists of the points x (if any) where the linear function $\langle \cdot, x^* \rangle$ achieves its maximum over C.*

PROOF. Take $f = \delta(\cdot \mid C)$ in the theorem, so that f^* is the support function $\delta^*(\cdot \mid C)$. Invoke the equivalence of (a*) and (b). ‖

COROLLARY 23.5.4. *Let K be a non-empty closed convex cone. Then $x^* \in \partial\delta(x \mid K)$ if and only if $x \in \partial\delta(x^* \mid K^\circ)$. These conditions are equivalent to having*

$$x \in K, \qquad x^* \in K^\circ, \qquad \langle x, x^* \rangle = 0.$$

PROOF. Take $f = \delta(\cdot \mid K), f^* = \delta(\cdot \mid K^\circ)$, in the theorem and invoke the equivalence of (a), (a*) and (d). ‖

We have shown that the support function $\delta^*(\cdot \mid \partial f(x))$ can be obtained by closing the directional derivative function $f'(x; \cdot)$. However, discrepancies can exist between the values of these two functions at certain relative boundary points of their effective domains. These discrepancies have a dual meaning which is disclosed by the study of "approximate subgradients."

Let f be any convex function finite at x. A vector x^* is called an ε-*subgradient* of f at x (where $\varepsilon > 0$) if

$$f(z) \geq (f(x) - \varepsilon) + \langle x^*, z - x \rangle, \qquad \forall z.$$

The set of all such ε-gradients is denoted by $\partial_\varepsilon f(x)$.

Insight into the nature of ε-subgradients can be obtained from the function

$$h(y) = f(x + y) - f(x),$$

whose conjugate is given by

$$h^*(x^*) = f^*(x^*) + f(x) - \langle x, x^* \rangle.$$

Observe that h^* is a non-negative closed convex function on R^n, and that the set of points where h^* vanishes is $\partial f(x)$ (Theorem 23.5). We have $x^* \in \partial_\varepsilon f(x)$ if and only if

$$\varepsilon \geq \langle x^*, y \rangle - h(y), \qquad \forall y.$$

The supremum of $\langle x^*, y \rangle - h(y)$ in y is $h^*(x^*)$; thus

$$\partial_\varepsilon f(x) = \{x^* \mid h^*(x^*) \leq \varepsilon\}.$$

In particular, therefore, $\partial_\varepsilon f(x)$ is a closed convex set. As ε decreases, $\partial_\varepsilon f(x)$ gets smaller, if anything, and the intersection of the nest of sets $\partial_\varepsilon f(x)$, $\varepsilon > 0$, is $\partial f(x)$.

Although $\partial_\varepsilon f(x)$ decreases to $\partial f(x)$ as ε decreases to 0, the supremum $\delta^*(y \mid \partial_\varepsilon f(x))$ of a linear function $\langle \cdot, y \rangle$ on $\partial_\varepsilon f(x)$ does not necessarily decrease all the way to its supremum $\delta^*(y \mid \partial f(x))$ over $\partial f(x)$. This discrepancy corresponds exactly to the possible discrepancy between $f'(x; y)$ and $\delta^*(y \mid \partial f(x))$, as we shall now demonstrate.

THEOREM 23.6. *Let f be a closed proper convex function, and let x be a point where f is finite. Then*

$$f'(x;y) = \lim_{\varepsilon \downarrow 0} \delta^*(y \mid \partial_\varepsilon f(x)).$$

PROOF. Setting $h(y) = f(x + y) - f(x)$ as above, we can express $\partial_\varepsilon f(x)$ as the level set $\{x^* \mid h^*(x^*) - \varepsilon \leq 0\}$. Since $h^* - \varepsilon$ is the conjugate of $h + \varepsilon$, it follows from Theorem 13.5 that $\delta^*(\cdot \mid \partial_\varepsilon f(x))$ is the closure of the positively homogeneous convex function generated by $h + \varepsilon$. Since $h + \varepsilon$ is finite and positive at the origin, the positively homogeneous function generated by $h + \varepsilon$ is itself closed according to Theorem 9.7, and its value at y is the infimum of

$$((h + \varepsilon)\lambda)(y) = \lambda[f(x + \lambda^{-1}y) - f(x) + \varepsilon]$$

over $\lambda > 0$. Replacing λ by its reciprocal, we get the formula

$$\delta^*(y \mid \partial_\varepsilon f(x)) = \inf_{\lambda > 0} \frac{f(x + \lambda y) - f(x) + \varepsilon}{\lambda}.$$

As $\varepsilon \downarrow 0$, this decreases to

$$\inf_{\lambda > 0} \frac{f(x + \lambda y) - f(x)}{\lambda},$$

which is $f'(x; y)$ by Theorem 23.1.   ‖

To illustrate Theorem 23.6, we consider the function

$$f = \operatorname{conv} \{f_1, f_2\}$$

on R^2, where for each $x = (\xi_1, \xi_2)$

$$f_1(x) = \begin{cases} 0 & \text{if } \xi_1^2 + (\xi_2 - 1)^2 \le 1, \\ +\infty & \text{otherwise,} \end{cases}$$

$$f_2(x) = \begin{cases} 1 & \text{if } \xi_1 = 1, \\ +\infty & \text{otherwise.} \end{cases}$$

It can be seen that, for $y = (\eta_1, \eta_2)$,

$$f'(0; y) = \begin{cases} 0 & \text{if } \eta_2 > 0, \text{ or if } \eta_1 = 0 = \eta_2, \\ \eta_1 & \text{if } \eta_1 > 0 \text{ and } \eta_2 = 0, \\ +\infty & \text{if } \eta_2 < 0, \text{ or if } \eta_1 < 0 \text{ and } \eta_2 = 0. \end{cases}$$

The closure of the function $f'(0; \cdot)$, on the other hand, has the value 0 when $\eta_2 \ge 0$ and the value $+\infty$ when $\eta_2 < 0$. The dual meaning of the discrepancies between $f'(0; \cdot)$ and its closure may be seen from an inspection of the sets

$$\partial_\varepsilon f(0) = \{x^* \mid f(0) + f^*(x^*) - \langle 0, x^* \rangle \le \varepsilon\}$$
$$= \{x^* \mid f^*(x^*) \le \varepsilon\}, \qquad \varepsilon > 0.$$

By Theorem 16.5, f^* is the pointwise maximum of f_1^* and f_2^*. We have by direct calculation

$$f_1^*(x^*) = (\xi_1^{*2} + \xi_2^{*2})^{1/2} + \xi_2^*,$$
$$f_2^*(x^*) = \xi_1^* - 1,$$

and therefore $\partial_\varepsilon f(0)$ consists of all the vectors $x^* = (\xi_1^*, \xi_2^*)$ which satisfy

$$\max \{(\xi_1^{*2} + \xi_2^{*2})^{1/2} + \xi_2^*, \xi_1^* - 1\} \le \varepsilon.$$

In other words, $\partial_\varepsilon f(0)$ is for $\varepsilon > 0$ the intersection of the "parabolic" convex set

$$\{x^* \mid \xi_2^* \le (\varepsilon/2) - (\xi_1^{*2}/2\varepsilon)\}$$

and the closed half-space

$$\{x^* \mid \xi_1^* \le 1 + \varepsilon\},$$

whereas

$$\partial f(0) = \partial_0 f(0) = \{x^* \mid \xi_1^* = 0, \xi_2^* \le 0\}.$$

For $y_1 = (1, 0)$, the supremum of $\langle \cdot, y_1 \rangle$ over $\partial_\varepsilon f(0)$ is 1 for all $\varepsilon > 0$, but the supremum over $\partial f(0)$ is just 0. This corresponds to the fact that the values of $f'(0; \cdot)$ and its closure at y_1 are 1 and 0, respectively. Similarly, for $y_2 = (-1, 0)$ the supremum of $\langle \cdot, y_2 \rangle$ over $\partial_\varepsilon f(0)$ is $+\infty$ for all $\varepsilon > 0$, but the supremum over $\partial f(0)$ is 0, and this corresponds to the fact that the values of $f'(0; \cdot)$ and its closure at y_2 are $+\infty$ and 0, respectively.

In classical analysis, one generally expects the gradient of f at x to be orthogonal to the level surface of f through x. An analogous result for subgradients may be stated in terms of normals to convex sets.

THEOREM 23.7. *Let f be a proper convex function. Let x be a point such that f is subdifferentiable at x but f does not achieve its minimum at x. Then the normal cone to $C = \{z \mid f(z) \leq f(x)\}$ at x is the closure of the convex cone generated by $\partial f(x)$.*

PROOF.   The set $\{z \mid f(z) < f(x)\}$ has the same closure as C by Theorem 7.6, since $f(x) > \inf f$ by hypothesis. Hence, for x^* to be normal to C at x, it is necessary and sufficient that $\langle z - x, x^* \rangle \leq 0$ whenever $f(z) < f(x)$. Now the vectors y of form $\lambda(z - x)$ with $\lambda > 0$ and $f(z) < f(x)$ are precisely those such that $f'(x; y) < 0$ (Theorem 23.1). The normal cone K_0 to C at x is thus the polar of the (non-empty) convex cone

$$K = \{y \mid f'(x; y) < 0\}.$$

We have (by Theorem 7.6 and Theorem 23.2)

$$\text{cl } K = \{y \mid \text{cl}_y f'(x; y) \leq 0\} = \{y \mid \delta^*(y \mid \partial f(x)) \leq 0\}$$

$$= \{y \mid \langle y, x^* \rangle \leq 0, \forall x^* \in \partial f(x)\} = K_1^\circ,$$

where K_1 is the convex cone generated by $\partial f(x)$ (consisting of all non-negative multiples of elements of $\partial f(x)$). Thus

$$K_0 = K^\circ = (\text{cl } K)^\circ = K_1^{\circ\circ} = \text{cl } K_1,$$

and this is what we wanted to prove.   ‖

COROLLARY 23.7.1.   *Let f be a proper convex function, and let x be an interior point of dom f such that $f(x)$ is not the minimum of f. A vector x^* is then normal to $C = \{z \mid f(z) \leq f(x)\}$ at x if and only if there exists a $\lambda \geq 0$ such that $x^* \in \lambda \, \partial f(x)$.*

PROOF.   The hypothesis implies by Theorem 23.4 that $\partial f(x)$ is a non-empty closed bounded convex set not containing the origin. In this case, the closure of the convex cone generated by $\partial f(x)$ is simply the union of the sets $\lambda \, \partial f(x)$ for $\lambda \geq 0$ (Corollary 9.6.1).   ‖

It is immediate from the definition of subgradient that

$$\partial(\lambda f)(x) = \lambda \, \partial f(x), \qquad \forall x, \forall \lambda > 0.$$

This formula is also valid trivially for $\lambda = 0$, provided that $\partial f(x) \neq \emptyset$.

A more surprising fact is that the formula

$$\partial(f_1 + \cdots + f_m)(x) = \partial f_1(x) + \cdots + \partial f_m(x), \qquad \forall x,$$

is valid when $f_1, \ldots, f_m$ are proper convex functions whose effective domains overlap sufficiently.

THEOREM 23.8. *Let $f_1, \ldots, f_m$ be proper convex functions on R^n, and let $f = f_1 + \cdots + f_m$. Then*

$$\partial f(x) \supset \partial f_1(x) + \cdots + \partial f_m(x), \qquad \forall x.$$

If the convex sets ri (dom f_i), $i = 1, \ldots, m$, *have a point in common, then actually*

$$\partial f(x) = \partial f_1(x) + \cdots + \partial f_m(x), \qquad \forall x.$$

This condition for equality can be weakened slightly if certain of the functions, say $f_1, \ldots, f_k$, are polyhedral: then it is enough if the sets dom f_i, $i = 1, \ldots, k$, *and* ri (dom f_i), $i = k + 1, \ldots, m$, *have a point in common.*

PROOF. If $x^* = x_1^* + \cdots + x_m^*$, where $x_i^* \in \partial f_i(x)$, we have for every z

$$f(z) = f_1(z) + \cdots + f_m(z) \geq f_1(x) + \langle z - x, x_1^* \rangle + \cdots + f_m(x) + \langle z - x, x_m^* \rangle$$

$$= f(x) + \langle z - x, x_1^* + \cdots + x_m^* \rangle = f(x) + \langle z - x, x^* \rangle,$$

and hence $x^* \in \partial f(x)$. This proves the general inclusion. Assuming the ri (dom f_i) have a point in common, we have f^* given by the last formula in Theorem 16.4. Hence, by Theorem 23.5, $x^* \in \partial f(x)$ if and only if

$$\langle x, x^* \rangle = f_1(x) + \cdots + f_m(x)$$

$$+ \inf \{ f_1^*(x_1^*) + \cdots + f_m^*(x_m^*) \mid x_1^* + \cdots + x_m^* = x^* \},$$

where for each x^* the infimum is attained by some $x_1^*, \ldots, x_m^*$. Thus $\partial f(x)$ consists of the vectors of the form $x_1^* + \cdots + x_m^*$ such that

$$\langle x, x_1^* \rangle + \cdots + \langle x, x_m^* \rangle = f_1(x) + \cdots + f_m(x) + f_1^*(x_1^*) + \cdots + f_m^*(x_m^*).$$

But one always has $\langle x, x_i^* \rangle \leq f_i(x) + f_i^*(x_i^*)$, with equality if and only if $x_i^* \in \partial f_i(x)$. Thus $\partial f(x)$ is the same as $\partial f_1(x) + \cdots + \partial f_m(x)$. In the case where some of the functions are polyhedral, one invokes Theorem 20.1 in place of Theorem 16.4. ‖

COROLLARY 23.8.1. *Let $C_1, \ldots, C_m$ be convex sets in R^n whose relative interiors have a point in common. Then the normal cone to $C_1 \cap \cdots \cap C_m$ at any given point x is $K_1 + \cdots + K_m$, where K_i is the normal cone to C_i*

at x. If certain of the sets, say $C_1, \ldots, C_k$, *are polyhedral, the conclusion holds if merely the sets* $C_1, \ldots, C_k$, ri $C_{k+1}, \ldots$, ri C_m *have a point in common.*

PROOF.   Apply the theorem to the indicator functions $f_i = \delta(\cdot \mid C_i)$.   ‖

Because of the importance of Theorem 23.8 in various applications, it seems worthwhile to give a second proof which does not cover the final assertion (concerning polyhedral convexity), but which invokes only separation theory, rather than Theorem 16.4 or Theorem 20.1.

ALTERNATIVE PROOF.   Carrying over the above proof of the general inclusion in Theorem 23.8, we proceed to show that, when the sets ri (dom f_i), $i = 1, \ldots, m$, have a point in common, then for any $\bar{x}$

$$\partial(f_1 + \cdots + f_m)(\bar{x}) \subset \partial f_1(\bar{x}) + \cdots + \partial f_m(\bar{x}).$$

We shall consider only the case where $m = 2$, since the general case will follow from this by induction (upon application of Theorem 6.5 to the sets dom f_i). Thus, given any $\bar{x}$ and $\bar{x}^*$ such that

$$\bar{x}^* \in \partial(f_1 + f_2)(\bar{x}),$$

we shall show that

$$\bar{x}^* \in \partial f_1(\bar{x}) + \partial f_2(\bar{x}).$$

Replacing f_1 and f_2 by the proper convex functions

$$g_1(x) = f_1(x + \bar{x}) - f_1(\bar{x}) - \langle x, \bar{x}^* \rangle,$$
$$g_2(x) = f_2(x + \bar{x}) - f_2(\bar{x})$$

if necessary, we can reduce the argument to the case where

$$\bar{x} = 0, \quad \bar{x}^* = 0, \quad f_1(0) = 0 = f_2(0),$$

and consequently (since $\bar{x}^* \in \partial(f_1 + f_2)(\bar{x})$ by assumption)

$$\min_x (f_1 + f_2)(x) = (f_1 + f_2)(0) = 0.$$

Let us consider now the convex sets

$$C_1 = \{(x, \mu) \in R^{n+1} \mid \mu \geq f_1(x)\},$$
$$C_2 = \{(x, \mu) \in R^{n+1} \mid \mu \leq -f_2(x)\}.$$

According to Lemma 7.3, we have

$$\text{ri } C_1 = \{(x, \mu) \mid x \in \text{ri } (\text{dom} f_1), \mu > f_1(x)\},$$
$$\text{ri } C_2 = \{(x, \mu) \mid x \in \text{ri } (\text{dom} f_2), \mu < -f_2(x)\},$$

and since the minimum of $f_1 + f_2$ is 0 it follows that

$$\text{ri } C_1 \cap \text{ri } C_2 = \emptyset.$$

Hence C_1 and C_2 can be separated properly by some hyperplane in R^{n+1} (Theorem 11.3). The separating hyperplane cannot be vertical, for if it were its image under the projection $(x, \mu) \to x$ would be a hyperplane in R^n separating $\text{dom} f_1$ and $\text{dom} f_2$ properly, and this is impossible because

$$\text{ri} \,(\text{dom} f_1) \cap \text{ri} \,(\text{dom} f_2) \neq \emptyset$$

(Theorem 11.3). The separating hyperplane must therefore be the graph of an affine function on R^n, in fact a linear function since C_1 and C_2 have the origin of R^{n+1} in common. Thus there exists an $x^* \in R^n$ such that

$$\mu \geq \langle x, x^* \rangle, \qquad \forall (x, \mu) \in C_1.$$

$$\mu \leq \langle x, x^* \rangle, \qquad \forall (x, \mu) \in C_2.$$

The latter conditions can be expressed respectively as

$$f_1(x) \geq f_1(0) + \langle x - 0, x^* \rangle, \qquad \forall x \in R^n,$$

$$f_2(x) \geq f_2(0) + \langle x - 0, -x^* \rangle, \qquad \forall x \in R^n,$$

or in other words,

$$x^* \in \partial f_1(0) \quad \text{and} \quad -x^* \in \partial f_2(0).$$

It follows from this that

$$0 \in \partial f_1(0) + \partial f_2(0),$$

and the proof is complete. ‖

Here is another result which is useful in the calculation of subgradients.

THEOREM 23.9. *Let $f(x) = h(Ax)$, where h is a proper convex function on R^m and A is a linear transformation from R^n to R^m. Then*

$$\partial f(x) \supset A^* \partial h(Ax), \qquad \forall x.$$

If the range of A contains a point of ri $(\text{dom } h)$, or if h is polyhedral and the range of A merely contains a point of $\text{dom } h$, then

$$\partial f(x) = A^* \partial h(Ax), \qquad \forall x.$$

PROOF. If $x^* \in A^* \partial h(Ax)$, then $x^* = A^* y^*$ for some $y^* \in \partial h(Ax)$. For every $z \in R^n$ we have

$$f(z) = h(Az) \geq h(Ax) + \langle y^*, Az - Ax \rangle = f(x) + \langle x^*, z - x \rangle,$$

and hence $x^* \in \partial f(x)$. On the other hand, suppose that the range of A contains a point of ri $(\text{dom } h)$. Then f is proper and

$$f^*(x^*) = \inf \{h^*(y^*) \,|\, A^* y^* = x^*\}$$

by Theorem 16.3, where the infimum is attained by some y^* for each x^*

such that $f^*(x^*) \neq +\infty$. Given any $x^* \in \partial f(x)$, we have

$$f(x) + f^*(x^*) = \langle x, x^* \rangle$$

by Theorem 23.5, and hence there exists a vector y^* such that $A^*y^* = x^*$ and

$$f(x) + h^*(y^*) = \langle x, A^*y^* \rangle.$$

This condition says that

$$h(Ax) + h^*(y^*) = \langle Ax, y^* \rangle,$$

in other words that $y^* \in \partial h(Ax)$ by Theorem 23.5. Thus $x^* \in A^* \partial h(Ax)$. If h is polyhedral, the same proof is valid if merely $Ax \in \text{dom } h$ for some x, because the formula for f^* in terms of h^* can be obtained still from Theorem 16.3 via Corollary 19.3.1. ‖

For polyhedral convex functions, the theory of directional derivatives and subdifferentials is considerably simplified by the following theorem.

THEOREM 23.10. *Let f be a polyhedral convex function, and let x be a point where f is finite. Then f is subdifferentiable at x, and $\partial f(x)$ is a polyhedral convex set. The directional derivative function $f'(x; \cdot)$ is a proper polyhedral convex function, and it is the support function of $\partial f(x)$.*

PROOF.   The polyhedral convex set

$$(\text{epi } f) - (x, f(x))$$

contains the origin, so the convex cone it generates is polyhedral and in particular closed (Corollary 19.7.1). This cone just is the epigraph of $f'(x; \cdot)$, so $f'(x; \cdot)$ is a polyhedral convex function. Since $f'(x; 0) = 0$, $f'(x; \cdot)$ is proper. (A polyhedral convex function which has the value $-\infty$ somewhere cannot have any finite values at all.) It follows in particular that $f'(x; \cdot)$ coincides with the support function of $\partial f(x)$ (Theorem 23.2). This implies that $\partial f(x)$ is a non-empty polyhedral convex set (Corollary 19.2.1). ‖

A polyhedral convex function whose subdifferential appears very often in extremum problems is the indicator function f of the non-negative orthant of R^n:

$$f(x) = \delta(x \mid x \geq 0) = \begin{cases} 0 & \text{if } \quad \xi_1 \geq 0, \ldots, \xi_n \geq 0, \\ +\infty & \text{if not,} \end{cases}$$

where $(\xi_1, \ldots, \xi_n) = x$. The subgradients x^* of f at x form the normal cone to the non-negative orthant at x, so

$$\partial f(x) = \{x^* = (\xi_1^*, \ldots, \xi_n^*) \mid x^* \leq 0, \langle x, x^* \rangle = 0\}.$$

In other words, for this f the relation $x^* \in \partial f(x)$ is equivalent to n complementary slackness conditions:

$$\xi_j \geq 0, \xi_j^* \leq 0, \xi_j \xi_j^* = 0, \qquad i = 1, \ldots, n.$$

Differential Continuity and

Monotonicity

Let f be a closed proper convex function on R^n. The subdifferential mapping ∂f defined in the preceding section assigns to each $x \in R^n$ a certain closed convex subset $\partial f(x)$ of R^n. The *effective domain* of ∂f, which is the set

$$\text{dom } \partial f = \{x \mid \partial f(x) \neq \emptyset\},$$

is not necessarily convex, but it differs very little from being convex, in the sense that

$$\text{ri } (\text{dom} f) \subset \text{dom } \partial f \subset \text{dom} f$$

(Theorem 23.4). The *range* of ∂f as a multivalued mapping is defined by

$$\text{range } \partial f = \cup \{\partial f(x) \mid x \in R^n\}.$$

The range of ∂f is the effective domain of ∂f^* by Corollary 23.5.1, so

$$\text{ri } (\text{dom} f^*) \subset \text{range } \partial f \subset \text{dom} f^*.$$

Certain continuity and monotonicity properties of ∂f and the set

$$\text{graph } \partial f = \{(x, x^*) \in R^{2n} \mid x^* \in \partial f(x)\}$$

will be proved in this section. These properties correspond to continuity and monotonicity properties of the directional derivatives of f, and they imply Lipschitz properties of f itself. Necessary and sufficient conditions will be given in order that a multivalued mapping be the subdifferential mapping of a closed proper convex function.

The one-dimensional case will be treated first, because it is much simpler and it helps to motivate the general results.

THEOREM 24.1. *Let f be a closed proper convex function on R. For convenience, extend the right and left derivative functions f'_+ and f'_- beyond the interval $\text{dom} f$ by setting both $= +\infty$ for points lying to the right of* dom f and both $= -\infty$ *for points lying to the left. Then* f'_+ *and* f'_- *are non-decreasing functions on* R, *finite on the interior of* dom f, *such that*

$$f'_+(z_1) \le f'_-(x) \le f'_+(x) \le f'_-(z_2) \quad \text{when} \quad z_1 < x < z_2.$$

Moreover, for every x *one has*

$$\lim_{z \downarrow x} f'_+(z) = f'_+(x), \qquad \lim_{z \uparrow x} f'_+(z) = f'_-(x),$$

$$\lim_{z \downarrow x} f'_-(z) = f'_+(x), \qquad \lim_{z \uparrow x} f'_-(z) = f'_-(x).$$

PROOF.   For any $x \in \text{dom} f$, we have by definition

$$f'_+(x) = \lim_{z \downarrow x} \frac{f(z) - f(x)}{z - x} = \lim_{\lambda \downarrow 0} \frac{f(x + \lambda) - f(x)}{\lambda} = f'(x; 1),$$

$$f'_-(x) = \lim_{z \uparrow x} \frac{f(z) - f(x)}{z - x} = \lim_{\lambda \downarrow 0} \frac{f(x - \lambda) - f(x)}{-\lambda} = -f'(x; -1).$$

According to Theorem 23.1, these limits exist in the monotone decreasing and monotone increasing sense, respectively, and $f'_-(x) \le f'_+(x)$. (The latter inequality also holds by definition if $x \notin \text{dom} f$.) It is clear, from the monotonicity of the difference quotients, that $f'_+(x) < +\infty$ if and only if x lies to the left of the right endpoint of cl $(\text{dom} f)$, and $f'_-(x) > -\infty$ if and only if x lies to the right of the left endpoint. Thus the points where f'_+ and f'_- are both finite are precisely those in int $(\text{dom} f)$. If y and z are both in dom f and $y < z$, we have

$$f'_+(y) \le \frac{f(z) - f(y)}{z - y} = \frac{f(y) - f(z)}{y - z} \le f'_-(z).$$

If y and z are not both in dom f, and $y < z$, then $f'_+(y) \le f'_-(z)$ by definition. The triple inequality in the theorem now follows. This inequality implies in particular that f'_+ and f'_- are non-decreasing. It further implies that

$$f'_+(x) \le \lim_{z \downarrow x} f'_-(z) \le \lim_{z \downarrow x} f'_+(z).$$

To prove that equality really holds, it suffices to show that the second limit is no greater than $f'_+(x)$ in the case where dom f contains the interval $(x, x + \varepsilon)$ for some $\varepsilon > 0$. (Otherwise equality holds by the extended definition of f'_+ and f'_-.) In this case the limit of $f(z)$ as $z \downarrow x$ is $f(x)$ by Corollary 7.5.1, so that for $x < y < x + \varepsilon$ we have

$$\frac{f(y) - f(x)}{y - x} = \lim_{z \downarrow x} \frac{f(y) - f(z)}{y - z} \ge \lim_{z \downarrow x} f'_+(z).$$

Therefore

$$\lim_{z \downarrow x} f'_+(z) \le \lim_{y \downarrow x} \frac{f(y) - f(x)}{y - x} = f'_+(x).$$

The other two limit formulas in the theorem are proved similarly. ‖
Under the hypothesis of Theorem 24.1, we have

$$\partial f(x) = \{x^* \in R \mid f'_-(x) \le x^* \le f'_+(x)\}$$

for every x, as already pointed out after Theorem 23.2. For example, let

$$f(x) = \begin{cases} |x| - 2(1 - x)^{1/2} & \text{if } -3 \le x \le 1, \\ +\infty & \text{otherwise.} \end{cases}$$

This f is a closed proper convex function on R. We have

$$f'_+(x) = \begin{cases} +\infty & \text{if } x \ge 1, \\ 1 + (1 - x)^{-1/2} & \text{if } 0 \le x < 1, \\ -1 + (1 - x)^{-1/2} & \text{if } -3 \le x < 0, \\ -\infty & \text{if } x < -3, \end{cases}$$

$$f'_-(x) = \begin{cases} +\infty & \text{if } x \ge 1, \\ 1 + (1 - x)^{-1/2} & \text{if } 0 < x < 1, \\ -1 + (1 - x)^{-1/2} & \text{if } -3 < x \le 0, \\ -\infty & \text{if } x \le -3, \end{cases}$$

so that

$$\partial f(x) = \begin{cases} \emptyset & \text{if } x \ge 1, \\ \{1 + (1 - x)^{-1/2}\} & \text{if } 0 < x < 1, \\ [0, 2] & \text{if } x = 0, \\ \{-1 + (1 - x)^{-1/2}\} & \text{if } -3 < x < 0, \\ (-\infty, -1/2] & \text{if } x = -3, \\ \emptyset & \text{if } x < -3. \end{cases}$$

Observe that when the graph of ∂f is drawn it takes the form of a "continuous infinite curve." We shall see in Theorem 24.3 that the graphs of the subdifferential mappings of the closed proper convex functions on R can be characterized in fact as the "complete non-decreasing curves" in R^2.
To see that the limit formulas in Theorem 24.1 can fail when f is not

closed, consider the case where

$$f(x) = \begin{cases} 0 & \text{if } x > 0, \\ 1 & \text{if } x = 0, \\ +\infty & \text{if } x < 0. \end{cases}$$

In this case

$$f'_+(x) = \begin{cases} 0 & \text{if } x > 0, \\ -\infty & \text{if } x \leq 0, \end{cases}$$

and therefore f'_+ is not right-continuous at 0.

When f is closed and proper, each of the functions f'_+ and f'_- determines the other by the limit formulas in the theorem. Indeed, let φ be any function from R to $[-\infty, +\infty]$ such that

$$f'_-(x) \leq \varphi(x) \leq f'_+(x), \qquad \forall x \in R,$$

and let

$$\varphi_+(x) = \lim_{z \downarrow x} \varphi(z), \qquad \varphi_-(x) = \lim_{z \uparrow x} \varphi(z).$$

Then φ is non-decreasing by Theorem 24.1, and one has $f'_+ = \varphi_+$ and $f'_- = \varphi_-$. Thus φ determines ∂f completely. The next theorem shows how φ determines f itself up to an additive constant. (Note that φ can be taken to be finite on the (non-empty) interval $I = $ domain ∂f. Outside of I, φ is necessarily infinite, while on int I it is necessarily finite.)

THEOREM 24.2. *Let $a \in R$, and let φ be a non-decreasing function from R to $[-\infty, +\infty]$ such that $\varphi(a)$ is finite. Let φ_+ and φ_- be the right and left limits of φ as above. Then the function f given by*

$$f(x) = \int_a^x \varphi(t)\, dt$$

is a well-defined closed proper convex function on R such that

$$f'_- = \varphi_- \leq \varphi \leq \varphi_+ = f'_+.$$

Moreover, if g is any other closed proper convex function on R such that $g'_- \leq \varphi \leq g'_+$, then $g = f + \alpha$ for some $\alpha \in R$.

PROOF. Let J be the interval where φ is finite. Since φ is non-decreasing, $f(x)$ is well defined and finite as a Riemann integral for $x \in J$. At finite endpoints of cl J, $f(x)$ is well-defined as a limit of Riemann integrals (or as a Lebesgue integral), while for $x \notin$ cl J the integral is unambiguously $+\infty$. We shall show that f is convex on J. It will then follow from the continuity of the integral on cl J that f is a closed proper convex function on R. Let x and y be points of J, $x < y$, and let $z = (1 - \lambda)x + \lambda y$ with $0 < \lambda < 1$. Then $\lambda = (z - x)/(y - x)$ and $(1 - \lambda) = (y - z)/(y - x)$. We have

$$f(z) - f(x) = \int_x^z \varphi(t)\, dt \leq (z - x)\varphi(z),$$

$$f(y) - f(z) = \int_z^y \varphi(t)\, dt \geq (y - z)\varphi(z).$$

Therefore

$$(1 - \lambda)[f(z) - f(x)] + \lambda[f(z) - f(y)]$$
$$\leq [(1 - \lambda)(z - x) - \lambda(y - z)]\varphi(z) = 0,$$

and we have

$$f(z) \leq (1 - \lambda)f(x) + \lambda f(y).$$

This proves the convexity of f. For any $x \in J$, we have

$$\frac{f(z) - f(x)}{z - x} = \frac{1}{z - x} \int_x^z \varphi(t)\, dt \geq \varphi(x), \qquad \forall z > x,$$

so that $f'_+(x) \geq \varphi(x)$. Similarly, $\varphi(x) \geq f'_-(x)$ for $x \in J$. These two in-equalities also hold trivially when $x \notin J$, so we must have $f'_+ = \varphi_+$ and $f'_- = \varphi_-$ as explained just before the theorem. Now if g is any other closed proper convex function on R such that $g'_- \leq \varphi \leq g'_+$, we also have $g'_+ = \varphi_+$ and $g'_- = \varphi_-$, and hence $g'_+ = f'_+$ and $g'_- = f'_-$. Then

$$\operatorname{ri}(\operatorname{dom} g) = \operatorname{ri}(\operatorname{dom} f) = \operatorname{ri} J$$

by the finiteness properties of left and right derivatives in Theorem 24.1 and the fact that

$$J \subset \operatorname{dom} f \subset \operatorname{cl} J.$$

Inasmuch as f and g are closed, their values on R are completely deter-mined by their values on $\operatorname{ri} J$. Thus we need only show that $g = f + \operatorname{const}$. on $\operatorname{ri} J$. This is trivial if J consists of a single point, so we suppose $\operatorname{ri} J = \operatorname{int} J \neq \emptyset$. On $\operatorname{int} J$, the left and right derivatives of f and g are finite by Theorem 24.1. By the additivity of limits, the left and right derivatives of the function $h = f - g$ on $\operatorname{int} J$ exist and

$$h'_+(x) = f'_+(x) - g'_+(x) = 0,$$

$$h'_-(x) = f'_-(x) - g'_-(x) = 0.$$

Thus the two-sided derivative of h on $\operatorname{int} J$ exists and is identically 0. This implies $f - g = \operatorname{const}$. on $\operatorname{int} J$. $\parallel$

COROLLARY 24.2.1. *Let* f *be a finite convex function on a non-empty open real interval I. Then*

$$f(y) - f(x) = \int_x^y f'_+(t)\, dt = \int_x^y f'_-(t)\, dt$$

for any x and y in I.

PROOF. Extend f to be a closed proper convex function on R, and apply the theorem with $\varphi = f'_+$ or $\varphi = f'_-$. ‖

A *complete non-decreasing curve* is a subset of R^2 of the form

$$\Gamma = \{(x, x^*) \mid x \in R, x^* \in R, \varphi_-(x) \leq x^* \leq \varphi_+(x)\},$$

where φ is some non-decreasing function from R to $[-\infty, +\infty]$ which is not everywhere infinite. Such a set Γ resembles the graph of a continuous non-decreasing function on the interval

$$I = \{x \mid (x, x^*) \in \Gamma \text{ for some } x^*\},$$

except that it may contain vertical segments as well as horizontal segments. It is an elementary exercise to show that, for any complete non-decreasing curve Γ, the mapping $(x, x^*) \to x + x^*$ is one-to-one from Γ onto R and continuous in both directions. Thus Γ is a true curve and is "unbounded at both ends."

The complete non-decreasing curves can be characterized as the *maximal totally ordered* subsets of R^2 with respect to the coordinatewise partial ordering. (In this ordering, a subset Γ of R^2 is totally ordered if and only if, for any two pairs (x_0, x_0^*) and (x_1, x_1^*) in Γ, one has $x_0 \leq x_1$ and $x_0^* \leq x_1^*$, or one has $x_0 \geq x_1$ and $x_0^* \geq x_1^*$. A maximal totally ordered subset is one which is not properly contained in any other totally ordered subset.)

The results in this section furnish the following simple characterization of subdifferential mappings from R to R.

THEOREM 24.3. *The graphs of the subdifferential mappings ∂f of the closed proper convex functions f on R are precisely the complete non-decreasing curves Γ in R^2. Moreover f is uniquely determined by Γ up to an additive constant.*

PROOF. Immediate from Theorem 24.1 and Theorem 24.2. ‖

If Γ is a complete non-decreasing curve, then so is

$$\Gamma^* = \{(x^*, x) \mid (x, x^*) \in \Gamma\}.$$

In fact, if f is a closed proper convex function on R such that $\Gamma = \text{graph } \partial f$, then $\Gamma^* = \text{graph } \partial f^*$ by Theorem 23.5. By the same theorem, Γ consists of the points where the non-negative lower semi-continuous function

$$h(x, x^*) = f(x) + f^*(x^*) - xx^*$$

on R^2 vanishes.

In the general n-dimensional case, the nature of subdifferential mappings is not so easy to picture. Before characterizing such mappings abstractly, we shall establish some fundamental continuity results.

THEOREM 24.4. *Let f be a closed proper convex function on R^n. If $x_1, x_2, \ldots$, and $x_1^*, x_2^*, \ldots$, are sequences such that $x_i^* \in \partial f(x_i)$, where x_i converges to x and x_i^* converges to x^*, then $x^* \in \partial f(x)$. In other words, the graph of ∂f is a closed subset of $R^n \times R^n$.*

PROOF. By Theorem 23.5,

$$\langle x_i, x_i^* \rangle \geq f(x_i) + f^*(x_i^*), \qquad \forall i.$$

Taking the "lim inf" as $i \to \infty$, and using the fact that f and f^* are closed, we get

$$\langle x, x^* \rangle \geq f(x) + f^*(x^*)$$

and hence $x^* \in \partial f(x)$. ‖

THEOREM 24.5. *Let f be a convex function on R^n, and let C be an open convex set on which f is finite. Let $f_1, f_2, \ldots$, be a sequence of convex functions finite on C and converging pointwise to f on C. Let $x \in C$, and let $x_1, x_2, \ldots$, be a sequence of points in C converging to x. Then, for any $y \in R^n$ and any sequence $y_1, y_2, \ldots$, converging to y, one has*

$$\limsup_{i \to \infty} f_i'(x_i; y_i) \leq f'(x; y).$$

Moreover, given any $\varepsilon > 0$, there exists an index i_0 such that

$$\partial f_i(x_i) \subset \partial f(x) + \varepsilon B, \qquad \forall i \geq i_0,$$

where B is the Euclidean unit ball of R^n.

PROOF. Given any $\mu > f'(x; y)$, there exists a $\lambda > 0$ such that $x + \lambda y \in C$ and

$$[f(x + \lambda y) - f(x)]/\lambda < \mu.$$

By Theorem 10.8, $f_i(x_i + \lambda y_i)$ tends to $f(x + \lambda y)$ and $f_i(x_i)$ tends to $f(x)$. Hence, for all sufficiently large indices i, one has

$$[f_i(x_i + \lambda y_i) - f_i(x_i)]/\lambda < \mu.$$

Since

$$f_i'(x_i; y_i) \leq [f_i(x_i + \lambda y_i) - f_i(x_i)]/\lambda,$$

it follows that

$$\limsup_{i \to \infty} f_i'(x_i; y_i) \leq \mu.$$

This is true for any $\mu > f'(x; y)$, so the "lim sup" inequality in the theorem is valid. We may conclude in particular (by taking $y_i = y$ for every i) that

$$\limsup_{i \to \infty} f'_i(x_i; y) \leq f'(x; y), \qquad \forall y \in R^n.$$

The convex functions $f'_i(x_i; \cdot)$ and $f'(x; \cdot)$ are the support functions of the non-empty closed bounded convex sets $\partial f_i(x_i)$ and $\partial f(x)$, respectively (Theorem 23.4), and hence they are finite throughout R^n. Therefore, given any $\varepsilon > 0$, there exists by Corollary 10.8.1 an index i_0 such that

$$f'_i(x_i; y) \leq f'(x; y) + \varepsilon, \qquad \forall y \in B, \qquad \forall i \geq i_0.$$

By positive homogeneity we have

$$f'_i(x_i; y) \leq f'(x; y) + \varepsilon |y|, \qquad \forall y \in R^n, \qquad \forall i \geq i_0,$$

in other words

$$\delta^*(y \mid \partial f_i(x_i)) \leq \delta^*(y \mid \partial f(x)) + \varepsilon \delta^*(y \mid B)$$
$$= \delta^*(y \mid \partial f(x) + \varepsilon B), \qquad \forall y \in R^n, \qquad \forall i \geq i_0.$$

This implies that

$$\partial f_i(x_i) \subset \partial f(x) + \varepsilon B, \qquad \forall i \geq i_0$$

(Corollary 13.1.1). ‖

COROLLARY 24.5.1.   *If f is a proper convex function on R^n, $f'(x; y)$ is an upper semi-continuous function of*

$$(x, y) \in [\text{int } (\text{dom} f) \times R^n].$$

Moreover, given any $x \in$ int (dom f) and any $\varepsilon > 0$, there exists a $\delta > 0$ such that

$$\partial f(z) \subset \partial f(x) + \varepsilon B, \qquad \forall z \in (x + \delta B),$$

where B is the Euclidean unit ball of R^n.

PROOF.   Take $C = $ int (dom f) and $f_i = f$ for every i. ‖

The fact that one generally has only a "lim sup" relation in Theorem 24.5 is illustrated by the case where $C = R$, $f(x) = |x|$ and

$$f_i(x) = |x|^{p_i}, \qquad p_i > 1, \qquad p_i \to 1.$$

The right derivatives $f'_i(0; 1)$ are here all 0, but $f'(0; 1)$ itself is 1. It is clear also from the one-dimension case, of course, that the upper semi-continuity in Corollary 24.5.1 cannot be strengthened in general to continuity (although one does have continuity in y for each fixed $x \in$ int (dom f), since $f'(x; \cdot)$ is a finite convex function on R^n). It will be seen in Theorem 25.4 that the question of the continuity of $f'(x; y)$ in x is closely related to that of the existence of two-sided directional derivatives.

Suppose that f is a proper convex function, and let $x \in$ int (dom f).

Let $x_1, x_2, \ldots$, be a sequence of vectors tending to x, and let $x_i^* \in \partial f(x_i)$ for each i. According to Corollary 24.5.1, the sequence $x_1^*, x_2^*, \ldots$, tends toward the (non-empty closed bounded) set $\partial f(x)$, but it need not actually have a limit, unless $\partial f(x)$ consists of just one vector. More can be said, however, if the sequence $x_1, x_2, \ldots$, approaches x from a single direction, i.e. if the sequence is asymptotic to the half-line emanating from x in the direction of a certain vector y. In this event, according to the theorem below, $x_1^*, x_2^*, \ldots$, must tend toward the portion of the boundary of $\partial f(x)$ consisting of the points x^* at which y is normal to $\partial f(x)$. If there is only one such x^* (and this is true for almost all vectors y, as we shall see in the next section), then $x_1^*, x_2^*, \ldots$, must converge to x^*.

For any $x \in \mathrm{dom}\, f$ and any y such that $f'(x; y)$ is finite, we shall denote the directional derivatives of the convex function $f'(x; \cdot)$ at y by $f'(x; y; \cdot)$. Thus

$$f'(x; y; z) = \lim_{\lambda \downarrow 0} [f'(x; y + \lambda z) - f'(x; y)]/\lambda.$$

Observe that, by the positive homogeneity of $f'(x; \cdot)$,

$$f'(x; y + \lambda z) \leq f'(x; y) + \lambda f'(x; z)$$

and hence

$$f'(x; y; z) \leq f'(x; z), \qquad \forall z.$$

THEOREM 24.6. *Let f be a closed proper convex function and let $x \in \mathrm{dom}\, f$. Let $x_1, x_2, \ldots$, be a sequence in $\mathrm{dom}\, f$ converging to x but distinct from x, and suppose that*

$$\lim_{i \to \infty} |x_i - x|^{-1}(x_i - x) = y,$$

where $f'(x; y) > -\infty$ and the half-line $\{x + \lambda y \mid \lambda \geq 0\}$ meets $\mathrm{int}\,(\mathrm{dom}\, f)$. Then

$$\limsup_{i \to \infty} f'(x_i; z) \leq f'(x; y; z), \qquad \forall z.$$

Moreover, given any $\varepsilon > 0$ there exists an index i_0 such that

$$\partial f(x_i) \subset \partial f(x)_y + \varepsilon B, \qquad \forall i \geq i_0,$$

where B is the Euclidean unit ball and $\partial f(x)_y$ consists of the points $x^* \in \partial f(x)$ such that y is normal to $\partial f(x)$ at x^*.

PROOF. Let $\alpha > 0$ be such that $x + \alpha y$ belongs to $\mathrm{int}\,(\mathrm{dom}\, f)$. We can find a simplex S such that $y \in \mathrm{int}\, S$ and $x + \alpha S \subset \mathrm{int}\,(\mathrm{dom}\, f)$. Let P be the convex hull of x and $x + \alpha S$. Then P is a polytope in $\mathrm{dom}\, f$. Fix any vector z, and choose $\lambda > 0$ so small that $y + \lambda z \in \mathrm{int}\, S$. Set $\varepsilon_i = |x_i - x|$, $y_i = |x_i - x|^{-1}(x_i - x)$ and $u_i = y_i + \lambda z$. By our hypothesis, ε_i tends to 0 and y_i tends to y. It is possible to choose an index i_1 so large that, for every $i \geq i_1$, one has $y_i \in \text{int } S$, $u_i \in \text{int } S$ and $\varepsilon_i < \alpha$. Then, for every $i \geq i_1$, the vectors $x_i = x + \varepsilon_i y_i$ and $x_i + \varepsilon_i \lambda z = x + \varepsilon_i u_i$ belong to P, and we have

$$0 = \varepsilon_i^{-1}[f(x + \varepsilon_i y_i) - f(x)] + \varepsilon_i^{-1}[f(x + \varepsilon_i u_i) - f(x + \varepsilon_i y_i)]$$
$$+ \varepsilon_i^{-1}[f(x) - f(x + \varepsilon_i u_i)]$$
$$\geq f'(x; y_i) + f'(x + \varepsilon_i y_i; u_i - y_i) + f'(x + \varepsilon_i u_i; -u_i).$$

Since $u_i - y_i = \lambda z$, it follows from the relations in Theorem 23.1 that

$$f'(x; y_i) + \lambda f'(x_i; z) \leq -f'(x + \varepsilon_i u_i; -u_i)$$
$$\leq f'(x + \varepsilon_i u_i; u_i) \leq [f(x + \varepsilon_i u_i + \beta u_i) - f(x + \varepsilon_i u_i)]/\beta,$$

where β is an arbitrary number in the open interval $(0, \alpha)$. We take the "lim sup" of both sides of this inequality as $i \to \infty$. Since $u_i \to y + \lambda z$ and $\varepsilon_i \downarrow 0$, we have $x + \varepsilon_i u_i + \beta u_i$ in P, as well as $x + \varepsilon_i u_i$ in P, for sufficiently large indices i. Since the polytope P is a locally simplicial set (Theorem 20.5), f is continuous relative to P (Theorem 10.2). Thus

$$\lim_{i \to \infty} f(x + \varepsilon_i u_i) = f(x),$$

$$\lim_{i \to \infty} f(x + \varepsilon_i u_i + \beta u_i) = f(x + \beta(y + \lambda z)).$$

The vector $x + \alpha y$ belongs to the interior of $\text{dom} f$, so y belongs to the interior of $\text{dom} f'(x; \cdot)$. Hence $f'(x; \cdot)$ is continuous at y (Theorem 10.1), and

$$\lim_{i \to \infty} f'(x; y_i) = f'(x; y).$$

Therefore

$$f'(x; y) + \lambda \limsup_{i \to \infty} f'(x_i; z) \leq [f(x + \beta(y + \lambda z)) - f(x)]/\beta$$

for $0 < \beta < \alpha$. Taking the limit as $\beta \downarrow 0$, we get

$$f'(x; y) + \lambda \limsup_{i \to \infty} f'(x_i; z) \leq f'(x; y + \lambda z).$$

By hypothesis $f'(x; y) > -\infty$, and it follows that

$$\limsup_{i \to \infty} f'(x_i; z) \leq [f'(x; y + \lambda z) - f'(x; y)]/\lambda.$$

This inequality holds for any sufficiently small $\lambda > 0$, and the limit of the difference quotient as λ tends to 0 is $f'(x; y; z)$ by definition. This proves the first assertion of the theorem. The second assertion follows by exactly the same argument as in the proof of Theorem 24.5, with $f'(x; \cdot)$ replaced by $f'(x; y; \cdot)$. By Theorem 23.4 the latter function is finite everywhere (because the convex function $f'(x; \cdot)$ has y in the interior of its effective domain and is finite at y), and it is the support function of the closed convex set $\partial f(x)_y$ described in the theorem (Corollary 23.5.3). ‖

The next theorem describes a boundedness property of ∂f and relates it to Lipschitz properties of f which were established in §10.

THEOREM 24.7. *Let f be a closed proper convex function, and let S be a non-empty closed bounded subset of* int (dom f). *Then the set*

$$\partial f(S) = \cup \{\partial f(x) \mid x \in S\}$$

is non-empty, closed and bounded. The real number

$$\alpha = \sup \{|x^*| \mid x^* \in \partial f(S)\} < \infty$$

has the property that

$$f'(x; z) \leq \alpha |z|, \quad \forall x \in S, \quad \forall z,$$

$$|f(y) - f(x)| \leq \alpha |y - x|, \quad \forall y \in S, \quad \forall x \in S.$$

PROOF. We shall show first that $\partial f(S)$ is bounded. For each $x \in S$, $\partial f(x)$ is non-empty, bounded and has $f'(x; \cdot)$ as its support function (Theorem 23.4). Hence

$$\alpha = \sup_{x^* \in \partial f(S)} \sup_{|z|=1} \langle x^*, z \rangle$$

$$= \sup_{|z|=1} \sup_{x \in S} \sup_{x^* \in \partial f(x)} \langle x^*, z \rangle$$

$$= \sup_{|z|=1} \sup_{x \in S} f'(x; z).$$

Since S is closed and bounded and $f'(x; z)$ is upper semi-continuous in x on S (Corollary 24.5.1), the quantity

$$g(z) = \sup \{f'(x; z) \mid x \in S\}$$

is finite for each z. The function g is the pointwise supremum of a collection of convex functions. Thus g is a finite convex function and must be continuous (Theorem 10.1). It follows that

$$\infty > \sup \{g(z) \mid |z| = 1\} = \alpha,$$

and hence that $\partial f(S)$ is bounded.

To see that $\partial f(S)$ is closed, consider any sequence $x_1^*, x_2^*, \ldots$, in $\partial f(S)$ converging to a point x^*. Choose $x_i \in S$ such that $x_i^* \in \partial f(x_i)$. Since S is closed and bounded, we can suppose (extracting subsequences if necessary) that the sequence $x_1, x_2, \ldots$, converges to a point $x \in S$. Then $x^* \in \partial f(x)$ by Theorem 24.4, so $x^* \in \partial f(S)$ and the closedness of

$\partial f(S)$ is established. For any points x and y in S, $x \neq y$, we have

$$f(y) - f(x) \geq f'(x; y - x) \geq -f'(x; x - y),$$

(Theorem 23.1), and hence

$$f(x) - f(y) \leq f'(x; x - y) = |x - y| \cdot f'(x; z),$$

where $z = |x - y|^{-1}(x - y)$. This z has $|z| = 1$, so $f'(x; z) \leq \alpha$. Thus

$$f(x) - f(y) \leq \alpha |x - y|$$

for any x and y in S, and the theorem follows. ‖

The subdifferentials of convex functions on R^n will now be characterized in terms of a monotonicity property. A multivalued mapping ρ from R^n to R^n will be called *cyclically monotone* if one has

$$\langle x_1 - x_0, x_0^* \rangle + \langle x_2 - x_1, x_1^* \rangle + \cdots + \langle x_0 - x_m, x_m^* \rangle \leq 0$$

for any set of pairs (x_i, x_i^*), $i = 0, 1, \ldots, m$ (m arbitrary) such that $x_i^* \in \rho(x_i)$. A *maximal* cyclically monotone mapping is one whose graph is not properly contained in the graph of any other cyclically monotone mapping.

If f is a proper convex function, ∂f is cyclically monotone. Indeed, if $x_i^* \in \partial f(x_i)$ for $i = 0, \ldots, m$, we have

$$\langle x_1 - x_0, x_0^* \rangle \leq f(x_1) - f(x_0),$$

and so forth, for each of the inner products in the sum in the definition of "cyclically monotone," so that the sum is majorized by

$$[f(x_1) - f(x_0)] + [f(x_2) - f(x_1)] + \cdots + [f(x_0) - f(x_m)] = 0.$$

THEOREM 24.8. *Let ρ be a multivalued mapping from R^n to R^n. In order that there exists a closed proper convex function f on R^n such that $\rho(x) \subset \partial f(x)$ for every x, it is necessary and sufficient that ρ be cyclically monotone.*

PROOF. The necessity is clear, since a subdifferential mapping ∂f is itself cyclically monotone. Suppose on the other hand that ρ is cyclically monotone. Fix any pair (x_0, x_0^*) in the graph of ρ (which can be supposed to be non-empty), and define f on R^n by

$$f(x) = \sup \{\langle x - x_m, x_m^* \rangle + \cdots + \langle x_1 - x_0, x_0^* \rangle\},$$

where the supremum is taken over all finite sets of pairs (x_i, x_i^*), $i = 1$, $\ldots, m$, in the graph of ρ. Since f is the supremum of a certain collection of affine functions (one for each choice of $(x_1, x_1^*), \ldots, (x_m, x_m^*)$), f is a closed convex function. The cyclic monotonicity of ρ implies that $f(x_0) = 0$ and hence that f is proper. Now let x and x^* be any vectors such that $x^* \in \rho(x)$. We shall show that $x^* \in \partial f(x)$. It is enough to show that, for any $\alpha < f(x)$ and any $y \in R^n$, we have

$$f(y) > \alpha + \langle y - x, x^* \rangle.$$

Given $\alpha < f(x)$, there exist (by the definition of f) certain pairs (x_i, x_i^*), $i = 1, \ldots, m$, such that $x_i^* \in \rho(x_i)$ and

$$\alpha < \langle x - x_m, x_m^* \rangle + \cdots + \langle x_1 - x_0, x_0^* \rangle.$$

Setting $x_{m+1} = x$ and $x_{m+1}^* = x^*$, we have

$$f(y) \geq \langle y - x_{m+1}, x_{m+1}^* \rangle + \langle x_{m+1} - x_m, x_m^* \rangle + \cdots + \langle x_1 - x_0, x_0^* \rangle$$
$$> \langle y - x, x^* \rangle + \alpha$$

by the definition of f. This proves that $\rho \subset \partial f$. ‖

THEOREM 24.9. *The subdifferential mappings of the closed proper convex functions on R^n are the maximal cyclically monotone mappings from R^n to R^n. The function is uniquely determined by its subdifferential mapping up to an additive constant.*

PROOF. If ρ is a maximal cyclically monotone mapping, there exists by Theorem 24.8 some closed proper convex function f such that $\rho \subset \partial f$. Since ∂f is itself cyclically monotone, we must actually have $\rho = \partial f$. On the other hand, let f be any closed proper convex function, and let ρ be a cyclically monotone mapping such that $\partial f \subset \rho$. By Theorem 24.8, $\rho \subset \partial g$ for a certain closed proper convex function g. Then $\partial f(x) \subset \partial g(x)$ for every x. To prove the theorem, it will be enough to show that this implies $g = f + \text{const.}$ From the relation $\partial f \subset \partial g$, we have

$$\text{ri} \, (\text{dom} f) \subset \text{dom} \, \partial f \subset \text{dom} \, \partial g \subset \text{dom} \, g$$

(Theorem 23.4). For any $x \in \text{ri} \, (\text{dom} f)$ and $y \in R^n$,

$$f'(x; y) = \sup_{x^* \in \partial f(x)} \langle x^*, y \rangle \leq \sup_{x^* \in \partial g \, (x)} \langle x^*, y \rangle \leq g'(x; y)$$

(Theorem 23.4 and Theorem 23.2). It follows that, for any x_1 and x_2 in $\text{ri} \, (\text{dom} f)$, the convex functions h and k defined by

$$h(\lambda) = f((1 - \lambda)x_1 + \lambda x_2), \qquad k(\lambda) = g((1 - \lambda)x_1 + \lambda x_2),$$

have the property that

$$k'_-(\lambda) \leq h'_-(\lambda) \leq h'_+(\lambda) \leq k'_+(\lambda), \qquad 0 \leq \lambda \leq 1.$$

By Theorem 6.4, the interval $I = \text{int} \, (\text{dom} \, h)$ is non-empty, and we have

$$[0, 1] \subset I = \{\lambda \mid (1 - \lambda)x_1 + \lambda x_2 \in \text{ri} \, (\text{dom} f)\}$$
$$\subset \{\lambda \mid (1 - \lambda)x_1 + \lambda x_2 \in \text{dom} \, g\} = \text{dom} \, k.$$

Hence by Corollary 24.2.1

$$f(x_2) - f(x_1) = h(1) - h(0) = \int_0^1 h'_+(\lambda) \, d\lambda = k(1) - k(0) = g(x_2) - g(x_1).$$

Thus there exists a real constant α such that $g(x) = f(x) + \alpha$ for every $x \in \text{ri}(\text{dom} f)$. Since f and g are closed, we must actually have $g(x) = f(x) + \alpha$ for every $x \in \text{cl}(\text{dom} f)$ by Corollary 7.5.1. For $x \notin \text{cl}(\text{dom} f)$, $f(x) = +\infty$ so that $g(x) \leq f(x) + \alpha$ trivially; equality must be proved. We shall make use of a dual argument. For the conjugate functions f^* and g^*, we have

$$\partial f^*(x^*) = (\partial f)^{-1}(x^*) \subset (\partial g)^{-1}(x^*) = \partial g^*(x^*),$$

so there exists a real constant α^* such that $g^*(x^*) \leq f^*(x^*) + \alpha^*$ for every x^*, with equality for $x^* \in \text{cl}(\text{dom} f^*)$. For any vectors x and x^* such that $x^* \in \partial f(x)$, we have $x^* \in \partial g(x)$ too and hence

$$f(x) + f^*(x^*) = \langle x, x^* \rangle = g(x) + g^*(x^*);$$

moreover $x \in \text{dom} f$ and $x^* \in \text{dom} f^*$, so this implies $\alpha^* = -\alpha$. Thus $g^* \leq f^* - \alpha = (f + \alpha)^*$. Since the conjugacy correspondence is order-inverting, we may conclude that $g \geq f + \alpha$. But we already had $g \leq f + \alpha$. Thus $g = f + \alpha$. ‖

A multivalued mapping ρ from R^n to R^n is said to be *monotone* if

$$\langle x_1 - x_0, x_1^* - x_0^* \rangle \geq 0$$

for every (x_0, x_0^*) and (x_1, x_1^*) in the graph of ρ. This condition corresponds to the case where $m = 1$ in the definition of cyclic monotonicity; thus every cyclically monotone mapping is in particular a monotone mapping.

When $n = 1$, the monotone mappings are simply the mappings whose graphs are totally ordered in R^2 with respect to the coordinatewise partial ordering, so that the *maximal* monotone mappings correspond to the complete non-decreasing curves Γ. It follows from Theorem 24.3 and Theorem 24.9 that, when $n = 1$, the monotone mappings and the cyclically monotone mappings are the same. However, when $n > 1$ there exist monotone mappings which are not cyclically monotone. For example, when ρ is the (single-valued) linear transformation from R^n to R^n corresponding to an $n \times n$ matrix Q, ρ is cyclically monotone if and only if Q is symmetric and positive semi-definite (as may be deduced from Theorem 24.9 as an exercise). Yet ρ is monotone if merely

$$\langle x_1 - x_0, Q(x_1 - x_0) \rangle \geq 0, \qquad \forall x_0, x_1,$$

i.e. if the symmetric part $(1/2)(Q + Q^*)$ of Q is positive semi-definite.

It will be proved in §31 (Corollary 31.5.2) that the subdifferential mapping ∂f of any closed proper convex function f is also a maximal monotone mapping. (Note that this is *not* immediate from Theorem 24.9 and the fact that every cyclically monotone mapping is a monotone mapping.) Other examples of maximal monotone mappings will be constructed in §37 from the subdifferential mappings of saddle-functions.

Differentiability of Convex Functions

Let f be a function from R^n to $[-\infty, +\infty]$, and let x be a point where f is finite. According to the usual definition, f is *differentiable* at x if and only if there exists a vector x^* (necessarily unique) with the property that

$$f(z) = f(x) + \langle x^*, z - x \rangle + o(|z - x|),$$

or in other words

$$\lim_{z \to x} \frac{f(z) - f(x) - \langle x^*, z - x \rangle}{|z - x|} = 0.$$

Such an x^*, if it exists, is called the *gradient* of f at x and is denoted by $\nabla f(x)$.

Suppose that f is differentiable at x. Then by definition, for any $y \neq 0$,

$$0 = \lim_{\lambda \downarrow 0} \frac{f(x + \lambda y) - f(x) - \langle \nabla f(x), \lambda y \rangle}{\lambda |y|}$$

$$= [f'(x; y) - \langle \nabla f(x), y \rangle]/|y|.$$

Therefore $f'(x; y)$ exists and is a linear function of y:

$$f'(x; y) = \langle \nabla f(x), y \rangle, \qquad \forall y.$$

In particular, for $j = 1, \ldots, n$,

$$\langle \nabla f(x), e_j \rangle = \lim_{\lambda \to 0} \frac{f(x + \lambda e_j) - f(x)}{\lambda} = \frac{\partial f}{\partial \xi_j}(x),$$

where e_j is the vector forming the jth row of the $n \times n$ identity matrix, and ξ_j denotes the jth component of x. It follows that

$$\nabla f(x) = \left(\frac{\partial f}{\partial \xi_1}(x), \ldots, \frac{\partial f}{\partial \xi_n}(x) \right),$$

so that for any $y = (\eta_1, \ldots, \eta_n)$

$$f'(x; y) = \frac{\partial f}{\partial \xi_1}(x)\eta_1 + \cdots + \frac{\partial f}{\partial \xi_n}(x)\eta_n.$$

In the case where f is convex, one may ask how the concept of "gradient" is related to concept of "subgradient" which has been developed in §23 and §24. The relationship turns out to be very simple.

THEOREM 25.1. *Let f be a convex function, and let x be a point where f is finite. If f is differentiable at x, then $\nabla f(x)$ is the unique subgradient of f at x, so that in particular*

$$f(z) \geq f(x) + \langle \nabla f(x), z - x \rangle, \qquad \forall z.$$

Conversely, if f has a unique subgradient at x, then f is differentiable at x.

PROOF. Suppose first that f is differentiable at x. Then $f'(x; \cdot)$ is the linear function $\langle \nabla f(x), \cdot \rangle$. By Theorem 23.2, the subgradients at x are the vectors x^* such that

$$\langle \nabla f(x), y \rangle \geq \langle x^*, y \rangle, \qquad \forall y,$$

and this condition is satisfied if and only if $x^* = \nabla f(x)$. Thus $\nabla f(x)$ is the unique subgradient of f at x. Suppose on the other hand that f has a unique subgradient x^* at x. The convex function g defined by

$$g(y) = f(x + y) - f(x) - \langle x^*, y \rangle$$

then has 0 as its unique subgradient at the origin. We must show that this implies

$$\lim_{v \to 0} \frac{g(y)}{|y|} = 0.$$

The closure of $g'(0; \cdot)$ is the support function of $\partial g(0)$, which here is the constant function 0 (Theorem 23.2). Therefore $g'(0; \cdot)$ itself is identically 0, since $g'(0; \cdot)$ cannot differ from its closure other than at boundary points of its effective domain, and we have

$$0 = g'(0; u) = \lim_{\lambda \downarrow 0} [g(\lambda u) - g(0)]/\lambda, \qquad \forall u.$$

Here $g(0) = 0$, and the difference quotient is a non-decreasing function of λ. The convex functions h_λ, where

$$h_\lambda(u) = g(\lambda u)/\lambda, \qquad \lambda > 0,$$

thus decrease pointwise to the constant function 0 as λ decreases to 0. Let B be the Euclidean unit ball, and let $\{a_1, \ldots, a_m\}$ be any finite collection of points whose convex hull includes B. Each $u \in B$ can be expressed as a convex combination

$$u = \lambda_1 a_1 + \cdots + \lambda_m a_m,$$

and one then has

$$0 \leq h_\lambda(u) \leq \sum_{i=1}^{m} \lambda_i h_\lambda(a_i)$$

$$\leq \max \{h_\lambda(a_i) \mid i = 1, \ldots, m\}.$$

Since $h_\lambda(a_i)$ decreases to 0 for each i as $\lambda \downarrow 0$, we may conclude that $h_\lambda(u)$ decreases to 0 uniformly in $u \in B$ as $\lambda \downarrow 0$. Given any $\varepsilon > 0$, there exists therefore a $\delta > 0$ such that

$$g(\lambda u)/\lambda \leq \varepsilon, \qquad \forall \lambda \in (0, \delta], \qquad \forall u \in B.$$

Since each vector y such that $0 < |y| \leq \delta$ can be expressed as λu with $\lambda = |y|$ and $u \in B$, we have $g(y)/|y| \leq \varepsilon$ whenever $0 < |y| \leq \delta$. This proves that the limit of $g(y)/|y|$ is 0 as claimed. ‖

COROLLARY 25.1.1. *Let f be a convex function. If f is finite and differentiable at a given point x, then f is proper and $x \in$ int (dom f).*

PROOF. The inequality in the theorem implies that $f(z) > -\infty$ for every z, and hence that f is proper. It is obvious from the definition of differentiability that, if f is differentiable at x, f must be finite in some neighborhood of x. ‖

We notice from Corollary 25.1.1 that the gradient mappings ∇f and $\nabla(\text{cl} f)$ coincide, inasmuch as f and $\text{cl} f$ coincide on int (dom f).

COROLLARY 25.1.2. *Let f be a proper convex function on R^n. Then the exposed points of the convex set epi f^* in R^{n+1} are the points of the form $(x^*, f^*(x^*))$ such that, for some x, f is differentiable at x and $\nabla f(x) = x^*$.*

PROOF. Since $(\text{cl} f)^* = f^*$, and $\nabla(\text{cl} f) = \nabla f$ as just remarked, we can assume f is closed. By definition, (x^*, μ^*) is an exposed point of epi f^* if and only if there is a supporting hyperplane H to epi f^* which meets epi f^* only at (x^*, μ^*). Such an H has to be non-vertical, and μ^* must be $f^*(x^*)$. In fact H must be the graph of an affine function $\langle x, \cdot \rangle - \mu$ such that $x \in \partial f^*(x^*)$, and $x \notin \partial f^*(z^*)$ for every $z^* \neq x^*$. By Theorem 23.5, this condition means that x^* is the unique element of $\partial f(x)$. Thus the exposed points of epi f^* are of the form $(x^*, f^*(x^*))$ where, for some x, x^* is the sole element of $\partial f(x)$. Apply the theorem. ‖

COROLLARY 25.1.3. *Let C be a non-empty closed convex set, and let g be any positively homogeneous proper convex function such that*

$$C = \{z \mid \langle y, z \rangle \leq g(y), \forall y\}.$$

(In particular, g may be taken to be the support function of C.) Then z is an exposed point of C if and only if there exists a point y such that g is differentiable at y and $\nabla g(y) = z$.

PROOF. The indicator function of C is g^* by Corollary 13.2.1. Apply the preceding corollary to g. ‖

THEOREM 25.2. *Let f be a convex function on R^n, and let x be a point at which f is finite. A necessary and sufficient condition for f to be differentiable at x is that the directional derivative function $f'(x; \cdot)$ be linear. Moreover, this condition is satisfied if merely the n two-sided partial derivatives $\partial f(x)/\partial \xi_j$ exist at x and are finite.*

PROOF. If the function $f'(x; \cdot)$ is linear, it is a closed convex function and hence directly equal to the support function of $\partial f(x)$ (Theorem 23.2). Then $\partial f(x)$ must consist of a single point, implying by Theorem 25.1 that f is differentiable at x. To complete the proof, we need only show that the existence and finiteness of the $\partial f/\partial \xi_j$ at x implies $f'(x; \cdot)$ is linear. Let e_j be the vector forming the jth row of the $n \times n$ identity matrix. We have

$$f'(x; e_j) = \frac{\partial f}{\partial \xi_j}(x) = -f'(x; -e_j), \qquad j = 1, \ldots, n.$$

The effective domain of $f'(x; \cdot)$ therefore contains the $2n$ vectors $\pm e_j$, and consequently it contains all positive multiples of the $\pm e_j$ by positive homogeneity. Since the effective domain is convex, it must be all of R^n. It follows that $f'(x; \cdot)$ is proper, for otherwise it would be identically $-\infty$ (Theorem 7.2). The linearity is assured by Theorem 4.8. ‖

Results about the existence of two-sided directional derivatives and gradients may be deduced from the continuity theorems of §24, as will be demonstrated next.

THEOREM 25.3. *Let f be a finite convex function on an open interval I of the real line. Let D be the subset of I where the (ordinary two-sided) derivative f' exists. Then D contains all but perhaps countably many points of I (so that in particular D is dense in I), and f' is continuous and non-decreasing relative to D.*

PROOF. Extend f to be a closed proper convex function on R. By Theorem 24.1, $f'_+(x) = f'_-(x)$ if and only if f'_+ is continuous at x. Thus D consists of the points of I where f'_+ is continuous. The points of I not in D are those where the non-decreasing function f'_+ has a jump, and there can be only countably many such jumps. Since f' agrees with f'_+ on D, f' is continuous and non-decreasing relative to D. ‖

THEOREM 25.4. *Let f be a proper convex function on R^n. For a given $y \neq 0$, let D be the set of points x in int (dom f) where $f'(x; y) = -f'(x; -y)$, i.e. where the ordinary two-sided directional derivative*

$$\lim_{\lambda \to 0} \frac{f(x + \lambda y) - f(x)}{\lambda}$$

exists. Then D consists precisely of the points of int (dom f) where f'(x; y) is continuous as a function of x. Moreover, D is dense in int (dom f). In fact, the complement of D in int (dom f) is a set of measure zero, and it can be expressed as a countable union of sets S_k closed relative to int (dom f), such that no bounded interval of a line in the direction of y contains more than finitely many points of any one S_k.

PROOF. In view of Corollary 24.5.1, the points of int (dom f) where $f'(x; y)$ is continuous as a function of x are the same as those where it is lower semi-continuous as a function of x. We claim, however, that

$$\liminf_{z \to x} f'(z; y) = -f'(x; -y), \qquad \forall x \in \text{int (dom } f).$$

Proving this relation will establish the continuity assertion in the theorem. In the first place, $\geq$ must hold, because $f'(z; y) \geq -f'(z; -y)$ for every z in dom f by Theorem 23.1, and $f'(z; -y)$ is upper semi-continuous in z on int (dom f). On the other hand, $\leq$ must hold because, for the one-dimensional convex function $g(\lambda) = f(x + \lambda y)$, one has

$$\lim_{\lambda \uparrow 0} f'(x + \lambda y; y) = \lim_{\lambda \uparrow 0} g'_+(\lambda) = g'_-(0) = -f'(x; -y)$$

(Theorem 24.1). Thus the "lim inf" relation holds as claimed. We demonstrate next how the complement of D in int (dom f) can be expressed in the manner described in the theorem. The complement consists of the points x of int (dom f) where

$$0 < f'(x; y) + f'(x; -y) = h(x).$$

Hence it is the union of the increasing sequence of sets

$$S_k = \{x \in \text{int (dom } f) \mid h(x) \geq 1/k\}, \qquad k = 1, 2, \ldots$$

As the sum of two upper semi-continuous functions of x, h is itself upper semi-continuous on int (dom f). Thus each S_k is closed relative to int (dom f) (and hence is a measurable set). Given any $x \in R^n$, let L_x be the line through x in the direction of y. Suppose L_x meets S_k. By restricting f to L_x one gets a one-dimensional convex function g as above, and the points $z = x + \lambda y$ in $L_x \cap S_k$ correspond to the values of λ such that

$$g'_+(\lambda) - g'_-(\lambda) \geq 1/k.$$

The inequality in Theorem 24.1 ensures that there cannot be more than a finite number of such points in any bounded interval. This proves that each S_k has the required intersection property, which implies that S_k has measure zero. (The measure of S_k can be obtained by integrating the measure of $L_x \cap S_k$ as a function of $x \in S'_k$, where S'_k is the projection of

S_k on the subspace of R^n orthogonal to y.) Since the complement of D in int (dom f) is the union of the sets S_k, it too must have measure zero. In particular, then, this complement can have no interior points, so that D is dense in int (dom f). ‖

The main theorem about the gradient mapping of a convex function is the following.

THEOREM 25.5. *Let f be a proper convex function on R^n, and let D be the set of points where f is differentiable. Then D is a dense subset of* int (dom f), *and its complement in* int (dom f) *is a set of measure zero. Furthermore, the gradient mapping $\nabla f: x \to \nabla f(x)$ is continuous on D.*

PROOF. Let $e_1, \ldots, e_n$ be the rows of the $n \times n$ identity matrix. Applying Theorem 25.4 to $y = e_j$, we see that the subset D_j of int (dom f) where $\partial f/\partial \xi_j$ exists has complement of zero measure in int (dom f). The union of these complements for $j = 1, \ldots, n$ likewise has measure zero. It is the complement of $D_1 \cap \cdots \cap D_n$, and the latter set is D by Theorem 25.2. In particular, the complement of D in int (dom f) has no interior, i.e. D is dense in int (dom f). Each partial derivative function $\partial f/\partial \xi_j$ is continuous on the corresponding D_j by Theorem 25.4, so all n partial derivatives are continuous on D. Since $\nabla f(x)$ is the vector of first partial derivatives where it exists, ∇f is continuous on D. ‖

COROLLARY 25.5.1. *Let f be a finite convex function on an open convex set C. If f is differentiable on C, then f is actually continuously differentiable on C.*

The set D in Theorem 25.5 is topologically a G_δ, i.e. the intersection of a countable collection of open sets. Indeed, the proof shows that D is the intersection of sets $D_1, \ldots, D_n$, each of which is a G_δ by Theorem 25.4.

We shall now show how the entire subdifferential mapping ∂f of a convex function f can be constructed from the gradient mapping ∇f, when f is closed and ∇f is not vacuous.

THEOREM 25.6. *Let f be a closed proper convex function such that* dom f *has a non-empty interior. Then*

$$\partial f(x) = \text{cl (conv } S(x)) + K(x), \qquad \forall x,$$

where $K(x)$ is the normal cone to dom f *at x (empty if $x \notin$* dom f) *and $S(x)$ is the set of all limits of sequences of the form $\nabla f(x_1), \nabla f(x_2), \ldots$, such that f is differentiable at x_i and x_i tends to x.*

PROOF. For each x we have $S(x) \subset \partial f(x)$, because the graph of ∂f is closed (Theorem 24.4). Since $\partial f(x)$ is a closed convex set, this implies cl (conv $S(x)$) is included in $\partial f(x)$. We observe next that, for any x such that $\partial f(x) \neq \emptyset$ (and hence $x \in$ dom f), $K(x)$ is the recession cone of

$\partial f(x)$. Indeed, given any $x^* \in \partial f(x)$, the recession cone of $\partial f(x)$ consists of the vectors y^* such that

$$x^* + \lambda y^* \in \partial f(x), \qquad \forall \lambda \geq 0,$$

(Theorem 8.3), i.e. such that

$$f(z) \geq f(x) + \langle x^* + \lambda y^*, z - x \rangle, \qquad \forall z, \forall \lambda \geq 0.$$

This condition is satisfied if and only if

$$\langle y^*, z - x \rangle \leq 0, \qquad \forall z \in \text{dom} f,$$

which means by definition that $y^* \in K(x)$. It follows that

$$\text{cl (conv } S(x)) + K(x) \subset \partial f(x) + K(x) = \partial f(x).$$

The opposite inclusion must now be proved. Since int $(\text{dom} f)$ is not empty, $K(x)$ contains no lines, and this implies that $\partial f(x)$ itself contains no lines, because $x^* + K(x)$ is included in $\partial f(x)$ for every $x^* \in \partial f(x)$. Hence $\partial f(x)$ is the convex hull of its extreme points and extreme directions by Theorem 18.5. Every extreme point of $\partial f(x)$ is a limit of exposed points by Theorem 18.6. On the other hand, every vector whose direction is an extreme direction of $\partial f(x)$ belongs (by Theorem 8.3, since $\partial f(x)$ is closed) to the recession cone of $\partial f(x)$, i.e. to the convex cone $K(x)$. Thus

$$\partial f(x) \subset \text{conv (cl } E) + K(x),$$

where E is the set of all exposed points of $\partial f(x)$. Of course

$$\text{conv (cl } E) \subset \text{cl (conv } E),$$

since cl $(\text{conv } E)$ is a convex set containing cl E. Therefore, to prove that

$$\partial f(x) \subset \text{cl (conv } S(x)) + K(x)$$

it suffices to prove that $E \subset S(x)$, i.e. that every exposed point of $\partial f(x)$ can be expressed as the limit of a sequence of gradients $\nabla f(x_i)$ with x_i tending to x.

Given any exposed point x^* of $\partial f(x)$, there exists by definition a supporting hyperplane to $\partial f(x)$ which meets $\partial f(x)$ only at x^*. Thus there exists a vector y with $|y| = 1$ such that y is normal to $\partial f(x)$ at x^*, but y is not normal to $\partial f(x)$ at any other point, i.e.

$$\langle y, x^* \rangle > \langle y, z^* \rangle, \forall z^* \in \partial f(x), z^* \neq x^*.$$

Since $K(x)$ is the recession cone of $\partial f(x)$, the latter condition on y implies in particular that

$$\langle y, y^* \rangle < 0, \qquad \forall y^* \in K(x), y^* \neq 0.$$

Hence (since $K(x)$ is also the normal cone to dom f at x) there does not exist a vector $y^* \neq 0$ such that

$$\langle z, y^* \rangle \leq \langle x, y^* \rangle \leq \langle x + \alpha y, y^* \rangle$$

for every $z \in \text{dom} f$ and every $\alpha \geq 0$. In other words, the half-line $\{x + \alpha y \mid \alpha \geq 0\}$ cannot be separated from $\text{dom} f$. It follows from Theorem 11.3 that this half-line must meet the (non-empty) interior of $\text{dom} f$. Thus (by Theorem 6.1 and the fact that $x \in \text{dom} f$) there exists an $\alpha > 0$ such that $x + \varepsilon y \in \text{int} (\text{dom} f)$ when $0 < \varepsilon \leq \alpha$. Choose any sequence $\varepsilon_1, \varepsilon_2, \ldots$, tending to 0 such that $0 < \varepsilon_i \leq \alpha$ for all i. Since f is differentiable on a dense subset of $\text{int} (\text{dom} f)$ by Theorem 25.5, there exists for each i an $x_i \neq x$ such that f is differentiable at x_i and

$$|x_i - (x + \varepsilon_i y)| < \varepsilon_i^2.$$

We have

$$\lim_{i \to \alpha} x_i = x,$$

$$\lim_{i \to \infty} |x_i - x|^{-1}(x_i - x) = y,$$

and this implies by Theorem 24.6 that, given any $\varepsilon > 0$, we have

$$\partial f(x_i) \subset \partial f(x)_y + \varepsilon B$$

for all sufficiently large indices i, where B is the closed unit Euclidean ball and $\partial f(x)_y$ is the set of all points of $\partial f(x)$ at which y is a normal vector. Here $\partial f(x_i)$ consists of just $\nabla f(x_i)$ (Theorem 25.1), while $\partial f(x)_y$ consists of just x^*. Thus, given any $\varepsilon > 0$, we have

$$|\nabla f(x_i) - x^*| < \varepsilon$$

for all sufficiently large indices i. This shows that

$$\lim_{i \to \infty} \nabla f(x_i) = x^*,$$

and since x^* was an arbitrary exposed point of $\partial f(x)$, the proof of the theorem is complete.   ‖

Ordinarily, of course, if $f_1, f_2, \ldots$, is a sequence of differentiable functions on an open interval I converging pointwise to a differentiable f on I, the sequence of derivatives $f_1', f_2', \ldots$, need not converge to f' and may diverge wildly. It is a remarkable fact, however, that if the functions are convex $f_1', f_2', \ldots$, not only converges to f' but converges uniformly on each closed bounded subinterval of I. This is a corollary of the following theorem.

THEOREM 25.7.   *Let C be an open convex set, and let f be a convex function which is finite and differentiable on C. Let $f_1, f_2, \ldots$, be a sequence of convex functions finite and differentiable on C such that $\lim_{i \to \infty} f_i(x) = f(x)$ for every $x \in C$. Then*

$$\lim_{i \to \infty} \nabla f_i(x) = \nabla f(x), \qquad \forall x \in C.$$

In fact, the mappings ∇f_i converge to ∇f uniformly on every closed bounded subset of C.

PROOF.   Let S be a closed bounded subset of C. To prove the theorem, it is enough to prove that the partial derivatives of f_i converge to those of f uniformly on S. Thus it is enough to prove that, given any vector y and any $\varepsilon > 0$, there exists an index i_0 such that

$$|\langle \nabla f_i(x), y \rangle - \langle \nabla f(x), y \rangle| \leq \varepsilon, \qquad \forall i \geq i_0, \qquad \forall x \in S.$$

This inequality can be written as a pair of inequalities

$$\langle \nabla f_i(x), y \rangle \leq \langle \nabla f(x), y \rangle + \varepsilon,$$
$$\langle \nabla f_i(x), -y \rangle \leq \langle \nabla f(x), -y \rangle + \varepsilon.$$

We shall show that there exists an index i_1 such that the first inequality holds for every $i \geq i_1$ and every $x \in S$. An index i_2 can be produced similarly for the second inequality, and the desired i_0 is then obtained by taking the larger of i_1 and i_2. Arguing by contradiction, we suppose that there is no i_1 with the specified properties. Then there are infinitely many indices i for which one can select a corresponding vector $x_i \in S$ such that

$$\langle \nabla f_i(x_i), y \rangle > \langle \nabla f(x_i), y \rangle + \varepsilon.$$

Passing to subsequences if necessary, we can suppose that this holds for every index i, and that the selected sequence $x_1, x_2, \ldots$, converges to a point x of S. For any $\lambda > 0$ small enough that $x + \lambda y \in C$, we have $x_i + \lambda y \in C$ for all sufficiently large indices i and

$$\langle \nabla f_i(x_i), y \rangle \leq [f_i(x_i + \lambda y) - f_i(x_i)]/\lambda.$$

The functions f_i converge to f uniformly on closed bounded subsets of C (Theorem 10.8), and since f is continuous on C this implies that $f_i(x_i)$ tends to $f(x)$ and $f_i(x_i + \lambda y)$ tends to $f(x + \lambda y)$. Since ∇f is continuous by Theorem 25.5, $\nabla f(x_i)$ tends to $\nabla f(x)$. Therefore

$$\langle \nabla f(x), y \rangle + \varepsilon = \lim_{i \to \infty} \langle \nabla f(x_i), y \rangle + \varepsilon$$
$$\leq \limsup_{i \to \infty} \langle \nabla f_i(x_i), y \rangle$$
$$\leq \lim_{i \to \infty} [f_i(x_i + \lambda y) - f_i(x_i)]/\lambda$$
$$= [f(x + \lambda y) - f(x)]/\lambda.$$

This is supposed to hold for every sufficiently small $\lambda > 0$. But

$$\langle \nabla f(x), y \rangle = f'(x; y) = \lim_{\lambda \downarrow 0} [f(x + \lambda y) - f(x)]/\lambda,$$

so this situation is impossible.   ‖

It may be remarked that, in the hypothesis of Theorem 25.7, $f_i(x)$ need only converge to $f(x)$ for every $x \in C'$, where C' is some dense subset of C. This implies by Theorem 10.8 and the continuity of finite convex functions on C that $f_i(x)$ converges to $f(x)$ for every $x \in C$.

The Legendre Transformation

The classical Legendre transformation for differentiable functions defines a correspondence which, for convex functions, is intimately connected with the conjugacy correspondence. The Legendre transformation will be investigated here in the light of the general differential theory of convex functions. We shall show that the case where it is well-defined and involutory is essentially the case where the subdifferential mapping of the convex function is single-valued and in fact one-to-one.

A multivalued mapping ρ which assigns to each $x \in R^n$ a set $\rho(x) \subset R^n$ is said to be *single-valued*, of course, if $\rho(x)$ contains *at most* one element x^* for each x. (Thus ρ is to reduce to an ordinary function on dom $\rho = \{x \mid \rho(x) \neq \emptyset\}$, but dom ρ is not required to be all of R^n.) If ρ and ρ^{-1} are both single-valued, ρ is said to be *one-to-one*. Here ρ^{-1} denotes the inverse of ρ, which in the sense of multivalued mappings is defined by

$$\rho^{-1}(x^*) = \{x \mid x^* \in \rho(x)\}.$$

Thus ρ is one-to-one if and only if the set

$$\text{graph } \rho = \{(x, x^*) \in R^{2n} \mid x^* \in \rho(x)\}$$

does not contain two different pairs with the same x component, or two with the same x^* component.

An extended-real-valued function f on R^n is said to be *smooth*, of course, only if f is actually finite and differentiable throughout R^n. However, we shall call a proper convex function f *essentially smooth* if it satisfies the following three conditions for $C = \text{int (dom } f)$:

(a) C is not empty;
(b) f is differentiable throughout C;
(c) $\lim_{i \to \infty} |\nabla f(x_i)| = +\infty$ whenever $x_1, x_2, \ldots,$ is a sequence in C converging to a boundary point x of C.

Note that a smooth convex function on R^n is in particular essentially smooth (since (c) holds vacuously).

THEOREM 26.1. *Let f be a closed proper convex function. Then ∂f is a single-valued mapping if and only if f is essentially smooth. In this case, ∂f*

251

reduces to the gradient mapping ∇f, i.e. $\partial f(x)$ consists of the vector $\nabla f(x)$ alone when $x \in$ int $(\mathrm{dom}\, f)$, while $\partial f(x) = \emptyset$ when $x \notin$ int $(\mathrm{dom}\, f)$.

PROOF. From Theorem 25.1, we see that the mapping ∂f is single-valued if and only if it reduces everywhere to ∇f. The criterion for this is just that $\partial f(x)$ be empty whenever f is not differentiable. Since $\partial f(x) \neq \emptyset$ when $x \in$ ri $(\mathrm{dom}\, f)$ (Theorem 23.4), this condition implies that f is differentiable throughout ri $(\mathrm{dom}\, f)$. All points where f is differentiable belong to int $(\mathrm{dom}\, f)$, however. Thus ∂f is single-valued if and only if the above conditions (a) and (b) hold for $C = $ int $(\mathrm{dom}\, f)$, and $\partial f(x) = \emptyset$ when $x \notin C$. Of course $\partial f(x)$ is always empty for $x \notin \mathrm{dom}\, f$. It will be enough therefore to show (assuming (a) and (b)) that condition (c) fails for a given boundary point x of C if and only if $\partial f(x) \neq \emptyset$. Now, if (c) fails, there exists a sequence $x_1, x_2, \ldots$, converging to x such that the sequence $\nabla f(x_1), \nabla f(x_2), \ldots$, is bounded. Extracting a subsequence if necessary, we can assume that $\nabla f(x_i)$ converges to a certain vector x^*. This x^* must belong to $\partial f(x)$ (Theorem 24.4), so $\partial f(x) \neq \emptyset$. Conversely, suppose $\partial f(x) \neq \emptyset$. Then $\partial f(x)$ contains the limit of some sequence $\nabla f(x_1), \nabla f(x_2), \ldots$, by Theorem 25.6, so (c) fails. $\|$

The definition of essentially smooth may be expressed in terms of directional derivatives instead of norms of gradients:

LEMMA 26.2. *Condition* (c) *in the definition of essentially smooth is equivalent to the following condition (assuming* (a) *and* (b)):

(c′) $f'(x + \lambda(a - x); a - x) \downarrow -\infty$ *as $\lambda \downarrow 0$ for any $a \in C$ and any boundary point x of C.*

PROOF. Both (c) and (c′) involve the behavior of f only on the open convex set C, so there is no loss of generality if we suppose the proper convex function f to be closed. Let $a \in C$ and let x be a boundary point of C. As demonstrated in the proof of Theorem 26.1, (c) fails for x if and only if $\partial f(x) \neq \emptyset$. On the other hand, according to Theorem 23.3, $\partial f(x) \neq \emptyset$ if and only if $f(x) < \infty$ and $f'(x; y) > -\infty$ for every y. The last property is implied simply by $f'(x; a - x) > -\infty$ (Theorem 7.2), because $f'(x; \cdot)$ is a convex function with $a - x$ in the interior of its effective domain. Thus (c) fails for x if and only if $f(x) < \infty$ and $f'(x; a - x) > -\infty$. We claim, however, that the latter holds if and only if (c′) fails for x. Consider the closed proper convex function g on R defined by $g(\lambda) = f(x + \lambda(a - x))$. By Theorem 24.1,

$$\lim_{\lambda \downarrow 0} f'(x + \lambda(a - x); a - x) = \lim_{\lambda \downarrow 0} g'_+(\lambda) = g'_+(0),$$

where (by the extended definition of g'_+ in Theorem 24.1)

$$g'_+(0) = \begin{cases} f'(x; a - x) & \text{if } 0 \in \mathrm{dom}\, g, \text{ i.e. } x \in \mathrm{dom}\, f, \\ -\infty & \text{if } 0 \notin \mathrm{dom}\, g, \text{ i.e. } x \notin \mathrm{dom}\, f. \end{cases}$$

The limit in (c′) thus fails to be $-\infty$ if and only if $x \in \mathrm{dom}\, f$ and $f'(x; a - x) > -\infty$. ‖

Theorem 26.1 will now be dualized with respect to the conjugacy correspondence.

A real-valued function f on a convex set C is said to be *strictly convex* on C if

$$f((1 - \lambda)x_1 + \lambda x_2) < (1 - \lambda)f(x_1) + \lambda f(x_2), \qquad 0 < \lambda < 1,$$

for any two different points x_1 and x_2 in C. A proper convex function f on R^n will be called *essentially strictly convex* if f is strictly convex on every convex subset of

$$\{x \mid \partial f(x) \neq \emptyset\} = \mathrm{dom}\, \partial f.$$

Since by Theorem 23.4

$$\mathrm{ri}\,(\mathrm{dom}\, f) \subset \mathrm{dom}\, \partial f \subset \mathrm{dom}\, f,$$

this condition implies f is strictly convex on $\mathrm{ri}\,(\mathrm{dom}\, f)$. (As demonstrated in §23, $\mathrm{dom}\, \partial f$ itself is not always a convex set.)

A closed proper convex function f which is essentially strictly convex need not be strictly convex on the entire convex set $\mathrm{dom}\, f$, as is shown by

$$f(x) = \begin{cases} (\xi_2^2/2\xi_1) - 2\xi_2^{1/2} & \text{if } \xi_1 > 0, \xi_2 \geq 0, \\ 0 & \text{if } \xi_1 = 0 = \xi_2, \\ +\infty & \text{otherwise, where } x = (\xi_1, \xi_2). \end{cases}$$

Here $\mathrm{dom}\, \partial f$ is an open convex set, namely the positive quadrant of R^2, and f is strictly convex on $\mathrm{dom}\, \partial f$, but along the non-negative ξ_1-axis f is identically zero and hence not strictly convex. Observe incidentally that this f happens to be, not only essentially strictly convex, but essentially smooth.

It is also possible for a closed proper convex function f to be strictly convex on $\mathrm{ri}\,(\mathrm{dom}\, f)$ but fail to be strictly convex on some other convex subset of $\mathrm{dom}\, \partial f$ (and therefore fail to be an essentially strictly convex function on R^n). An example of this behavior is

$$f(x) = \begin{cases} (\xi_2^2/2\xi_1) + \xi_2^2 & \text{if } \xi_1 > 0, \xi_2 \geq 0, \\ 0 & \text{if } \xi_1 = 0 = \xi_2, \\ +\infty & \text{otherwise, where } x = (\xi_1, \xi_2). \end{cases}$$

For this function, $\mathrm{ri}\,(\mathrm{dom}\, f)$ is the positive quadrant of R^2, and f is strictly convex on this set, but $\mathrm{dom}\, \partial f$ also includes the entire non-negative ξ_1-axis, which is a convex set on which f is constant.

THEOREM 26.3. *A closed proper convex function is essentially strictly convex if and only if its conjugate is essentially smooth.*

PROOF. Let f be a closed proper convex function. According to Theorem 23.5, the subdifferential mapping of the conjugate function f^* is $(\partial f)^{-1}$, and by Theorem 26.1 this mapping is single-valued if and only if f^* is essentially smooth. Thus it suffices to show that f is essentially strictly convex if and only if $\partial f(x_1) \cap \partial f(x_2) = \emptyset$ whenever $x_1 \neq x_2$.

Suppose first that f is not essentially strictly convex. Then there exist two different points x_1 and x_2 such that, for a certain $x = (1 - \lambda)x_1 + \lambda x_2$, $0 < \lambda < 1$, one has $\partial f(x) \neq \emptyset$ and

$$f(x) = (1 - \lambda)f(x_1) + \lambda f(x_2).$$

Take any $x^* \in \partial f(x)$, and let H be the graph of the affine function $h(z) = f(x) + \langle x^*, z - x \rangle$. This H is a supporting hyperplane to epi f at $(x, f(x))$ Now $(x, f(x))$ is a relative interior point of the line segment in epi f joining $(x_1, f(x_1))$ and $(x_2, f(x_2))$, so the points $(x_1, f(x_1))$ and $(x_2, f(x_2))$ must belong to H. Thus $x^* \in \partial f(x_1)$ and $x^* \in \partial f(x_2)$, implying

$$\partial f(x_1) \cap \partial f(x_2) \neq \emptyset.$$

Suppose conversely that x^* is an element of $\partial f(x_1) \cap \partial f(x_2)$, where $x_1 \neq x_2$. The graph of $h(z) = \langle x^*, z \rangle - \mu$ for a certain μ (namely $\mu = f^*(x^*)$) is then a non-vertical supporting hyperplane H to epi f containing $(x_1, f(x_1))$ and $(x_2, f(x_2))$. The line segment joining these points belongs to H, so f cannot be strictly convex along the line segment joining x_1 and x_2. Every x in this line segment has $x^* \in \partial f(x)$. Hence f is not an essentially strictly convex function. ‖

COROLLARY 26.3.1. *Let f be a closed proper convex function. Then ∂f is a one-to-one mapping if and only if f is strictly convex on* int (dom f) *and essentially smooth.*

PROOF. We have $(\partial f)^{-1} = \partial f^*$ by Corollary 23.5.1. Thus, by Theorem 26.1, ∂f is one-to-one if and only if f and f^* are both essentially smooth. Since f is the conjugate of f^*, the essential smoothness of f^* is equivalent to the essential strict convexity of f. When f is essentially smooth, essential strict convexity reduces to strict convexity on int (dom f) by Theorem 26.1. ‖

Various results about the preservation of essential smoothness may be derived from Theorem 26.3.

COROLLARY 26.3.2. *Let f_1 and f_2 be closed proper convex functions on R^n such that f_1 is essentially smooth and*

$$\text{ri } (\text{dom } f_1^*) \cap \text{ri } (\text{dom } f_2^*) \neq \phi.$$

Then $f_1 \,\square\, f_2$ is essentially smooth.

PROOF. By Theorem 26.3, f_1^* is essentially strictly convex. Furthermore,

$$f_1 \,\square\, f_2 = (f_1^* + f_2^*)^*$$

by Theorem 16.4 and

$$\partial(f_1^* + f_2^*)(x^*) = \partial f_1^*(x^*) + \partial f_2^*(x^*), \qquad \forall x^*$$

by Theorem 23.8. The latter implies in particular that

$$\text{dom } \partial(f_1^* + f_2^*) \subset \text{dom } \partial f_1^*,$$

and it follows from this and the essential strict convexity of f_1^* that $f_1^* + f_2^*$ is essentially strictly convex. Therefore $(f_1^* + f_2^*)^*$ is essentially smooth by Theorem 26.3. ∥

COROLLARY 26.3.3. *Let f be a closed proper convex function on R^n which is essentially smooth, and let A be a linear transformation from R^n onto R^m. If there exists a $y^* \in R^m$ such that $A^*y^* \in \text{ri } (\text{dom } f^*)$, then the convex function Af on R^m is essentially smooth.*

PROOF. By Theorem 26.3, f^* is essentially strictly convex. Furthermore,

$$Af = (f^*A^*)^*$$

by Theorem 16.3, and

$$\partial(f^*A^*)(y^*) = A \, \partial f^*(A^*y^*)$$

by Theorem 23.9, so that

$$\text{dom } \partial(f^*A^*) = A^{*-1} \text{dom } \partial f^*.$$

Here A^{*-1} is single-valued (inasmuch as A maps R^n onto R^m), and it follows therefore from the essential strict convexity of f^* that f^*A^* is strictly convex. Hence $(f^*A^*)^*$ is essentially smooth by Theorem 26.3. ∥

Corollary 26.3.2 implies, for instance, that if C is any non-empty closed convex set in R^n and

$$f(x) = \inf \{|x - y|^p \mid y \in C\}, \qquad p > 1,$$

then f is a differentiable convex function on R^n (hence *continuously* differentiable by Corollary 25.5.1). Namely $f = f_1 \,\square\, f_2$, where

$$f_1(x) = |x|^p, \qquad f_2(x) = \delta(x \mid C),$$

and dom f_1^* is all of R^n (Corollary 13.3.1).

Corollary 26.3.3 implies that, if f is any (finite) differentiable convex function on R^n and A is a linear transformation from R^n onto R^m such that

$$Ax = 0 \quad \text{and} \quad x \neq 0 \quad \text{imply} \quad (f0^+)(x) > 0,$$

then Af is a differentiable convex function on R^m. (The condition here on $f0^+$ implies by Corollary 16.2.1 that the range of A^* meets ri $(\text{dom } f^*)$.) In particular, taking A to be a projection of the form

$$(\xi_1, \ldots, \xi_m, \xi_{m+1}, \ldots, \xi_n) \rightarrow (\xi_1, \ldots, \xi_m),$$

we see that if f is a differentiable convex function whose recession cone contains no non-zero vectors of the form

$$(0, \ldots, 0, \xi_{m+1}, \ldots, \xi_n),$$

then the convex function g defined by

$$g(\xi_1, \ldots, \xi_m) = \inf_{\xi_{m+1}, \ldots, \xi_n} f(\xi_1, \ldots, \xi_m, \xi_{m+1}, \ldots, \xi_n)$$

is (continuously) differentiable throughout R^m. Needless to say, differentiability could hardly be expected to be preserved under this kind of construction if f were not convex.

Let f be a differentiable real-valued function on an open subset C of R^n. The *Legendre conjugate* of the pair (C, f) is defined to be the pair (D, g), where D is the image of C under the gradient mapping ∇f, and g is the function on D given by the formula

$$g(x^*) = \langle (\nabla f)^{-1}(x^*), x^* \rangle - f((\nabla f)^{-1}(x^*)).$$

It is not actually necessary to have ∇f one-to-one on C in order that g be well-defined (i.e. single-valued). It suffices if

$$\langle x_1, x^* \rangle - f(x_1) = \langle x_2, x^* \rangle - f(x_2)$$

whenever $\nabla f(x_1) = \nabla f(x_2) = x^*$. Then the value of $g(x^*)$ can be obtained unambiguously from the formula by replacing $(\nabla f)^{-1}(x^*)$ by any of the vectors it contains.

Passing from (C, f) to the Legendre conjugate (D, g), if the latter is well-defined, is called the *Legendre transformation*.

In the case where f and C are convex, we can extend f to be a closed convex function on all of R^n with C as the interior of its effective domain. The Legendre conjugate of (C, f) is then related to the (ordinary) conjugate of the extended f as follows.

THEOREM 26.4. *Let f be any closed proper convex function such that the set $C = \mathrm{int}\,(\mathrm{dom}\,f)$ is non-empty and f is differentiable on C. The Legendre conjugate (D, g) of (C, f) is then well-defined. Moreover, D is a subset of $\mathrm{dom}\,f^*$ (namely the range of ∇f), and g is the restriction of f^* to D.*

PROOF. On C, ∂f reduces to ∇f (Theorem 25.1). For a given x^* in the range of ∇f, the vectors x such that $\nabla f(x) = x^*$ are the vectors in C where the function $\langle \cdot, x^* \rangle - f$ happens to attain its supremum $f^*(x^*)$ (Theorem 23.5). Thus, no matter which x we choose in $(\nabla f)^{-1}(x^*)$, we get the same value for $\langle x, x^* \rangle - f(x)$, namely $f^*(x^*)$. The formula for $g(x^*)$ is therefore unambiguous, and it gives $f^*(x^*)$. ‖

COROLLARY 26.4.1. *Let f be any essentially smooth closed proper convex function, and let $C =$ int $(\text{dom} f)$. (In particular, f may be any differentiable convex function on R^n, in which case $C = R^n$.) Then the Legendre conjugate (D, g) of (C, f) is well-defined. One has*

$$D = \{x^* \mid \partial f^*(x^*) \neq \emptyset\},$$

so that D is almost convex in the sense that

$$\text{ri} \, (\text{dom} f^*) \subset D \subset \text{dom} f^*.$$

Furthermore, g is the restriction of f^* to D, and g is strictly convex on every convex subset of D.

PROOF. The hypothesis implies that $\partial f = \nabla f$ (Theorem 26.1), so that $(\nabla f)^{-1} = \partial f^*$ (Theorem 23.5). The range D of ∇f therefore consists of the set of points x^* such that $\partial f^*(x^*) \neq \emptyset$. This set lies between dom f^* and ri $(\text{dom} f^*)$ according to Theorem 23.4. The strict convexity of g follows from the fact that f^* is essentially strictly convex (Theorem 26.3). ‖

Corollary 26.4.1 says, among other things, that the conjugate of an essentially smooth convex function f may be obtained from the Legendre conjugate (D, g) of (C, f) merely by extending g to be a closed proper convex function. We have $f^* = g$ on D, and at any boundary point x^* of D the value of f^* can be obtained as the limit of g along any line segment joining x^* with a point of ri D (Theorem 7.5). Outside of cl D, we have $f^*(x^*) = +\infty$.

Although the Legendre conjugate in Theorem 26.4 is well-defined, one can not always invert ∇f explicitly to get a workable formula in a given case. Notice, however, that the mapping ∇f, which is continuous from C onto D by Theorem 25.5, provides a natural parameterization of D. Under the (nonlinear) change of variables $x^* = \nabla f(x)$, we have

$$f^*(\nabla f(x)) = \langle x, \nabla f(x) \rangle - f(x).$$

In this sense, the Legendre conjugate of (C, f) can be treated as a (non-convex) function on C itself.

If f is a differentiable convex function on a non-empty open convex set C such that condition (c) of the definition of essentially smooth is not satisfied, then the domain D of the Legendre conjugate might not be "almost convex." For example, let C be the open upper half-plane in R^2, and let

$$f(\xi_1, \xi_2) = \xi_1^2/4\xi_2$$

on C. Then f is differentiable and convex, but the image D of C under ∇f is not convex, in fact D is the parabola

$$P = \{(\xi_1^*, \xi_2^*) \mid \xi_2^* = -(\xi_1^*)^2\}.$$

Condition (c) fails for f at the origin.

In general, the Legendre conjugate of a differentiable convex function need not be differentiable or convex, and we cannot speak of the Legendre conjugate of the Legendre conjugate. The Legendre transformation does, however, yield a symmetric one-to-one correspondence in the class of all pairs (C, f) such that C is an open convex set and f is a strictly convex function on C satisfying conditions (a), (b) and (c) (or (c′)) of the definition of essentially smooth. This is shown in the theorem below. For convenience, a pair (C, f) in the class just described will be called a *convex function of Legendre type*. By Corollary 26.3.1, a closed proper convex function f has ∂f one-to-one if and only if the restriction of f to $C = \mathrm{int}\,(\mathrm{dom}\,f)$ is a convex function of Legendre type.

THEOREM 26.5. *Let f be a closed convex function. Let $C = \mathrm{int}\,(\mathrm{dom}\,f)$ and $C^* = \mathrm{int}\,(\mathrm{dom}\,f^*)$. Then (C, f) is a convex function of Legendre type if and only if (C^*, f^*) is a convex function of Legendre type. When these conditions hold, (C^*, f^*) is the Legendre conjugate of (C, f), and (C, f) is in turn the Legendre conjugate of (C^*, f^*). The gradient mapping ∇f is then one-to-one from the open convex set C onto the open convex set C^*, continuous in both directions, and $\nabla f^* = (\nabla f)^{-1}$.*

PROOF. Since $\partial f^* = (\partial f)^{-1}$, we have ∂f one-to-one if and only if we have ∂f^* one-to-one. The first assertion of the theorem thus follows from Corollary 26.3.1. The rest of the theorem is then immediate from Theorem 26.1 and Corollary 26.4.1, except for the continuity of ∇f and ∇f^*, which is guaranteed by Theorem 25.5. ‖

To illustrate Theorem 26.5, we return to an example considered earlier in this section:

$$f(x) = \begin{cases} (\xi_2^2/2\xi_1) - 2\xi_2^{1/2} & \text{if } \xi_1 > 0, \xi_2 \geq 0, \\ 0 & \text{if } \xi_1 = 0 = \xi_2, \\ +\infty & \text{otherwise, where } x = (\xi_1, \xi_2). \end{cases}$$

As already remarked, this f is both essentially strictly convex and essentially smooth. Thus (C, f) is a convex function Legendre type, where

$$C = \mathrm{int}\,(\mathrm{dom}\,f) = \{x = (\xi_1, \xi_2) \mid \xi_1 > 0, \xi_2 > 0\}.$$

For $x \in C$ and $x^* = (\xi_1^*, \xi_2^*)$, we have $x^* = \nabla f(x)$ if and only if

$$\xi_1^* = -\xi_2^2/2\xi_1^2,$$

$$\xi_2^* = (\xi_2/\xi_1) - (1/\xi_2^{1/2}).$$

These nonlinear equations can easily be solved explicitly for ξ_1 and ξ_2 in terms of ξ_1^* find ξ_2^* (in most examples, unfortunately, this would not be the case), and we see that the equation $x = (\nabla f)^{-1}(x^*)$ is expressed by

$$\xi_1 = 1/(-2\xi_1^*)^{1/2}[(-2\xi_1^*)^{1/2} - \xi_2^*]^2,$$
$$\xi_2 = 1/[(-2\xi_1^*)^{1/2} - \xi_2^*]^2$$

for $x^* \in C^*$, the range of ∇f, where

$$C^* = \{x^* = (\xi_1^*, \xi_2^*) \mid \xi_1^* < 0, \xi_2^* < (-2\xi_1^*)^{1/2}\}.$$

According to Theorem 26.5, we actually have

$$C^* = \text{int } (\text{dom } f^*),$$

and from the formula

$$f^*(x^*) = \langle (\nabla f)^{-1}(x^*), x^* \rangle - f((\nabla f)^{-1}(x^*)),$$

which is valid for $x^* \in C^*$, we obtain

$$f^*(x^*) = 1/[(-2\xi_1^*)^{1/2} - \xi_2^*].$$

The Legendre conjugate of (C, f) is (C^*, f^*), and (C^*, f^*) is therefore another convex function of Legendre type. It may be verified as an exercise that the Legendre conjugate of this (C^*, f^*) is indeed (C, f) again.

The values of the conjugate function f^* on the whole space can always be determined from the values on ri (dom f^*) by a closure construction, and in this example we obtain, from the knowledge of (C^*, f^*), the formula

$$f^*(x^*) = \begin{cases} 1/[(-2\xi_1^*)^{1/2} - \xi_2^*] & \text{if } \xi_1^* \leq 0, \xi_2^* < (-2\xi_1^*)^{1/2}, \\ +\infty & \text{otherwise.} \end{cases}$$

To finish off this section, we shall describe the case where the Legendre transformation and the conjugacy correspondence coincide completely. Recall that, according to the definition given in §13, a finite convex function f on R^n is said also to be *co-finite* if epi f contains no non-vertical half-lines, and this is equivalent (by Corollary 8.5.2.) to the condition that

$$+\infty = (f0^+)(y) = \lim_{\lambda \to \infty} f(\lambda y)/\lambda, \qquad \forall y \neq 0.$$

THEOREM 26.6. *Let f be a (finite) differentiable convex function on R^n. In order that ∇f be a one-to-one mapping from R^n onto itself, it is necessary and sufficient that f be strictly convex and co-finite. When these conditions hold, f^* is likewise a differentiable convex function on R^n which is strictly convex and co-finite, and f^* is the same as the Legendre conjugate of f, i.e.*

$$f^*(x^*) = \langle (\nabla f)^{-1}(x^*), x^* \rangle - f((\nabla f)^{-1}(x^*)), \qquad \forall x^*.$$

The Legendre conjugate of f^* is then in turn f.

PROOF.   This is obvious from Corollary 26.3.1, Theorem 26.5 and the fact that dom $f^* = R^n$ if and only if f is co-finite (Corollary 13.3.1).   ‖

The following characterization of co-finiteness is helpful in applying Theorem 26.6.

LEMMA 26.7.   *Let f be a differentiable convex function on R^n. In order that f be co-finite, it is necessary and sufficient that*

$$\lim_{i \to \infty} |\nabla f(x_i)| = +\infty$$

for every sequence $x_1, x_2, \ldots$, such that

$$\lim_{i \to \infty} |x_i| = +\infty.$$

PROOF.   Since dom $f^* = R^n$ if and only if f is co-finite, it suffices to show that int $(\text{dom} f^*) \neq R^n$ if and only if there exists an unbounded sequence $x_1, x_2, \ldots$, such that the sequence $\nabla f(x_1), \nabla f(x_2), \ldots$, is convergent. Suppose the latter holds. Let $x_i^* = \nabla f(x_i)$ for $i = 1, 2, \ldots$, and let $x^* = \lim_{i \to \infty} x_i^*$. Then $x_i \in \partial f^*(x_i^*)$ for every i. If x^* were an interior point of dom f^*, $\partial f^*(x^*)$ would be bounded (Theorem 23.4), and by Corollary 24.5.1 there would exist an index i_0 such that

$$\partial f^*(x_i^*) \subset \partial f^*(x^*) + B, \quad i \geq i_0,$$

(B = unit ball). This would contradict the unboundedness of the sequence $x_1, x_2, \ldots$, so we may conclude that $x^* \notin \text{int} (\text{dom} f^*)$. Suppose now conversely that int $(\text{dom} f^*) \neq R^n$. Let x^* be any boundary point of dom f^*. Either $\partial f^*(x^*)$ is unbounded or $\partial f^*(x^*) = \emptyset$. If $\partial f^*(x^*)$ is unbounded, it contains an unbounded sequence $x_1, x_2, \ldots$, and for each x_i we have $x^* \in \partial f(x_i)$, i.e. $x^* = \nabla f(x_i)$, so that $\nabla f(x_1), \nabla f(x_2), \ldots$, is trivially a convergent sequence. On the other hand, if $\partial f^*(x^*) = \emptyset$ let $x_1^*, x_2^*, \ldots$, be any sequence in ri $(\text{dom} f^*)$ converging to x^*. For each i, choose an element $x_i \in \partial f^*(x_i^*)$, which is possible since f^* is subdifferentiable on ri $(\text{dom} f^*)$ by Theorem 23.4. Then $x_i^* = \nabla f(x_i)$ for every i, so that $\nabla f(x_i)$ tends to x^*. The sequence $x_1, x_2, \ldots$, must be unbounded, for if not some subsequence would converge to a point x, and we would have $x \in \partial f^*(x^*)$ by Theorem 24.4, contrary to $\partial f^*(x^*) = \emptyset$.   ‖

Part VI · Constrained Extremum Problems

The Minimum of a Convex Function

The great importance of extremum problems and variational principles in applied mathematics leads one to the general study of the minimum or maximum (or of certain minimax extrema) of a function h over a set C. When a sufficient amount of convexity is present, the study is greatly simplified, and many significant theorems can be established, particularly as regards duality and characterizations of the points where the extrema are attained.

In this section we shall study the minimum of a convex function h over a convex set C in R^n. There is no real loss of generality in assuming h to be a proper convex function on R^n. Minimizing h over C is of course equivalent to minimizing

$$f(x) = h(x) + \delta(x \mid C) = \begin{cases} h(x) & \text{if } x \in C, \\ +\infty & \text{if } x \notin C, \end{cases}$$

over all of R^n. We therefore begin with a discussion of the (unconstrained) minimum of a (possibly infinity-valued) convex function f on R^n and then specialize to the case where $f = h + \delta(\cdot \mid C)$. In §28 we shall consider in detail the case where C is the set of solutions to some system of inequalities.

In what follows, our attention will be focused on the properties of the parameterized nest of level sets

$$\text{lev}_\alpha f = \{x \mid f(x) \leq \alpha\}, \qquad \alpha \in R,$$

belonging to a given proper convex function f. The sets $\text{lev}_\alpha f$ are convex, and if f is closed (a sensible regularity assumption) they are all closed. The union of the $\text{lev}_\alpha f$ for $\alpha \in R$ is dom f. Minimizing f over R^n is the same as minimizing f over the convex set dom f.

Let $\inf f$ denote the infimum of $f(x)$ as x ranges over R^n. In terms of level sets, $\inf f$ is characterized by the property that $\text{lev}_\alpha f = \emptyset$ for $\alpha < \inf f$. For $\alpha = \inf f$, $\text{lev}_\alpha f$ consists of the points x where the infimum of f is attained; we call this level set the *minimum set* of f. Obviously it is of great importance in a given case to know whether the minimum set is empty or non-empty, or whether it consists of a unique point. Certainly it cannot contain more than one point if f is strictly convex on dom f. In any event, *the minimum set of f is a certain convex subset of R^n, closed if f is closed.*

The manner in which the sets $\text{lev}_\alpha f$ decrease to the minimum set as $\alpha \downarrow \inf f$ is important in situations where one is concerned with the convergence of vector sequences $x_1, x_2, \ldots$, such that $f(x_i)$ decreases to $\inf f$.

A necessary and sufficient condition for a given point x to belong to the minimum set of f is that $0 \in \partial f(x)$, i.e. that $x^* = 0$ be a subgradient of f at x. Indeed, this is true simply by the definition of "subgradient." What makes the condition $0 \in \partial f(x)$ useful and significant is the general differential theory of convex functions, particularly the results in §23 relating subgradients and directional derivatives, and the formulas for computing subgradients in various situations.

According to Theorem 23.2, one has $0 \in \partial f(x)$ if and only if f is finite at x and

$$f'(x; y) \geq 0, \ \forall y.$$

Of course, the one-sided directional derivatives $f'(x; y)$ depend only on the values of f in an arbitrarily small neighborhood of x. It follows that, if x is a point where f has a finite *local* (relative) minimum, i.e. if $x \in \text{dom} f$ and $f(z) \geq f(x)$ for every z within a certain positive distance ε of x, then $0 \in \partial f(x)$, so that x is a point where f has its *global* minimum. This is one of the most striking consequences of convexity, and it is one of the main technical justifications for assuming convexity in the first attempts at analyzing a new class of minimum problems.

The theory of the minimum of a convex function is rich with duality, and a great deal will be said on this topic in the following sections. At the heart of this duality is the fact that there is an extensive correspondence between properties of the nest of level sets $\text{lev}_\alpha f$ and properties of the conjugate function f^* at the origin. The correspondence has been established bit by bit in previous sections, and it is appropriate to summarize it here for convenience.

THEOREM 27.1.   *The following statements are valid for any closed proper convex function f.*

(a) $\inf f = -f^*(0)$. *Thus f is bounded below if and only if $0 \in \text{dom} f^*$.*

(b) *The minimum set of f is $\partial f^*(0)$. Thus the infimum of f is attained if and only if f^* is subdifferentiable at 0. This condition is satisfied in particular when $0 \in \text{ri}\,(\text{dom} f^*)$; moreover, one has $0 \in \text{ri}\,(\text{dom} f^*)$ if and only if every direction of recession of f is a direction in which f is constant.*

(c) *For the infimum of f to be finite but unattained, it is necessary and sufficient that $f^*(0)$ be finite and $f^{*\prime}(0; y) = -\infty$ for some y.*

(d) *The minimum set of f is a non-empty bounded set if and only if* $0 \in \text{int} (\text{dom} f^*)$. *This holds if and only if f has no directions of recession.*

(e) *The minimum set of f consists of a unique vector x if and only if* f^* *is differentiable at 0 and* $x = \nabla f^*(0)$.

(f) *The non-empty sets among the sets* $\text{lev}_\alpha f$ *(including the minimum set of f, if that is non-empty) all have the same recession cone. This coincides with the recession cone of f. It is the polar of the convex cone generated by* $\text{dom} f^*$.

(g) *For each* $\alpha \in R$, *the support function of* $\text{lev}_\alpha f$ *is the closure of the positively homogeneous convex function generated by* $f^* + \alpha$. *If f is bounded below, the support function of the minimum set of f is the closure of the directional derivative function* $f^{*\prime}(0; \cdot)$.

(h) *If* $\inf f$ *is finite, one has*

$$\lim_{\alpha \downarrow \inf f} \delta^*(y \mid \text{lev}_\alpha f) = f^{*\prime}(0; y), \quad \forall y.$$

(i) *One has* $0 \in \text{cl} (\text{dom} f^*)$ *if and only if* $(f 0^+)(y) \geq 0$ *for every y. Thus* $0 \notin \text{cl} (\text{dom} f^*)$ *if and only if there exists a vector* $y \neq 0$ *and a real number* $\varepsilon > 0$ *such that*

$$f(x + \lambda y) \leq f(x) - \lambda\varepsilon, \quad \forall \lambda \geq 0, \quad \forall x \in \text{dom} f.$$

PROOF. (a): By the definition of $f^*(0)$ in §12. (b): Theorem 23.5, Theorem 23.4 and Corollary 13.3.4. (c): By (a), (b) and Theorem 23.3.' (d): By (b), Theorem 23.4 and Corollary 13.3.4. (e): By (b) and Theorem 25.1. (f): Theorem 8.7 and Theorem 14.2. (g): Theorem 13.5 applied to $f - \alpha$, Theorem 23.2. (h): By (a) and Theorem 23.6, the set $\partial_\varepsilon f^*(0)$ being $\text{lev}_\alpha f$ for $\alpha = \inf f + \varepsilon$. (i): Corollary 13.3.4 and Theorem 8.5. ‖

The directions of recession of f are by definition the directions of the non-zero vectors y (if any) such that $f(x + \lambda y)$ is a non-increasing function of λ for every choice of x. If such a direction exists, there will obviously exist *unbounded* sequences $x_1, x_2, \ldots$, such that $f(x_i) \downarrow \inf f$, and hence the infimum of f might not be finite or attained. It is a remarkable fact about closed convex functions that such behavior is possible *only* if a direction of recession exists.

THEOREM 27.2. *Let f be a closed proper convex function which has no direction of recession. The infimum of f is then finite and attained. Moreover, for every* $\varepsilon > 0$ *there exists a* $\delta > 0$ *such that every vector x satisfying* $f(x) \leq \inf f + \delta$ *lies within the distance* ε *of the minimum set of f (i.e.* $|z - x| < \varepsilon$ *for at least one z such that* $f(z) = \inf f$), *the minimum set here being a non-empty closed bounded convex set.*

PROOF. That the infimum is finite and attained is immediate from Theorem 27.1(d). Also, since there are no directions of recession, all the (closed, convex) sets $\text{lev}_\alpha f$ are bounded (Theorem 27.1(f) and Theorem 8.4). Let M denote the minimum set of f, and let B denote the unit Euclidean ball. Fix any $\varepsilon > 0$. The set $M + \varepsilon(\text{int } B)$ is open, because it is a union of translates of the open set $\varepsilon (\text{int } B)$. For each $\delta > 0$, let S_δ denote the intersection of the complement of $M + \varepsilon (\text{int } B)$ with $\text{lev}_\alpha f$, where $\alpha = \inf f + \delta$. The sets S_δ form a nest of closed bounded subsets of R^n. If every S_δ were non-empty, there would exist a point x common to every S_δ. Such an x would have the contradictory property that $f(x) \leq \inf f + \delta$ for every $\delta > 0$ (hence $x \in M$), yet $x \notin M + \varepsilon$ (int B). Thus S_δ must be empty for some $\delta > 0$. For this δ, the level set $\{x \,|\, f(x) \leq \inf f + \delta\}$ lies entirely in $M + \varepsilon$ (int B) as required.   ‖

COROLLARY 27.2.1.   *Let f be a closed proper convex function which has no direction of recession. Let $x_1, x_2, \ldots,$ be any sequence such that*

$$\lim_{i \to \infty} f(x_i) = \inf f.$$

Then $x_1, x_2, \ldots,$ is a bounded sequence, and all its cluster points belong to the minimum set of f.

COROLLARY 27.2.2.   *Let f be a closed proper convex function which attains its infimum at a unique point x. If $x_1, x_2, \ldots,$ is any sequence of vectors such that $f(x_1), f(x_2), \ldots,$ converges to $\inf f$, then $x_1, x_2, \ldots,$ converges to x.*

PROOF.   If the minimum set (a certain $\text{lev}_\alpha f$) consists of a single point, f cannot have any directions of recession. Apply the theorem.   ‖

A closed proper convex function on the real line attains its infimum if it is neither a non-increasing function nor a non-decreasing function. This is the one-dimensional case of Theorem 27.2. In the n-dimensional case, the theorem says that a closed proper convex function f attains its infimum if the restriction of f to each line in R^n is a one-dimensional convex function of the sort just described (or the constant function $+\infty$). It suffices actually if each restriction *which is not a constant function* is of the sort described. This follows from part (b) of Theorem 27.1.

A reasonable conjecture is that, if f is a closed proper convex function on R^n which attains its infimum relative to each line in R^n (i.e. the restriction of f to each line is a function whose infimum is attained), then f attains its infimum on R^n. Here is an example which shows that this conjecture is false. Let P be the "parabolic" convex set in R^2 defined by

$$P = \{(\xi_1, \xi_2) \,|\, \xi_2 \geq \xi_1^2\}.$$

For each $x \in R^2$, let $f_0(x)$ be the square of the distance of x from P, i.e.

$$f_0(x) = \inf \{|x - y|^2 \,|\, y \in P\} = (f_1 \,\square\, f_2)(x)$$

where $f_1(x) = |x|^2$ and $f_2(x) = \delta(x \mid P)$. Let

$$f(x) = f(\xi_1, \xi_2) = f_0(\xi_1, \xi_2) - \xi_1.$$

Then f is a finite convex function on R^2. (In fact, it can be shown that f is continuously differentiable.) Along any line which is not parallel to the ξ_2-axis, the limit of $f(x)$ is $+\infty$ in both directions, so the infimum of f relative to such a line is attained. Along any line parallel to the ξ_2-axis, $f(x)$ is non-increasing as a function of ξ_2 and constant for large positive values of ξ_2 and hence attains its infimum. Thus f satisfies the hypothesis of the conjecture. But f does not attain its infimum on R^2. Along the parabola $\xi_2 = \xi_1^2$, the value of $f(\xi_1, \xi_2)$ is $-\xi_1$, so f is not even bounded below!

In particular one can have $(f0^+)(y) \geq 0$ for every y, and yet $\inf f = -\infty$. This corresponds to the case where $0 \in \text{cl} \, (\text{dom} \, f^*)$ but $0 \notin \text{dom} \, f^*$ (see parts (a) and (i) of Theorem 27.1).

We shall now take up the special case where f is explicitly of the form $h + \delta(\cdot \mid C)$, i.e. where a convex function h is to be minimized over a convex set C not necessarily equal to dom h. Properties of the infimum are to be described in terms of the relationship between h and C.

THEOREM 27.3. *Let h be a closed proper convex function, and let C be a non-empty closed convex set over which h is to be minimized. If h and C have no direction of recession in common (which is trivially true if either h or C has no direction of recession at all), then h attains its infimum over C. In the case where C is polyhedral, h achieves its infimum over C under the weaker hypothesis that every common direction of recession of h and C is a direction in which h is constant.*

PROOF. Let $f(x) = h(x) + \delta(x \mid C)$. The infimum of h over C is the same as the infimum of f over R^n. If f is identically $+\infty$, the infimum is trivially attained throughout C. If f is not identically $+\infty$, f is a closed proper convex function whose directions of recession are the common directions of recession of h and C. By Theorem 27.2, f attains its infimum when there are no such directions. This establishes the non-polyhedral case of the theorem. A different argument is needed to get the refinement for C polyhedral. Setting

$$\beta = \inf \{h(x) \mid x \in C\} < +\infty,$$

we consider the collection of closed convex sets consisting of C and the sets lev$_\alpha \, h$, $\alpha > \beta$. By hypothesis, C is polyhedral, and the only directions in which all the sets in the collection recede are directions in which all the sets other than C are linear (Theorem 27.1(f)). Helly's Theorem, in the form of Theorem 21.5, is applicable to such a collection. Every finite subcollection has a non-empty intersection by the choice of β, so the whole collection has a non-empty intersection. The points of this intersection are the points where the infimum of h relative to C is attained.   ‖

COROLLARY 27.3.1.   *Let h be a closed proper convex function such that every direction of recession of h is a direction in which h is affine. (This condition is satisfied of course if h is an affine or quadratic convex function, or merely if* dom h^* *is an affine set (Corollary 13.3.2).) Then h attains its infimum relative to any polyhedral convex set C on which it is bounded below.*

PROOF.   Under the hypothesis, any common direction of recession of h and C must be a direction in which h is affine. Thus, if y is a vector with such a direction, one has, for any $x \in C$,

$$x + \lambda y \in C \quad \text{and} \quad h(x + \lambda y) = h(x) + \nu\lambda, \qquad \forall \lambda \geq 0,$$

where

$$\nu = (h0^+)(y) = -(h0^+)(-y) \leq 0.$$

When h is bounded below on C, this condition implies that $\nu = 0$, so that every common direction of recession of h and C is actually a direction in which h is constant, and the theorem is applicable.   ‖

The hypothesis of Corollary 27.3.1 is also satisfied by any *polynomial convex function*, i.e. a convex function h such that $h(\xi_1, \ldots, \xi_n)$ can be expressed as a polynomial in the variables $\xi_1, \ldots, \xi_n$. (Then $h(x + \lambda y)$ is a polynomial convex function of the single real variable λ, no matter what x and y are chosen, and such a function of λ must either be affine or have limit $+\infty$ as $|\lambda| \to \infty$.)

COROLLARY 27.3.2.   *A polyhedral (or equivalently: finitely generated) convex function h attains its infimum relative to any polyhedral convex set C on which it is bounded below.*

PROOF.   Let D be the intersection of epi h with the "vertical prism" in R^{n+1} consisting of the points (x, μ) such that $x \in C$. Since C and h are polyhedral, D is the intersection of two polyhedral convex sets and hence is polyhedral. Minimizing h over C is equivalent to minimizing the linear function $(x, \mu) \to \mu$ over D. This infimum is attained, if it is not $-\infty$, by the preceding corollary.   ‖

COROLLARY 27.3.3.   *Let f_0 and f_i be closed proper convex functions on R^n for $i \in I$, where I is an arbitrary index set (finite or infinite). Assume that the system of constraints*

$$f_i(x) \leq 0, \qquad \forall i \in I,$$

is consistent. If there is no direction of recession common to f_0 and all the functions f_i, then the infimum of f_0 subject to these constraints is attained. More generally, the infimum is attained if there exists a finite subset I_0 of I

such that f_i is polyhedral (or affine, for example) for $i \in I_0$, and such that the only directions of recession common to f_0 and all the functions f_i are directions in which f_0 and all the functions f_i, for $i \in I \setminus I_0$, are constant.

PROOF. In the non-polyhedral case, let $h = f_0$, and let C be the set of vectors satisfying the constraints; the theorem is then applicable. For the polyhedral refinement, let

$$h(x) = \begin{cases} f_0(x) & \text{if } f_i(x) \leq 0, \quad \forall i \in I \setminus I_0, \\ +\infty & \text{if not.} \end{cases}$$

Let C be the polyhedral convex set consisting of the vectors x satisfying $f_i(x) \leq 0$ for every $i \in I_0$. The polyhedral case of the theorem is then applicable. ‖

For an illustration of Corollary 27.3.3, consider the problem of minimizing $f_0(x)$ subject to the constraints

$$f_1(x) \leq 0, \ldots, f_m(x) \leq 0, \qquad x \geq 0,$$

where f_i is of the form

$$f_i(x) = (1/p_i)\langle x, Q_i x \rangle^{p_i/2} + \langle a_i, x \rangle + \alpha_i$$

for $i = 0, \ldots, s$ with $p_i > 1$ and Q_i an $n \times n$ symmetric positive semi-definite matrix, while f_i is of the form

$$f_i(x) = \langle a_i, x \rangle + \alpha_i$$

for $i = s + 1, \ldots, m$. (The convexity of f_i for $i = 0, 1, \ldots, s$ follows from the fact that the function

$$g_i(x) = \langle x, Q_i x \rangle^{1/2}$$

is a gauge; see the example following Corollary 15.3.2.) The condition $x \geq 0$ can be written, of course, as the system

$$f_{m+1}(x) \leq 0, \ldots, f_{m+n}(x) \leq 0,$$

where

$$f_{m+j}(x) = -\xi_j \quad \text{for} \quad x = (\xi_1, \ldots, \xi_n).$$

To get a sufficient criterion for the attainment of the minimum in this case, we can apply the last part of Corollary 27.3.3 with

$$I = \{1, \ldots, m + n\}, \qquad I_0 = \{s + 1, \ldots, m + n\}.$$

The directions of recession of f_i are by definition the directions of the vectors $y \neq 0$ such that $(f_i 0^+)(y) \leq 0$, and by the formula in Corollary 8.5.2 we have

$$(f_i 0^+)(y) = \begin{cases} \langle a_i, y \rangle & \text{if } Q_i y = 0, \\ +\infty & \text{if } Q_i y \neq 0, \end{cases}$$

for $i = 0, 1, \ldots, s$,

$$(f_i 0^+)(y) = \langle a_i, y \rangle$$

for $i = s + 1, \ldots, m$, and

$$(f_{m+j} 0^+)(y) = -\eta_j \quad \text{for} \quad y = (\eta_1, \ldots, \eta_n).$$

The existence criterion furnished by Corollary 27.3.3 is that the system

$$(f_i 0^+)(y) \leq 0 \quad \text{for} \quad i = 1, \ldots, m + n$$

should not be satisfied by any y which does not also satisfy

$$(f_i 0^+)(-y) \leq 0 \quad \text{for} \quad i = 0, 1, \ldots, s.$$

In other words, $f_0(x)$ attains its minimum subject to the given constraints if every solution y to the system

$$y \geq 0, \quad \langle a_i, y \rangle \leq 0 \quad \text{for} \quad i = 0, 1, \ldots, m,$$
$$Q_i y = 0 \quad \text{for} \quad i = 0, 1, \ldots, s,$$

actually satisfies

$$\langle a_i, y \rangle = 0 \quad \text{for} \quad i = 0, 1, \ldots, s.$$

The points at which a constrained infimum is attained may be characterized by means of subdifferential theory. Suppose, for example, that we want to minimize a function of the form

$$f = \lambda_1 f_1 + \cdots + \lambda_m f_m$$

on R^n, where $f_1, \ldots, f_m$ are proper convex functions and $\lambda_1, \ldots, \lambda_m$ are non-negative real numbers. (Some of the functions may be indicator functions.) A necessary and sufficient condition for the infimum to be attained at the point x is that $0 \in \partial f(x)$. Now, under certain mild restrictions specified in Theorem 23.8, the formula

$$\partial f(x) = \lambda_1 \partial f_1(x) + \cdots + \lambda_m \partial f_m(x), \quad \forall x,$$

is valid. In this case, then, we get a necessary and sufficient subdifferential condition:

$$0 \in [\lambda_1 \partial f_1(x) + \cdots + \lambda_m \partial f_m(x)],$$

which can be analyzed further according to the nature of $f_1, \ldots, f_m$. The following theorem exemplifies this approach.

THEOREM 27.4. *Let h be a proper convex function, and let C be a nonempty convex set. In order that x be a point where the infimum of h relative to C is attained, it is sufficient that there exist a vector $x^* \in \partial h(x)$ such that $-x^*$ is normal to C at x. This condition is necessary, as well as sufficient,* *if* ri (dom *h*) *intersects* ri *C, or if C is polyhedral and* ri (dom *h*) *merely intersects C.*

PROOF. We want to minimize $h + \delta(\cdot \mid C)$ on R^n. By Theorem 23.8, the condition

$$0 \in [\partial h(x) + \partial\delta(x \mid C)]$$

is always sufficient for the infimum to be attained at x, and it is a necessary condition under the given assumptions about the intersection of dom h and C. The set $\partial\delta(x \mid C)$ is just the normal cone to C at x. ‖

The condition in Theorem 27.4 can also be derived easily from separation theory without invoking Theorem 23.8. The argument, which is a specialization of the alternative proof of Theorem 23.8 given in §23, may be sketched as follows. Let α be the infimum of h over C, and consider in R^{n+1} the convex sets $C_1 = $ epi h and

$$C_2 = \{(x, \mu) \mid x \in C, \mu \leq \alpha\}.$$

These sets can be separated by a non-vertical hyperplane, i.e. the graph of some affine function $\langle \cdot, x^* \rangle + \beta$. If x is any point where the infimum of h over C is attained, then this x^* belongs to $\partial h(x)$, and $-x^*$ is normal to C at x. The details of this proof can be developed as an exercise.

If h is actually differentiable at x in Theorem 27.4, $\partial h(x)$ reduces to the single vector $\nabla h(x)$ (Theorem 25.1). The minimality condition is then that $-\nabla h(x)$ be normal to C at x. In the case where C is a subspace L, say, this condition would say that $x \in L$ and $\nabla h(x) \perp L$.

As an application of Theorem 27.4, consider the problem of finding the nearest point of a convex set C to a given point a. This is the same as the problem of minimizing the differentiable convex function

$$h(x) = (1/2) |x - a|^2$$

over C. The intersection hypothesis of Theorem 27.4 is satisfied trivially, so a necessary and sufficient condition for x to be the point of C nearest to a is that the vector

$$-\nabla h(x) = a - x$$

be normal to C at x.

For another application, consider the problem of minimizing a function of the form

$$h(x) = f_1(\xi_1) + \cdots + f_n(\xi_n) \quad \text{for} \quad x = (\xi_1, \ldots, \xi_n)$$

over a subspace L of R^n, where f_j is a closed proper convex function on R

for $j = 1, \ldots, n$. Here $\partial h(x)$ is a generalized rectangle:

$$\partial h(x) = \{x^* = (\xi_1^*, \ldots, \xi_n^*) \mid \xi_j^* \in \partial f_j(\xi_j) \quad \text{for} \quad j = 1, \ldots, n\}$$

$$= \{x^* \in R^n \mid f'_-(\xi_j) \le \xi_j^* \le f'_+(\xi_j) \quad \text{for} \quad j = 1, \ldots, n\},$$

and for any $x \in \text{dom } h$ and $z = (\zeta_1, \ldots, \zeta_n)$ one has

$$h'(x; z) = f'_1(\xi_1; \zeta_1) + \cdots + f'_n(\xi_n; \zeta_n)$$

$$= \sup \{\zeta_1 \xi_1^* + \cdots + \zeta_n \xi_n^* \mid \xi_j^* \in \partial f_j(\xi_j), \quad j = 1, \ldots, n\},$$

where (since each $\partial f_j(\xi_j)$ is a closed interval) the latter supremum is attained unless it is infinite. In particular, for any $x \in \text{dom } h$ and $z \in R^n$,

$$h'(x; z) = \sup \{\langle z, x^* \rangle \mid x^* \in \partial h(x)\}.$$

Suppose that L contains at least one element of ri $(\text{dom } h)$, i.e. an x such that

$$\xi_j \in \text{ri } (\text{dom } f_j) \quad \text{for} \quad j = 1, \ldots, n.$$

Then by Theorem 27.4 the infimum of h over L is attained at a given point x if and only if $x \in L$ and there exists an $x^* \in L^\perp$ satisfying the inequality system

$$\alpha_j \le \xi_j^* \le \beta_j, \quad j = 1, \ldots, n,$$

where $\alpha_j = f'_-(\xi_j)$ and $\beta_j = f'_+(\xi_j)$.

Observe incidentally that, if such an x^* does *not* exist, then there exists by Theorem 22.6 an *elementary* vector $z = (\zeta_1, \ldots, \zeta_n)$ of L such that

$$\zeta_1 \, \partial f_1(\xi_1) + \cdots + \zeta_n \, \partial f_n(\xi_n) < 0,$$

or in other words, in view of the above formula for $h'(x; z)$,

$$h'(x; z) < 0.$$

Thus, given any $x \in L \cap \text{dom } h$, either the infimum of h over L is already attained at x, or one can move away from x in the direction of some elementary vector of L (and there are only a *finite* number of such "elementary directions" according to Corollary 22.4.1) to a point $x' \in L \cap \text{dom } h$ with $h(x') < h(x)$. In situations where the elementary vectors of L are convenient to work with, as for example when L is the space of all circulations (or the space of all tensions) in some directed graph as discussed in §22, this observation leads to an efficient algorithm for minimizing h over L (at least when the functions f_j are suitably polyhedral, i.e. piecewise linear).

Further remarks on the preceding example will be made following Corollary 31.4.3.

Ordinary Convex Programs and Lagrange Multipliers

The theory of Lagrange multipliers tells how to transform certain constrained extremum problems into extremum problems involving fewer constraints but more variables. Here we shall present the branch of the theory which is applicable to problems of minimizing convex functions subject to "convex" constraints.

By an *ordinary convex program* (P) (as opposed to a "generalized" convex program, to be defined in §29), we shall mean a problem of the following form: minimize $f_0(x)$ over C subject to constraints

$$f_1(x) \leq 0, \ldots, f_r(x) \leq 0, \qquad f_{r+1}(x) = 0, \ldots, f_m(x) = 0,$$

where C is a non-empty convex set in R^n, f_i is a finite convex function on C for $i = 0, 1, \ldots, r$, and f_i is an affine function on C for $i = r + 1, \ldots, m$. Included here are the special cases where $r = m$ (i.e. no equality constraints) or $r = 0$(i.e. no inequality constraints).

Of course, to define (P) as a "problem" is rather vague, and it might lead to misunderstandings. For the sake of mathematical rigor, we must say that what we really mean by an ordinary convex program (P) is an $(m + 3)$-tuple $(C, f_0, \ldots, f_m, r)$ satisfying the conditions just given. Technically speaking, the fact that we have constrained minimization in mind is only to be inferred from the concepts we define in terms of (P) and the theorems we prove concerning them.

Only the values of the functions f_i on C itself are actually involved in the definition of (P). However, *we shall assume for convenience that each f_i is defined on all of R^n in such a way that* (a) f_0 *is a proper convex function with* $\mathrm{dom}\, f_0 = C$, (b) $f_1, \ldots, f_r$ *are proper convex functions with*

$$\mathrm{ri}\,(\mathrm{dom}\, f_i) \supset \mathrm{ri}\, C, \qquad \mathrm{dom}\, f_i \supset C,$$

and (c) f_i *is affine throughout R^n for every $i \neq 0$ such that f_i is affine on C.* There is no loss of generality in this assumption. (One can always arrange for (a) and (b) to be satisfied by setting $f_i(x) = +\infty$ when $x \notin C$, for $i = 0, 1, \ldots, r$. As for (c), one need only recall that the graph of each affine function f_i on C is contained in at least one non-vertical hyperplane, and such a hyperplane is the graph of an affine extension of f_i to all of R^n.)

A vector x will be called a *feasible solution* to (P) if $x \in C$ and x satisfies the m constraints of (P). In other words, the set of feasible solutions to (P) is defined to be the (possibly empty) convex set

$$C_0 = C \cap C_1 \cap \cdots \cap C_m,$$

where

$$C_i = \{x \mid f_i(x) \leq 0\}, \qquad i = 1, \ldots, r,$$

$$C_i = \{x \mid f_i(x) = 0\}, \qquad i = r + 1, \ldots, m.$$

The convex function f on R^n defined by

$$f(x) = f_0(x) + \delta(x \mid C_0) = \begin{cases} f_0(x) & \text{if } x \in C_0, \\ +\infty & \text{if } x \notin C_0, \end{cases}$$

will be called the *objective function* for (P). Note that f has C_0 as its effective domain, and that f is closed when $f_0, f_1, \ldots, f_r$ are closed. Minimizing f over R^n is the same as minimizing $f_0(x)$ over all feasible solutions x. The infimum of f (which may be finite or $-\infty$ or $+\infty$) will be called the *optimal value* in (P). The points where the infimum of f is attained will be called the *optimal solutions* to (P), provided that f is not identically $+\infty$, i.e. that $C_0 \neq \emptyset$. The set of all optimal solutions is thus a (possibly empty) convex subset of the set of all feasible solutions.

Some theorems about the existence of optimal solutions can be obtained by applying the results in §27, most notably Corollary 27.3.3. There is no need to discuss such results any further here. We shall focus our attention rather on various characterizations of optimal solutions.

It should be emphasized that, by our definitions, two ordinary convex programs can have the same objective function (and hence the same feasible solutions, optimal value and optimal solutions), and yet be significantly different. An ordinary convex program has structure not reflected by its objective function alone, since its definition requires the specification of the values of the functions f_i throughout C, and it is this further structure which is all-important where Lagrange multipliers are concerned.

We define $(\lambda_1, \ldots, \lambda_m) \in R^m$ to be a *vector of Kuhn-Tucker coefficients* for (P), or simply a *Kuhn-Tucker vector* for (P), if $\lambda_i \geq 0$ for $i = 1, \ldots, r$ and the infimum of the proper convex function

$$f_0 + \lambda_1 f_1 + \cdots + \lambda_m f_m$$

(whose effective domain is C) is finite and equal to the optimal value in (P). (This terminology is discussed on p. 429.)

One reason for theoretical interest in Kuhn-Tucker coefficients is that such coefficients λ_i, if known, would make it possible to commute operations in (P). Instead of first determining the feasible solutions to (P) and then minimizing f_0 over these, one would be able first to minimize $f_0 + \lambda_1 f_1 + \cdots + \lambda_m f_m$ over R^n and then eliminate from the minimum set the points which fail to satisfy certain constraints. This is explained in the following theorem.

THEOREM 28.1. *Let (P) be an ordinary convex program. Let $(\lambda_1, \ldots, \lambda_m)$ be a Kuhn–Tucker vector for (P), and let*

$$h = f_0 + \lambda_1 f_1 + \cdots + \lambda_m f_m.$$

Let D be the set of points where h attains its infimum over R^n. Let I be the set of indices i such that $1 \le i \le r$ and $\lambda_i = 0$, and let J be the complement of I in $\{1, \ldots, m\}$. Let D_0 be the set of points $\bar{x} \in D$ such that

$$f_i(\bar{x}) = 0, \qquad \forall i \in J,$$

$$f_i(\bar{x}) \le 0, \qquad \forall i \in I.$$

Then D_0 is the set of all optimal solutions to (P).

PROOF. By hypothesis $\inf h = \inf f$, where f is the objective function for (P) and $\inf f$ is finite. For any feasible solution x to (P), we have

$$\lambda_i f_i(x) \le 0, \qquad i = 1, \ldots, m,$$

so that

$$f_0(x) + \lambda_1 f_1(x) + \cdots + \lambda_m f_m(x) \le f_0(x).$$

Thus $h(x) \le f(x)$ for every x, with equality if and only if x is a feasible solution such that

$$\lambda_i f_i(x) = 0, \qquad i = 1, \ldots, m.$$

It follows that the minimum set of f is contained in the minimum set of h and is in fact D_0. But the minimum set of f is the set of optimal solutions to (P). ∥

The fact that the D_0 in Theorem 28.1 can be a proper subset of the minimum set D is clear from the case where $C = R^n$ and every f_i is affine. In this case h, being an affine function which is bounded below on R^n, must be constant; thus $D = R^n$, while $D_0 \subset C_0$. However, there is another important case where D_0 does coincide with D, so that no further conditions have to be checked after a point has been determined which minimizes h:

COROLLARY 28.1.1.   *Let (P) be an ordinary convex program, and let* $(\lambda_1, \ldots, \lambda_m)$ *be a Kuhn-Tucker vector for (P). Assume that the functions* f_i *are all closed. If the infimum of*

$$h = f_0 + \lambda_1 f_1 + \cdots + \lambda_m f_m$$

is attained at a unique point $\bar{x}$, *this* $\bar{x}$ *is the unique optimal solution to (P).*

PROOF.   The hypothesis implies that h and the objective function f of (P) are closed. Suppose that the infimum of h is attained at a unique point $\bar{x}$. The corollary will follow from the theorem if we show that (P) has at least one optimal solution, i.e. that the infimum of f is attained somewhere. Since the minimum set of h consists of $\bar{x}$ alone, h has no directions of recession, i.e. the closed convex set epi h contains no "horizontal" half-lines. We have $f \geq h$ by the proof of the theorem so epi f likewise contains no "horizontal" half-lines. Thus f has no directions of recession, and by Theorem 27.2 the minimum set of f is not empty.   ‖

It should be noted in Corollary 28.1.1 that, if f_0 is strictly convex on C, then h is strictly convex on C, so that the infimum of h is attained at a unique point if attained at all.

Kuhn–Tucker coefficients can be interpreted heuristically as "equilibrium prices," and this is an important motivating idea. For each $u = (v_1, \ldots, v_m)$† in R^m, let $p(u) = p(v_1, \ldots, v_m)$ denote the infimum of $f_0(x)$ over C subject to the constraints

$$f_i(x) \leq v_i, \qquad i = 1, \ldots, r,$$

$$f_i(x) = v_i, \qquad i = r + 1, \ldots, m.$$

(The infimum is $+\infty$ by convention if these constraints cannot be satisfied.) Of course, $p(0)$ is the optimal value in (P), and in general $p(u)$ is the optimal value in the ordinary convex program (P_u) obtained by replacing f_i by $f_i - v_i$ for $i = 1, \ldots, m$. Thinking of the vectors u as representing "perturbations" of (P), we call p the *perturbation function* for (P) and direct our attention to the properties of p around $u = 0$.

Let us assume that $f_0(x)$ can be interpreted as the "cost" of x; thus in (P) we want to minimize cost subject to certain constraints. It may be possible, however, to modify the constraints to our advantage by buying perturbations u. Specifically, let us assume that we are allowed to change (P) to any (P_u) that we please, except that we must pay for the change, the price being v_i^* per unit of perturbation variable v_i. Then, for any perturbation u, the minimum cost we can achieve in the perturbed problem

---

† Here the symbol v, to be distinguished from the italic v, is the Greek letter upsilon.

(P_u), plus the cost of u, will be

$$p(v_1, \ldots, v_m) + v_1^* v_1 + \cdots + v_m^* v_m = p(u) + \langle u^*, u \rangle.$$

A perturbation will be "worth buying" if and only if this quantity is less than $p(0, \ldots, 0)$, the optimal value in the unperturbed problem.

Now here is where Kuhn–Tucker coefficients come in. We claim that, when the optimal value in (P) is finite, $(\lambda_1, \ldots, \lambda_m)$ is a Kuhn–Tucker vector for (P) if and only if, at the prices $v_i^* = \lambda_i$, no perturbation whatsoever would be worth buying (so that one would be content with the constraints as given, an "equilibrium" situation). Indeed, the infimum in u of the cost $p(u) + \langle u^*, u \rangle$ is the same as the infimum of

$$f_0(x) + v_1^* v_1 + \cdots + v_m^* v_m$$

in u and x subject to $v_i \geq f_i(x)$ for $i = 1, \ldots, r$ and $v_i = f_i(x)$ for $i = r + 1, \ldots, m$. The latter is

$$\inf_x \{ f_0(x) + v_1^* f_1(x) + \cdots + v_m^* f_m(x) \}$$

if $v_i^* \geq 0$ for $i = 1, \ldots, r$, but $-\infty$ otherwise. Thus, when $p(0, \ldots, 0)$ is finite and $v_i^* = \lambda_i$, the inequality

$$p(v_1, \ldots, v_m) + \lambda_1 v_1 + \cdots + \lambda_m v_m \geq p(0, \ldots, 0)$$

holds for every $u = (v_1, \ldots, v_m)$ if and only if $\lambda_i \geq 0$ for $i = 1, \ldots, r$ and

$$\inf_x \{ f_0(x) + \lambda_1 f_1(x) + \cdots + \lambda_m f_m(x) \} = p(0, \ldots, 0).$$

This condition means that $(\lambda_1, \ldots, \lambda_m)$ is a Kuhn–Tucker vector for (P).

The next theorem shows that Kuhn–Tucker coefficients can usually be expected to exist.

THEOREM 28.2. *Let (P) be an ordinary convex program, and let I be the set of indices $i \neq 0$ such that f_i is not affine. Assume that the optimal value in (P) is not $-\infty$, and that (P) has at least one feasible solution in* ri C *which satisfies with strict inequality all the inequality constraints for $i \in I$. Then a Kuhn–Tucker vector (not necessarily unique) exists for (P).*

PROOF. We shall first treat the case where there are no equality constraints, i.e. $r = m$. Let the indices in I be $1, \ldots, k$ for convenience, and let the optimal value in (P) be denoted by α. By hypothesis, the system

$$f_1(x) < 0, \ldots, f_k(x) < 0, \qquad f_{k+1}(x) \leq 0, \ldots, f_m(x) \leq 0,$$

has at least one solution in ri C. However, by the definition of α, the system

$$f_0(x) - \alpha < 0, \qquad f_1(x) < 0, \ldots, f_k(x) < 0,$$
$$f_{k+1}(x) \leq 0, \ldots, f_m(x) \leq 0,$$

has no solution in C. The second system satisfies the hypothesis of Theorem 21.2, so there exist non-negative real numbers $\lambda_0, \lambda_1, \ldots, \lambda_m$, not all zero, such that

$$\lambda_0(f_0(x) - \alpha) + \lambda_1 f_1(x) + \cdots + \lambda_m f_m(x) \geq 0, \qquad \forall x \in C.$$

Actually λ_0 itself must be positive, for if $\lambda_0 = 0$ we would have $\lambda_1 f_1 + \cdots + \lambda_m f_m$ non-negative on C with at least one of the coefficients $\lambda_1, \ldots, \lambda_k$ positive, and this would contradict the assumed existence of a solution to the first inequality system. Dividing all the coefficients λ_i by λ_0 if necessary, we can suppose that $\lambda_0 = 1$. The function

$$h = f_0 + \lambda_1 f_1 + \cdots + \lambda_m f_m$$

then satisfies $h(x) \geq \alpha$ for all $x \in C$ and $h(x) = +\infty$ for all $x \notin C$, so that $\inf h \geq \alpha$. On the other hand, for any feasible solution x one has $h(x) \leq f_0(x)$ (since $\lambda_i \geq 0$ and $f_i(x) \leq 0$ for $i = 1, \ldots, m$), and hence $\inf h$ cannot be greater than the infimum of f_0 over the set of feasible solutions, which is α. Thus $\inf h = \alpha$, and $(\lambda_1, \ldots, \lambda_m)$ is a Kuhn–Tucker vector for (P). The theorem is now established for the case where there are no equality constraints.

When equality constraints are present, i.e. $r < m$, the corresponding functions $f_{r+1}, \ldots, f_m$ are affine by the definition of (P). Each constraint $f_i(x) = 0$ can be replaced by two constraints

$$f_i(x) \leq 0, \qquad (-f_i)(x) \leq 0,$$

to obtain an "equivalent" ordinary convex program (P') with only inequality constraints. The part of the theorem which has already been proved is applicable to (P'). Kuhn–Tucker coefficients for (P') are non-negative real numbers $\lambda_1, \ldots, \lambda_r, \lambda'_{r+1}, \ldots, \lambda'_m, \lambda''_{r+1}, \ldots, \lambda''_m$ such that the infimum of the function

$$f_0 + \sum_{i=1}^r \lambda_i f_i + \sum_{i=r+1}^m \lambda'_i f_i + \sum_{i=r+1}^m \lambda''_i(-f_i)$$

is finite and equal to the optimal value in (P'), which is the same as the optimal value in (P). Setting $\lambda_i = \lambda'_i - \lambda''_i$ for $i = r + 1, \ldots, m$, one obtains from such coefficients a Kuhn–Tucker vector $(\lambda_1, \ldots, \lambda_m)$ for (P). $\|$

COROLLARY 28.2.1. *Let (P) be an ordinary convex program with only inequality constraints, i.e. with $r = m$. Assume that the optimal value in (P) is not $-\infty$, and that there exists at least one $x \in C$ such that*

$$f_1(x) < 0, \ldots, f_m(x) < 0.$$

Then a Kuhn–Tucker vector exists for (P).

PROOF. If $x \in C$ satisfies $f_i(x) < 0$ for $i = 1, \ldots, m$ and y is any point of ri C (and hence of ri $(\mathrm{dom}\, f_i)$ for $i = 1, \ldots, m$ by assumption (b) at the beginning of this section), then the point

$$z = (1 - \lambda)x + \lambda y$$

for $\lambda > 0$ sufficiently small satisfies $f_i(z) < 0$ for $i = 1, \ldots, m$ and $z \in$ ri C (Theorem 7.5 and Theorem 6.1). Thus the hypothesis of the theorem is satisfied. ‖

Corollary 28.2.1 can also be given a more direct proof which, unlike the proof of Theorem 28.2 via Theorem 21.2, uses only the separation results in §11 and no facts about polyhedral convexity. The argument is almost the same as in the first half of the proof of Theorem 28.2 (under the assumption that there are only inequality constraints), except that one uses the more elementary Theorem 21.1 instead of Theorem 21.2; the details of this proof make an easy exercise.

Another important special case of Theorem 28.2, whose proof does depend on the theory of polyhedral convexity, is the following.

COROLLARY 28.2.2. *Let (P) be an ordinary convex program with only linear constraints, i.e. with*

$$f_i(x) = \langle a_i, x \rangle - \alpha_i, \qquad i = 1, \ldots, m.$$

If the optimal value in (P) is not $-\infty$ *and (P) has a feasible solution in* ri C, *then a Kuhn–Tucker vector exists for (P).*

For an example of an ordinary convex program for which Kuhn–Tucker coefficients do *not* exist, let $C = R^2$, $f_0(\xi_1, \xi_2) = \xi_1$, $f_1(\xi_1, \xi_2) = \xi_2$, $f_2(\xi_1, \xi_2) = \xi_1^2 - \xi_2$, $r = 2$. The only $x = (\xi_1, \xi_2)$ satisfying the constraints

$$f_1(\xi_1, \xi_2) \leq 0, \qquad f_2(\xi_1, \xi_2) \leq 0,$$

is $x = (0, 0)$. This program therefore has $(0, 0)$ as its unique optimal solution and 0 as its optimal value. However, if (λ_1, λ_2) were a Kuhn–Tucker vector we would have $\lambda_1 \geq 0$, $\lambda_2 \geq 0$ and

$$0 \leq f_0(\xi_1, \xi_2) + \lambda_1 f_1(\xi_1, \xi_2) + \lambda_2 f_2(\xi_1, \xi_2)$$

$$= \xi_1 + (\lambda_1 - \lambda_2)\xi_2 + \lambda_2 \xi_1^2, \qquad \forall \xi_1, \xi_2,$$

which is impossible. The hypothesis of Theorem 28.2 is not satisfied here, since there is no (ξ_1, ξ_2) such that $f_1(\xi_1, \xi_2) \leq 0$ and $f_2(\xi_1, \xi_2) < 0$.

This example can be modified to show that something like the relative interior condition in Theorem 28.2 and Corollary 28.2.2 is needed, even when f_0 is linear on C and all the constraints are linear equations. Let

$$C = \{(\xi_1, \xi_2) \in R^2 \mid \xi_1^2 - \xi_2 \leq 0\},$$

$f_0(\xi_1, \xi_2) = \xi_1$, $f_1(\xi_1, \xi_2) = \xi_2$, $r = 0$. In the ordinary convex program given by this choice of elements, $x = (0, 0)$ is again the unique optimal solution and 0 is the optimal value. A Kuhn–Tucker vector would consist of a single real number λ_1 such that

$$0 \leq f_0(\xi_1, \xi_2) + \lambda_1 f_1(\xi_1, \xi_2) = \xi_1 + \lambda_1 \xi_2, \qquad \forall (\xi_1, \xi_2) \in C.$$

But no such λ_1 exists.

Kuhn–Tucker coefficients can be characterized in terms of the directional derivatives of the perturbation function p of (P) at $u = 0$, as we shall show in a more general setting in §29. It will follow from this characterization that Kuhn–Tucker coefficients always exist, except for certain situations where their existence would be highly unnatural from the heuristic "equilibrium price" point of view.

We shall now show how Kuhn–Tucker coefficients and optimal solutions in an ordinary convex program (P) can be characterized in terms of the "saddle-point" extrema of a certain concave-convex function on $R^m \times R^n$.

The *Lagrangian* of (P) is the function L on $R^m \times R^n$ defined by

$$L(u^*, x) = \begin{cases} f_0(x) + v_1^* f_1(x) + \cdots + v_m^* f_m(x) & \text{if } u^* \in E_r, x \in C, \\ -\infty & \text{if } u^* \notin E_r, x \in C, \\ +\infty & \text{if } x \notin C, \end{cases}$$

where

$$E_r = \{u^* = (v_1^*, \ldots, v_m^*) \in R^m \mid v_i^* \geq 0, i = 1, \ldots, r\}.$$

The variable v_i^* is known as the *Lagrange multiplier* associated with the *i*th constraint in (P).

If v_i^* is interpreted heuristically as the price per unit of perturbation variable v_i as above, L has a natural meaning. For any given $u^* \in R^m$ and $x \in R^n$, we have

$$L(u^*, x) = \inf \{f_0(x) + v_1^* v_1 + \cdots + v_m^* v_m \mid u \in U_x\},$$

where U_x is the set of perturbations $u = (v_1, \ldots, v_m)$ such that $v_i \geq f_i(x)$ for $i = 1, \ldots, r$ and $v_i = f_i(x)$ for $i = r + 1, \ldots, m$, i.e. such that x satisfies the constraints in the perturbed problem (P_u). Thus $L(u^*, x)$ can be interpreted as the minimum cost at which x can be obtained when the price for perturbations is u^*.

Observe that L is concave in u^* for each x and convex in x for each u^*. Moreover, L reflects *all* the structure of (P), because the $(m + 3)$-tuple $(C, f_0, \ldots, f_m, r)$ can be recovered completely from L. (Namely, C and r are uniquely determined by L because the set of points where L is finite is $E_r \times C$. The values of the functions $f_0, f_1, \ldots, f_m$ on C can be obtained from L by the formulas

$$f_0(x) = L(0, x), \qquad x \in C,$$

$$f_i(x) = L(e_i, x) - L(0, x) \quad \text{for} \quad i = 1, \ldots, m, \qquad x \in C,$$

where e_i is the vector forming the ith row of the $m \times m$ identity matrix.)
There is thus a one-to-one correspondence between ordinary convex programs and their Lagrangians.

A vector pair $(\bar{u}^*, \bar{x})$ is said to be a *saddle-point* of L (with respect to maximizing in u^* and minimizing in x) if

$$L(u^*, \bar{x}) \leq L(\bar{u}^*, \bar{x}) \leq L(\bar{u}^*, x), \qquad \forall u^*, \qquad \forall x.$$

THEOREM 28.3. *Let (P) be an ordinary convex program in the notation above. Let $\bar{u}^*$ and $\bar{x}$ be vectors in R^m and R^n, respectively. In order that $\bar{u}^*$ be a Kuhn–Tucker vector for (P) and $\bar{x}$ be an optimal solution to (P), it is necessary and sufficient that $(\bar{u}^*, \bar{x})$ be a saddle-point of the Lagrangian L of (P). Moreover, this condition holds if and only if $\bar{x}$ and the components λ_i of $\bar{u}^*$ satisfy*

(a) $\lambda_i \geq 0, f_i(\bar{x}) \leq 0$ and $\lambda_i f_i(\bar{x}) = 0, i = 1, \ldots, r,$
(b) $f_i(\bar{x}) = 0$ for $i = r + 1, \ldots, m,$
(c) $0 \in [\partial f_0(\bar{x}) + \lambda_1 \, \partial f_1(\bar{x}) + \cdots + \lambda_m \, \partial f_m(\bar{x})].$ (*Omit terms with $\lambda_1 = 0$.*)

PROOF. By the definition of "saddle-point," $(\bar{u}^*, \bar{x})$ is a saddle-point of L if and only if

$$\sup_{u^*} L(u^*, \bar{x}) = L(\bar{u}^*, \bar{x}) = \inf_x L(\bar{u}^*, x).$$

However, the inequality

$$\sup_{u^*} L(u^*, \bar{x}) \geq L(\bar{u}^*, \bar{x}) \geq \inf_x L(\bar{u}^*, x)$$

always holds. Thus $(\bar{u}^*, \bar{x})$ is a saddle-point if and only if

$$\sup_{u^*} L(u^*, \bar{x}) = \inf_x L(\bar{u}^*, x).$$

No matter what the choice of $\bar{x}$, we have

$$\sup_{u^*} L(u^*, \bar{x}) = \sup \{ f_0(\bar{x}) + v_1^* f_1(\bar{x}) + \cdots + v_m^* f_m(\bar{x}) \,|\, (v_1^*, \ldots, v_m^*) \in E_r \}$$

$$= f_0(\bar{x}) + \delta(\bar{x} \,|\, C_0) > -\infty,$$

where C_0 is the set of all feasible solutions to (P). On the other hand, for any given $\bar{u}^* = (\lambda_1, \ldots, \lambda_m)$ we have

$$+\infty > \inf_x L(\bar{u}^*, x) = \begin{cases} \inf h & \text{if} \quad \bar{u}^* \in E_r, \\ -\infty & \text{if} \quad \bar{u}^* \notin E_r, \end{cases}$$

where $h = f_0 + \lambda_1 f_1 + \cdots + \lambda_m f_m$. Thus $(\bar{u}^*, \bar{x})$ is a saddle-point of L

if and only if

(d) $\bar{u}^* \in E_r$, $\bar{x} \in C_0$, $\inf h = f_0(\bar{x})$.

Condition (d) is satisfied when $\bar{u}^*$ is a Kuhn–Tucker vector and $\bar{x}$ is an optimal solution, since then $\inf h = \alpha$ and $f_0(\bar{x}) = \alpha$, where α is the optimal value in (P) and is finite. On the other hand, suppose (d) is satisfied. For any $x \in C_0$ we have

$$\lambda_i f_i(x) \le 0, \qquad i = 1, \ldots, m,$$

and consequently $h(x) \le f_0(x)$. Therefore

$$\inf h = \inf_{x \in C_0} h(x) \le \inf_{x \in C_0} f_0(x) = \alpha \le f_0(\bar{x}),$$

and it follows that

$$\inf h = \alpha = f_0(\bar{x}).$$

Thus (d) implies that $\bar{u}^*$ is a Kuhn–Tucker vector and $\bar{x}$ is an optimal solution.

Of course, from the argument just given we have

$$\inf h \le h(\bar{x}) \le f_0(\bar{x})$$

when $\bar{u}^* \in E_r$ and $\bar{x} \in C_0$, where the second inequality is strict unless

$$\lambda_i f_i(\bar{x}) = 0, \qquad i = 1, \ldots, r.$$

Thus (d) implies (a), (b) and the condition:

(c') $h(\bar{x}) = \inf h$.

Conversely, (a), (b) and (c') imply (d), because $h(\bar{x}) = f_0(\bar{x})$ under (a) and (b). To complete the proof of the theorem, we need only show that (c') is equivalent to (c), assuming $\bar{u}^* \in E_r$. By definition, h attains its infimum at $\bar{x}$ if and only if 0 is a subgradient of h at $\bar{x}$, i.e. $0 \in \partial h(\bar{x})$. Since

$$\bigcap_{i=0}^{m} \mathrm{ri}\,(\mathrm{dom}\, f_i) = \mathrm{ri}\, C \ne \emptyset$$

(by our blanket assumption on p. 273), we have

$$\partial h(x) = \partial f_0(x) + \partial(\lambda_1 f_1)(x) + \cdots + \partial(\lambda_m f_m)(x)$$
$$= \partial f_0(x) + \lambda_1\, \partial f_1(x) + \cdots + \lambda_m\, \partial f_m(x)$$

for every x by Theorem 23.8. Thus the condition $0 \in \partial h(\bar{x})$ is equivalent to (c). ‖

Conditions (a), (b) and (c) are known as the *Kuhn–Tucker conditions* for (P). When the functions f_i are actually differentiable at $\bar{x}$, $\partial f_i(\bar{x})$ reduces of course to the gradient $\nabla f_i(\bar{x})$ of f_i at $\bar{x}$ (Theorem 25.1), and (c) becomes the gradient equation

$$\nabla f_0(\bar{x}) + \lambda_1 \nabla f_1(\bar{x}) + \cdots + \lambda_m \nabla f_m(\bar{x}) = 0.$$

It follows from Theorem 28.3 that, in circumstances where Kuhn–Tucker vectors are sure to exist, solving the constrained minimization problem in (P) is equivalent to solving the unconstrained (or rather: more simply constrained) extremum problem of finding a saddle-point of L.

COROLLARY 28.3.1 (Kuhn–Tucker Theorem). *Let (P) be an ordinary convex program satisfying the hypothesis of Theorem 28.2. In order that a given vector $\bar{x}$ be an optimal solution to (P), it is necessary and sufficient that there exist a vector $\bar{u}^*$ such that $(\bar{u}^*, \bar{x})$ is a saddle-point of the Lagrangian L of (P). Equivalently, $\bar{x}$ is an optimal solution if and only if there exist Lagrange multiplier values λ_i which, together with $\bar{x}$, satisfy the Kuhn–Tucker conditions for (P).*

It is interesting to see that the Kuhn–Tucker conditions for an ordinary convex program could also be derived in a different and rather instructive way directly from the theory of subdifferentiation. For simplicity of exposition, we shall assume that $C = R^n$, $r = m$, and that the system

$$f_1(x) < 0, \dots, f_m(x) < 0,$$

has at least one solution. (The argument could be extended to the general case, however.) Setting as above

$$C_i = \{x \mid f_i(x) \le 0\}, \qquad i = 1, \dots, m,$$

we may express the objective function f for (P) by

$$f(x) = f_0(x) + \delta(x \mid C_1) + \cdots + \delta(x \mid C_m).$$

The optimal solutions to (P) are the vectors $\bar{x}$ such that $0 \in \partial f(\bar{x})$. Our assumption about the system $f_i(x) < 0$, $i = 1, \dots, m$, implies (by the continuity of finite convex functions on R^n) that

$$\text{int } C_1 \cap \cdots \cap \text{int } C_m \ne \emptyset.$$

Of course, C_i is the effective domain of $\delta(\cdot \mid C_i)$ for $i = 1, \dots, m$, and $R^n = \text{dom} f_0$. It follows from Theorem 23.8 that

$$\partial f(x) = \partial f_0(x) + \partial \delta(x \mid C_1) + \cdots + \partial \delta(x \mid C_m).$$

Furthermore, $\partial \delta(x \mid C_i)$ is the normal cone to C_i at x, and according to Corollary 23.7.1 this is given by

$$\partial \delta(x \mid C_i) = \begin{cases} \bigcup \{\lambda_i \, \partial f_i(x) \mid \lambda_i \ge 0\} & \text{if } f_i(x) = 0, \\ \{0\} & \text{if } f_i(x) < 0, \\ \emptyset & \text{if } f_i(x) > 0. \end{cases}$$

It follows that $\partial f(x)$ is non-empty if and only if x satisfies $f_i(x) \le 0$ for

$i = 1, \ldots, m$, in which case $\partial f(x)$ is the union of

$$\partial f_0(x) + \lambda_1 \, \partial f_1(x) + \cdots + \lambda_m \, \partial f_m(x)$$

over all choices of coefficients $\lambda_i \geq 0$ such that

$$\lambda_i f_i(x) = 0 \quad \text{for} \quad i = 1, \ldots, m.$$

Thus $0 \in \partial f(\bar{x})$ if and only if there exist coefficients $\lambda_1, \ldots, \lambda_m$ which along with $\bar{x}$ satisfy the Kuhn–Tucker conditions.

Theorem 28.3 shows how the optimal solutions and Kuhn–Tucker vectors for (P) can be characterized in terms of the Lagrangian L of (P). The following theorem shows how the optimal value in (P) can likewise be characterized in terms of the Lagrangian L.

THEOREM 28.4.   *Let* (P) *be an ordinary convex program with Lagrangian* L. *If* $\bar{u}^*$ *is a Kuhn–Tucker vector for* (P) *and* $\bar{x}$ *is an optimal solution, the saddle-value* $L(\bar{u}^*, \bar{x})$ *is the optimal value in* (P). *More generally,* $\bar{u}^*$ *is a Kuhn–Tucker vector for* (P) *if and only if*

$$-\infty < \inf_x L(\bar{u}^*, x) = \sup_{u^*} \inf_x L(u^*, x) = \inf_x \sup_{u^*} L(u^*, x),$$

in which case the common extremum value in the latter equation is the optimal value in (P).

PROOF.   If $\bar{u}^* = (\lambda_1, \ldots, \lambda_m)$ is a Kuhn–Tucker vector and $\bar{x}$ is an optimal solution, we have

$$L(\bar{u}^*, \bar{x}) = f_0(\bar{x}) + \lambda_1 f_1(\bar{x}) + \cdots + \lambda_m f_m(\bar{x}) = f_0(\bar{x})$$

by the Kuhn–Tucker conditions in Theorem 28.3, and hence $L(\bar{u}^*, \bar{x})$ is the optimal value in (P). Now in general, as was shown in the proof of Theorem 28.3,

$$\sup_{u^*} L(u^*, x) = f_0(x) + \delta(x \mid C_0) = f(x),$$

where f is the objective function for (P). Thus

$$\inf_x \sup_{u^*} L(u^*, x) = \alpha,$$

where α is the optimal value in (P). For any $\bar{u}^* = (\lambda_1, \ldots, \lambda_m)$ we have

$$\sup_{u^*} L(u^*, x) \geq L(\bar{u}^*, x), \qquad \forall x,$$

so that

$$\alpha \geq \inf_x L(\bar{u}^*, x).$$

Moreover, according to the proof of Theorem 28.3,

$$\inf_x L(\bar{u}^*, x) = \begin{cases} \inf (f_0 + \lambda_1 f_1 + \cdots + \lambda_m f_m) & \text{if} \quad \bar{u}^* \in E_r, \\ -\infty & \text{if} \quad \bar{u}^* \notin E_r. \end{cases}$$

Therefore $\bar{u}^*$ is a Kuhn–Tucker vector if and only if the supremum of the function

$$g = \inf_x L(\cdot, x)$$

over R^m is $\alpha > -\infty$ and is attained at $\bar{u}^*$. This is the assertion of the theorem. ‖

COROLLARY 28.4.1. *Let* (P) *be an ordinary convex program having at least one Kuhn–Tucker vector, e.g. an ordinary convex program satisfying the hypothesis of Theorem 28.2. Let g be the concave function defined by*

$$g(u^*) = \inf_x L(u^*, x),$$

where L is the Lagrangian of (P). *The Kuhn–Tucker vectors for* (P) *are then precisely the points $\bar{u}^*$ where g attains its supremum over R^m.*

The concavity of g in Corollary 28.4.1 is immediate, of course, from the fact that g is the pointwise infimum of the concave functions $L(\cdot, x)$, $x \in R^n$. Note that, as a matter of fact, g is the pointwise infimum of the *affine* functions of the form

$$u^* = (v_1^*, \ldots, v_m^*) \to f_0(x) + v_1^* f_1(x) + \cdots + v_m^* f_m(x)$$
$$+ v_1^* \zeta_1 + \cdots + v_r^* \zeta_r,$$

where $x \in C$ and $\zeta_i \geq 0$ for $i = 1, \ldots, r$.

Corollary 28.4.1 shows that the problem of determining a Kuhn–Tucker vector for a given (P) can be reduced to the numerical problem of maximizing a certain concave function g on R^m. In some cases the latter problem is computationally feasible, since an explicit representation is known for g as the pointwise infimum of a collection of affine functions, even though no "analytic" formula for g may be known. It is interesting to note that

$$g(u^*) = -p^*(-u^*),$$

where p is the perturbation function for (P) introduced earlier in this section. (In terms of the conjugacy correspondence for concave functions defined at the beginning of §30, one has $g = (-p)^*$. The function p is convex, as will be shown in Theorem 29.1.)

An important *decomposition principle* can be derived from the Lagrange multiplier theory for ordinary convex programs. Suppose that the functions f_i in (P) can be expressed in the form

$$f_i(x) = f_{i1}(x_1) + \cdots + f_{is}(x_s), \qquad i = 0, 1, \ldots, m,$$

where each f_{ik} is a proper convex function on R^{n_k} (affine for $i > r$) and

$$x = (x_1, \ldots, x_s), \qquad x_k \in R^{n_k}, \qquad n_1 + \cdots + n_s = n.$$

Let

$$C^k = \text{dom } f_{0k} \subset R^{n_k}, \qquad k = 1, \ldots, s,$$

so that (by our convention that dom $f_0 = C$)

$$C = \{x = (x_1, \ldots, x_s) \mid x_k \in C^k, k = 1, \ldots, s\}.$$

Then (P) can be described as the problem of minimizing

$$f_{01}(x_1) + \cdots + f_{0s}(x_s)$$

subject to

$$x_k \in C^k \quad \text{for} \quad k = 1, \ldots, s,$$

$$f_{i1}(x_1) + \cdots + f_{is}(x_s) \leq 0 \quad \text{for} \quad i = 1, \ldots, r$$

$$f_{i1}(x_1) + \cdots + f_{is}(x_s) = 0 \quad \text{for} \quad i = r + 1, \ldots, m.$$

Heuristically, we may think of such a problem as arising when s different problems of the form

$$\text{minimize} \quad f_{0k} \quad \text{over} \quad C^k, \qquad k = 1, \ldots, s$$

(where the convex set C^k may given by some system of inequalities or equations in R^{n_k}) become interdependent through the introduction of a few joint constraints. The decomposition principle asserts that, when a Kuhn–Tucker vector $(\lambda_1, \ldots, \lambda_m)$ exists for (P), it is possible to break (P) down again into s independent problems over the sets C^k by appropriate modification of the functions f_{0k}.

Specifically, given the coefficients λ_i, we can reduce (P), as explained in Theorem 28.1 (or Corollary 28.1.1) to the problem of minimizing h over C, where

$$h = f_0 + \lambda_1 f_1 + \cdots + \lambda_m f_m.$$

In view of the given expressions for $f_0, \ldots, f_m$, however, we have

$$h(x) = h_1(x_1) + \cdots + h_s(x_s), \qquad x_k \in R^{n_k},$$

where

$$h_k = f_{0k} + \lambda_1 f_{1k} + \cdots + \lambda_m f_{mk}, \qquad k = 1, \ldots, s,$$

so that the problem of minimizing h over C is equivalent to the s independent problems:

$$\text{minimize} \quad h_k \quad \text{over} \quad C^k, \qquad k = 1, \ldots, s.$$

Note that the latter extremum problems are in the spaces R^{n_k}, whereas by Corollary 28.4.1 the problem of determining a Kuhn–Tucker vector $(\lambda_1, \ldots, \lambda_m)$ is an extremum problem in R^m. Thus, by means of the decomposition principle, the extremum problem (P) of dimensionality n can be replaced by $s + 1$ extremum problems of (possibly much lower) dimensionalities $n_1, \ldots, n_s$, and m. In many cases such a reduction in dimensionality makes possible the numerical solution of problems which would otherwise be hopelessly large.

A good illustration of the decomposition principle is provided by the problem of minimizing

$$q(x) = q_1(\xi_1) + \cdots + q_n(\xi_n)$$

subject to

$$x = (\xi_1, \ldots, \xi_n) \geq 0, \qquad \xi_1 + \cdots + \xi_n = 1,$$

where each q_k is a proper convex function on R such that

$$\text{dom } q_k \supset [0, 1].$$

To express this problem in the above form, we set

$$f_{0k}(\xi_k) = \begin{cases} q_k(\xi_k) & \text{if } \xi_k \geq 0, \\ +\infty & \text{if } \xi_k < 0, \end{cases}$$

for $k = 1, \ldots, n$ and

$$f_{1k}(\xi_k) = \xi_k, \qquad k = 1, \ldots, n - 1,$$
$$f_{1n}(\xi_n) = \xi_n - 1.$$

In terms of these functions, the problem is to minimize

$$f_0(x) = f_{01}(\xi_1) + \cdots + f_{0n}(\xi_n)$$

subject to the linear constraint

$$f_1(x) = f_{11}(\xi_1) + \cdots + f_{1n}(\xi_n) = 0.$$

Thus we have an ordinary convex program to which the decomposition principle is applicable, namely the (P) given by $(C, f_0, f_1, 0)$, where

$$C = \text{dom } f_0 = \{x \mid \xi_k \in \text{dom } f_{0k}, k = 1, \ldots, n\}.$$

The infimum in (P) is finite (since the functions q_k, being finite and convex on $[0, 1]$, are all bounded on $[0, 1]$), and the interior of C contains points x such that $f_1(x) = 0$. Hence by Corollary 28.2.2 a Kuhn–Tucker vector, consisting here of just a single Kuhn–Tucker coefficient λ_1, exists for (P). If such a λ_1 can be calculated, we can replace the original problem by n problems of one dimension:

$$\text{minimize } f_{0k}(\xi_k) + \lambda_1 f_{1k}(\xi_k) \text{ in } \xi_k.$$

Having determined for $k = 1, \ldots, n$ the real interval I_k consisting of the points where the latter minimum is attained, we can get all the optimal solutions to the original problem by taking all the vectors $x = (\xi_1, \ldots, \xi_n)$ such that $\xi_k \in I_k$ for $k = 1, \ldots, n$ and $\xi_1 + \cdots + \xi_n = 1$ (Theorem 28.1).

To see how a Kuhn–Tucker coefficient λ_1 may be calculated, we apply

Corollary 28.4.1. The Lagrangian of (P) is given simply by

$$L(v_1^*, x) = -v_1^* + \sum_{k=1}^{n} [f_{0k}(\xi_k) + v_1^* \xi_k]$$

for every $v_1^* \in R$ and $x \in R^n$, where v_1^* is the Lagrange multiplier associated with the constraint $f_1(x) = 0$, and hence the g in Corollary 28.4.1 is given by

$$g(v_1^*) = -v_1^* + \sum_{k=1}^{n} \inf \{f_{0k}(\xi_k) + v_1^* \xi_k | \xi_k \in R\}$$
$$= -v_1^* - \sum_{k=1}^{n} f_{0k}^*(-v_1^*).$$

Thus λ_1 is a Kuhn–Tucker coefficient for (P) if and only if $v_1^* = \lambda_1$ minimizes the quantity

$$-g(v_1^*) = v_1^* + f_{01}^*(-v_1^*) + \cdots + f_{0n}^*(-v_1^*).$$

Computing the minimum of the convex function $-g$ is, of course, a relatively easy matter, since only one real variable v_1^* is involved and the conjugate functions f_{0k}^* are fairly simple to determine. Thus the decomposition principle allows us in this example to replace a problem in essentially $n - 1$ real variables (one of the original variables ξ_k being redundant because of the constraint $\xi_1 + \cdots + \xi_n = 1$) by $n + 1$ problems, each in a single real variable.

A more general example of an application of the decomposition principle is the following. For $k = 1, \ldots, s$, let f_{0k} be a proper convex function on R^{n_k} and let A_k be an $m \times n_k$ real matrix. We shall consider the problem of minimizing

$$f_{01}(x_1) + \cdots + f_{0s}(x_s), \qquad x_k \in R^{n_k},$$

subject to

$$A_1 x_1 + \cdots + A_s x_s = a,$$

where a is a given element of R^m. Here each f_{0k} might itself be the objective function in some ordinary convex program (P_k); in particular, the effective domain C^k of f_{0k} might be given by some further system of constraints. However, at the moment we are only concerned with the constraints which make $x_1, \ldots, x_s$ interdependent, and we are supposing that these constraints are linear.

For $i = 1, \ldots, m$ and $k = 1, \ldots, s$ let a_{ik} denote the vector in R^{n_k} forming the ith row of the matrix A_k, and let α_i be the ith component of a. Let

$$f_{ik}(x_k) = \langle a_{ik}, x_k \rangle \quad \text{for} \quad k = 1, \ldots, s - 1,$$
$$f_{is}(x_s) = \langle a_{is}, x_s \rangle - \alpha_i,$$

and let

$$f_i(x) = f_{i1}(x_1) + \cdots + f_{is}(x_s),$$

where

$$x = (x_1, \ldots, x_s), \qquad x_k \in R^{n_k}.$$

The ordinary convex program (P) we want to consider is the one given by $(C, f_0, \ldots, f_m, 0)$, where

$$C = \mathrm{dom}\, f_0 = \{x \mid x_k \in C^k, k = 1, \ldots, s\}.$$

Observe incidentally that, if each C^k happens to be the set of feasible solutions to some (P_k), the interior of C could well be empty; it is for the sake of situations like this that we have formulated Theorem 28.2 and Corollary 28.2.2 in terms of ri C.

According to Corollary 28.2.2, if the infimum in (P) is finite and it is possible to choose vectors $x_k \in$ ri C^k such that

$$A_1 x_1 + \cdots + A_s x_s = a,$$

then a Kuhn–Tucker vector

$$\bar{u}^* = (\lambda_1, \ldots, \lambda_m)$$

exists for (P). Given such a $\bar{u}^*$, we could replace (P) by the s independent problems in which the function h_k is minimized over R^{n_k} for $k = 1, \ldots, s$, where

$$
\begin{aligned}
h_k &= f_{0k} + \lambda_1 f_{1k} + \cdots + \lambda_m f_{mk} \\
&= \begin{cases} f_{0k} + \langle\, \cdot\, , A_k^* \bar{u}^* \rangle & \text{if} \quad k = 1, \ldots, s - 1, \\ f_{0s} + \langle\, \cdot\, , A_s^* \bar{u}^* \rangle - \langle a, \bar{u}^* \rangle & \text{if} \quad k = s \end{cases}
\end{aligned}
$$

(A_k^* being the transpose of A_k). The set of optimal solutions to (P) would then consist of the vectors $x = (x_1, \ldots, x_s)$ such that $x_k \in D_k$ and $\sum_{k=1}^s A_k x_k = a$, where D_k is the set of points where h_k attains its minimum (Theorem 28.1).

The Lagrangian in this example is

$$L(u^*, x) = -\langle a, u^* \rangle + \sum_{k=1}^s [f_{0k}(x_k) + \langle x_k, A_k^* u^* \rangle]$$

so that in Corollary 28.4.1 we have

$$
\begin{aligned}
g(u^*) &= -\langle a, u^* \rangle - \sum_{k=1}^s \sup_{x_k} \{\langle x_k, -A_k^* u^* \rangle - f_{0k}(x_k)\} \\
&= -\langle a, u^* \rangle - \sum_{k=1}^s f_{0k}^*(-A_k^* u^*).
\end{aligned}
$$

The Kuhn–Tucker vectors for (P) can therefore be obtained by minimizing the convex function

$$w(u^*) = \langle a, u^* \rangle + f_{01}^*(-A_1^* u^*) + \cdots + f_{0s}^*(-A_s^* u^*)$$

on R^m.

The problem of minimizing w is not necessarily easy, but it is worth noting that it may be tractable even in certain cases where the conjugate functions f_{0k}^* can not be written down explicitly. Let us assume for simplicity that each of the functions f_{0k} is co-finite (this being true in particular if f_{0k} is closed and C^k is bounded), so that f_{0k}^* is finite throughout R^m. Then w is finite throughout R^m, and by Theorems 23.8 and 23.9 the subgradients of w are given by the formula

$$\partial w(u^*) = a - A_1 \,\partial f_{01}^*(-A_1^* u^*) - \cdots - A_s \,\partial f_{0s}^*(-A_s^* u^*).$$

On the other hand, by Theorem 23.5 we have

$$x_k \in \partial f_{0k}^*(-A_k^* u^*)$$

if and only if x_k minimizes the function

$$f_{0k} + \langle \,\cdot\,, A_k^* u^* \rangle,$$

the minimum value itself being, of course, $f_{0k}^*(-A_k^* u^*)$. Thus, given any $u^* \in R^m$, it is possible to calculate $w(u^*)$ and $\partial w(u^*)$ by solving the s problems

$$\text{minimize} \quad f_{0k}(x_k) + \langle x_k, A_k^* u^* \rangle \quad \text{in} \quad x_k.$$

It follows that, in cases where the latter problems are relatively easy to solve (as for example when every f_{0k} is of the form

$$f_{0k}(x_k) = \begin{cases} g_k(x_k) & \text{if} \quad x_k \geq 0,\ B_k x_k = b_k, \\ +\infty & \text{otherwise,} \end{cases}$$

with g_k some finite differentiable convex function on R^{n_k}), one could minimize w by any method which demanded only the ability to calculate $w(u^*)$ and an element of $\partial w(u^*)$ for any given u^*. Note in particular that, if every f_{0k} is actually strictly convex on C^k, then every f_{0k}^* is differentiable (Theorem 26.3) and $\partial w(u^*)$ reduces to $\nabla w(u^*)$, so that gradient methods could be considered.

Bifunctions and Generalized Convex Programs

In an ordinary convex program (P), one is interested in minimizing a certain convex function on R^n, the objective function for (P), whose effective domain is the set of all feasible solutions to (P). But there is more to (P) than just this abstract minimization problem. Another ordinary convex program can have the same objective function as (P) and yet have a different Lagrangian and different Kuhn–Tucker coefficients. If one is to have a full generalization of the concept of "convex program," one must somehow take this fact into account.

The Kuhn–Tucker vectors corresponding to an ordinary convex program can be characterized in terms of a certain class of perturbations of the objective function for the program, as we have shown in §28, and this is the key to the generalization which will be developed below. A convex program will be defined in effect as an (extended-real-valued) convex "objective function" together with a particular class of "convex" perturbations of this "objective function." For such generalized programs a Lagrangian theory will be given in which Kuhn–Tucker vectors can be interpreted in terms of "equilibrium prices" for the perturbations, much as in §28.

In order to express the dependence of the objective function in a minimization problem on a vector u corresponding to a perturbation, we find it convenient to introduce the concept of a "bifunction," as a generalization of a multivalued mapping. This is not so much a new concept as a different way of treating an old concept, the distinction between "variables" and "parameters." Actually, there is nothing in the present section which would compel us to introduce bifunction terminology. All the results could just as well be stated in more conventional terms. But the concept of a bifunction will be increasingly useful as this book progresses, so we might as well begin exploiting it now.

We define a *bifunction* from R^m to R^n to be a mapping F which assigns to each $u \in R^m$ a function Fu on R^n with values in $[-\infty, +\infty]$. The value of *Fu* at a point $x \in R^n$ will be denoted by $(Fu)(x)$. The function

$$(u, x) \to (Fu)(x), \qquad (u, x) \in R^m \times R^n = R^{m+n},$$

will be called the *graph function* of *F*. (The concept of a bifunction could be developed in more general terms, but the definition given here will suffice for present purposes.)

It is clear that each extended-real-valued function f on R^{m+n} is the graph function of exactly one bifunction from R^m to R^n, namely the F defined by

$$Fu = f(u, \cdot), \qquad \forall u \in R^m.$$

Thus a bifunction can simply be regarded as the first stage of a function broken down into two stages:

$$F: u \to Fu: x \to (Fu)(x).$$

The one-to-one correspondence between bifunctions from R^m to R^n and extended-real-valued functions on R^{m+n} is analogous to the one-to-one correspondence between multivalued mappings from R^m to R^n and subsets of R^{m+n} (the graphs of the mappings). The terminology of bifunctions is useful in the same contexts where the terminology of multivalued mappings is useful, i.e. where one wants to stress analogies with notions familiar for single-valued mappings from R^m to R^n.

For heuristic purposes, it is helpful to think of a bifunction as a generalization of a multivalued mapping in the following way. Let F be any bifunction from R^m to R^n such that $(Fu)(x)$ is never $-\infty$, and for each $u \in R^m$ let Su be the set of points $x \in R^n$ such that $(Fu)(x) < +\infty$. To specify F completely, it is enough to specify for each u the set Su and a certain real-valued function on Su (the restriction of Fu), since F can be reconstructed from this information by a $+\infty$ extension. Thus F may be identified heuristically with a correspondence which assigns to each $u \in R^m$ a set Su equipped with a distinguished real-valued function (a "valuation" giving the "cost," say, of each element x of Su). This correspondence reduces to the multivalued mapping $S: u \to Su$ if the distinguished function is identically zero on Su for every u, i.e. if F is the $(+\infty)$ *indicator bifunction* of S:

$$(Fu)(x) = \begin{cases} 0 & \text{if } x \in Su, \\ +\infty & \text{if } x \notin Su. \end{cases}$$

(Here we have invoked $+\infty$ extensions and excluded $-\infty$ as a possible value of Fu, but in later sections, where concave bifunctions as well as convex bifunctions appear, the opposite situation—where the roles of $+\infty$ and $-\infty$ are reversed—will also have to be kept in mind. In particular, we will sometimes make use of indicator bifunctions which have $-\infty$ in place of $+\infty$.)

A bifunction F from R^m to R^n will be called *convex* if its graph function is convex on R^{m+n}. This implies in particular that Fu is a convex function on R^n for each $u \in R^m$. A convex bifunction will be said to be *closed* or *proper* according to whether its graph function is closed or proper, respectively.

The *graph domain* of a convex bifunction F from R^m to R^n is defined to be the effective domain of the graph function of F (a certain convex set in R^{m+n}). The *effective domain* of F, denoted by dom F, is defined to be the set of all vectors $u \in R^m$ such that Fu is not the constant function $+\infty$. Thus dom F is the projection on R^m of the graph domain of F in R^{m+n} and hence is a convex set in R^m. If F is proper, dom F consists of the vectors u such that the convex function Fu is proper.

A simple example of a convex bifunction which will be very important to us theoretically, although not in conjunction with generalized convex programs, is the $(+\infty)$ *indicator bifunction of a linear transformation A* from R^m to R^n, i.e. the F defined by

$$(Fu)(x) = \delta(x \mid Au) = \begin{cases} 0 & \text{if } x = Au, \\ +\infty & \text{if } x \neq Au. \end{cases}$$

This F is convex, because its graph function is the indicator function of the graph of A, which happens to be a convex set (a subspace) in R^{m+n}. Observe that F is closed and proper, and dom $F = R^m$. This example, as will be seen later, provides a useful bridge between linear algebra and the theory of convex bifunctions.

The main example for present purposes is the following. Let (P) be an ordinary convex program in the notation of §28. For each $u = (v_1, \ldots, v_m) \in R^m$, let Su denote the subset of R^n consisting of the vectors x such that

$$f_1(x) \leq v_1, \ldots, f_r(x) \leq v_r, \qquad f_{r+1}(x) = v_{r+1}, \ldots, f_m(x) = v_m.$$

Define the bifunction F from R^m to R^n by

$$Fu = f_0 + \delta(\cdot \mid Su), \qquad \forall u.$$

We shall call F the *convex bifunction associated with the ordinary convex program* (P). The convexity of F follows from the fact that the graph function of F can be expressed as a sum of functions g_i on R^{m+n}, each of which is obviously convex:

$$(Fu)(x) = g_0(u, x) + g_1(u, x) + \cdots + g_m(u, x).$$

where

$$g_i(u, x) = \begin{cases} f_0(x) & \text{for} \quad i = 0, \\ \delta(u, x \mid f_i(x) \leq v_i) & \text{for} \quad i = 1, \dots, r, \\ \delta(u, x \mid f_i(x) = v_i) & \text{for} \quad i = r + 1, \dots, m. \end{cases}$$

(The notation here means that, for $i = 1, \dots, r, g_i(u, x)$ is 0 when the ith component v_i of u satisfies $v_i \geq f_i(x)$, and otherwise $g_i(u, x) = +\infty$. Thus, for $i = 1, \dots, r, g_i$ is the indicator function of a copy of the convex set epi f_i. Analogously for $i = r + 1, \dots, m$.) We have

$$\operatorname{dom} F = \{u \in R^m \mid Su \cap C \neq \emptyset\},$$

where $C = \operatorname{dom} f_0$. The convex set $\operatorname{dom} F$ is not empty, because it contains the vector

$$(f_1(x), \dots, f_m(x))$$

for each $x \in C$. Since $\operatorname{dom} F \neq \emptyset$ and $(Fu)(x)$ is never $-\infty$, F is proper. If the convex functions $f_0, f_1, \dots, f_r$ are closed, F is closed. (Recall that, in the notation of §28, it is assumed that $f_{r+1}, \dots, f_m$ are affine functions on R^n and hence closed.)

It is important to realize that an ordinary convex program (P) is uniquely determined by its associated bifunction F. The $(m + 3)$-tuple $(C, f_0, \dots, f_m, r)$ can be reconstructed from F as follows. In the first place,

$$C = \{x \in R^n \mid \exists u \in R^m, (Fu)(x) < +\infty\}.$$

For any $x \in C$ and any u such that $(Fu)(x) < +\infty$, one has

$$f_0(x) = (Fu)(x).$$

This fixes f_0. For any $x \in C$, one has

$$\{u \in R^m \mid (Fu)(x) < +\infty\} = (f_1(x), \dots, f_m(x)) + K,$$

where K is the convex cone in R^m consisting of the vectors $y = (\eta_1, \dots, \eta_m)$ such that $\eta_i \geq 0$ for $i = 1, \dots, r$ and $\eta_i = 0$ for $i = r + 1, \dots, m$. This characterizes the integer r and determines the values of the functions $f_1, \dots, f_m$ on C.

Thus, instead of defining an ordinary convex program on R^n formally in terms of a certain $(m + 3)$-tuple, we could just as well define it in terms of a certain convex bifunction from R^m to R^n. This is the approach which we shall now take in a more general setting.

Let F be any convex bifunction from R^m to R^n. We define the *(generalized) convex program* (P) associated with F to be the "minimization problem with perturbations" in which the function $F0$ is to be minimized over R^n and the given perturbations are those that replace $F0$ by Fu for different choices of $u \in R^m$. (The rigorous definition of (P) is that (P) is simply F itself; cf. the remarks at the beginning of §28. However, the introduction of (P) and of terminology tied to (P) rather than to F is useful, even if redundant, because there are contexts in which we shall want to use the notion of a convex bifunction without evoking the picture of a particular "minimization problem with perturbations.")

The convex function $F0$ will be called the *objective function* for (P), and its infimum (over R^n) will be called the *optimal value* in (P). The vectors in the convex set

$$\text{dom } F0 = \{x \in R^n \mid (F0)(x) < +\infty\}$$

will be called the *feasible solutions* to (P), and (P) will be said to be *consistent* if at least one such vector exists. (Thus (P) is consistent if and only if the optimal value in (P) is $< +\infty$.) We define an *optimal solution* to (P) to be a vector $x \in R^n$ such that $(F0)(x)$ is finite and equal to the optimal value in (P). (Thus we do not speak of optimal solutions to (P) when (P) is inconsistent, even though in that case $(F0)(x)$ is the optimal value $+\infty$ in (P) for every x.)

The set of all optimal solutions to (P) is empty unless $F0$ is proper; when $F0$ is proper it is the minimum set of $F0$, a (possibly empty) convex subset of the set of all feasible solutions to (P).

A general condition for the existence of optimal solutions to (P) may be obtained by applying Theorem 27.2 to $F0$. In what follows we shall be concerned not so much with optimal solutions as with generalized Kuhn–Tucker vectors.

Thinking of the convex function Fu on R^n as the objective function for (P) perturbed by the amount u, we define the *perturbation function* for (P) to be the (extended-real-valued) function $\inf F$ on R^m given by

$$(\inf F)(u) = \inf Fu = \inf_x (Fu)(x).$$

Note that the value of $\inf F$ at $u = 0$ is just the optimal value in (P).

We define a vector $u^* \in R^m$ to be a *Kuhn–Tucker vector* for (P) if the quantity

$$\inf_u \{\langle u^*, u \rangle + \inf Fu\} = \inf_{u,x} \{\langle u^*, u \rangle + (Fu)(x)\}$$

is finite and equal to the optimal value $\inf F0$ in (P). Since $\langle u^*, u \rangle + \inf Fu$ equals $\inf F0$ when $u = 0$, this condition on u^* is equivalent to the condition that ($\inf F0$ be finite and)

$$\inf Fu + \langle u^*, u \rangle \geq \inf F0, \qquad \forall u.$$

Kuhn–Tucker vectors for (P) can therefore be interpreted heuristically as equilibrium price vectors exactly as in §28, and this is one of the main motivations for studying them.

The function L on $R^m \times R^n$ given by

$$L(u^*, x) = \inf_u \{\langle u^*, u \rangle + (Fu)(x)\}, \qquad \forall u^*, \forall x,$$

will be called the *Lagrangian* of (P). Since

$$\inf_{u,x} \{\langle u^*, u \rangle + (Fu)(x)\} = \inf_x L(u^*, x),$$

the definition of a Kuhn–Tucker vector for (P) can be stated in terms of L just as well as in terms of inf F: u^* is a Kuhn–Tucker vector if and only if the infimum of the function $L(u^*, \cdot)$ on R^n is finite and equal to the optimal value in (P). We shall demonstrate below that Kuhn–Tucker vectors and optimal solutions under the generalized definitions correspond to saddle-points of the Lagrangian L, just as in the case of ordinary convex programs in Theorem 28.3, at least when F is closed and proper.

If (P) is an ordinary convex program, the concepts just defined do, of course, reduce to those defined earlier. The function inf F becomes the perturbation function p in the discussion of "equilibrium prices" in §28. The formula for the Lagrangian of (P) reduces to

$$L(u^*, x) = \inf \{v_1^* v_1 + \cdots + v_m^* v_m + f_0(x) \mid u \in U_x\},$$

where U_x is the set of vectors $u = (v_1, \ldots, v_m) \in R^m$ such that $v_i \geq f_i(x)$ for $i = 1, \ldots, r$ and $v_i = f_i(x)$ for $i = r + 1, \ldots, m$, and therefore

$$L(u^*, x) = \begin{cases} f_0(x) + v_1^* f_1(x) + \cdots + v_m^* f_m(x) & \text{if } u^* \in E_r, x \in C \\ -\infty & \text{if } u^* \notin E_r, x \in C, \\ +\infty & \text{if } x \notin C, \end{cases}$$

as in §28, where E_r is the set of vectors $u^* = (v_1^*, \ldots, v_m^*)$ such that $v_i^* \geq 0$ for $i = 1, \ldots, r$. The definition of a Kuhn–Tucker vector u^* in terms of $L(u^*, \cdot)$ becomes, for this L, the definition in §28.

For a miscellaneous but illuminating example of a generalized convex program which is not an ordinary convex program, consider the bifunction $F: R^2 \to R^n$ defined by

$$(Fu)(x) = \begin{cases} [\langle x, Qx \rangle/(1 + v_1)] + \langle a, x \rangle & \text{if } v_1 > -1 \text{ and } x \in B + v_2 e, \\ 0 & \text{if } v_1 = -1, Qx = 0 \text{ and } x \in B + v_2 e, \\ +\infty & \text{otherwise,} \end{cases}$$

where $u = (v_1, v_2)$, Q is a symmetric $n \times n$ positive semi-definite matrix, B is the Euclidean unit ball of R^n, and a and e are elements of R^n with $|e| = 1$. This F is a closed proper convex bifunction, as is easily seen from the fact that

$$(Fu)(x) = f_1(1 + v_1, x) + f_2(v_2, x)$$

where f_1 is given in terms of the quadratic convex function

$$q(x) = \langle x, Qx \rangle + \langle a, x \rangle$$

by the formula

$$f_1(\lambda, x) = \begin{cases} (q\lambda)(x) & \text{if} \quad \lambda > 0, \\ (q0^+)(x) & \text{if} \quad \lambda = 0, \\ +\infty & \text{if} \quad \lambda < 0 \end{cases}$$

(see the remarks preceding Corollary 8.5.2), and f_2 is the indicator of the convex set

$$\{(v_2, x) \mid |x - v_2 e| \leq 1\}.$$

The objective function in the convex program (P) associated with F is

$$F0 = q + \delta(\cdot \mid B).$$

Thus in (P) we want to minimize q over the Euclidean unit ball B, and B is the set of all feasible solutions. We are also interested in what happens when this minimization problem is perturbed through right scalar multiplication of q and translation of B in the direction of the vector e (or the opposite direction). Specifically, with each $u = (v_1, v_2)$ with $v_1 > -1$ we associate the problem of minimizing

$$Fu = q \cdot (1 + v_1) + \delta(\cdot \mid B + v_2 e)$$

over R^n (or equivalently $q \cdot (1 + v_1)$ over $B + v_2 e$), and we study the minimum in this problem (the quantity inf Fu) as a function of the perturbation variables v_1 and v_2 in a neighborhood of $v_1 = 0$, $v_2 = 0$. The Lagrangian L in this generalized convex program is easily calculated from the formula in the above definition: for $u^* = (v_1^*, v_2^*)$ one has

$$L(u^*, x) = 2[v_1^* \langle x, Qx \rangle]^{1/2} - v_1^* - \langle x, a - v_2^* e \rangle$$
$$- |v_2^*| [1 - |x - \langle x, e \rangle e|^2]^{1/2}$$

if $v_1^* \geq 0$ and $|x - \langle x, e \rangle e| \leq 1$, whereas

$$L(u^*, x) = \begin{cases} -\infty & \text{if} \quad v_1^* < 0 \quad \text{and} \quad |x - \langle x, e \rangle e| \leq 1, \\ +\infty & \text{if} \quad |x - \langle x, e \rangle e| > 1. \end{cases}$$

Further examples of generalized convex programs involving perturbation by right scalar multiplication or translation will be investigated in §30 and §31.

The reader should note that, in a certain sense, "generalized" convex programs are really no more general than ordinary convex programs. In fact they can be expressed in a roundabout way as ordinary convex programs with linear equations as the only explicit constraints. Let F be any proper convex bifunction from R^m to R^n. Let

$$D = \{(u, x) \mid (Fu)(x) < +\infty\} \subset R^{m+n},$$
$$g_0(u, x) = (Fu)(x),$$
$$g_i(u, x) = g_i(v_1, \ldots, v_m, x) = v_i, \qquad i = 1, \ldots, m.$$

Let (Q) be the ordinary convex program in which $g_0(u, x)$ is minimized over D subject to the constraints

$$g_i(u, x) = 0, \qquad i = 1, \ldots, m.$$

The objective function g for (Q) is essentially the same as the one for the convex program (P) associated with F, namely

$$g(u, x) = \begin{cases} (F0)(x) & \text{if } u = 0, \\ +\infty & \text{if } u \neq 0, \end{cases}$$

and the perturbations associated with (Q) correspond in a direct way to those in (P). Moreover, the Kuhn–Tucker vectors $u^* = (v_1^*, \ldots, v_m^*)$ for (Q) can be identified with those of (P). To a certain extent, therefore, the theory of (P) could be expressed equivalently in terms of (Q). However, this would not be very natural (cf. the case where (P) is itself an ordinary convex program with inequality constraints). The Lagrangian of (Q) involves u, as well as u^* and x, and it therefore differs from the Lagrangian of (P) in an essential way. An attempt to make ordinary convex programs with linear equation constraints the basic model for everything, rather than "generalized" convex programs, would consequently lead to a seriously restrictive theory of Lagrangians and duality. We shall show in §36 that the saddle-point problems corresponding to the Lagrangians of generalized convex programs are in effect the most general ("regularized") concave-convex minimax problems.

The fundamental fact about the perturbation function of any convex program, ordinary or generalized, is the following.

THEOREM 29.1.   *Let F be any convex bifunction from R^m to R^n. Then the perturbation function* inf F *in the convex program (P) associated with F is a convex function on R^m whose effective domain is* dom F. *When the optimal value in (P) is finite, the Kuhn–Tucker vectors for (P) are precisely the vectors $u^* \in R^m$ such that $-u^*$ is a subgradient of* inf F *at $u = 0$, i.e.*

$$-u^* \in \partial(\text{inf } F)(0).$$

PROOF. Let $f(u, x) = (Fu)(x)$, and let A be the linear transformation $(u, x) \rightarrow u$. Then f is a convex function on R^{m+n} and $Af = \inf F$, whence it follows from Theorem 5.7 that inf F is a convex function. Since the value of inf F at a given point u is $+\infty$ only if Fu is the constant function $+\infty$, we have

$$\text{dom } (\inf F) = \text{dom } F.$$

By definition, u^* is a Kuhn–Tucker vector for (P) if and only if inf F is finite at 0 and

$$(\inf F)(u) \geq (\inf F)(0) + \langle -u^*, u \rangle, \forall u.$$

This inequality says that $-u^* \in \partial (\inf F)(0)$. ‖

The importance of Theorem 29.1 is that it enables us to bring to bear on the study of perturbations and Kuhn–Tucker vectors the whole theory of continuity and differentiability of convex functions. We shall state some of the principal results as corollaries to Theorem 29.1.

COROLLARY 29.1.1. *Let F be any convex bifunction from R^m to R^n. Suppose that the optimal value in the convex program (P) associated with F is finite. Then the one-sided directional derivative*

$$(\inf F)'(0; u) = \lim_{\lambda \downarrow 0} \frac{(\inf F)(\lambda u) - (\inf F)(0)}{\lambda}$$

exists for every $u \in R^m$ and is a positively homogeneous convex function of u. The Kuhn–Tucker vectors u^* for (P) form a closed convex set in R^m whose support function is the closure of the function

$$u \rightarrow (\inf F)'(0; -u).$$

PROOF. Apply Theorems 23.1 and 23.2 to inf F. ‖

COROLLARY 29.1.2. *Let F be any convex bifunction from R^m to R^n. Suppose that the optimal value in the convex program (P) associated with F is finite. A Kuhn–Tucker vector then fails to exist for (P) if and only if there exists a vector $u \in R^m$ such that the two-sided directional derivative*

$$\lim_{\lambda \to 0} \frac{(\inf F)(\lambda u) - (\inf F)(0)}{\lambda}$$

exists and equals $-\infty$.

PROOF. Apply Theorem 23.3 to inf F. ‖

Corollary 29.1.1 makes possible a complete interpretation of Kuhn–Tucker vectors in terms of rates of change of the optimal value in (P) with respect to the given perturbations of the objective function for (P). From the "equilibrium price" point of view, Corollary 29.1.2 is the definitive result about the existence of Kuhn–Tucker vectors. It says that a convex program with a finite optimal value has at least one Kuhn–Tucker vector, unless the program is unstable in a certain sense which obviously precludes all possibility of an "equilibrium." If there exists a vector u with the property in Corollary 29.1.2 then u gives a direction of perturbation in which the optimal value in the program drops off infinitely steeply. Perturbation in this direction is "infinitely advantageous" in the sense of the heuristic remarks in §28, and therefore no equilibrium price vector $u^* = (v_1^*, \ldots, v_m^*)$ can exist, because no finite prices can compensate for an infinite marginal improvement in minimum cost.

The question of the uniqueness of Kuhn–Tucker vectors has a satisfying answer:

COROLLARY 29.1.3. *Let F be any convex bifunction from R^m to R^n. Suppose that the optimal value in the convex program (P) associated with F is finite. Then (P) has a unique Kuhn–Tucker vector $u^* = (v_1^*, \ldots, v_m^*)$ if and only if the function inf F is differentiable at $u = 0$, in which case u^* is given by the formula*

$$v_i^* = -\frac{\partial}{\partial v_i} (\inf F)\bigg|_{u=0}.$$

PROOF. This is immediate from Theorem 29.1 and Theorem 25.1. ‖

For example, consider an ordinary convex program (P) in the notation of §28 with $r = m$. Assume that (P) has unique Kuhn–Tucker coefficients $v_1^*, \ldots, v_m^*$. By definition, $(\inf F)(v_1, 0, \ldots, 0)$ is the infimum of $f_0(x)$ subject to the constraints

$$f_1(x) \leq v_1, \qquad f_2(x) \leq 0, \ldots, f_m(x) \leq 0.$$

The derivative of this function of v_1 is $-v_1^*$ at $v_1 = 0$, according to Corollary 29.1.3.

We have already defined a convex program (P) to be *consistent* if it has feasible solutions, i.e. if $0 \in$ dom F. To aid in stating further corollaries of Theorem 29.1, we shall call (P) *strongly consistent* if $0 \in$ ri (dom F) and *strictly consistent* if $0 \in$ int (dom F). When (P) is an ordinary convex program in the notation of §28, (P) is strongly consistent if and only if there exists an $x \in$ ri C such that

$$f_1(x) < 0, \ldots, f_r(x) < 0, \qquad f_{r+1}(x) = 0, \ldots, f_m(x) = 0.$$

This may be proved by the reader as an exercise. It is obvious that, when (P) is an ordinary convex program with $r = m$, i.e. with only inequality constraints, (P) is strictly consistent if and only if there exists an $x \in C$ such that

$$f_1(x) < 0, \ldots, f_m(x) < 0.$$

In general, a convex program (P) is strictly consistent if and only if, for every $u \in R^m$, there exists a $\lambda > 0$ such that $\lambda u \in \text{dom } F$, i.e. such that $F(\lambda u)$ on R^n is not the constant function $+\infty$ (Corollary 6.4.1). Thus, informally speaking, a consistent convex program is strictly consistent unless there is some direction of perturbation in which the set of feasible solutions immediately becomes vacuous.

COROLLARY 29.1.4. *Let F be any convex bifunction from R^m to R^n. Suppose that the optimal value in the convex program (P) associated with F is finite and that (P) is strongly (or strictly) consistent. Let U^* denote the set of all Kuhn–Tucker vectors for (P). Then $U^* \neq \emptyset$ and*

$$(\inf F)'(0; u) = -\inf\{\langle u^*, u \rangle \mid u^* \in U^*\}, \quad \forall u \in R^m.$$

PROOF. Apply Theorem 23.4 to inf F. (Here inf F is necessarily proper by Theorem 7.2, since by hypothesis it is finite at 0, and 0 is a relative interior point of its effective domain.) ‖

COROLLARY 29.1.5. *Let F be any convex bifunction from R^m to R^n. Suppose that the optimal value in the convex program (P) associated with F is finite and that (P) is strictly consistent. Then there is an open convex neighborhood of 0 in R^m on which inf F is finite and continuous. Moreover, the Kuhn–Tucker vectors for (P) form a non-empty closed bounded convex subset of R^m.*

PROOF. By hypothesis, inf F is finite at 0 and $0 \in \text{int } (\text{dom } F)$. Moreover, dom F is the effective domain of inf F by Theorem 29.1. Therefore inf F is finite and continuous on int (dom F) (Theorems 7.2 and 10.1). By Theorem 23.4, $\partial(\inf F)(0)$ is a non-empty closed bounded convex set, and hence so is the set of all Kuhn–Tucker vectors for (P). ‖

COROLLARY 29.1.6. *Let F be any convex bifunction from R^m to R^n. If there exists a vector $u \in R^m$ such that inf $Fu = -\infty$, then inf $Fu = -\infty$ for every $u \in \text{ri } (\text{dom } F)$ (whereas inf $Fu = +\infty$ for every $u \notin \text{dom } F$).*

PROOF. Apply Theorem 7.2 to inf F. ‖

When (P) is an ordinary convex program, Corollary 29.1.4 does not provide quite as broad a criterion for the existence of Kuhn–Tucker vectors as does Theorem 28.2, because Theorem 28.2 takes into special account the fact that certain of the inequality constraints may be affine. Corollary 29.1.4 and other results for general convex programs can be refined to some extent, however, by invoking polyhedral convexity.

A convex bifunction F will be called *polyhedral* if its graph function is polyhedral. The convex programs associated with such bifunctions will be called *polyhedral convex programs*.

As important examples of polyhedral convex programs, we mention the *linear programs*; these we define to be the ordinary convex programs in which (in the notation of §28) the functions $f_0, f_1, \ldots, f_m$ are all affine on C and C is of the form

$$\{x = (\xi_1, \ldots, \xi_n) \mid \xi_j \geq 0, j = 1, \ldots, s\}$$

for some integer s, $0 \leq s \leq n$. The Lagrangians of linear programs are thus the functions L on $R^m \times R^n$ of the form

$$L(u^*, x) = \begin{cases} K(u^*, x) & \text{if } (\xi_1, \ldots, \xi_s) \geq 0, \quad (v_1^*, \ldots, v_r^*) \geq 0, \\ -\infty & \text{if } (\xi_1, \ldots, \xi_s) \geq 0, \quad (v_1^*, \ldots, v_r^*) \not\geq 0, \\ +\infty & \text{if } (\xi_1, \ldots, \xi_s) \not\geq 0, \end{cases}$$

where K is a *bi-affine* function on $R^m \times R^n$, i.e. a function of the form

$$K(u^*, x) = \langle u^*, Ax \rangle + \langle u^*, a \rangle + \langle a^*, x \rangle + \alpha.$$

(The reader may wonder why we do not simply define a linear program to be an ordinary convex program in which every f_i is affine and $C = R^n$, since conditions like $\xi_j \geq 0$ can always be assumed to be represented explicitly among the constraints $f_i(x) \leq 0$. The reason is that, in the theory of §28, constraints corresponding to the functions f_i have Lagrange multipliers assigned to them, whereas constraints incorporated into the minimization problem by the specification of the set C do not. It should be recalled that, from our point of view, two convex programs can involve exactly the same minimization problem and yet be different, because they are associated with different convex bifunctions. The bifunction chosen in a given case depends on what perturbations and Kuhn–Tucker vectors one is interested in. Different bifunctions yield different Lagrangians and furthermore, as will be seen in §30, lead to different dual programs.)

An example of a polyhedral convex program which is not just a linear program may be obtained, for instance, from the problem of minimizing

$$\|x\|_\infty = \max \{|\xi_j| \mid j = 1, \ldots, n\}, \quad x = (\xi_1, \ldots, \xi_n),$$

over the polytope

$$\text{conv } \{a_1, \ldots, a_s\},$$

where the a_k's are given points in R^n. To get a convex program from a minimization problem, one needs to introduce some suitable class of perturbations; let us take the perturbations here to be translations of the points a_k in the directions of certain vectors $b_1, \ldots, b_m$. Specifically, let us consider the bifunction $F: R^m \to R^n$ defined by

$$Fu = \|\cdot\|_\infty + \delta(\cdot \mid Cu),$$

where for each $u = (v_1, \ldots, v_m)$

$$Cu = \bigcap_{i=1}^m \text{conv } \{a_1 + v_i b_i, \ldots, a_s + v_i b_i\}.$$

This F is a (proper) polyhedral convex bifunction, since the graph function of F on R^{m+n} is the sum of the polyhedral convex function

$$(u, x) \to \|x\|_\infty$$

and the indicators of the m polyhedral convex sets

$$\{(u, x) \mid x - v_i b_i \in \text{conv }\{a_1, \ldots, a_s\}\}, \qquad i = 1, \ldots, m$$

(see the theorems in the second half of §19). The convex program (P) associated with F is by definition polyhedral, and the objective function in (P) is

$$F0 = \|\cdot\|_\infty + \delta(\cdot \mid \text{conv }\{a_1, \ldots, a_r\}),$$

so that the minimization problem in (P) is the given problem.

Polyhedral convex programs have many special properties. The most important of these are listed in the next theorem.

THEOREM 29.2. *Let F be a polyhedral convex bifunction from R^m to R^n. The objective function F0 and the perturbation function inf F in the poly-hedral convex program (P) associated with F are then polyhedral convex functions. If the optimal value in (P) is finite, (P) has at least one optimal solution and at least one Kuhn–Tucker vector. Moreover, the set of all optimal solutions and the set of all Kuhn–Tucker vectors are polyhedral convex sets.*

PROOF. The graph function $f(u, x) = (Fu)(x)$ is a polyhedral convex function on R^{m+n}, and hence $(F0)(x)$ is a polyhedral convex function of $x \in R^n$. As was shown in the proof of Theorem 29.1, inf F is the image of f under a certain linear transformation A. Since linear transformations preserve polyhedral convexity (Corollary 19.3.1) we may conclude from this that inf F is polyhedral. Assume now that the optimal value in (P) is finite. Then $F0$ is bounded below on R^n, and by Corollary 27.3.2 the infimum of $F0$ is attained. Of course, the minimum set of $F0$ is polyhedral, being a level set of the form $\{x \mid (F0)(x) \leq \alpha\}$. Thus (P) has optimal solutions, and these form a polyhedral convex set. Since inf F is polyhedral and $(\text{inf } F)(0)$ is finite, $\partial (\text{inf } F)(0)$ is a non-empty polyhedral convex set (Theorem 23.10). It follows from Theorem 29.1 that the Kuhn–Tucker vectors for (P) form a non-empty polyhedral convex set. ‖

We turn now to the Lagrangian characterization of Kuhn–Tucker vectors and optimal solutions.

THEOREM 29.3. *Let F be a closed proper convex bifunction from R^m to R^n, and let L be the Lagrangian of the convex program (P) associated with F. Let $\bar{u}^*$ and $\bar{x}$ be vectors in R^m and R^n, respectively. In order that $\bar{u}^*$ be a Kuhn–Tucker vector for (P) and $\bar{x}$ be an optimal solution to (P), it is necessary and* *sufficient that* $(\bar{u}^*, \bar{x})$ *be a saddle-point of* L, *i.e.*

$$L(u^*, \bar{x}) \le L(\bar{u}^*, \bar{x}) \le L(\bar{u}^*, x), \; \forall u^*, \; \forall x.$$

PROOF. As observed at the beginning of the proof of Theorem 28.3, the saddle-point condition is equivalent to

$$\sup_{u^*} L(u^*, \bar{x}) = \inf_x L(\bar{u}^*, x).$$

We have already pointed out, in connection with the definition of L, that

$$\inf_x L(\bar{u}^*, x) = \inf_u \{\langle \bar{u}^*, u \rangle + \inf Fu\} \le \inf F0.$$

On the other hand, in terms of the convex function $h(u) = (Fu)(\bar{x})$, we have

$$L(u^*, \bar{x}) = \inf_u \{\langle u^*, u \rangle + h(u)\}$$

$$= -\sup \{\langle -u^*, u \rangle - h(u)\} = -h^*(-u^*)$$

and hence

$$\sup_{u^*} L(u^*, \bar{x}) = \sup_{u^*} \{\langle 0, -u^* \rangle - h^*(-u^*)\}$$

$$= h^{**}(0) = (\operatorname{cl} h)(0).$$

Since F is closed by hypothesis, we have $\operatorname{cl} h = h$ and

$$\sup_{u^*} L(u^*, \bar{x}) = (F0)(\bar{x}) \ge \inf F0.$$

The properness of F implies that

$$\inf_u \{\langle \bar{u}^*, u \rangle + \inf Fu\} < +\infty, \qquad (F0)(\bar{x}) > -\infty.$$

Therefore $(\bar{u}^*, \bar{x})$ is a saddle-point of L if and only if

$$\inf_u \{\langle \bar{u}^*, u \rangle + \inf Fu\} = \inf F0 = (F0)(\bar{x}) \in R.$$

This condition means by definition that $\bar{u}^*$ is a Kuhn–Tucker vector for (P) and $\bar{x}$ is an optimal solution to (P). ‖

A generalization of the Kuhn–Tucker Theorem follows at once:

COROLLARY 29.3.1. *Let F be a closed proper convex bifunction from R^m to R^n. Suppose that the convex program (P) associated with F is strongly (or strictly) consistent, or that (P) is polyhedral and merely consistent. In order that a given vector $\bar{x} \in R^n$ be an optimal solution to (P), it is necessary and sufficient that there exist a vector $\bar{u}^* \in R^m$ such that $(\bar{u}^*, \bar{x})$ is a saddle-point of the Lagrangian L of (P).*

PROOF. Under the given consistency hypothesis, if (P) has an optimal solution $\bar{x}$ it also has at least one Kuhn–Tucker vector $\bar{u}^*$ (Corollary 29.1.4 and Theorem 29.2). ‖

Generalized Kuhn–Tucker conditions may be formulated in terms of the "subgradients" of the Lagrangian L, which is actually a concave-convex function on $R^m \times R^n$. We refer the reader to §36, where the general nature of the Lagrangian saddle-point problem corresponding to (P) is explained in detail. Another form of Corollary 29.3.1, the general Kuhn–Tucker Theorem, is stated as Theorem 36.6.

Theorem 28.4 can likewise be generalized from ordinary convex programs to the convex programs associated with arbitrary closed proper convex bifunctions. However, we shall leave this as an exercise at present, since the result will be obvious from the theory of dual programs.

In order to apply Theorem 29.3 and the duality theory which will be given in §30, it is sometimes necessary to regularize a given convex program by "closing" its associated bifunction. If F is any convex bifunction from R^m to R^n, the *closure* cl F of F is defined to be the bifunction from R^m to R^n whose graph function is the closure of the graph function of F. Thus cl F is a closed convex bifunction, proper if and only if F is proper. The following theorem and corollary describe the relationship between the convex programs associated with F and cl F.

THEOREM 29.4. *Let F be a convex bifunction from R^m to R^n. For each $u \in$ ri (dom F), one has*

$$(\text{cl } F)u = \text{cl } (Fu),$$

$$\inf (\text{cl } F)u = \inf Fu.$$

Moreover, assuming that F is proper, one has

$$\text{dom } F \subset \text{dom } (\text{cl } F) \subset \text{cl } (\text{dom } F).$$

PROOF. For $f(u, x) = (Fu)(x)$, one has

$$(\text{cl } f)(u, x) = ((\text{cl } F)u)(x)$$

by definition. Since dom F is the projection on R^m of dom f, ri (dom F) is the projection of ri (dom f) (Theorem 6.6). Hence, given any $u \in$ ri (dom F), there exists some x such that $(u, x) \in$ ri (dom f). In particular such an x belongs to ri (dom Fu) (Theorem 6.4), and if $f(u, x) > -\infty$ we have f and Fu proper (Theorem 7.2) and

$$((\text{cl } F)u)(y) = \lim_{\lambda \uparrow 1} f(u, (1 - \lambda)x + \lambda y) = (\text{cl}(Fu))(y), \forall y$$

(Theorem 7.5). If $f(u, x) = -\infty$, of course

$$((\text{cl } F)u)(y) = -\infty = (\text{cl } (Fu))(y), \forall y.$$

Therefore, at all events, $(\text{cl } F)u = \text{cl } (Fu)$ for every $u \in$ ri (dom F). The convex functions Fu and cl (Fu) have the same infimum on R^n, so Fu and $(\text{cl } F)u$ have the same infimum when $u \in$ ri (dom F). Thus the functions inf F and inf (cl F) agree on ri (dom F). When F is proper, its graph function f is proper and

$$\text{dom} f \subset \text{dom} (\text{cl} f) \subset \text{cl} (\text{dom} f).$$

Projecting these sets on R^m, we get the effective domain inclusions in the theorem.   ‖

COROLLARY 29.4.1.   *Let F be a convex bifunction from R^m to R^n. Let* (P) *be the convex program associated with F, and let* (cl P) *be the convex program associated with* cl F. *Assume that* (P) *is strongly consistent. Then* (cl P) *is strongly consistent. The objective function for* (cl P) *is the closure of the objective function for* (P), *so that* (P) *and* (cl P) *have the same optimal value and every optimal solution to* (P) *is an optimal solution to* (cl P). *The perturbation functions for* (P) *and* (cl P) *agree on a neighborhood of 0, so that the Kuhn–Tucker vectors for the two programs are the same.*

Adjoint Bifunctions and Dual Programs

A fundamental fact about generalized convex programs is that each such "minimization problem with perturbations" has a dual, which is a certain "maximization problem with perturbations," a generalized concave program. In most cases, two programs dual to each other have the same optimal-value, and the optimal solutions to one are the Kuhn–Tucker vectors for the other.

The duality theory for convex programs is based on a concept of the "adjoint" of a convex bifunction. The adjoint operation for bifunctions may be regarded as a generalization of the adjoint operation for linear transformations, and a considerable "convex algebra" parallel to linear algebra may be built around it, as will be shown in §33 and §38.

Only minimization problems have been discussed in preceding sections, but here we shall need to deal on an equal footing with maximization problems in which the objective function is an extended-real-valued concave function. The changes in passing from minimization to maximization and from convexity to concavity are essentially trivial and obvious. The roles of $+\infty$, $\leq$ and "inf" are everywhere interchanged with those of $-\infty$, $\geq$ and "sup." We shall summarize the most important alterations.

A function g from R^n to $[-\infty, +\infty]$ is *concave* if $-g$ is convex. For a concave function g, one defines

$$\text{epi}\, g = \{(x, \mu) \mid x \in R^n, \mu \in R, \mu \leq g(x)\},$$

$$\text{dom}\, g = \{x \mid g(x) > -\infty\}.$$

One says that g is *proper* if $g(x) > -\infty$ for at least one x and $g(x) < +\infty$ for every x, i.e. if $-g$ is proper. The *closure* cl g of g is the pointwise infimum of all the affine functions h such that $h \geq g$, i.e. it is $-(\text{cl}\,(-g))$. If g is proper, or if $x \in \text{cl}\,(\text{dom}\, g)$, one has

$$(\text{cl}\, g)(x) = \limsup_{y \to x} g(y).$$

If g is the constant function $-\infty$, then cl $g = g$; but if g is an improper concave function which has the value $+\infty$ somewhere, then cl g is the constant function $+\infty$. One says that g is *closed* if cl $g = g$ (i.e. if $-g$ is closed). If g is proper, g is closed if and only if it is upper semi-continuous, i.e. if and only if the convex sets

$$\{x \mid g(x) \geq \alpha\}, \qquad \alpha \in R,$$

are all closed.

The *conjugate* of a concave function g is defined by

$$g^*(x^*) = \inf_x \{\langle x, x^* \rangle - g(x)\},$$

and one has $g^{**} = $ cl g. *Caution*: in general,

$$g^* \neq -(-g)^*.$$

For the convex function $f = -g$, one has, not $g^*(x^*) = -f^*(x^*)$, but

$$g^*(x^*) = -f^*(-x^*).$$

The set $\partial g(x)$ consists by definition of the vectors x^* such that

$$g(z) \leq g(x) + \langle x^*, z - x \rangle, \forall z.$$

We shall call such vectors x^* *subgradients* of g at x, and the mapping $x \rightarrow \partial g(x)$ the *subdifferential* of g, for simplicity, even though terms like "supergradients" and "superdifferential" might be more appropriate. One has

$$\partial g(x) = -\partial(-g)(x).$$

If g is proper, one has

$$g(x) + g^*(x^*) \leq \langle x, x^* \rangle, \forall x, \forall x^*,$$

with equality holding if and only if $x^* \in \partial g(x)$. If g is closed, one has $x^* \in \partial g(x)$ if and only if $x \in \partial g^*(x^*)$.

A bifunction from R^m to R^n is said to be *concave* if its graph function is concave, and so forth. For a concave bifunction G, dom G is defined to be the set of vectors $u \in R^m$ such that Gu is not the constant function $-\infty$ on R^n. The *concave program* (Q) associated with G is defined to be the "maximization problem with perturbations" in which the concave function $G0$ is to be maximized over R^n and the given perturbations are those in which $G0$ is replaced by Gu for different choices of $u \in R^m$. One calls $G0$ the *objective function* for (Q). The vectors x such that $(G0)(x) > -\infty$ (i.e. those in dom $G0$) are the *feasible solutions* to (Q). The supremum of $G0$ over R^n is the *optimal value* in (Q), and if this supremum is finite the points where it is attained (if any) are called *optimal solutions* to (Q). The *perturbation function* for (Q) is the function sup G on R^m defined by

$$(\sup G)(u) = \sup Gu = \sup_x (Gu)(x).$$

This is an (extended-real-valued) concave function on R^m, and its effective domain is dom G. A vector $u^* \in R^m$ is said to be a *Kuhn–Tucker vector* for (Q) if the quantity

$$\sup_u \{\langle u^*, u \rangle + \sup Gu\} = \sup_{u,x} \{\langle u^*, u \rangle + (Gu)(x)\}$$

is finite and equal to the optimal value in (Q). This condition holds if and only if sup G is finite at 0 and

$$-u^* \in \partial(\sup G)(0).$$

The *Lagrangian L* of (Q) is defined by

$$L(u^*, x) = \sup_u \{\langle u^*, u \rangle + (Gu)(x)\}.$$

Consistency, *strong consistency* and *strict consistency* for a concave program mean (as for a convex program) that $0 \in$ dom G, $0 \in$ ri (dom G) and $0 \in$ int (dom G), respectively.

So much for the terminology of concave functions and concave programs. The results in §29 can be translated into this terminology without difficulty.

For any convex bifunction F from R^m to R^n,

$$F: u \to Fu: x \to (Fu)(x),$$

the *adjoint* of F is defined as the bifunction

$$F^*: x^* \to F^*x^*: u^* \to (F^*x^*)(u^*)$$

from R^n to R^m given by

$$(F^*x^*)(u^*) = \inf_{u,x} \{(Fu)(x) - \langle x, x^* \rangle + \langle u, u^* \rangle\}.$$

The adjoint of a concave bifunction is defined in the same way, except with the infimum replaced by a supremum.

The adjoint correspondence for bifunctions is actually just a modification of the conjugacy correspondences for convex and concave functions. Let f be the graph function of the convex bifunction F. Regarding

$$\langle u, -u^* \rangle + \langle x, x^* \rangle$$

as the inner product of the vectors (u, x) and $(-u^*, x^*)$ in R^{m+n}, we have

$$(F^*x^*)(u^*) = \inf_{u,x} \{f(u, x) - \langle x, x^* \rangle + \langle u, u^* \rangle\}$$

$$= -\sup_{u,x} \{\langle u, -u^* \rangle + \langle x, x^* \rangle - f(u, x)\} = -f^*(-u^*, x^*),$$

where f^* is the conjugate of f. The graph function of F^* is thus a closed concave function, i.e. F^* is a closed concave bifunction. By definition, then, the adjoint F^{**} of F^* is the bifunction from R^m to R^n given by

$$(F^{**}u)(x) = \sup_{x^*,u^*} \{(F^*x^*)(u^*) - \langle u^*, u \rangle + \langle x^*, x \rangle\}$$

$$= \sup_{x^*,u^*} \{\langle u, -u^* \rangle + \langle x, x^* \rangle - f^*(-u^*, x^*)\}$$

$$= f^{**}(u, x) = (\text{cl } f)(u, x).$$

But the convex bifunction from R^m to R^n whose graph function is cl f is, by definition, the closure cl F of F. The fundamental facts about the adjoint operation may therefore be summarized as follows.

THEOREM 30.1. *Let F be any convex or concave bifunction from R^m to R^n. Then F^* is a closed bifunction of the opposite type from R^n to R^m, proper if and only if F is proper, and*

$$F^{**} = \text{cl } F.$$

In particular, $F^{**} = F$ if F is closed. Thus the adjoint operation establishes a one-to-one correspondence between the closed proper convex (resp. concave) bifunctions from R^m to R^n and the closed proper concave (resp. convex) bifunctions from R^n to R^m. If F is polyhedral, so is F^*.

PROOF. By Theorem 12.2 and the preceding remarks. The fact that the adjoint operation preserves polyhedral convexity is immediate from Theorem 19.2. ‖

As a first example of the adjoint operation for bifunctions, let F be the *convex indicator bifunction of a linear transformation A from R^m to R^n,* i.e.

$$(Fu)(x) = \delta(x \mid Au) = \begin{cases} 0 & \text{if } x = Au, \\ +\infty & \text{if } x \neq Au. \end{cases}$$

Calculating F^* directly, we get

$$(F^*x^*)(u^*) = \inf_{u,x} \{\delta(x \mid Au) - \langle x, x^* \rangle + \langle u, u^* \rangle\}$$

$$= \inf_u \{-\langle Au, x^* \rangle + \langle u, u^* \rangle\} = \inf_u \langle u, u^* - A^*x^* \rangle$$

$$= \begin{cases} 0 & \text{if } u^* = A^*x^*, \\ -\infty & \text{if } u^* \neq A^*x^*. \end{cases}$$

Thus F^* is the *concave indicator bifunction of the adjoint transformation A^** from R^n to R^m,

$$F^*x^* = -\delta(\cdot \mid A^*x^*), \qquad \forall x^* \in R^n.$$

This shows that the adjoint operation for bifunctions can rightly be viewed as a generalization of the adjoint operation for linear transformations; see also §33.

Other notable examples of adjoint bifunctions will be given shortly. However, it is clear from the relationship between the graph functions of F and F^*, as explained before Theorem 30.1, that numerous examples of convex and concave bifunctions adjoint to each other could be generated just from the examples in Part III of convex functions conjugate to each other.

The explicit calculation of the adjoint of a given convex bifunction is, of course, not always an easy task, but the applicability of the general formulas in §16 should not be overlooked. Further formulas of use in this connection will be derived in §38 for the adjoints of bifunctions constructed by means of certain natural operations which are analogous to addition and multiplication of linear transformations.

Let F be a convex bifunction from R^m to R^n, and let (P) be the associated convex program. The concave program associated with the (concave) adjoint bifunction F^* will be called the program *dual* to (P) and will be denoted by (P^*).

In (P) we minimize $F0$ as a function of $x \in R^n$, and we perturb $F0$ by replacing it by Fu for different choices of $u \in R^m$. In the dual program (P^*), we maximize F^*0 as a function of $u^* \in R^m$, and we perturb F^*0 by replacing it by F^*x^* for various choices of $x^* \in R^n$. The optimal value in (P) is inf $F0$, and the optimal value in (P^*) is sup F^*0. A Kuhn–Tucker vector for (P) is a $u^* \in R^m$ such that the quantity

$$\inf_u \{\langle u, u^* \rangle + \inf Fu\} = \inf_{u,x} \{\langle u, u^* \rangle + (Fu)(x)\}$$

is finite and equal to the optimal value in (P), while a Kuhn–Tucker vector for (P^*) is an $x \in R^n$ such that the quantity

$$\sup_{x^*} \{\langle x, x^* \rangle + \sup F^*x^*\} = \sup_{x^*,u^*} \{\langle x, x^* \rangle + (F^*x^*)(u^*)\}$$

is finite and equal to the optimal value in (P^*).

In the case where F is closed and proper (the only case really of interest), we have $F^{**} = F$ by Theorem 30.1, so that program dual to (P^*) is in turn (P).

The relationship between a general dual pair of programs will be analyzed in a moment, but first we want to display the classic dual pair of linear programs as an example. Let A be a linear transformation from R^n to R^m, and let a and a^* be fixed vectors in R^m and R^n, respectively. The linear program (P) which we want to consider is the ordinary convex program in which

$$f_0(x) = \langle a^*, x \rangle + \delta(x \mid x \geq 0)$$

is to be minimized subject to the m linear constraints $f_i(x) \leq 0$ expressed by the system

$$a - Ax \leq 0.$$

The bifunction associated with (P) is the polyhedral proper convex bifunction F from R^m to R^n defined by

$$(Fu)(x) = \langle a^*, x \rangle + \delta(x \mid x \geq 0, a - Ax \leq u).$$

Here, of course, we are employing the convention that a vector inequality like $z \geq z'$ is to be interpreted componentwise, i.e. $\zeta_j \geq \zeta'_j$ for every index j. The indicator function notation has the obvious meaning: for $C = \{x \mid x \geq 0\}$, we simply write $\delta(x \mid x \geq 0)$ instead of $\delta(x \mid C)$, etc. Thus $\delta(x \mid x \geq 0)$ is the function of x which has the value 0 when $x \geq 0$ and $+\infty$ when $x \not\geq 0$.

To determine the program (P^*) dual to (P), we calculate the adjoint of F. By definition,

$$(F^*x^*)(u^*) = \inf_{u,x} \{\langle a^*, x \rangle + \delta(x \mid x \geq 0, a - Ax \leq u) - \langle x, x^* \rangle + \langle u, u^* \rangle\}$$

$$= \inf_{x \geq 0, v \geq 0} \{\langle x, a^* - x^* \rangle + \langle a - Ax + v, u^* \rangle\}$$

$$= \inf_{x \geq 0, v \geq 0} \{\langle a, u^* \rangle + \langle v, u^* \rangle + \langle x, a^* - x^* - A^*u^* \rangle\}$$

$$= \langle a, u^* \rangle + \inf_{v \geq 0} \langle v, u^* \rangle + \inf_{x \geq 0} \langle x, a^* - x^* - A^*u^* \rangle$$

$$= \begin{cases} \langle a, u^* \rangle & \text{if} \quad u^* \geq 0 \quad \text{and} \quad a^* - x^* - A^*u^* \geq 0, \\ -\infty & \text{otherwise.} \end{cases}$$

In other words,

$$(F^*x^*)(u^*) = \langle a, u^* \rangle - \delta(u^* \mid u^* \geq 0, a^* - A^*u^* \geq x^*),$$

and (P^*) is the linear program (of the maximizing type) in which one maximizes

$$g_0(u^*) = \langle a, u^* \rangle - \delta(u^* \mid u^* \geq 0)$$

subject to the n linear constraints $g_j(u^*) \geq 0$ expressed by the system

$$a^* - A^*u^* \geq 0.$$

In (P^*), the perturbations which receive attention are those which replace a^* by $a^* - x^*$ for different choices of $x^* \in R^n$, whereas in (P) they are those that replace a by $a - u$ for different choices of $u \in R^m$.

As a further illustration consider the problem which was examined at the end of §28 in connection with the decomposition principle: minimize

$$f_0(x) = f_{01}(x_1) + \cdots + f_{0s}(x_s)$$

subject to the linear constraints expressed by the vector equation

$$A_1 x_1 + \cdots + A_s x_s = a,$$

where a is an element of R^m, A_k is a linear transformation from R^{n_k} to R^m, f_{0k} is a proper convex function on R^{n_k} and

$$x = (x_1, \ldots, x_s) \in R^n, \qquad x_k \in R^{n_k}, \qquad n_1 + \cdots + n_s = n.$$

Here we are interested in an ordinary convex program (P) whose associated convex bifunction $F: R^m \to R^n$ is given by

$$(Fu)(x) = f_0(x) + \delta(x \mid Ax = a + u),$$

where

$$Ax = A_1 x_1 + \cdots + A_s x_s.$$

The adjoint bifunction F^* may be calculated as follows, where $x^* = (x_1^*, \ldots, x_s^*)$:

$$(F^* x^*)(u^*) = \inf_{u, x} \{f_0(x) + \delta(x \mid Ax = a + u) - \langle x, x^* \rangle + \langle u, u^* \rangle\}$$

$$= \inf_{x_1, \ldots, x_s} \{-\langle a, u^* \rangle + \textstyle\sum_{k=1}^s [f_{0k}(x) - \langle x_k, x_k^* \rangle + \langle A_k x_k, u^* \rangle]\}$$

$$= -\langle a, u^* \rangle - \textstyle\sum_{k=1}^s \sup_{x_k} \{\langle x_k, x_k^* - A_k^* u^* \rangle - f_{0k}(x_k)\}$$

$$= -\langle a, u^* \rangle - \textstyle\sum_{k=1}^s f_{0k}^*(x_k^* - A_k^* u^*)$$

The objective function in the concave program (P^*) dual to (P) is therefore given by

$$(F^* 0)(u^*) = -\langle a, u^* \rangle - f_{01}^*(-A_1^* u^*) - \cdots - f_{0s}^*(-A_s^* u^*),$$

and the problem of maximizing this expression in u^* is to be perturbed by translating each conjugate function f_{0k}^* on R^{n_k} by an amount $-x_k^* \in R^{n_k}$. The components x_k of the Kuhn–Tucker vectors $x = (x_1, \ldots, x_s)$ for (P^*) will measure (in the sense of the analogue of Theorem 29.1 and its corollaries for the case of concave programs) the differential effect such translations would have on the optimal value in (P^*).

Observe in the example just given that the objective function in (P^*) is the function whose maximization yields the Kuhn–Tucker vectors for (P), when such vectors exist, as explained at the end of §28. Thus, if a Kuhn–Tucker vector exists for (P), the set of such vectors is the same as the set of all optimal solutions to (P^*). We shall see below that the same thing is actually true for *any* convex or concave program and its dual.

Duals of ordinary convex programs with inequality constraints will be discussed at the end of this section, and further examples of dual programs will be considered in §31. We proceed now, however, with the development of the general theory of such programs.

Almost everything, about the general relationship between a convex program and its dual, hinges on the following fact.

THEOREM 30.2. *Let F be a convex bifunction from R^m to R^n, and let (P) be the convex program associated with F. The objective function F^*0 in the concave program (P^*) dual to (P) is then the conjugate of the concave function $-\inf F$ in (P), i.e. one has*

$$(-\inf F)^* = F^*0, \qquad (F^*0)^* = -\text{cl} \, (\inf F).$$

If F is closed, the objective function $F0$ in (P) is the conjugate of the convex function $-\sup F^*$ in (P^*), i.e. one has

$$(-\sup F^*)^* = F0, \qquad (F0)^* = -\text{cl} \, (\sup F^*).$$

PROOF. By definition,

$$(F^*0)(u^*) = \inf_{u,x} \{(Fu)(x) - \langle x, 0 \rangle + \langle u, u^* \rangle\}$$

$$= \inf_{u} \{\langle u, u^* \rangle + \inf_{x} (Fu)(x)\}$$

$$= \inf_{u} \{\langle u, u^* \rangle - (-\inf F)(u)\} = (-\inf F)^*(u^*).$$

On the other hand, if F is closed we have $F^{**} = F$ and hence

$$(F0)(x) = (F^{**}0)(x) = \sup_{x^*,u^*} \{(F^*x^*)(u^*) - \langle 0, u^* \rangle + \langle x, x^* \rangle\}$$

$$= \sup_{x^*} \{\langle x, x^* \rangle + \sup_{u^*} (F^*x^*)(u^*)\}$$

$$= \sup_{x^*} \{\langle x, x^* \rangle - (-\sup F^*)(x^*)\} = (-\sup F^*)^*(x).$$

The formulas for $(F^*0)^*$ and $(F0)^*$ are then consequences of the basic properties of the conjugacy correspondences. ‖

Observe from the preceding proof that the objective function in (P^*) is given by

$$(F^*0)(u^*) = \inf_x L(u^*, x), \, \forall u^*,$$

where L is the Lagrangian of (P). If F is closed, the objective function in (P), on the other hand, is given by

$$(F0)(x) = \sup_{u^*} L(u^*, x), \, \forall x,$$

as was shown in the proof of Theorem 29.3. Thus the optimal value in (P^*) is

$$\sup_{u^*} \inf_x L(u^*, x),$$

whereas, assuming that F is closed, the optimal value in (P) is

$$\inf_x \sup_{u^*} L(u^*, x).$$

The role of Lagrangians in the theory of dual programs will be discussed further towards the end of §36; see also Corollary 30.5.1 below.

The formulas in Theorem 30.2 lead to immediate results about the relationship between the optimal value in (P) and the optimal value in (P^*) and in particular to results about the consistency of (P) and (P^*).

COROLLARY 30.2.1. *Let F be a closed convex bifunction from R^m to R^n, and let (P) be the convex program associated with F. The dual program (P^*) is inconsistent if and only if there is a vector $u \in R^m$ such that Fu has no lower bound on R^n. On the other hand, (P) is inconsistent if and only if there is a vector $x^* \in R^n$ such that F^*x^* has no upper bound on R^m.*

PROOF. The inconsistency of (P) means that the function $F0$ is identically $+\infty$. Since $F0$ is the conjugate of the convex function $-\sup F^*$ by the theorem, this happens if and only if $-\sup F^*$ has the value $-\infty$ somewhere, i.e.

$$+\infty = (\sup F^*)(x^*) = \sup (F^*x^*)$$

for some $x^* \in R^n$. Dually for the inconsistency of (P^*). ‖

Corollary 30.2.1 should be considered in conjunction with Corollary 29.1.6.

COROLLARY 30.2.2. *Let F be a closed convex bifunction from R^m to R^n, and let (P) be the convex program associated with F. Then the optimal value inf $F0$ in (P) and the optimal value sup F^*0 in (P^*) satisfy*

$$(\text{cl} (\inf F))(0) = (\sup F^*)(0) = \sup F^*0,$$

$$(\text{cl} (\sup F^*))(0) = (\inf F)(0) = \inf F0.$$

In particular, one always has

$$\inf F0 \geq \sup F^*0.$$

PROOF. According to Theorem 30.2,

$$(\text{cl} (\inf F))(0) = -(F^*0)^*(0) = -\inf_{u^*} \{\langle 0, u^* \rangle - (F^*0)(u^*)\}$$

$$= \sup_{u^*} (F^*0)(u^*) = \sup F^*0.$$

Similarly for the other formula. ‖

COROLLARY 30.2.3. *Let F be a closed convex bifunction from R^m to R^n, and let (P) be the associated convex program. Except in the case where* *neither (P) nor (P*) is consistent, one has*

$$\lim_{u \to 0} \inf (\inf Fu) = \sup F^*0,$$

$$\lim_{x^* \to 0} \sup (\sup F^*x^*) = \inf F0.$$

PROOF. From what we know in general about the closure operation for convex functions, the formula

$$(\text{cl } (\inf F))(0) = \lim_{u \to 0} \inf (\inf F)(u)$$

holds except in cases where the left side is $-\infty$ and the right is $+\infty$. The left side equals sup F^*0 by the preceding corollary, and this is $-\infty$ only when (P^*) is inconsistent. When the right side is $+\infty$, we have inf $F0 = +\infty$ in particular, so that (P) is inconsistent. Thus the first formula in the corollary holds unless both (P) and (P^*) are inconsistent. Similarly for the second formula. ‖

Let us agree to call a convex program (P) *normal* if its perturbation function inf F is closed at $u = 0$, i.e.

$$(\text{cl } (\inf F))(0) = (\inf F)(0).$$

If (P) is consistent, or if merely $0 \in \text{cl } (\text{dom } F)$, this condition is equivalent to inf Fu being a lower semi-continuous function of u at $u = 0$. This is a natural property to demand of a convex program, for without this lower semi-continuity there would exist some $v \in R^m$ such that the limit of the convex function $h(\lambda) = \inf F(\lambda v)$ as $\lambda \downarrow 0$ is strictly less than $h(0) = \inf F0$ (Theorem 7.5). The program would thus be very unstable with respect to a certain direction of perturbation. If $0 \notin \text{cl } (\text{dom } F)$, (P) is normal trivially, except in the situation described in Corollary 29.1.6.

Normality is defined analogously for concave programs. Thus the dual of a convex program is normal if and only if

$$(\text{cl } (\sup F^*))(0) = (\sup F^*)(0);$$

this implies the upper semi-continuity of sup F^* at $x^* = 0$.

THEOREM 30.3.   *Let F be a closed convex bifunction from R^m to R^n, and let (P) be the convex program associated with F. Then the following conditions are equivalent:*

(a) *(P) is normal;*

(b) *(P^*) is normal;*

(c) *inf $F0 = \sup F^*0$, i.e. the optimal value in (P) equals the optimal value in (P^*).*

PROOF.   This is immediate from Corollary 30.2.2. ‖

We shall say simply that *normality holds* for a dual pair of programs if the three equivalent conditions in Theorem 30.3 are satisfied. Normality does hold "normally," as the next theorem shows.

THEOREM 30.4. *Let F be a closed convex bifunction from R^m to R^n, and let (P) be the convex program associated with F. Then any one of the following conditions is sufficient to guarantee that normality holds for (P) and (P^*):*

(a) *(P) is strongly (or strictly) consistent;*

(b) *(P^*) is strongly (or strictly) consistent;*

(c) *The optimal value in (P) is finite, and a Kuhn–Tucker vector exists for (P);*

(d) *The optimal value in (P^*) is finite, and a Kuhn–Tucker vector exists for (P^*);*

(e) *(P) is polyhedral and consistent;*

(f) *(P^*) is polyhedral and consistent;*

(g) *$\{x \mid (F0)(x) \le \alpha\}$ is non-empty and bounded for some α;*

(h) *$\{u^* \mid (F^*0)(u^*) \ge \alpha\}$ is non-empty and bounded for some α;*

(i) *(P) has a unique optimal solution, or the optimal solutions to (P) form a non-empty bounded set;*

(j) *(P^*) has a unique optimal solution, or the optimal solutions to (P^*) form a non-empty bounded set.*

PROOF. Under (a), 0 belongs to the relative interior of the effective domain of inf F (Theorem 29.1), so inf F agrees with cl (inf F) at 0 (Theorems 7.2, 7.4). Under (c), inf F is subdifferentiable at 0 (Theorem 29.1), and this too implies inf F is closed at 0 (Corollary 23.5.2). Under (e), inf F is a polyhedral convex function with 0 in its effective domain (Theorem 29.2). A polyhedral convex function always agrees with its closure on its effective domain. Thus (a), (c) and (e) imply that normality holds. Dually, (b), (d) and (f) imply that normality holds. Condition (g) is equivalent by Theorem 27.1 (d) to having $0 \in$ int (dom $(F0)^*$), i.e. (P^*) strictly consistent. Thus (g) is a special case of (b), and similarly (h) is a special case of (a). Of course, (i) and (j) are contained in (g) and (h). ‖

Theorem 30.4 can be applied, for example, to the dual pair of linear programs described earlier in this section. These are polyhedral convex programs, so it follows that the optimal value in (P) and the optimal value in (P^*) are equal, unless both programs are inconsistent. This result is known as the Gale–Kuhn–Tucker Duality Theorem. Of course, a polyhedral convex or concave program has an optimal solution by Theorem 29.2 when its optimal value is finite.

There do exist convex programs which are not normal, although such programs are necessarily rather freakish and are not of much interest in themselves, as is clear from Theorem 30.4. For an example of abnormality consider the closed proper convex bifunction F from R to R defined by

$$(Fu)(x) = \begin{cases} \exp\left(-\sqrt{ux}\right) & \text{if } u \geq 0, x \geq 0, \\ +\infty & \text{otherwise.} \end{cases}$$

The function inf F is given by

$$\inf Fu = \begin{cases} 0 & \text{if } u > 0, \\ 1 & \text{if } u = 0, \\ +\infty & \text{if } u < 0. \end{cases}$$

Thus the optimal value in (P) is 1, whereas

$$(\text{cl} (\inf F))(0) = 0.$$

The optimal value in (P^*) must be 0, by Corollary 30.2.3. Note that (P) is not strongly consistent.

The reader can easily construct other, similar examples of abnormality in which $(\inf F)(0)$ is finite but $(\text{cl} (\inf F))(0) = -\infty$, or in which $(\inf F)(0) = +\infty$ but $(\text{cl} (\inf F))(0)$ is finite or $-\infty$.

The next theorem describes a remarkable duality between Kuhn–Tucker vectors and optimal solutions.

THEOREM 30.5.   *Let F be a closed convex bifunction from R^m to R^n, and let (P) be the convex program associated with F. Suppose that normality holds for (P) and (P^*). Then u^* is a Kuhn–Tucker vector for (P) if and only if u^* is an optimal solution to (P^*). Dually, x is a Kuhn–Tucker vector for (P^*) if and only if x is an optimal solution to (P).*

PROOF.   As we know from Theorem 29.1, u^* is a Kuhn–Tucker vector for (P) if and only if inf $F0$ is finite and $-u^*$ belongs to $\partial(\inf F)(0)$. Since normality holds by assumption, inf F agrees with cl $(\inf F)$ at 0, and hence inf F and cl $(\inf F)$ have the same subgradients at 0 (see Theorem 23.5). Moreover, $-\text{cl} (\inf F) = (F^*0)^*$ by Theorem 30.2. Thus u^* is a Kuhn–Tucker vector for (P) if and only if $(F^*0)^*(0)$ is finite and

$$u^* \in \partial(F^*0)^*(0),$$

i.e. (by Theorem 27.1 in the concave case) if and only if the supremum of F^*0 is finite and attained at u^*. Thus the Kuhn–Tucker vectors u^* for (P) are the optimal solutions to (P^*). The proof of the dual assertion of the theorem is parallel. $\parallel$

COROLLARY 30.5.1.   *Let F be a closed convex bifunction from R^m to R^n, and let (P) be the convex program associated with F. Then the following conditions on a pair of vectors $\bar{x} \in R^n$ and $\bar{u}^* \in R^m$ are equivalent:*

(a) *normality holds, and $\bar{x}$ and $\bar{u}^*$ are optimal solutions to (P) and (P^*) respectively*;

(b) *$(\bar{u}^*, \bar{x})$ is a saddle-point of the Lagrangian L of (P);*

(c) *$(F0)(\bar{x}) \leq (F^*0)(\bar{u}^*)$ (in which case equality must actually hold).*

PROOF. The equivalence of (a) and (b) is immediate from Theorem 29.3, since the existence of a Kuhn–Tucker vector implies normality by Theorem 30.4. The equivalence of (a) and (c) is by the definition of the phrase "normality holds." ‖

An example of a normal convex program (P), such that (P) has an optimal solution but (P^*) has no optimal solution, is obtained when F is the closed convex bifunction from R to R given by

$$(Fu)(x) = \begin{cases} x & \text{if } x^2 \leq u, \\ +\infty & \text{if } x^2 > u. \end{cases}$$

The perturbation function is then given by

$$\inf Fu = \begin{cases} -u^{1/2} & \text{if } u \geq 0, \\ +\infty & \text{if } u < 0. \end{cases}$$

This function is lower semi-continuous at $u = 0$, but it has derivative $-\infty$ there. Thus (P) is normal but has no Kuhn–Tucker "vector" u^* (Corollary 29.1.2). No optimal solution can exist for (P^*), in view of Theorem 30.5, although $x = 0$ is trivially an optimal solution for (P).

Existence theorems for optimal solutions to (P) can be deduced from Theorem 30.5 by applying to (P^*) the various existence theorems for Kuhn–Tucker vectors.

COROLLARY 30.5.2. *Let F be a closed convex bifunction from R^m to R^n, and let (P) be the convex program associated with F. If (P) is consistent and (P^*) is strongly consistent, then (P) has an optimal solution. Dually, if (P) is strongly consistent and (P^*) is consistent, then (P^*) has an optimal solution.*

PROOF. If (P^*) is strongly consistent, normality holds (Theorem 30.4), so that the optimal values in (P) and (P^*) are equal. The common value cannot be $-\infty$ (because (P^*) is consistent) nor $+\infty$ (because (P) is consistent), and hence it is finite. At least one Kuhn–Tucker vector x exists for (P^*) by Corollary 29.1.4, and this x is an optimal solution to (P) by Theorem 30.5. ‖

Other duality results are clearly possible, in view of the conjugacy relationship between objective functions and perturbation functions in Theorem 30.2 and the correspondences listed in Theorem 27.1. Generally speaking, any property of the perturbation function in (P) is dual to some property of the objective function in (P^*), and any property of the perturbation function in (P^*) is dual to some property of the objective function in (P).

The rest of this section is devoted to a discussion of the dual of an ordinary convex program. The discussion will be limited, for notational simplicity, to the case where all the (explicit) constraints are inequalities.

Let (P) be the ordinary convex program in which $f_0(x)$ is to be minimized over C subject to

$$f_1(x) \leq 0, \ldots, f_m(x) \leq 0,$$

where $f_0, f_1, \ldots, f_m$ are proper convex functions on R^n such that $\operatorname{dom} f_0 = C$ and

$$\operatorname{dom} f_i \supset C, \quad \operatorname{ri}(\operatorname{dom} f_i) \supset \operatorname{ri} C, \quad i = 1, \ldots, m.$$

The convex bifunction F from R^m to R^n associated with (P) is given by

$$(Fu)(x) = f_0(x) + \delta(x \,|\, f_i(x) \leq v_i, \quad i = 1, \ldots, m),$$

where $u = (v_1, \ldots, v_m)$. The adjoint F^* of F may be calculated as follows, with $z = (\zeta_1, \ldots, \zeta_m)$:

$$(F^*x^*)(u^*) = \inf_{u,x} \{(Fu)(x) - \langle x, x^* \rangle + \langle u, u^* \rangle\}$$

$$= \inf_{x \in R^n} \inf_{u \in R^m} \{f_0(x) + \delta(x \,|\, f_i(x) \leq v_i, \quad i = 1, \ldots, m)$$
$$- \langle x, x^* \rangle + v_1^* v_1 + \cdots + v_m^* v_m\}$$

$$= \inf_{x \in C} \inf_{z \geq 0} \left\{ f_0(x) - \langle x, x^* \rangle + \sum_{i=1}^m v_i^*(f_i(x) + \zeta_i) \right\}$$

$$= \inf_{x \in C} \left\{ f_0(x) + \sum_{i=1}^m v_i^* f_i(x) - \langle x, x^* \rangle \right\} + \inf_{z \geq 0} \langle u^*, z \rangle$$

$$= -\sup_{x \in C} \left\{ \langle x, x^* \rangle - f_0(x) - \sum_{i=1}^m v_i^* f_i(x) \right\} - \delta(u^* \,|\, u^* \geq 0).$$

If $u^* \not\geq 0$, this expression is $-\infty$, whereas if $u^* \geq 0$ it is

$$-\sup_{x \in R^n} \{\langle x, x^* \rangle - (f_0 + v_1^* f_1 + \cdots + v_m^* f_m)(x)\}$$
$$= -(f_0 + v_1^* f_1 + \cdots + v_m^* f_m)^*(x^*).$$

By Theorem 16.4 and Theorem 16.1, the latter is

$$-(f_0^* \,\square\, (v_1^* f_1)^* \,\square\, \cdots \,\square\, (v_m^* f_m)^*)(x^*)$$

$$= -(f_0^* \,\square\, f_1^* v_1^* \,\square\, \cdots \,\square\, f_m^* v_m^*)(x^*)$$

$$= -\inf \left\{ f^*(x_0^*) + \sum_{i=1}^m (f_i^* v_i^*)(x_i^*) \,\Big|\, \sum_{i=0}^m x_i^* = x^* \right\},$$

where the infimum is attained and

$$(f_i^* v_i^*)(x_i^*) = \begin{cases} v_i^* f_i^*(v_i^{*-1} x_i^*) & \text{if } v_i^* > 0, \\ \delta(x_i^* \mid 0) & \text{if } v_i^* = 0. \end{cases}$$

Therefore F^* is given by

$$(F^* x^*)(u^*) = \begin{cases} -(f_0^* \,\square\, f_1^* v_1^* \,\square\, \cdots \,\square\, f_m^* v_m^*)(x^*) & \text{if } u^* \geq 0, \\ -\infty & \text{if } u^* \not\geq 0. \end{cases}$$

By definition, in the (generalized) concave program (P^*) dual to (P) one maximizes the concave function F^*0 over R^m, and one perturbs F^*0 by replacing it by $F^* x^*$ for different choices of $x^* \in R^n$. Since the effective domain of

$$f_0^* \,\square\, f_1^* v_1^* \,\square\, \cdots \,\square\, f_m^* v_m^*$$

(for $v_i^* \geq 0$) is the convex set

$$C_0^* + v_1^* C_1^* + \cdots + v_m^* C_m^* \subseteq R^n,$$

where

$$C_i^* = \operatorname{dom} f_i^*, \qquad i = 0, 1, \ldots, m,$$

the feasible solutions to (P^*) are the vectors $u^* = (v_1^*, \ldots, v_m^*)$ such that

$$v_1^* \geq 0, \ldots, v_m^* \geq 0,$$

$$0 \in (C_0^* + v_1^* C_1^* + \cdots + v_m^* C_m^*).$$

Over the convex subset of R^m consisting of all such vectors u^*, one is to maximize the (finite) concave function

$$(v_1^*, \ldots, v_m^*) \to -(f_0^* \,\square\, f_1^* v_1^* \,\square\, \cdots \,\square\, f_m^* v_m^*)(0)$$

$$= -\inf \{ f_0^*(z_0^*) + v_1^* f_1^*(z_1^*) + \cdots + v_m^* f_m^*(z_m^*) \mid$$

$$z_i^* \in C_i^*, \quad i = 0, 1, \ldots, m, \quad z_0^* + v_1^* z_1^* + \cdots + v_m^* z_m^* = 0 \}$$

(where infimum is attained by some $z_0^*, \ldots, z_m^*$ for each feasible u^*). The perturbation corresponding to a given $x^* \in R^n$ changes this problem by replacing 0 by x^* in the constraints

$$0 \in (C_0^* + v_1^* C_1^* + \cdots + v_m^* C_m^*),$$

$$z_0^* + v_1^* z_1^* + \cdots + v_m^* z_m^* = 0.$$

As the above calculations show, the objective function in (P^*) can also be expressed by

$$(F^*0)(u^*) = \begin{cases} \inf \left(f_0 + v_1^* f_1 + \cdots + v_m^* f_m \right) & \text{if } u^* \geq 0, \\ -\infty & \text{if } u^* \not\geq 0. \end{cases}$$

Hence the feasible solutions to (P^*) can also be described as the vectors $u^* \geq 0$ such that the infimum of the proper convex function

$$f_0 + v_1^* f_1 + \cdots + v_m^* f_m$$

on R^n is not $-\infty$. If the infimum is attained whenever it is not $-\infty$, and if the functions f_i are differentiable throughout R^n, then the feasible solutions are the vectors $u^* \geq 0$ such that, for some $x \in R^n$,

$$\nabla f_0(x) + v_1^* \nabla f_1(x) + \cdots + v_m^* \nabla f_m(x) = 0;$$

for any such u^* and x one has

$$(F^*0)(u^*) = f_0(x) + v_1^* f_1(x) + \cdots + v_m^* f_m(x).$$

It may seem strange that the objective function for the dual of an ordinary convex program should itself need to be expressed in terms of an extremum problem. In a certain sense, this is due to the fact that the perturbations which are naturally associated with an ordinary convex program are not enough to compensate for the non-linearity of the constraint functions. A more explicit dual program can be obtained by replacing the given ordinary convex program (P) by a generalized convex program (Q) having the same objective function as (P) but more perturbations.

Specifically, given (P) as above, let (Q) be the convex program associated with the convex bifunction G from R^k to R^n defined by

$$(Gw)(x) = f_0(x - x_0) + \delta(x \,|\, f_i(x - x_i) \leq v_i, \quad i = 1, \ldots, m),$$

where

$$w = (u, x_0, \ldots, x_m) \in R^m \times R^n \times \cdots \times R^n = R^k,$$

$k = m + (m + 1)n$. In (Q), as in (P), we minimize $f_0(x)$ subject to $f_i(x) \leq 0$, $i = 1, \ldots, m$, but in (Q) the given class of perturbations is larger: we perturb the constraint $f_i(x) \leq 0$ to $f_i(x) \leq v_i$ as in (P), but we also perturb each function f_i by translating it by an amount $x_i \in R^n$. Setting

$$w^* = (u^*, x_0^*, \ldots, x_m^*) \in R^k,$$

we can calculate G^* to determine the concave program dual to (Q). Making use of the initial steps in the calculation of F^* above, we get, for $u^* \geq 0$,

$$(G^* x^*)(w^*) = \inf_{w, x} \left\{ (Gw)(x) - \langle x, x^* \rangle + \langle u, u^* \rangle + \sum_{i=0}^{m} \langle x_i, x_i^* \rangle \right\}$$

$$= \inf_{x, x_i} \left\{ f_0(x - x_0) + \sum_{i=1}^{m} v_i^* f_i(x - x_i) - \langle x, x^* \rangle + \sum_{i=0}^{m} \langle x_i, x_i^* \rangle \right\}.$$

Upon substitution of $y_i = x - x_i$, this becomes

$$\inf_{x, \nu_i} \left\{ f_0(y_0) + \sum_{i=1}^{m} v_i^* f_i(y_i) - \left\langle x, x^* - \sum_{i=0}^{m} x_i^* \right\rangle - \sum_{i=0}^{m} \langle y_i, x_i^* \rangle \right\}$$

$$= -\sup_x \left\{ \left\langle x, x^* - \sum_{i=0}^{m} x_i^* \right\rangle \right\} - \sup_{y_0} \{ \langle y_0, x_0^* \rangle - f_0(y_0) \}$$

$$- \sum_{i=1}^{m} \sup_{\nu_i} \{ \langle y_i, x_i^* \rangle - (v_i^* f_i)(y_i) \}$$

$$= - \left[\delta\left(x^* - \sum_{i=0}^{m} x_i^* \mid 0 \right) + f_0^*(x_0^*) + \sum_{i=1}^{m} (v_i^* f_i)^*(x_i^*) \right]$$

$$= - \left[\delta\left(\sum_{i=0}^{m} x_i^* \mid x^* \right) + f_0^*(x_0^*) + \sum_{i=1}^{m} (f_i^* v_i^*)(x_i^*) \right].$$

If $u^* \not\geq 0$, we get $(G^* x^*)(w^*) = -\infty$ as in the calculation of F^*. It follows that

$$(G^* x^*)(w^*) = \begin{cases} -f_0^*(x_0^*) - (f_1^* v_1^*)(x_1^*) - \cdots - (f_m^* v_m^*)(x_m^*) \\ \qquad \text{if} \quad u^* \geq 0 \quad \text{and} \quad x_0^* + \cdots + x_m^* = x^*, \\ -\infty \quad \text{otherwise.} \end{cases}$$

The feasible solutions to the dual program (Q^*) are thus the vectors $w^* = (u^*, x_0^*, \ldots, x_m^*)$ such that

$$u^* \geq 0, \qquad x_0^* + x_1^* + \cdots + x_m^* = 0,$$

$$x_0^* \in C_0^*, \qquad x_1^* \in v_1^* C_1^*, \ldots, x_m^* \in v_m^* C_m^*,$$

where $C_i^* = \operatorname{dom} f_i^*$ as above. Over the set of these vectors w^*, we want to maximize the concave function

$$-[f_0^*(x_0^*) + (f_1^* v_1^*)(x_1^*) + \cdots + (f_m^* v_m^*)(x_m^*)].$$

The perturbation in (Q^*) corresponding to a given $x^* \in R^n$ alters the constraint $x_0^* + \cdots + x_m^* = 0$ to $x_0^* + \cdots + x_m^* = x^*$.

Of course, x_i^* is related to the z_i^* in the expression of (P^*) above by

$$x_0^* = z_0^*, \qquad x_i^* = v_i^* z_i^* \quad \text{for} \quad i = 1, \ldots, m.$$

According to the general duality theory, the x_i^* components of the optimal solutions w^* to (Q^*) describe the rates of change of

$$\inf \{ f_0(x) \mid f_i(x) \leq 0, \quad i = 1, \ldots, m \}$$

with respect to translating the functions f_i by amounts x_i.

If there are directions in which certain of the functions f_i are affine, the corresponding convex sets $C_i^* = \operatorname{dom} f_i^*$ are less than n-dimensional

(Theorem 13.4). The dual problem (Q^*) is then degenerate, in the sense that there is a proper subspace of R^k which contains dom G^*x^* for every x^*. In this case, in passing from (P) to (Q) one has overcompensated for the nonlinearity of the f_i by introducing redundant degrees of perturbation: a translation of f_i by an amount x_i in a direction in which f_i is affine merely alters f_i by a constant, i.e. it has the same effect as a perturbation corresponding to the variable v_i.

It is sometimes desirable in such cases to consider convex programs "intermediate" between (P) and (Q) in which the perturbations are chosen more carefully, so as to match the particular nonlinearity at hand. Suppose that each f_i is expressed in the form

$$f_i(x) = h_i(A_i x + a_i) + \langle a_i^*, x \rangle + \alpha_i,$$

where h_i is a closed proper convex function on R^{n_i}, A_i is a linear transformation from R^n to R^{n_i}, a_i and a_i^* are vectors in R^{n_i} and R^n respectively, and α_i is a real number. (The vectors in the null space of A_i then give directions in which f_i is affine.) Let (R) be the convex program associated with the convex bifunction H from R^d to R^n defined by

$$(Hw)(x) = \begin{cases} h_0(A_0 x + a_0 - p_0) + \langle a_0^*, x \rangle + \alpha_0 & \text{if} \\ \quad h_i(A_i x + a_i - p_i) + \langle a_i^*, x \rangle + \alpha_i \leq v_i \\ \quad \text{for} \quad i = 1, \ldots, m, \\ +\infty \quad \text{otherwise}, \end{cases}$$

where $d = m + n_0 + \cdots + n_m$ and

$$w = (u, p_0, \ldots, p_m), \qquad u \in R^m, \qquad p_i \in R^{n_i}.$$

The objective function in (R) is again the same as in the ordinary convex program (P), i.e. in (R) one minimizes $f_0(x)$ over C subject to $f_i(x) \leq 0$ for $i = 1, \ldots, m$. The adjoint of H may be determined by direct calculation as above. One finds that

$$(H^*x^*)(w^*) = \begin{cases} \alpha_0 + \langle a_0, p_0^* \rangle - h_0^*(p_0^*) \\ \quad + \sum_{i=1}^m [\alpha_i v_i^* + \langle a_i, p_i^* \rangle - (h_i^* v_i^*)(p_i^*)] \\ \quad \text{if} \quad u^* \geq 0 \quad \text{and} \\ \qquad a_0^* + \sum_{i=1}^m v_i^* a_i^* + \sum_{i=0}^m A_i^* p_i^* = x^*, \\ -\infty \quad \text{otherwise}, \end{cases}$$

where

$$w^* = (u^*, p_0^*, \ldots, p_m^*), \qquad u^* \in R^m, \qquad p_i^* \in R^{n_i}.$$

Thus, in the dual program (R^*), one maximizes

$$\alpha_0 + \langle a_0, p_0^* \rangle - h_0^*(p_0^*) + \sum_{i=1}^m [\alpha_i v_i^* + \langle a_i, p_i^* \rangle - (h_i^* v_i^*)(p_i^*)]$$

subject to

$$p_0^* \in D_0^*, \qquad p_i^* \in v_i^* D_i^* \quad \text{and} \quad v_i^* \geq 0 \quad \text{for} \quad i = 1, \ldots, m,$$

$$a_0^* + \sum_{i=1}^m v_i^* a_i^* + \sum_{i=0}^m A_i^* p_i^* = 0,$$

where $D_i^* = \text{dom } h_i^*$ for $i = 0, \ldots, m$. If the expression of each f_i is such that there are no directions in which h_i is affine, the convex set D_i^* will be of dimension n_i and hence will have a non-empty interior in R^{n_i}. (It is easily seen that, when C is n-dimensional, such an expression of f_i always exists with $n_i = \text{rank } f_i$, unless f_i is affine, in which event h_i, A_i, a_i, p_i and p_i^* can simply be omitted from all the formulas.)

The Lagrangian L of (R) may be calculated from the defining relation

$$L(w^*, x) = \inf_w \{\langle w, w^* \rangle + (Hw)(x)\}.$$

One obtains

$$L(w^*, x) = \alpha_0 + \langle a_0, p_0^* \rangle - h_0^*(p_0^*)$$

$$+ \sum_{i=1}^m [\alpha_i v_i^* + \langle a_i, p_i^* \rangle - (h_i^* v_i^*)(p_i^*)]$$

$$+ \langle a_0^* + \sum_{i=1}^m v_i^* a_i^* + \sum_{i=0}^m A_i^* p_i^*, x \rangle$$

if $u^* \geq 0$, whereas $L(w^*, x) = -\infty$ if $u^* \not\geq 0$.

As an illustration of programs (R) and (R^*), consider the important case where

$$f_i(x) = \log \left(\sum_{r=1}^{n_i} \exp \left(\alpha_{r0}^i + \sum_{j=1}^n \alpha_{rj}^i \xi_j \right) \right)$$

for $i = 1, \ldots, m$. Here f_i is a finite convex function on R^n (affine if $n_i = 1$); see the example preceding Theorem 16.5. (Note that, under the substitution $\tau_j = e^{\xi_j}$, each of the terms

$$\exp \left(\alpha_{r0}^i + \sum_{j=1}^n \alpha_{rj}^i \xi_j \right)$$

takes on the general form

$$\beta_0 \tau_1^{\beta_1} \tau_2^{\beta_2} \cdots \tau_n^{\beta_n},$$

where $\beta_0 > 0$.) The problem of minimizing $f_0(x)$ subject to $f_i(x) \leq 0$ for $i = 1, \ldots, m$ corresponds to (R) in the case where

$$h_i(\pi_{i1}, \ldots, \pi_{in_i}) = \log \left(\sum_{r=1}^{n_i} e^{\pi_{ir}} \right),$$

A_i is the linear transformation given by the $n_i \times n$ matrix (α_{rj}^i), a_i is the vector in R^{n_i} with components α_{r0}^i, and $a_i^* = 0$, $\alpha_i = 0$. From the calculation preceding Theorem 16.5, we have

$$h_i^*(\pi_{i1}^*, \ldots, \pi_{in_i}^*) = \begin{cases} \sum_{r=1}^{n_i} \pi_{ir}^* \log \pi_{ir}^* & \text{if} \quad \pi_{ir}^* \geq 0, \qquad \sum_{r=1}^{n_i} \pi_{ir}^* = 1, \\ +\infty & \text{otherwise.} \end{cases}$$

It follows that in (R^*) one maximizes the concave function

$$\sum_{i=0}^{m} \langle a_i, p_i^* \rangle - h_0^*(p_0^*) - \sum_{i=1}^{m} (h_i^* v_i^*)(p_i^*)$$
$$= \sum_{i=0}^{m} \sum_{r=1}^{n_i} (\pi_{ir}^* \alpha_{r0}^i - \pi_{ir}^* \log \pi_{ir}^*) + \sum_{i=1}^{m} v_i^* \log v_i^*$$

subject to the linear constraints

$$\pi_{ir}^* \geq 0 \quad \text{for} \quad i = 0, \ldots, m \quad \text{and} \quad r = 1, \ldots, n_i,$$

$$\sum_{r=1}^{n_0} \pi_{0r}^* = 1 \quad \text{and} \quad \sum_{r=1}^{n_i} \pi_{ir}^* = v_i^* \quad \text{for} \quad i = 1, \ldots, m,$$

$$\sum_{i=0}^{m} \sum_{r=1}^{n_i} \pi_{ir}^* \alpha_{rj}^i = 0 \quad \text{for} \quad j = 1, \ldots, n,$$

where

$$p_i = (\pi_{i1}, \ldots, \pi_{in_i}) \quad \text{for} \quad i = 0, 1, \ldots, m.$$

By the general duality theory, the components ξ_j of an optimal solution x to (R) describe the rates of change of the supremum in (R^*) with respect to perturbing the latter constraints to

$$\sum_{i=0}^{m} \sum_{r=1}^{n_i} \pi_{ir}^* \alpha_{rj}^i = \xi_j^* \quad \text{for} \quad j = 1, \ldots, n,$$

whereas the components v_i^* and π_{ir}^* of an optimal solution to (R^*) describe the rates of change of the infimum in (R) with respect to perturbing the functions f_i by subtracting certain constants v_i and performing certain translations. Optimal solutions to (R) and (R^*) correspond to saddle-points of the Lagrangian function

$$L(w^*, x) = \begin{cases} \sum_{i=0}^{m} \sum_{r=1}^{n_i} (\pi_{ir}^* \alpha_{r0}^i - \pi_{ir}^* \log \pi_{ir}^*) + \sum_{i=1}^{m} v_i^* \log v_i^* \\ \quad + \sum_{j=1}^{n} \sum_{i=0}^{m} \sum_{r=1}^{n_i} \pi_{ir}^* \alpha_{rj}^i \xi_j \quad \text{if} \quad w^* \in D^*, \\ -\infty \quad \text{if} \quad w^* \notin D^*, \end{cases}$$

where D^* is the set of vectors $w^* = (u^*, p_0^*, \ldots, p_m^*)$ such that

$$\pi_{ir}^* \geq 0 \quad \text{for} \quad i = 0, \ldots, m \quad \text{and} \quad r = 1, \ldots, n_i,$$

$$\sum_{r=1}^{n_0} \pi_{0r}^* = 1 \quad \text{and} \quad \sum_{r=1}^{n_i} \pi_{ir}^* = v_i^* \quad \text{for} \quad i = 1, \ldots, m.$$

Fenchel's Duality Theorem

Fenchel's duality theorem pertains to the problem of minimizing a difference $f(x) - g(x)$, where f is a proper convex function on R^n and g is a proper concave function on R^n. This problem includes, as a special case, the problem of minimizing f over a convex set C (take $g = -\delta(\cdot \mid C)$). In general, $f - g$ is a certain convex function on R^n. The duality resides in the connection between minimizing $f - g$ and maximizing the concave function $g^* - f^*$, where f^* is the (convex) conjugate of f and g^* is the (concave) conjugate of g. This duality is a special case of the general duality in §30, as we shall show, but it can also be deduced independently of the general theory by an elementary separation argument.

Note that the minimization of $f - g$ effectively takes place over the convex set

$$\text{dom} (f - g) = \text{dom} f \cap \text{dom} g,$$

whereas the maximization of $g^* - f^*$ effectively takes place over the convex set

$$\text{dom} (g^* - f^*) = \text{dom} g^* \cap \text{dom} f^*.$$

THEOREM 31.1 (Fenchel's Duality Theorem). *Let f be a proper convex function on R^n, and let g be a proper concave function on R^n. One has*

$$\inf_x \{f(x) - g(x)\} = \sup_{x^*} \{g^*(x^*) - f^*(x^*)\}$$

if either of the following conditions is satisfied:
  (a) ri (dom f) $\cap$ ri (dom g) $\neq \emptyset$;
  (b) f *and* g *are closed, and* ri (dom g^*) $\cap$ ri (dom f^*) $\neq \emptyset$.
Under (a) *the supremum is attained at some x^*, while under* (b) *the infimum is attained at some x; if* (a) *and* (b) *both hold, the infimum and supremum are necessarily finite.*

If g is actually polyhedral, ri (dom g) *and* ri (dom g^*) *can be replaced by* dom g *and* dom g^* *in* (a) *and* (b), *respectively (and the closure assumption in* (b) *is superfluous). Similarly if f is polyhedral. (Thus, if f and g are both polyhedral,* ri *can be omitted in all cases.)*

PROOF. For any x and x^* in R^n, we have

$$f(x) + f^*(x^*) \geq \langle x, x^* \rangle \geq g(x) + g^*(x^*)$$

by Fenchel's Inequality, and hence

$$f(x) - g(x) \geq g^*(x^*) - f^*(x^*).$$

Therefore

$$\inf (f - g) \geq \sup (g^* - f^*).$$

If the infimum is $-\infty$, the supremum is $-\infty$ too and is attained throughout R^n. Assume now that (a) holds and that $\alpha = \inf (f - g)$ is not $-\infty$. Then α is finite, and it is the greatest of the constants β such that $f \geq g + \beta$. To show that the supremum of $g^* - f^*$ is α and is attained, it is enough to show that there exists a vector x^* such that $g^*(x^*) - f^*(x^*) \geq \alpha$. Consider the epigraphs

$$C = \{(x, \mu) \mid x \in R^n, \mu \in R, \mu \geq f(x)\},$$

$$D = \{(x, \mu) \mid x \in R^n, \mu \in R, \mu \leq g(x) + \alpha\}.$$

These are convex sets in R^{n+1}. According to Lemma 7.3,

$$\text{ri } C = \{(x, \mu) \mid x \in \text{ri } (\text{dom} f), f(x) < \mu < \infty\}.$$

Since $f \geq g + \alpha$, ri C does not meet D. Hence there exists a hyperplane H in R^{n+1} which separates C and D properly (Theorem 11.3). If H were vertical, its projection on R^n would be a hyperplane separating the projections of C and D properly. But the projections of C and D on R^n are dom f and dom g, respectively, and these sets cannot be separated properly because of assumption (a) (Theorem 11.3). Therefore H is not vertical, i.e. H is the graph of a certain affine function h,

$$h(x) = \langle x, x^* \rangle - \alpha^*.$$

Since H separates C and D, we have

$$f(x) \geq \langle x, x^* \rangle - \alpha^* \geq g(x) + \alpha, \qquad \forall x.$$

The inequality on the left implies that

$$\alpha^* \geq \sup_x \{\langle x, x^* \rangle - f(x)\} = f^*(x^*),$$

while the inequality on the right implies that

$$\alpha^* + \alpha \leq \inf_x \{\langle x, x^* \rangle - g(x)\} = g^*(x^*).$$

It follows that $\alpha \leq g^*(x^*) - f^*(x^*)$ as desired.

When g is polyhedral, i.e. when D is a polyhedral convex set, we can sharpen the proof by invoking Theorem 20.2, instead of Theorem 11.3, to get a separating hyperplane H. By Theorem 20.2, H can be chosen so that it does not contain C. If H were vertical, its projection on R^n would be a hyperplane separating dom f and dom g and not containing all of dom f. When ri (dom f) and dom g (which is polyhedral) have a point in common, this situation is impossible by Theorem 20.2, so that H must be non-vertical and the proof goes through as before. Similarly when f, rather than g, is polyhedral.

When both f and g are polyhedral, a somewhat different argument, not involving relative interiors at all, can be used in the case where $\alpha = \inf (f - g)$ is finite to show the existence of an x^* such that $g^*(x^*) - f^*(x^*) \geq \alpha$. In this case, by the definition of α, the closure of the convex set $C - D$ in R^{n+1} contains the origin $(0, 0)$, but $C - D$ does not contain any (x, μ) with $x = 0$ and $\mu < 0$. Since C and D are polyhedral (by the fact that f and g are polyhedral), $C - D$ is polyhedral by Corollary 19.3.2 and hence is closed. Let $C - D$ be expressed as the intersection of a finite collection of closed half-spaces in R^{n+1}. These half-spaces all contain the origin of R^{n+1}, but at least one of them must be disjoint from the half-line $\{(0, \mu) \mid \mu < 0\}$, for otherwise this half-line would meet $C - D$, contrary to what we have just observed. At least one of these half-spaces must therefore be the epigraph of a linear function $\langle \cdot, x^* \rangle$ on R^n.

For this x^* we have

$$\mu_1 - \mu_2 \geq \langle x_1 - x_2, x^* \rangle$$

for every $(x_1, \mu_1) \in C$ and every $(x_2, \mu_2) \in D$, or in other words

$$\langle x_2, x^* \rangle - g(x_2) - \alpha \geq \langle x_1, x^* \rangle - f(x_1), \quad \forall x_1, \quad \forall x_2.$$

This implies that

$$g^*(x^*) - \alpha \geq f^*(x^*)$$

as required.

The part of the theorem concerning condition (b) follows by duality, since $f = f^{**}$ and $g = g^{**}$ when f and g are closed. Of course, f^* and g^* are polyhedral when f and g are polyhedral, respectively (Theorem 19.2). ‖

The next theorem shows how the extremum problems in Fenchel's Duality Theorem, in a generalized form, can be regarded as a dual pair of convex and concave programs in the sense of §30. The theorems in §29 and §30 can be applied to these programs, and in this way one can refine the conclusions of Fenchel's Duality Theorem and gain insight into their meaning.

THEOREM 31.2. *Let f be a proper convex function on R^n, let g be a proper concave function on R^m, and let A be a linear transformation from R^n to R^m. Let*

$$(Fu)(x) = f(x) - g(Ax + u), \quad \forall x \in R^n, \quad \forall u \in R^m.$$

Then F is a proper convex bifunction from R^m *to* R^n, *closed if f and g are closed. The optimal value in the convex program* (P) *associated with F is*

$$\inf_x \{f(x) - g(Ax)\} = \inf F0,$$

and (P) *is strongly consistent if and only if there exists a vector* $x \in$ ri (dom f) *such that* $Ax \in$ ri (dom g). *The adjoint of F is given by*

$$(F^*x^*)(u^*) = g^*(u^*) - f^*(A^*u^* + x^*), \qquad \forall u^* \in R^m, \qquad \forall x^* \in R^n.$$

The optimal value in the dual concave program (P*) *is*

$$\sup_{u^*} \{g^*(u^*) - f^*(A^*u^*)\} = \sup F^*0,$$

and (P*) *is strongly consistent if and only if there exists a vector* $u^* \in$ ri (dom g^*) *such that* $A^*u^* \in$ ri (dom f^*).

PROOF. It is obvious that $f(x) - g(Ax + u)$ is a proper convex function of (u, x), closed if f and g are closed. The assertions about F are thus valid. The optimal value in (P) is the infimum of $F0$ by definition. The function Fu is identically $+\infty$ unless there exists an x such that $f(x)$ and $g(Ax + u)$ are both finite, i.e. unless $Ax + u \in$ dom g for some $x \in$ dom f. Thus

$$\text{dom } F = \text{dom } g - A \text{ dom } f,$$

and by the calculus of relative interiors (Theorem 6.6, Corollary 6.6.2) we have

$$\text{ri (dom } F) = \text{ri (dom } g) - A \text{ (ri (dom } f)).$$

It follows that (P) is strongly consistent if and only if

$$0 \in [\text{ri (dom } g) - A \text{ (ri (dom } f))],$$

i.e. if and only if ri (dom g) and A (ri (dom f)) have a point in common. The formula for F^* is proved by direct calculation using the substitution $y = Ax + u$:

$$(F^*x^*)(u^*) = \inf_{x,u} \{(Fu)(x) - \langle x, x^* \rangle + \langle u, u^* \rangle\}$$

$$= \inf_{x,u} \{f(x) - g(Ax + u) - \langle x, x^* \rangle + \langle u, u^* \rangle\}$$

$$= \inf_{x,y} \{f(x) - g(y) - \langle x, x^* \rangle + \langle y, u^* \rangle - \langle Ax, u^* \rangle\}$$

$$= \inf_x \{f(x) - \langle x, A^*u^* + x^* \rangle\} + \inf_y \{\langle y, u^* \rangle - g(y)\}$$

$$= -f^*(A^*u^* + x^*) + g^*(u^*).$$

The statement about (P*) is justified like the statement about (P).   ‖

The convex and concave programs in Theorem 31.2 reduce to the extremum problems in Fenchel's Duality Theorem when $m = n$ and A is the identity transformation $I: R^n \to R^n$. Thus Fenchel's Duality Theorem is obtained from the problem of minimizing $f - g$ by introducing the perturbation which replaces g by a translate g_u for each u, where

$$g_u(x) = g(x + u).$$

The perturbation function in this convex program (P) is the (convex) function p given by

$$p(u) = \inf (f - g_u).$$

The duality between minimizing $f - g$ and maximizing $g^* - f^*$ has to do with the behavior of $p(u)$ around $u = 0$. Indeed, $g^* - f^*$ is the objective function in the dual concave program (P^*) (in which perturbation corresponds to translation of f^*), so $g^* - f^*$ is the concave conjugate of $-p$ (Theorem 30.2). Assuming that $\operatorname{dom} f \cap \operatorname{dom} g$ is not empty, or that $\operatorname{dom} g^* \cap \operatorname{dom} f^*$ is not empty (and assuming for simplicity in applying results of §30 that f and g are closed, so that F is closed by Theorem 31.2—this assumption is actually unnecessary in view of Corollary 29.4.1), we have

$$\sup (g^* - f^*) = \lim_{u \to 0} \inf p(u) \leq p(0) = \inf (f - g)$$

(Corollary 30.2.3). Condition (a) of Theorem 31.1 (which corresponds to (P) being strongly consistent, according to Theorem 31.2) is sufficient, as we have seen, for the existence of at least one vector x^* such that

$$g^*(x^*) - f^*(x^*) = \sup (g^* - f^*) = \inf (f - g).$$

When $\inf (f - g)$ is finite, such vectors x^* are precisely the Kuhn–Tucker vectors for (P) (Theorem 30.5), and a necessary as well as sufficient condition for their existence is that

$$p'(0; y) > -\infty, \quad \forall y$$

(Corollary 29.1.2). For there to exist exactly one such vector x^*, it is necessary and sufficient that p be finite and differentiable at $u = 0$, in which case the unique x^* is $-\nabla p(0)$ (Corollary 29.1.3).

Of course, the results in §29 and §30 can be applied in the same way to the more general programs in Theorem 31.2 to get conditions under which

$$\inf_x \{f(x) - g(Ax)\} = \sup_{u^*} \{g^*(u^*) - f^*(A^*u^*)\},$$

and so forth. In particular, Theorem 30.4 and Corollary 30.5.2 yield a generalization of Fenchel's Duality Theorem to these programs:

COROLLARY 31.2.1. *Let f be a closed proper convex function on R^n, let g be a closed proper concave function on R^m, and let A be a linear transformation from R^n to R^m. One has*

$$\inf_x \{f(x) - g(Ax)\} = \sup_{u^*} \{g^*(u^*) - f^*(A^*u^*)\}$$

if either of the following conditions is satisfied:
(a) *There exists an $x \in$ ri (dom f) such that $Ax \in$ ri (dom g);*
(b) *There exists a $u^* \in$ ri (dom g^*) such that $A^*u^* \in$ ri (dom f^*).*

Under (a) *the supremum is attained at some u^*, while under* (b) *the infimum is attained at some x.*

It can be shown that in Corollary 31.2.1, just as in Theorem 31.1, "ri" can be omitted whenever the corresponding function f or g is actually polyhedral. However, the proof will not be given here.

The convex programs in Theorem 31.2 have some interesting special cases. When

$$f(x) = \langle a^*, x \rangle + \delta(x \mid x \geq 0),$$

$$g(u) = -\delta(u \mid u \geq a),$$

for given vectors a and a^* in R^m and R^n, respectively, (P) is the linear program on p. 311, whose optimal value is

$$\inf \{\langle a^*, x \rangle \mid x \geq 0, Ax \geq a\}.$$

The conjugate functions in this case are given by

$$f^*(x^*) = \delta(x^* \mid x^* \leq a^*),$$

$$g^*(u^*) = \langle u^*, a \rangle - \delta(u^* \mid u^* \geq 0),$$

so that (P^*) is the dual linear program, whose optimal value is

$$\sup \{\langle u^*, a \rangle \mid u^* \geq 0, A^*u^* \leq a^*\}.$$

Another case worth noting is where f is an arbitrary positively homogeneous closed proper convex function on R^n (e.g. a norm) and $g(u) = -\delta(u \mid D)$, where D is a non-empty closed convex set in R^m. By Theorem 13.2, f is the support function of a certain non-empty closed convex set C in R^n, and $f^*(x^*) = \delta(x^* \mid C)$. The conjugate of g is given by

$$g^*(u^*) = \inf \{\langle u, u^* \rangle \mid u \in D\} = -\delta^*(-u^* \mid D),$$

and hence it is a positively homogeneous closed proper concave function. In (P) one minimizes $f(x)$ subject to the constraint $Ax \in D$, whereas in (P^*) one maximizes $g^*(u^*)$ subject to the constraint $A^*u^* \in C$.

The theory of subgradients can be employed to get conditions for the attainment of the extrema in Theorem 31.2. A necessary and sufficient condition for the infimum of $f - gA$ to be attained at x is trivially that

$$0 \in \partial(f - gA)(x).$$

In general, we have

$$\partial(f - gA)(x) \supset \partial f(x) - A^*\partial g(Ax)$$

by Theorems 23.8 and 23.9, with equality in particular when the image of ri (dom f) under A meets ri (dom g) (Theorem 6.7). The condition

$$0 \in (\partial f(x) - A^*\partial g(Ax))$$

is thus always sufficient and "usually" necessary for the infimum of $f - gA$ to be attained at x. Similarly, the condition

$$0 \in (\partial g^*(u^*) - A \, \partial f^*(A^*u^*))$$

is always sufficient and "usually" necessary for the supremum of $g^* - f^*A^*$ to be attained at u^*. When f and g are closed (so that $\partial f^* = (\partial f)^{-1}$ and $\partial g^* = (\partial g)^{-1}$ by Theorem 23.5), there is a remarkable duality between these two sufficient-and-usually-necessary conditions. This may be seen by considering the subdifferential relations

$$A^*u^* \in \partial f(x), \qquad Ax \in \partial g^*(u^*),$$

which we shall call the *Kuhn–Tucker conditions* for the programs in Theorem 31.2. (This terminology will be justified in §36, where Kuhn–Tucker conditions will be defined for arbitrary convex programs.) A vector x satisfies

$$0 \in (\partial f(x) - A^*\partial g(Ax))$$

if and only if there exists a vector u^* such that x and u^* together satisfy the Kuhn–Tucker conditions. On the other hand, a vector u^* satisfies

$$0 \in (\partial g^*(u^*) - A \, \partial f^*(A^*u^*))$$

if and only if there exists a vector x such that u^* and x satisfy the Kuhn–Tucker conditions. Thus the sufficient-and-usually-necessary condition for (P) can be satisfied if and only if the corresponding condition for (P^*) can be satisfied.

The significance of the Kuhn–Tucker conditions for the programs in Theorem 31.2 can also be stated and proved more directly, as follows.

THEOREM 31.3. *Let f be a closed proper convex function on R^n, let g be a closed proper concave function on R^m, and let A be a linear transformation from R^n to R^m. In order that x and u^* be vectors such that*

$$f(x) - g(Ax) = \inf (f - gA)$$
$$= \sup (g^* - f^*A^*) = g^*(u^*) - f^*(A^*u^*),$$

it is necessary and sufficient that x and u* satisfy the Kuhn–Tucker conditions:

$$A^*u^* \in \partial f(x), \qquad Ax \in \partial g^*(u^*).$$

PROOF. The Kuhn–Tucker conditions are equivalent to the conditions

$$f(x) + f^*(A^*u^*) = \langle x, A^*u^* \rangle,$$

$$g(Ax) + g^*(u^*) = \langle Ax, u^* \rangle$$

(Theorem 23.5), and these are in turn equivalent to

$$f(x) - g(Ax) = g^*(u^*) - f^*(A^*u^*),$$

in view of the general inequality

$$f(x) + f^*(A^*u^*) \geq \langle x, A^*u^* \rangle$$
$$= \langle Ax, u^* \rangle \geq g(Ax) + g^*(u^*).$$

The general inequality implies that

$$\inf (f - gA) \geq \sup (g^* - f^*A^*),$$

so the theorem follows. ‖

COROLLARY 31.3.1. *Assume the notation of the theorem. Assume also that the image of* ri (dom f) *under A meets* ri (dom g). *Then, in order that x be a vector at which the infimum of $f - gA$ is attained, it is necessary and sufficient that there exist a vector u* such that x and u* satisfy the Kuhn–Tucker conditions.*

PROOF. Apply Corollary 31.2.1. ‖

In the linear program example mentioned above we have

$$f(x) = \langle a^*, x \rangle + \delta(x \mid x \geq 0)$$

$$g^*(u^*) = \langle u^*, a \rangle - \delta(u^* \mid u^* \geq 0)$$

for certain vectors a and a^*. Calculating by the rule in Theorem 23.8, we get

$$\partial f(x) = a^* + \partial \delta(x \mid x \geq 0)$$

$$= \begin{cases} a^* + \{x^* \leq 0 \mid \langle x^*, x \rangle = 0\} & \text{if} \quad x \geq 0, \\ \emptyset & \text{if} \quad x \ngeq 0; \end{cases}$$

$$\partial g^*(u^*) = a - \partial \delta(u^* \mid u^* \geq 0)$$

$$= \begin{cases} a + \{u \geq 0 \mid \langle u^*, u \rangle = 0\} & \text{if} \quad u^* \geq 0, \\ \emptyset & \text{if} \quad u^* \ngeq 0. \end{cases}$$

The Kuhn–Tucker conditions in this case are therefore

$$x \geq 0, \qquad A^*u^* - a^* \leq 0, \qquad \langle x, A^*u^* - a^* \rangle = 0,$$
$$Ax - a \geq 0, \qquad u^* \geq 0, \qquad \langle Ax - a, u^* \rangle = 0.$$

In the example of "homogeneous" programs introduced following Corollary 31.2.1, we have

$$f(x) = \delta^*(x \mid C), \qquad g^*(u^*) = -\delta^*(-u^* \mid D),$$

where C and D are closed convex sets. The Kuhn–Tucker conditions then mean that x is normal to C at the point A^*u^*, and u^* is normal to D at the point Ax (Corollary 23.5.3).

In the case of the extremum problems in Fenchel's Duality Theorem, A is the identity transformation and the Kuhn–Tucker conditions reduce to

$$x^* \in \partial f(x), \qquad x \in \partial g^*(x^*).$$

Some important consequences of Fenchel's Duality Theorem will now be stated.

THEOREM 31.4. *Let f be a closed proper convex function on R^n, and let K be a non-empty closed convex cone in R^n. Let K^* be the negative of the polar of K, i.e.*

$$K^* = \{x^* \mid \langle x^*, x \rangle \geq 0, \quad \forall x \in K\}.$$

One has

$$\inf \{f(x) \mid x \in K\} = -\inf \{f^*(x^*) \mid x^* \in K^*\}$$

if either of the following conditions hold:
(a) ri $(\operatorname{dom} f) \cap$ ri $K \neq \emptyset$;
(b) ri $(\operatorname{dom} f^*) \cap$ ri $K^* \neq \emptyset$.
Under (a), the infimum of f^* over K^* is attained, while under (b) the infimum of f over K is attained.

If K is polyhedral, ri K and ri K^* can be replaced by K and K^* in (a) and (b).

In general, x and x^* satisfy

$$f(x) = \inf_{K} f = -\inf_{K^*} f^* = -f^*(x^*),$$

if and only if

$$x^* \in \partial f(x), \qquad x \in K, \qquad x^* \in K^*, \qquad \langle x, x^* \rangle = 0.$$

PROOF. Apply Theorem 31.1 with $g(x) = -\delta(x \mid K)$. The conjugate of $\delta(\cdot \mid K)$ is $\delta(\cdot \mid K^\circ)$ (Theorem 14.1), so we have

$$g^*(x^*) = -\delta(-x^* \mid K^\circ) = -\delta(x^* \mid K^*).$$

The Kuhn–Tucker conditions in Theorem 31.3 reduce to $x^* \in \partial f(x)$, $x \in \partial g^*(x^*)$. We have $x \in \partial g^*(x^*)$ if and only if

$$\langle x, x^* \rangle = g(x) + g^*(x^*) = -\delta(x \mid K) - \delta(x^* \mid K^*)$$

(Theorem 23.5), and this means that $x \in K$, $x^* \in K^*$ and $\langle x, x^* \rangle = 0$. ‖

COROLLARY 31.4.1. *Let f be a closed proper convex function on R^n. One has*

$$\inf \{f(x) \mid x \geq 0\} = -\inf \{f^*(x^*) \mid x^* \geq 0\}$$

if either of the following conditions holds:

(a) *There exists a vector $x \in \mathrm{ri}\,(\mathrm{dom}\,f)$ such that $x \geq 0$;*

(b) *There exists a vector $x^* \in \mathrm{ri}\,(\mathrm{dom}\,f^*)$ such that $x^* \geq 0$.*

Under (a), the second infimum is attained, while under (b) the first infimum is attained. In general, in order that the two infima be the negatives of each other and be attained at $x = (\xi_1, \ldots, \xi_n)$ and $x^* = (\xi_1^*, \ldots, \xi_n^*)$, respectively, it is necessary and sufficient that $x^* \in \partial f(x)$ and

$$\xi_j \geq 0, \qquad \xi_j^* \geq 0, \qquad \xi_j \xi_j^* = 0, \qquad j = 1, \ldots, n.$$

PROOF. Take K to be the non-negative orthant of R^n. ‖

COROLLARY 31.4.2. *Let f be a closed proper convex function on R^n, and let L be a subspace of R^n. One has*

$$\inf \{f(x) \mid x \in L\} = -\inf \{f^*(x^*) \mid x^* \in L^\perp\}$$

if either of the following conditions is satisfied:

(a) $L \cap \mathrm{ri}\,(\mathrm{dom}\,f) \neq \emptyset$;

(b) $L^\perp \cap \mathrm{ri}\,(\mathrm{dom}\,f^*) \neq \emptyset$.

Under (a) the infimum of f^* on $L^\perp$ is attained, while under (b) the infimum of f on L is attained. In general, x and x^* satisfy

$$f(x) = \inf{}_L f = -\inf{}_{L^\perp} f^* = -f^*(x^*)$$

if and only if $x \in L$, $x^* \in L^\perp$ and $x^* \in \partial f(x)$.

PROOF. Take $K = L$. ‖

If $f(x) = h(z + x) - \langle z^*, x \rangle$, where z and z^* are given vectors and h is any closed proper convex function, then

$$f^*(x^*) = h^*(z^* + x^*) - \langle z, x^* \rangle - \langle z, z^* \rangle$$

(Theorem 12.3). A remarkable duality between h and h^* is brought to light when Theorem 31.4 and its corollaries are applied to f with z and z^* regarded as parameters. For simplicity, we shall state this duality only in the case where h and h^* are both finite everywhere.

COROLLARY 31.4.3. *Let h be a convex function on R^n which is both finite and co-finite. Let K be any non-empty closed convex cone in R^n, and* *let $K^* = -K°$. Then for every z and z^* in R^n one has*

$$\inf_{x \in K} \{h(z + x) - \langle z^*, x \rangle\} + \inf_{x^* \in K^*} \{h^*(z^* + x^*) - \langle z, x^* \rangle\} = \langle z, z^* \rangle,$$

where the infima are both finite and attained.

PROOF. Since h is co-finite, h^* is finite everywhere (Corollary 13.3.1). The convex function

$$f(x) = h(z + x) - \langle z^*, x \rangle$$

and its conjugate f^* thus have $\text{dom} f = R^n$ and $\text{dom} f^* = R^n$. Apply Theorem 31.4 to f. ‖

If f is taken to be a partial affine function in Corollary 31.4.1, one obtains the Gale–Kuhn–Tucker Duality Theorem for linear programs. This is obvious from any Tucker representation of f and the corresponding Tucker representation of f^*; see §12. Duality theorems for "quadratic" programs can be derived similarly from Corollary 31.4.1 by taking f to be a partial quadratic function, and so forth.

The subspaces L and $L^\perp$ in Corollary 31.4.2 can, of course, be given various Tucker representations, as explained in §1, and in this way one can interpret the corollary as a result about extremal properties of "dual linear systems of variables." Observe that L can in particular be taken to be the space of all circulations in some directed graph G, in which event $L^\perp$ is the space of all tensions in G (see §22). Then the two problems dual to each other are, on the one hand, to find a circulation x in G which minimizes $f(x)$, and on the other hand to find a tension x^* in G which minimizes $f^*(x^*)$.

An especially important case of Corollary 31.4.2 is the one where f is *separable*, i.e.

$$f(x) = f(\xi_1, \ldots, \xi_n) = f_1(\xi_1) + \cdots + f_n(\xi_n),$$

where $f_1, \ldots, f_n$ are closed proper convex functions on R. Then f^* is separable too. In fact, as is easily verified,

$$f^*(x^*) = f^*(\xi_1^*, \ldots, \xi_n^*) = f_1^*(\xi_1^*) + \cdots + f_n^*(\xi_n^*),$$

where f_j^* is the conjugate of f_j. The extremum problems in Corollary 31.4.2 then become:

(I)   minimize $f_1(\xi_1) + \cdots + f_n(\xi_n)$ subject to $(\xi_1, \ldots, \xi_n) \in L$;

(II)   minimize $f_1^*(\xi_1^*) + \cdots + f_n^*(\xi_n^*)$ subject to $(\xi_1^*, \ldots, \xi_n^*) \in L^\perp$.

The Kuhn–Tucker conditions at the end of Corollary 31.4.2 become:

(III) $(\xi_1, \ldots, \xi_n) \in L$, $(\xi_1^*, \ldots, \xi_n^*) \in L^\perp$,

$$(\xi_j, \xi_j^*) \in \Gamma_j \text{ for } j = 1, \ldots, n,$$

where Γ_j is the graph of the subdifferential ∂f_j. (The fact that, in this separable case,

$$x^* \in \partial f(x) \quad \text{if and only if} \quad \xi_j^* \in \partial f_j(\xi_j) \quad \text{for} \quad j = 1, \ldots, n,$$

can be deduced, as an exercise, directly from the definition of "subgradient.")

The interesting thing about these Kuhn–Tucker conditions is that, according to Theorem 24.3, the sets Γ_j which can occur are precisely the complete non-decreasing curves in R^2. Thus, given any set of n complete non-decreasing curves Γ_j in R^2 and subspaces L and $L^\perp$ in R^n, Corollary 31.4.2 gives an extremal characterization of the solutions to system (III) in terms of problems (I) and (II), where each f_j is a closed proper convex function on R determined by Γ_j uniquely up to an additive constant. (In the case where L and $L^\perp$ are the circulation space and tension space, respectively, of a directed graph G as described in §22, the curve Γ_j can be interpreted as a specified "resistance" relation between the amount of flow ξ_j in the edge e_j and the potential difference ξ_j^* across e_j.)

Theorem 22.6 can be put to good use in the analysis of problems (I) and (II), because so many of the sets associated with the functions f_i, like $\text{dom} f_j$, $\text{dom } \partial f_j$ and $\partial f_j(\xi_j)$, are real intervals.

THEOREM 31.5 (Moreau). *Let f be a closed proper convex function on R^n, and let $w(z) = (\frac{1}{2}) |z|^2$. Then*

$$(f \,\square\, w) + (f^* \,\square\, w) = w,$$

i.e. for each $z \in R^n$ one has

$$\inf_x \{f(x) + w(z - x)\} + \inf_{x^*} \{f^*(x^*) + w(z - x^*)\} = w(z),$$

where both infima are finite and uniquely attained. The unique vectors x and x^* for which the respective infima are attained for a given z are the unique vectors x and x^* such that

$$z = x + x^*, \qquad x^* \in \partial f(x),$$

and they are given by

$$x = \nabla(f^* \,\square\, w)(z), \qquad x^* = \nabla(f \,\square\, w)(z).$$

PROOF.   Fix any z, and define g by

$$g(x) = -w(z - x).$$

Then g is a finite concave function on R^n, and by direct calculation

$$g^*(x^*) = \inf_x \{\langle x, x^* \rangle + w(z - x)\}$$
$$= -w(z - x^*) + w(z).$$

According to Fenchel's Duality Theorem,

$$\inf\{f - g\} + \inf\{f^* - g^*\} = 0,$$

where both infima are finite and attained. This proves the infimum formula in the theorem. The vectors x and x^* for which the respective infima are attained are unique, due to the strict convexity of w, and they are characterized as the solutions to the Kuhn–Tucker conditions

$$x^* \in \partial f(x), \qquad x = \nabla g^*(x^*) = z - x^*.$$

Since $\partial f^* = (\partial f)^{-1}$ (Corollary 23.5.1), it follows from the uniqueness that x and x^* satisfy these conditions if and only if

$$z - x \in \partial f(x), \qquad z - x^* \in \partial f^*(x^*).$$

The latter conditions can be written as

$$z \in [\partial f(x) + \nabla w(x)] = \partial(f + w)(x),$$

$$z \in [\partial f^*(x^*) + \nabla w(x^*)] = \partial(f^* + w)(x^*)$$

(Theorem 23.8), and hence as

$$x \in \partial(f + w)^*(z), \qquad x^* \in \partial(f^* + w)^*(z).$$

The uniqueness of x and x^* implies that ∂ can be replaced by ∇ (Theorem 25.1). Of course

$$(f + w)^* = f^* \,\square\, w^*, \qquad (f^* + w)^* = f \,\square\, w^*,$$

by Theorem 16.4, where $w^* = w$ by direct calculation. ‖

According to Theorem 31.5, given any closed proper convex function f on R^n, each $z \in R^n$ can be decomposed uniquely with respect to f into a sum

$$z = x + x^*$$

such that (x, x^*) belongs to the graph of ∂f. The component x in this decomposition, which is the unique x for which

$$\inf_x \{f(x) + (1/2) |z - x|^2\}$$

is attained, is denoted by prox $(z \,|\, f)$, and the mapping

$$z \to \text{prox}\,(z \,|\, f)$$

is called the *proximation* corresponding to f. The proximation corresponding to f^* is thus related to the proximation corresponding to f by the formula

$$\text{prox}\,(z \,|\, f^*) = z - \text{prox}\,(z \,|\, f), \qquad \forall z.$$

If f is the indicator function of a non-empty closed convex set C, prox $(z \mid f)$ is the point of C nearest to z. If $f = \delta(\cdot \mid K)$, where K is a non-empty closed convex cone, so that $f^* = \delta(\cdot \mid K°)$, the decomposition of z with respect to f yields a unique expression of z as a sum $z = x + x^*$ such that

$$x \in K, \qquad x^* \in K°, \qquad \langle x, x^* \rangle = 0.$$

This reduces to the familiar orthogonal decomposition of z with respect to a subspace L when $K = L$, $K° = L^{\perp}$.

Theorem 31.5 says that prox $(\cdot \mid f)$ is the gradient mapping of a certain differentiable convex function on R^n, namely $f^* \square w$. It follows then from Corollary 25.5.1 that prox $(\cdot \mid f)$ is a *continuous* mapping of R^n into itself. The range of prox $(\cdot \mid f)$ is of course dom ∂f, the image of the graph of ∂f under the projection $(x, x^*) \to x$.

The continuity of prox $(\cdot \mid f)$ also follows from the fact that prox $(\cdot \mid f)$ is a *contraction*, i.e.

$$|\text{prox} (z_1 \mid f) - \text{prox} (z_0 \mid f)| \leq |z_1 - z_0|, \qquad \forall z_0, z_1.$$

To verify the contraction property, observe that for

$$x_i = \text{prox} (z_i \mid f), \qquad i = 0, 1,$$

$$x_i^* = \text{prox} (z_i \mid f^*), \qquad i = 0, 1,$$

one has $z_i = x_i + x_i^*$, $i = 0, 1$, and consequently

$$|z_1 - z_0|^2 = |x_1 - x_0|^2 + 2\langle x_1 - x_0, x_1^* - x_0^* \rangle + |x_1^* - x_0^*|^2.$$

Furthermore, since $x_i^* \in \partial f(x_i)$, $i = 0, 1$, and ∂f is a monotone mapping (as explained at the end of §24), one has

$$\langle x_1 - x_0, x_1^* - x_0^* \rangle \geq 0.$$

Therefore

$$|z_1 - z_0|^2 \geq |x_1 - x_0|^2,$$

and $|x_1 - x_0| \leq |z_1 - z_0|$ as claimed.

The theory of proximations leads to two important conclusions about the geometric nature of the graphs of subdifferential mappings:

COROLLARY 31.5.1. *Let f be any closed proper convex function on R^n. The mapping*

$$(x, x^*) \to x + x^*$$

is then one-to-one from the graph of ∂f onto R^n, and it is continuous in both directions. (Thus the graph of ∂f is homeomorphic to R^n.)

COROLLARY 31.5.2. *If f is any closed proper convex function on R^n, ∂f is a maximal monotone mapping from R^n to R^n.*

PROOF. We already know from the end of §24 that ∂f is a monotone mapping. To prove maximality, we must show that, given any (y, y^*) not in the graph of ∂f, there exists some (x, x^*) in the graph of ∂f such that

$$\langle y - x, y^* - x^* \rangle < 0.$$

This is easy: by Theorem 31.5, there exists some (x, x^*) in the graph of ∂f such that

$$y + y^* = x + x^*,$$

and for this (x, x^*) we have

$$\langle y - x, y^* - x^* \rangle = -|y - x|^2 = -|y^* - x^*|^2.$$

The Maximum of a Convex Function

The theory of the maximum of a convex function relative to a convex set has an entirely different character from the theory of the minimum. For one thing, it is possible, even likely, in a given case that there are many local maxima besides the global maximum. This phenomenon is rather disastrous as far as computation is concerned, because once a local maximum has been found there is, more or less by definition, no local information to tell one how to proceed to a higher local maximum. In particular, there is no local criterion for deciding whether a given local maximum is really the global maximum. Generally speaking, one would have to make a list of all the local maxima and find the global maximum by comparison.

There is some consolation, however, that the global maximum of a convex function f relative to a convex set C, generally occurs, not at just any point of C, but at some extreme point. This will be seen below.

A good illustration of the difference between minimizing and maximizing a convex function is obtained by taking C to be a triangular convex set in R^2 and f to be a function of the form $f(x) = |x - a|$, where a is a point in R^2. Minimizing f over C is the same as looking for the point of C nearest to a. This problem always has a unique solution, which could lie anywhere in C, depending on the position of a. Maximizing f over C, on the other hand, is the same as looking for the point of C farthest from a. The farthest point can only be one of the three vertices of C, but local (non-global) maxima may well occur at these vertices.

The first fact to be established is a maximum principle resembling the one for analytic functions.

THEOREM 32.1. *Let f be a convex function, and let C be a convex set contained in* dom f. *If f attains its supremum relative to C at some point of* ri C, *then f is actually constant throughout C.*

PROOF. Suppose the relative supremum is attained at a point $z \in$ ri C. Let x be a point of C other than z. We must show that $f(x) = f(z)$. Since

342

$z \in$ ri C, there is a real number $\mu > 1$ such that the point $y = (1 - \mu)x + \mu z$ belongs to C. For $\lambda = \mu^{-1}$, one has

$$z = (1 - \lambda)x + \lambda y, \qquad 0 < \lambda < 1,$$

and the convexity of f implies that

$$f(z) \leq (1 - \lambda)f(x) + \lambda f(y).$$

At the same time, $f(x) \leq f(z)$ and $f(y) \leq f(z)$ because $f(z)$ is the supremum of f relative to C. If $f(x) \neq f(z)$, we would necessarily have $f(z) > f(x)$. Then $f(y)$ would have to be finite in the convexity inequality, (since otherwise $f(y) = -\infty$ and $f(z) = -\infty$), and we would deduce the impossible relation

$$f(z) < (1 - \lambda)f(z) + \lambda f(z) = f(z).$$

Therefore $f(x) = f(z)$. ‖

COROLLARY 32.1.1. *Let f be a convex function, and let C be a convex set contained in* dom f. *Let W be the set of points (if any) where the supremum of f relative to C is attained. Then W is a union of faces of C.*

PROOF. Let x be any point of W. There exists a unique face C' of C such that $x \in$ ri C' (Theorem 18.2). The supremum of f relative to C' is attained at x, so f must be constant on C' by the theorem. Thus $C' \subset W$. This demonstrates that W is a union of faces. ‖

Theorem 32.1 implies that a convex function f which attains its supremum relative to an affine set M in dom f must be constant on M. As a matter of fact, this conclusion holds even if the supremum is merely finite, as has already been noted in Corollary 8.6.2.

The convex hull operation is important in the study of maximization, according to the following theorem.

THEOREM 32.2. *Let f be a convex function, and let $C = $ conv S, where S is an arbitrary set of points. Then*

$$\sup \{f(x) \mid x \in C\} = \sup \{f(x) \mid x \in S\},$$

where the first supremum is attained only when the second (more restrictive) supremum is attained.

PROOF. This is obvious from the fact that a level set of the form $\{x \mid f(x) < \alpha\}$, being a convex set, contains C if and only if it contains S. ‖

COROLLARY 32.2.1. *Let f be a convex function, and let C be any closed convex set which is not merely an affine set or half of an affine set. The supremum of f relative to C is then the same as the supremum of f relative to the relative boundary of C, and the former is attained only when the latter is attained.*

PROOF.   Here C is the convex hull of its relative boundary by Theorem 18.4.   ‖

Theorem 32.2 can be applied to a given closed convex set C by representing C as the convex hull of its extreme points and extreme directions as in §18.

THEOREM 32.3.   *Let f be a convex function, and let C be a closed convex set contained in* dom f. *Suppose there are no half-lines in C on which f is unbounded above. Then*

$$\sup \{f(x) \mid x \in C\} = \sup \{f(x) \mid x \in E\},$$

where E is the subset of C consisting of the extreme points of $C \cap L^{\perp}$, L being the lineality space of C. The supremum relative to C is attained only when the supremum relative to E is attained.

PROOF.   The hypothesis implies that f is constant along every line in C (Corollary 8.6.2). The set $D = C \cap L^{\perp}$ is a closed convex set containing no lines, and $C = D + L$. Given any $x \in C$, the affine set $x + L$ in C intersects D, and on this affine set f is constant. Hence the supremum over C can be reduced to the supremum over D. Now, D is the convex hull of its extreme points and extreme directions (Theorem 18.5), so

$$D = K + \operatorname{conv} E$$

for a certain convex cone K. Every point of D which is not actually in conv E belongs to a half-line of the form

$$\{x + \lambda y \mid \lambda \geq 0\}, \qquad x \in \operatorname{conv} E, \qquad y \in K.$$

Along such a half-line, $f(x + \lambda y)$ is bounded above as a function of λ by hypothesis and hence is non-increasing as a function of λ (Theorem 8.6). The supremum of f relative to such a half-line is thus attained at the endpoint x. This demonstrates that the supremum over D can be reduced to the supremum over conv E. The desired conclusion follows then from Theorem 32.2.   ‖

COROLLARY 32.3.1.   *Let f be a convex function, and let C be a closed convex set contained in* dom f. *Suppose that C contains no lines. Then, if the supremum of f relative to C is attained at all, it is attained at some extreme point of C.*

PROOF.   If C contains no lines, then $L = \{0\}$ and $C \cap L^{\perp} = C$.   ‖

COROLLARY 32.3.2.   *Let f be a convex function, and let C be a non-empty closed bounded convex set contained in* ri (dom f). *Then the supremum of f relative to C is finite, and it is attained at some extreme point of C.*

PROOF.   Since $C \subset$ ri (dom f), f is continuous relative to C (Theorem 10.1). The supremum of f relative to C is then finite and attained, because

C is closed and bounded. By the preceding corollary, it is attained at some extreme point. ‖

COROLLARY 32.3.3. *Let f be a convex function, and let C be a non-empty polyhedral convex set contained in* dom f. *Suppose there are no half-lines in C on which f is unbounded above. Then the supremum of f relative to C is attained.*

PROOF. In this case, the set $C \cap L^{\perp}$ in the theorem is polyhedral, so that E is a finite set (Corollary 19.1.1). ‖

COROLLARY 32.3.4. *Let f be a convex function, and let C be a non-empty polyhedral convex set contained in* dom f. *Suppose that C contains no lines, and that f is bounded above on C. Then the supremum of f relative to C is attained at one of the (finitely many) extreme points of C.*

PROOF. This just combines Corollary 32.3.1 and Corollary 32.3.3. ‖

Corollary 32.3.4 applies in particular to the problem of maximizing an affine function over the set of solutions to a finite system of weak linear inequalities. This is a fact of fundamental importance in the computational theory for linear programs.

The condition $C \subset$ ri $(\text{dom} f)$ in Corollary 32.3.2 cannot be weakened to $C \subset$ dom f, even when f is closed, without a risk that the supremum of f relative to C might not be attained or might not be finite. This is illustrated by the following pair of examples.

In the first example, we take f to be the closed proper convex function on R^2 defined by

$$f(\xi_1, \xi_2) = \begin{cases} (\xi_1^2/\xi_2) - \xi_2 & \text{if } \xi_2 > 0, \\ 0 & \text{if } \xi_1 = \xi_2 = 0, \\ +\infty & \text{otherwise.} \end{cases}$$

(It can be seen that f is the support function of the parabolic convex set which consists of the points (ξ_1, ξ_2) such that

$$\xi_1^2 + 4\xi_2 + 4 \leq 0,$$

and this is one way to verify that f is convex and closed.) We take C to be the non-empty closed bounded convex subset of dom f defined by

$$C = \{(\xi_1, \xi_2) \mid \xi_1^2 \leq \xi_2 \leq 1\}.$$

Clearly $f(\xi_1, \xi_2) < 1$ throughout C. The value of $f(\xi_1, \xi_2)$ approaches 1 as (ξ_1, ξ_2) moves toward $(0, 0)$ along the boundary of C. Thus 1 is the supremum of f relative to C, and this supremum is not attained.

The second example is obtained from the same f with C replaced by the (non-empty closed bounded convex) set

$$D = \{(\xi_1, \xi_2) \mid \xi_1^4 \leq \xi_2 \leq 1\}.$$

Along the boundary curve $\xi_1^4 = \xi_2$ of D, the value of $f(\xi_1, \xi_2)$ is $\xi_1^{-2} - \xi_2$, and this rises to $+\infty$ as (ξ_1, ξ_2) moves toward the origin. Thus f is not even bounded above on D.

The theory of subgradients can be used to some extent to characterize the points where a relative supremum is attained.

THEOREM 32.4. *Let f be a convex function, and let C be a convex set on which f is finite but not constant. Suppose that the supremum of f relative to C is attained at a certain point $x \in \text{ri}(\text{dom} f)$. Then every $x^* \in \partial f(x)$ is a non-zero vector normal to C at x.*

PROOF. Here f must be proper by Theorem 7.2, since f is assumed to be finite at a point of $\text{ri}(\text{dom} f)$. Let the supremum be α, and let

$$D = \{z \mid f(z) \leq \alpha\}.$$

By hypothesis, C is contained in D and x is a point of C such that $f(x) = \alpha$. Since f is not constant on C, we have $\inf f < f(x)$ and hence $0 \notin \partial f(x)$. The set $\partial f(x)$ is non-empty, because $x \in \text{ri}(\text{dom} f)$ (Theorem 23.4). Every vector in $\partial f(x)$ is normal to D at x (Theorem 23.7) and hence in particular is normal to C at x. ∥

COROLLARY 32.4.1. *Let f be a proper convex function, and let S be a non-empty set on which f is not constant. Suppose the supremum of f relative to S is attained at a certain point $x \in \text{ri}(\text{dom} f)$. Then every $x^* \in \partial f(x)$ is a non-zero vector such that the linear function $\langle \cdot, x^* \rangle$ attains its supremum relative to S at x.*

PROOF. Let $C = \text{conv } S$. By Theorem 32.2, the supremum of f relative to C is the same as the supremum relative to S. The supremum is $f(x)$, which is finite because $x \in \text{ri}(\text{dom} f)$. The theorem can be applied to C. Thus every $x^* \in \partial f(x)$ is a non-zero vector normal to C at x. The normality means that the linear function $\langle \cdot, x^* \rangle$ attains its supremum relative to C (which is again the supremum relative to S) at x. ∥

A noteworthy case of Theorem 32.4 is where C is the unit Euclidean ball. The vectors normal to C at a boundary point x are then just the vectors of the form λx, $\lambda > 0$, so that the maximization of f over C leads to the "eigenvalue" condition

$$\lambda x \in \partial f(x), \qquad |x| = 1.$$

Part VII · Saddle-Functions and Minimax Theory

Saddle-Functions

Let C and D be subsets of R^m and R^n respectively, and let K be a function from $C \times D$ to $[-\infty, +\infty]$. We say that K is a *concave-convex* function if $K(u, v)$ is a concave function of $u \in C$ for each $v \in D$ and a convex function of $v \in D$ for each $u \in C$. *Convex-concave* functions are defined similarly. We speak of both kinds of functions as *saddle-functions*.

The theory of saddle-functions, like that of purely convex or concave functions, can be reduced conveniently to the case where the functions are everywhere defined but possibly infinity-valued. There are some ambiguities, however, which at first may seem awkward or puzzling.

Let K be a concave-convex function on $C \times D$. In extending $K(u, v)$ beyond D as a convex function of v for a fixed $u \in C$, we can set $K(u, v) = +\infty$. On the other hand, in extending $K(u, v)$ beyond C as a concave function of u for a fixed $v \in D$, we naturally set $K(u, v) = -\infty$. This leaves us in doubt as to how $K(u, v)$ should be extended to points (u, v) such that $u \notin C$ and $v \notin D$. It turns out that there is usually not one natural extension but two, or even more. The functions K_1 and K_2 defined by

$$K_1(u, v) = \begin{cases} K(u, v) & \text{if} \quad u \in C, \quad v \in D, \\ +\infty & \text{if} \quad u \in C, \quad v \notin D, \\ -\infty & \text{if} \quad u \notin C, \end{cases}$$

$$K_2(u, v) = \begin{cases} K(u, v) & \text{if} \quad u \in C, \quad v \in D, \\ -\infty & \text{if} \quad u \notin C, \quad v \in D, \\ +\infty & \text{if} \quad v \notin D, \end{cases}$$

are the simplest examples of concave-convex functions on $R^m \times R^n$ which agree with K on $C \times D$. We shall call K_1 the *lower simple extension* of K and K_2 the *upper simple extension* of K. Either K_1 or K_2 is adequate for most of the analysis of K. We shall therefore develop most of the theory of saddle-functions in terms of saddle-functions on all of $R^m \times R^n$, pointing out from time to time relationships with results about restricted saddle-functions.

Given a concave-convex function K on $R^m \times R^n$, we can apply the convex and concave closure operations to achieve some regularization. The function obtained by closing $K(u, v)$ as a convex function of v for each fixed u is called the *convex closure* of K and is denoted by $\mathrm{cl}_v\, K$ or $\mathrm{cl}_2\, K$. Similarly, the function obtained by closing $K(u, v)$ as a concave function of u for each fixed v is called the *concave closure* of K and is denoted by $\mathrm{cl}_u\, K$ or $\mathrm{cl}_1\, K$. We shall see in a moment that these closure operations preserve concavity-convexity. If K coincides with its convex closure, we say K is *convex-closed*, etc.

There is a surprising correspondence between saddle-functions and convex bifunctions which is at the heart of the theory of saddle-functions. This correspondence is a generalization of the classical correspondence between bilinear functions and linear transformations.

If A is any linear transformation from R^m to R^n, the function K defined by

$$K(u, x^*) = \langle Au, x^* \rangle$$

is a bilinear function on $R^m \times R^n$. Conversely, of course, any bilinear function K on $R^m \times R^n$ can be expressed this way for a unique linear transformation A from R^m to R^n. The analogous correspondence between saddle-functions K on $R^m \times R^n$ and bifunctions F from R^m to R^n is one-to-one modulo closure operations, and it is based on the conjugacy correspondence rather than the ordinary inner product.

For the sake of emphasizing the analogies with linear algebra, it is convenient to introduce an inner product notation for the conjugate of a convex or concave function f:

$$\langle f, x^* \rangle = \langle x^*, f \rangle = f^*(x^*).$$

(More general "inner products" $\langle f, g \rangle$, where f is a convex function and g is a concave function, will be defined in §38.) Note that $\langle f, x^* \rangle = \langle x, x^* \rangle$ when f is the indicator of the point x, i.e. when

$$f(z) = \begin{cases} 0 & \text{if } z = x, \\ +\infty & \text{if } z \neq x. \end{cases}$$

For any convex or concave bifunction F from R^m to R^n, we form

$$\langle Fu, x^* \rangle = \langle x^*, Fu \rangle = (Fu)^*(x^*)$$

as a function of (u, x^*) on $R^m \times R^n$. Thus, by definition,

$$\langle Fu, x^* \rangle = \sup_x \{ \langle x, x^* \rangle - (Fu)(x) \}$$

if F is convex, whereas

$$\langle Fu, x^* \rangle = \inf_x \{\langle x, x^* \rangle - (Fu)(x)\}$$

if F is concave.

If F is the convex indicator bifunction of a linear transformation A from R^m to R^n, i.e.

$$(Fu)(x) = \delta(x \mid Au),$$

we have

$$\langle Fu, x^* \rangle = \langle Au, x^* \rangle.$$

(Note: when the graph function of the bifunction F is actually affine, it would be possible to regard F as either convex or concave, so that $\langle Fu, x^* \rangle$ might be ambiguous. This causes no real technical difficulty, however, since it is always clear from the context how a given F is to be regarded. Such ambiguities could be eliminated rigorously by introducing a concept of an "oriented bifunction," meaning a bifunction paired with one of the symbols "sup" or "inf." For a "sup" oriented bifunction one would define $\langle Fu, x^* \rangle$ using "sup," while for an "inf" oriented bifunction one would define $\langle Fu, x^* \rangle$ using "inf." This device is not worth the effort in the present case, although we shall have occasion to employ it in a related situation in §39.)

THEOREM 33.1. *If F is any convex bifunction from R^m to R^n, then $\langle Fu, x^* \rangle$ is a concave-convex function of (u, x^*) which is convex-closed, and one has*

$$(\text{cl } (Fu))(x) = \sup_{x^*} \{\langle x, x^* \rangle - \langle Fu, x^* \rangle\}.$$

On the other hand, given any concave-convex function K on $R^m \times R^n$, define the bifunction F from R^m to R^n by

$$(Fu)(x) = \sup_{x^*} \{\langle x, x^* \rangle - K(u, x^*)\}.$$

Then F is convex, Fu is closed on R^n for each $u \in R^m$, and one has

$$\langle Fu, x^* \rangle = (\text{cl}_2 K)(u, x^*).$$

(Similarly for concave bifunctions F and convex-concave functions K.)

PROOF. Since $\langle Fu, \cdot \rangle$ is just $(Fu)^*$ by definition, it is a closed convex function of x^*, and its conjugate is cl (Fu) (Theorem 12.2). This proves the first part of the theorem, except for the fact that $\langle Fu, x^* \rangle$ is concave in u. To prove the concavity, fix any x^* in R^n. We have

$$-\langle Fu, x^* \rangle = \inf_x h(u, x), \qquad \forall u \in R^m,$$

where h is the convex function on R^{m+n} defined by

$$h(u, x) = (Fu)(x) - \langle x, x^* \rangle.$$

Thus $-\langle Fu, x^* \rangle$ as a function of u is Ah, where A is the projection $(u, x) \to u$. It follows that $-\langle Fu, x^* \rangle$ is convex in u (Theorem 5.7), and hence that $\langle Fu, x^* \rangle$ is concave in u.

Next consider the bifunction F defined in the theorem for a given concave-convex function K. For each x^*, the function

$$k_{x^*}(u, x) = \langle x, x^* \rangle - K(u, x^*)$$

is a (jointly) convex function of (u, x) on R^{m+n}. As the pointwise supremum of the collection of such functions, the graph function of F is a convex function on R^{m+n}. Thus F is a convex bifunction. Of course, the formula for Fu says that Fu is the conjugate of the convex function $K(u, \cdot)$ for each $u \in R^m$, so Fu is closed and $(Fu)^* = \langle Fu, \cdot \rangle$ is the closure of $K(u, \cdot)$. The latter is $(\mathrm{cl}_2\, K)(u, \cdot)$ by definition.   ‖

COROLLARY 33.1.1.   *If K is any concave-convex function on $R^m \times R^n$, then $\mathrm{cl}_1\, K$ and $\mathrm{cl}_2\, K$ are concave-convex functions such that $\mathrm{cl}_1\, K$ is concave-closed and $\mathrm{cl}_2\, K$ is convex-closed. (Similarly for convex-concave functions.)*

PROOF.   According to the theorem, $(\mathrm{cl}_2\, K)(u, x^*)$ is of the form $\langle Fu, x^* \rangle$ for a certain convex bifunction F, where $\langle Fu, x^* \rangle$ is concave in u and closed convex in x^*. Similarly in all the other cases.   ‖

The convex bifunctions F from R^m to R^n correspond one-to-one with their graph functions

$$f(u, x) = (Fu)(x),$$

which are just the convex functions on R^{m+n}. To obtain $\langle Fu, x^* \rangle$ from f, one takes the conjugate of $f(u, x)$ as a function of x for each u. This may be thought of as a *partial conjugacy* operation, as opposed to the ordinary conjugacy operation, where one takes

$$f^*(u^*, x^*) = \sup_{u, x}\, \{\langle u, u^* \rangle + \langle x, x^* \rangle - f(u, x)\}.$$

In this sense, Theorem 33.1 says that convex-closed saddle-functions are just the partial conjugates of (purely) convex functions.

Let us call a convex or concave bifunction F *image-closed* if the function Fu is closed for every u. (If F is closed, then F is in particular image-closed.) For such bifunctions, a one-to-one correspondence is implied by Theorem 33.1.

COROLLARY 33.1.2   *The relations*

$$K(u, x^*) = \langle Fu, x^* \rangle, \qquad Fu = K(u, \cdot)^*,$$

express a one-to-one correspondence between the convex-closed concave-convex functions K on $R^m \times R^n$ and the image-closed convex bifunctions F

from R^m to R^n. (Similarly for concave-closed saddle-functions and image-closed concave bifunctions.)

In the case of polyhedral convexity, the correspondence between saddle-functions and bifunctions is somewhat simpler.

COROLLARY 33.1.3. *Let F be a polyhedral convex bifunction from R^m to R^n. Then $\langle Fu, x^* \rangle$ is a polyhedral convex function of x^* for each u and a polyhedral concave function of u for each x^*. Moreover, assuming F is proper, F can be expressed in terms of $\langle Fu, x^* \rangle$ by the formula*

$$(Fu)(x) = \sup_{x^*} \{\langle x, x^* \rangle - \langle Fu, x^* \rangle\}.$$

PROOF. For each u, Fu is a polyhedral convex function. If F is proper, Fu nowhere has the value $-\infty$; since the epigraph of Fu is a closed set, this implies cl $(Fu) = Fu$. The conjugate $\langle Fu, \cdot \rangle$ of Fu is a polyhedral convex function by Theorem 19.2. Now in the proof of Theorem 33.1 it was shown that the function $u \to -\langle Fu, x^* \rangle$ was the image of a certain convex function h under a linear transformation A. When F is polyhedral, the h involved is actually polyhedral, so that the image Ah is not only convex but polyhedral (Corollary 19.3.1). Therefore $\langle Fu, x^* \rangle$ is polyhedral concave in u. ‖

By Corollary 33.1.2, the relations

$$L(u, x^*) = \langle u, Gx^* \rangle, \qquad Gx^* = L(\cdot, x^*)^*,$$

express a one-to-one correspondence between the *concave-closed* concave-convex functions L on $R^m \times R^n$ and the image-closed *concave* bifunctions G from R^n to R^m. Of course, if F is any convex bifunction from R^m to R^n,

$$F: u \to Fu: x \to (Fu)(x),$$

the adjoint F^* of F is a closed concave bifunction from R^m to R^n,

$$F^*: x^* \to F^*x^*: u^* \to (F^*x^*)(u^*).$$

It follows that $\langle u, F^*x^* \rangle$ is concave-convex and concave-closed.

The exact relationship between $\langle Fu, x^* \rangle$ and $\langle u, F^*x^* \rangle$ is explained by the next theorem and its corollaries.

THEOREM 33.2. *For any convex or concave bifunction F from R^m to R^n, one has*

$$\langle u, F^*x^* \rangle = \text{cl}_u \langle Fu, x^* \rangle,$$

$$\text{cl}_{x^*} \langle u, F^*x^* \rangle = \langle (\text{cl } F)u, x^* \rangle.$$

PROOF. Suppose that F is convex. By definition

$$(F^*x^*)(u^*) = \inf_{u,x} \{(Fu)(x) - \langle x, x^* \rangle + \langle u, u^* \rangle\}$$

$$= \inf_u \{\langle u, u^* \rangle - \sup_x \{\langle x, x^* \rangle - (Fu)(x)\}\}$$

$$= \inf_u \{\langle u, u^* \rangle - \langle Fu, x^* \rangle\}.$$

Thus, for the concave-convex function $K(u, x^*) = \langle Fu, x^* \rangle$, F^* is the bifunction from R^n to R^m obtained by taking the (concave) conjugate of $K(u, x^*)$ in u for each x^*. This situation is covered by Theorem 33.1 (with only differences in notation): one has

$$\langle u, F^*x^* \rangle = (\text{cl}_1 K)(u, x^*) = \text{cl}_u \langle Fu, x^* \rangle.$$

The same formula holds when F is concave, as is seen by interchanging "inf" and "sup." Applying this formula to F^* in place of F, one gets

$$\langle F^{**}u, x^* \rangle = \text{cl}_{x^*} \langle u, F^*x^* \rangle.$$

By Theorem 30.1, $F^{**} = \text{cl } F$. ‖

COROLLARY 33.2.1. *Let F be any convex or concave bifunction from R^m to R^n. If $u \in \text{ri (dom } F)$, one has*

$$\langle Fu, x^* \rangle = \langle u, F^*x^* \rangle$$

for every $x^* \in R^n$. On the other hand, if F is closed and $x^* \in \text{ri (dom } F^*)$ the same equation holds for every $u \in R^m$.

PROOF. Suppose F is convex. If $u \notin \text{dom } F$, Fu is identically $+\infty$ and $\langle Fu, x^* \rangle = -\infty$ for every x^*. If $u \in \text{dom } F$, Fu is not identically $+\infty$, so $\langle Fu, x^* \rangle > -\infty$ for every x^*. Thus, for each x^*, the effective domain of the concave function $u \to \langle Fu, x^* \rangle$ is dom F. A concave function agrees with its closure on the relative interior of its effective domain, and here the closure function is $\langle \cdot, F^*x^* \rangle$ by Theorem 33.2. Thus $\langle Fu, x^* \rangle$ and $\langle u, F^*x^* \rangle$ agree when $u \in \text{ri (dom } F)$. The argument is similar when F is concave. The second fact in the corollary is proved by applying the first fact to F^*. ‖

COROLLARY 33.2.2. *Let F be a proper polyhedral convex or concave bifunction. Then*

$$\langle Fu, x^* \rangle = \langle u, F^*x^* \rangle$$

holds, except when both $u \notin \text{dom } F$ and $x^* \notin \text{dom } F^*$. (In the exceptional case, one of the quantities is $+\infty$ and the other $-\infty$.)

PROOF. Since F is polyhedral, we have $\text{cl } F = F$. The function $u \to \langle Fu, x^* \rangle$ is polyhedral by Theorem 33.1, and hence it coincides with its closure on its effective domain (rather than just on the relative interior of this effective domain). The proof of Corollary 33.2.1 may be sharpened accordingly. ‖

The preceding results show that, for a convex or concave bifunction F, the *inner product equation*

$$\langle Fu, x^* \rangle = \langle u, F^*x^* \rangle$$

holds for "most" choices of u and x^*. This provides more justification for the "adjoint" terminology which we have introduced for F^*. Even though $\langle Fu, x^* \rangle$ and $\langle u, F^*x^* \rangle$ may differ for certain choices of u and x^*, Theorem 33.2 implies that, when cl $F = F$, the functions $\langle Fu, x^* \rangle$ and $\langle u, F^*x^* \rangle$ completely determine each other and determine F and F^* as well.

When F is the convex indicator bifunction of a linear transformation A from R^m to R^n, F^* is the concave indicator bifunction of the adjoint transformation A^*, and the inner product equation for F and F^* reduces to the classical relation

$$\langle Au, x^* \rangle = \langle u, A^*x^* \rangle.$$

The inner product equation for a convex bifunction F and its adjoint asserts (by definition) that

$$\sup_x \{\langle x, x^* \rangle - (Fu)(x)\} = \inf_{u^*} \{\langle u, u^* \rangle - (F^*x^*)(u^*)\}.$$

In other words, it asserts a certain relationship between a problem of maximizing a concave function of x and a problem of minimizing a convex function of u^*. There is a close connection between this and the duality theory for convex programs.

In the generalized convex program (P) associated with a closed convex bifunction F as in §29, one studies the function inf F around $u = 0$. By definition,

$$(\inf F)(u) = \inf Fu = -\sup_x \{\langle x, 0 \rangle - (Fu)(x)\} = -\langle Fu, 0 \rangle.$$

In the dual program (P^*), one studies sup F^* around $x^* = 0$, where

$$(\sup F^*)(x^*) = \sup F^*x^* = -\inf_{u^*} \{\langle 0, u^* \rangle - (F^*x^*)(u^*)\}$$
$$= -\langle 0, F^*x^* \rangle.$$

The condition that the optimal values in (P) and (P^*) be equal, i.e. that

$$\inf F0 = \sup F^*0,$$

is thus equivalent to the condition that

$$\langle F0, 0 \rangle = \langle 0, F^*0 \rangle.$$

More generally, fix any $u \in R^m$ and $x^* \in R^n$, and define the convex bifunction H from R^m to R^n by

$$(Hv)(y) = (F(u + v))(y) - \langle y, x^* \rangle.$$

The (concave) adjoint H^* of H is then given by

$$(H^*y^*)(v^*) = \inf_{v,y} \{(Hv)(y) - \langle y, y^* \rangle + \langle v, v^* \rangle\}$$

$$= \inf_{v,y} \{(F(u + v))(y) - \langle y, x^* + y^* \rangle + \langle v, v^* \rangle\}$$

$$= \inf_{w,y} \{(Fw)(y) - \langle y, x^* + y^* \rangle + \langle w - u, v^* \rangle\}$$

$$= (F(x^* + y^*))(v^*) - \langle u, v^* \rangle.$$

Thus in the convex program (Q) associated with H the optimal value is

$$\inf H0 = \inf_y \{(Fu)(y) - \langle y, x^* \rangle\} = -\langle Fu, x^* \rangle,$$

while in the dual concave program (Q^*) the optimal value is

$$\sup H^*0 = \sup_{v^*} \{(F^*x^*)(v^*) - \langle u, v^* \rangle\} = -\langle u, F^*x^* \rangle.$$

Therefore, in general, the question of whether

$$\langle Fu, x^* \rangle = \langle u, F^*x^* \rangle$$

for a certain u and x^* is equivalent to the question of whether normality holds for a certain dual pair of programs (Q) and (Q^*).

The fact that one can sometimes have $\langle Fu, x^* \rangle = -\infty$ and $\langle u, F^*x^* \rangle = +\infty$ corresponds to the fact that (Q) and (Q^*) can sometimes both be inconsistent. This extreme situation occurs when both $u \notin \text{dom } F$ and $x^* \notin \text{dom } F^*$ (because then Fu is the constant function $+\infty$, while F^*x^* is the constant function $-\infty$). Similarly, since there exist dual pairs of (abnormal) programs in which the optimal values are both finite but unequal, or in which one of the optimal values is finite and the other is infinite (see the examples in §30), it really is possible on some occasions to have $\langle Fu, x^* \rangle$ and $\langle u, F^*x^* \rangle$ both finite but unequal, or one finite and the other infinite. The possibility of such discrepancies will be analyzed thoroughly in §34, where explicit examples will be given.

A saddle-function K on $R^m \times R^n$ is said to be *fully closed* if it is both convex-closed and concave-closed. For example, K is fully closed if it is finite everywhere (inasmuch as a finite convex or concave function is continuous and hence closed). By Corollary 33.1.2 and Theorem 33.2, the fully closed concave-convex functions are the functions of the form $K(u, x^*) = \langle Fu, x^* \rangle$, where F is a convex bifunction such that

$$\langle Fu, x^* \rangle = \langle u, F^*x^* \rangle, \qquad \forall u, \qquad \forall x^*.$$

Clearly F has this property if dom $F = R^m$, or if F is closed and dom $F^* = R^n$, by Corollary 33.2.1. But F cannot have this property if dom $F \neq R^m$ and dom $F^* \neq R^n$, since in this case, as pointed out in the preceding paragraph, $\langle Fu, x^* \rangle$ and $\langle u, F^*x^* \rangle$ are oppositely infinite for certain choices of (u, x^*). The class of fully closed saddle-functions thus corresponds to only a special class of closed bifunctions. For many purposes, weaker notions of closedness are needed.

A concave-convex function K will be said to be *lower closed* if $cl_2 (cl_1 K) = K$ and *upper closed* if $cl_1 (cl_2 K) = K$. The way to remember which is which, is that lower closedness entails lower semi-continuity in the argument of K for which this is natural, the convex argument, whereas upper closedness entails upper semi-continuity in the concave argument. (If K is convex-concave, instead of concave-convex, we say K is lower closed if $cl_1 (cl_2 K) = K$ and upper closed if $cl_2 (cl_1 K) = K$.)

A saddle-function is fully closed if and only if it is both lower closed and upper closed.

THEOREM 33.3.   *The relations*

$$K(u, x^*) = \langle Fu, x^* \rangle, \qquad Fu = K(u, \cdot)^*,$$

define a one-to-one correspondence between the lower closed concave-convex functions K on $R^m \times R^n$ and the closed convex bifunctions F from R^m to R^n. Similarly for upper closed saddle-functions and closed concave bifunctions.

PROOF.   By Theorem 33.2, the convex bifunction cl F satisfies

$$\langle (cl\ F)u, x^* \rangle = cl_{x^*} \langle u, F^*x^* \rangle = cl_{x^*} cl_u \langle Fu, x^* \rangle.$$

Thus the saddle-function $K(u, x^*) = \langle Fu, x^* \rangle$ is lower closed if and only if

$$\langle (cl\ F)u, x^* \rangle = \langle Fu, x^* \rangle, \qquad \forall u, \qquad \forall x^*.$$

For image-closed convex bifunctions F, the latter condition is equivalent to cl $F = F$. The result thus follows from the correspondence already established in Corollary 33.1.2.   ∥

COROLLARY 33.3.1.   *Let $\underline{K}$ and $\bar{K}$ be concave-convex functions on $R^m \times R^n$. In order that there exist a closed convex bifunction F (necessarily unique) such that*

$$\underline{K}(u, x^*) = \langle Fu, x^* \rangle, \qquad \bar{K}(u, x^*) = \langle u, F^*x^* \rangle,$$

it is necessary and sufficient that $\underline{K}$ and $\bar{K}$ satisfy the relations

$$cl_1 \underline{K} = \bar{K}, \qquad cl_2 \bar{K} = \underline{K}.$$

These relations imply that $\underline{K}$ is lower closed, $\bar{K}$ is upper closed, and $\underline{K} \leq \bar{K}$.

PROOF.   The necessity of the condition is already known from Theorem 33.2. To prove the sufficiency, we observe that the closure relations imply

$$\mathrm{cl}_2\,(\mathrm{cl}_1\,\underline{K}) = \mathrm{cl}_2\,\bar{K} = \underline{K},$$

so that $\underline{K}$ is lower closed and $\underline{K}(u, x^*) = \langle Fu, x^* \rangle$ for a unique closed convex bifunction F. We then have

$$\bar{K}(u, x^*) = (\mathrm{cl}_1\,\underline{K})(u, x^*) = \mathrm{cl}_u\,\langle Fu, x^* \rangle = \langle u, F^*x^* \rangle$$

by Theorem 33.2, and everything follows.   ‖

COROLLARY 33.3.2.   *The relations*

$$\bar{K} = \mathrm{cl}_1\,\underline{K}, \qquad \underline{K} = \mathrm{cl}_2\,\bar{K},$$

define a one-to-one correspondence between the lower closed concave-convex functions $\underline{K}$ and the upper closed concave-convex functions $\bar{K}$ on $R^m \times R^n$.

PROOF.   This is immediate from Theorem 33.3, Theorem 33.2 and the fact that the adjoint correspondence for closed convex and concave bifunctions is one-to-one.   ‖

COROLLARY 33.3.3.   *Let C and D be non-empty closed convex sets in R^m and R^n, respectively, and let K be any finite continuous concave-convex function on $C \times D$. Let $\underline{K}$ and $\bar{K}$ be the lower and upper simple extensions of K to $R^m \times R^n$, respectively. Then $\underline{K}$ is lower closed, $\bar{K}$ is upper closed, and there exists a unique closed convex bifunction F from R^m to R^n such that*

$$\underline{K}(u, x^*) = \langle Fu, x^* \rangle, \qquad \bar{K}(u, x^*) = \langle u, F^*x^* \rangle.$$

The bifunctions F and F^* are expressed in terms of K by

$$(Fu)(x) = \begin{cases} \sup\,\{\langle x, x^* \rangle - K(u, x^*)\,|\,x^* \in D\} & \text{if } u \in C, \\ +\infty & \text{if } u \notin C, \end{cases}$$

$$(F^*x^*)(u^*) = \begin{cases} \inf\,\{\langle u, u^* \rangle - K(u, x^*)\,|\,u \in C\} & \text{if } x^* \in D, \\ -\infty & \text{if } x^* \notin D. \end{cases}$$

In particular, dom $F = C$ *and* dom $F^* = D$.

PROOF.   The continuity of K and the closedness of C and D ensure that $\mathrm{cl}_1\,\underline{K} = \bar{K}$ and $\mathrm{cl}_2\,\bar{K} = \underline{K}$. The result is then immediate from Corollary 33.3.1 and the definitions.   ‖

Closures and Equivalence Classes

A pairing has been established in §33 between the lower closed saddle-functions $\underline{K}$ and the upper closed saddle-functions $\bar{K}$ on $R^m \times R^n$, where each pair corresponds to a uniquely determined closed convex bifunction and its (closed concave) adjoint. This pairing will be extended below to an equivalence relation among *closed* saddle-functions, "closed" being a slightly weaker notion than "lower closed" or "upper closed." The structure of closed saddle-functions will be analyzed in detail. We shall show that each "proper" equivalence class of closed saddle-functions is uniquely determined by its "kernel," which is a finite saddle-function on a product of relatively open convex sets.

Let K be any saddle-function on $R^m \times R^n$. Having formed $\mathrm{cl}_1 K$ and $\mathrm{cl}_2 K$ (which are saddle-functions by Corollary 33.1.1), we can proceed to form $\mathrm{cl}_2 \mathrm{cl}_1 K$ and $\mathrm{cl}_1 \mathrm{cl}_2 K$. If K is concave-convex, $\mathrm{cl}_2 \mathrm{cl}_1 K$ is called the *lower closure* and $\mathrm{cl}_1 \mathrm{cl}_2 K$ the *upper closure* of K. If K is convex-concave, the terminology is reversed. By definition, then, K is lower closed if and only if it coincides with its lower closure, etc. It is not obvious that lower and upper closures are always lower and upper closed, respectively, i.e. that

$$\mathrm{cl}_2 \mathrm{cl}_1 \mathrm{cl}_2 \mathrm{cl}_1 K = \mathrm{cl}_2 \mathrm{cl}_1 K, \quad \forall K,$$
$$\mathrm{cl}_1 \mathrm{cl}_2 \mathrm{cl}_1 \mathrm{cl}_2 K = \mathrm{cl}_1 \mathrm{cl}_2 K, \quad \forall K,$$

but this is true, as we now demonstrate.

THEOREM 34.1. *If K is any saddle-function on $R^m \times R^n$, the lower closure of K is a lower closed saddle-function and the upper closure of K is an upper closed saddle-function.*

PROOF. We assume for definiteness that K is concave-convex. Let F be the bifunction from R^m to R^n defined by $Fu = K(u, \cdot)^*$. According to Theorem 33.1, F is convex and

$$\langle Fu, x^* \rangle = (\mathrm{cl}_2 K)(u, x^*).$$

When the closure operation in u is applied to both sides of this equation, we get

$$\langle u, F^*x^* \rangle = (\mathrm{cl}_1 \mathrm{cl}_2 K)(u, x^*)$$

by Theorem 33.2. Since F^* is a closed concave bifunction (Theorem 30.1), we may conclude from Theorem 33.3 that $\mathrm{cl}_1 \, \mathrm{cl}_2 \, K$ is an upper closed concave-convex function. The proof for the lower closure operation is analogous.   ‖

For reasons explained just prior to Theorem 33.3, one cannot hope to construct from an arbitrary given saddle-function K a saddle-function which is *both* lower closed and upper closed by repeated application of cl_1 and cl_2. In general, the lower and upper closure operations do not quite produce the same result:

$$\mathrm{cl}_2 \, \mathrm{cl}_1 \, K \neq \mathrm{cl}_1 \, \mathrm{cl}_2 \, K.$$

This discrepancy is a fundamental one, and it plays a crucial role in the theory of saddle-functions. The typical nature of the difference between $\mathrm{cl}_2 \, \mathrm{cl}_1 \, K$ and $\mathrm{cl}_1 \, \mathrm{cl}_2 \, K$ will be illustrated by examples.

The saddle-function in the first example will be concave-convex on the plane $R \times R$. Let C and D be the open unit interval $(0, 1)$. On the open square $C \times D$, let K be given by the formula

$$K(u, v) = u^v, \qquad 0 < u < 1, \qquad 0 < v < 1.$$

(Note that this formula does give a function which is concave in u and convex in v.) To get the values of K on the rest of $R \times R$, take either the lower simple extension or the upper simple extension of this function on $C \times D$. (It makes no difference which extension one takes.) The reader may verify (as a very good exercise for understanding the nature of the closure operations for saddle-functions) that

$$(\mathrm{cl}_1 \, \mathrm{cl}_2 \, K)(u, v) = \begin{cases} u^v & \text{if} \quad u \in [0, 1], \, v \in [0, 1], \, (u, v) \neq (0, 0), \\ 1 & \text{if} \quad (u, v) = (0, 0), \\ +\infty & \text{if} \quad u \in [0, 1], \, v \notin [0, 1], \\ -\infty & \text{if} \quad u \notin [0, 1], \, v \in [0, 1], \\ +\infty & \text{if} \quad u \notin [0, 1], \, v \notin [0, 1], \end{cases}$$

$$(\mathrm{cl}_2 \, \mathrm{cl}_1 \, K)(u, v) = \begin{cases} u^v & \text{if} \quad u \in [0, 1], \, v \in [0, 1], \, (u, v) \neq (0, 0), \\ 0 & \text{if} \quad (u, v) = (0, 0), \\ +\infty & \text{if} \quad u \in [0, 1], \, v \notin [0, 1], \\ -\infty & \text{if} \quad u \notin [0, 1], \, v \in [0, 1], \\ -\infty & \text{if} \quad u \notin [0, 1], \, v \notin [0, 1]. \end{cases}$$

Thus $\mathrm{cl}_1 \, \mathrm{cl}_2 \, K$ and $\mathrm{cl}_2 \, \mathrm{cl}_1 \, K$ differ in two places. The less significant place is where $u \notin [0, 1]$ and $v \notin [0, 1]$, one of the functions having the value $+\infty$ and the other $-\infty$. To some extent, this discrepancy is a just consequence of our conventions regarding $\pm \infty$, although it does have a natural meaning in minimax theory, as will be seen later. The really interesting discrepancy is at the origin, where one of the functions has the value 1 and the other the value 0. This reflects an intrinsic property of the function u^v on the unit square: there simply is no way to define 0^0 so as to have u^v both lower semi-continuous in v and upper semi-continuous in u at the $(0, 0)$ corner of the square. Any value between 0 and 1 can be assigned to 0^0 so as to make u^v concave-convex on the square, but there is no unique natural value.

As another example, let K be the lower or upper simple extension of the concave-convex function on the positive quadrant of $R \times R$ with values given by u/v. Then

$$(\operatorname{cl}_1 \operatorname{cl}_2 K)(u, v) = \begin{cases} u/v & \text{if} \quad u \geq 0, v > 0, \\ -\infty & \text{if} \quad u < 0, v > 0, \\ +\infty & \text{if} \quad v \leq 0, \end{cases}$$

$$(\operatorname{cl}_2 \operatorname{cl}_1 K)(u, v) = \begin{cases} u/v & \text{if} \quad u \geq 0, v > 0, \\ 0 & \text{if} \quad (u, v) = (0, 0), \\ +\infty & \text{if} \quad u \geq 0, v \leq 0 \quad \text{and} \quad (u, v) \neq (0, 0), \\ -\infty & \text{if} \quad u < 0. \end{cases}$$

Thus $\operatorname{cl}_1 \operatorname{cl}_2 K$ differs from $\operatorname{cl}_2 \operatorname{cl}_1 K$ when $u < 0$ and $v \leq 0$ ($\operatorname{cl}_1 \operatorname{cl}_2 K$ having the value $+\infty$ and $\operatorname{cl}_2 \operatorname{cl}_1 K$ the value $-\infty$), and when $(u, v) = (0, 0)$ ($\operatorname{cl}_1 \operatorname{cl}_2 K$ having the value $+\infty$ and $\operatorname{cl}_2 \operatorname{cl}_1 K$ the value 0). The notable feature of this example is that the set of points where $\operatorname{cl}_2 \operatorname{cl}_1 K$ is finite is not the same as the set where $\operatorname{cl}_1 \operatorname{cl}_2 K$ is finite, and it is not even a product set in $R \times R$.

In certain freakish cases, $\operatorname{cl}_1 \operatorname{cl}_2 K$ and $\operatorname{cl}_2 \operatorname{cl}_1 K$ differ so completely as to be almost unrelated to each other. Let K be the concave-convex function on $R \times R$ defined by

$$K(u, v) = \begin{cases} +\infty & \text{if} \quad uv > 0, \\ 0 & \text{if} \quad uv = 0, \\ -\infty & \text{if} \quad uv < 0. \end{cases}$$

Then trivially

$$(\operatorname{cl}_1 \operatorname{cl}_2 K)(u, v) = \begin{cases} 0 & \text{if} \quad u = 0, \\ -\infty & \text{if} \quad u \neq 0, \end{cases}$$

$$(\operatorname{cl}_2 \operatorname{cl}_1 K)(u, v) = \begin{cases} 0 & \text{if} \quad v = 0, \\ +\infty & \text{if} \quad v \neq 0. \end{cases}$$

Note that the set of points where $K(u, v)$ is finite is far from being a product of convex sets.

In the u^v example, one plainly has

$$\text{cl}_2 \, (\text{cl}_1 \, \text{cl}_2 \, K) = \text{cl}_2 \, \text{cl}_1 \, K,$$

$$\text{cl}_1 \, (\text{cl}_2 \, \text{cl}_1 \, K) = \text{cl}_1 \, \text{cl}_2 \, K.$$

Further application of cl_1 and cl_2 thus merely produces an oscillation between the lower and upper closures, and it accomplishes nothing. Indeed, the concave-convex functions $\bar{K} = \text{cl}_1 \, \text{cl}_2 \, K$ and $\underline{K} = \text{cl}_2 \, \text{cl}_1 \, K$ satisfy the relations

$$\text{cl}_1 \, \underline{K} = \bar{K}, \qquad \text{cl}_2 \, \bar{K} = \underline{K},$$

so that by Corollary 33.3.1 there exists a unique closed convex bifunction F from R to R such that

$$(\text{cl}_2 \, \text{cl}_1 \, K)(u, v) = \langle Fu, v \rangle,$$

$$(\text{cl}_1 \, \text{cl}_2 \, K)(u, v) = \langle u, F^*v \rangle.$$

The situation is entirely the same in the u/v example. In the freakish example, however, the functions $\text{cl}_1 \, \text{cl}_2 \, K$ and $\text{cl}_2 \, \text{cl}_1 \, K$ are fully closed, yet different, and no application of cl_1 or cl_2 turns either function into the other.

The concept of the effective domain of a saddle-function will be useful in describing the general structure of lower and upper closures. Given any concave-convex function K on $R^m \times R^n$, we define

$$\text{dom}_1 \, K = \{u \mid K(u, v) > -\infty, \forall v\},$$

$$\text{dom}_2 \, K = \{v \mid K(u, v) < +\infty, \forall u\}.$$

Observe that $\text{dom}_2 \, K$ is the intersection of the effective domains of the convex functions $K(u, \cdot)$ as u ranges over R^m, while $\text{dom}_1 \, K$ is the intersection of the effective domains of the concave functions $K(\cdot, v)$ as v ranges over R^n. In particular, $\text{dom}_1 \, K$ is a convex set in R^m and $\text{dom}_2 \, K$ is a convex set in R^n. The (convex) product set

$$\text{dom} \, K = \text{dom}_1 \, K \times \text{dom}_2 \, K$$

is called the *effective domain* of K. Since

$$-\infty < K(u, v) < +\infty$$

when $u \in \text{dom}_1 \, K$ and $v \in \text{dom}_2 \, K$, K is finite on $\text{dom} \, K$. However, there may also be certain points outside of $\text{dom} \, K$ where K is finite, as in the u/v example above. If $\text{dom} \, K \neq \emptyset$, K is said to be *proper*.

If K is the lower simple extension of a finite saddle-function on a non-empty convex set $C \times D$, one has $\text{dom}_1 K = C$ and $\text{dom}_2 K = D$, so that

$$\text{dom } K = C \times D$$

and K is proper. Similarly if K is an upper simple extension.

Two concave-convex functions K and L on $R^m \times R^n$ are said to be *equivalent* if $\text{cl}_1 K = \text{cl}_1 L$ and $\text{cl}_2 K = \text{cl}_2 L$. For example, the lower and upper simple extensions of a finite saddle-function on a convex $C \times D \neq \emptyset$ are equivalent. It is clear from the properties of the closure operations for convex and concave functions that equivalent saddle-functions must nearly coincide.

If $\text{cl}_1 K$ and $\text{cl}_2 K$ are both equivalent to K, K is said to be *closed*. In view of the fact that

$$\text{cl}_1 \text{cl}_1 K = \text{cl}_1 K, \qquad \text{cl}_2 \text{cl}_2 K = \text{cl}_2 K,$$

the conditions

$$\text{cl}_1 \text{cl}_2 K = \text{cl}_1 K, \qquad \text{cl}_2 \text{cl}_1 K = \text{cl}_2 K,$$

are necessary and sufficient for K to be a closed saddle-function. Trivially, if K is closed and L is equivalent to K, then L is closed.

THEOREM 34.2. *Given any closed convex bifunction F from R^m to R^n, let*

$$\underline{K}(u, x^*) = \langle Fu, x^* \rangle, \qquad \bar{K}(u, x^*) = \langle u, F^*x^* \rangle,$$

and let $\Omega(F)$ be the collection of all concave-convex functions K on $R^m \times R^n$ such that $\underline{K} \leq K \leq \bar{K}$. Then $\Omega(F)$ is an equivalence class (containing $\underline{K}$ and $\bar{K}$), and all the functions in $\Omega(F)$ are closed. Conversely, every equivalence class of closed concave-convex functions is of the form $\Omega(F)$ for a unique closed convex bifunction F.

For any K in $\Omega(F)$, one has

$$\text{cl}_1 K = \bar{K}, \qquad \text{cl}_2 K = \underline{K},$$
$$\text{dom } K = \text{dom } F \times \text{dom } F^*,$$
$$(Fu)(x) = \sup_{x^*} \{\langle x, x^* \rangle - K(u, x^*)\},$$
$$(F^*x^*)(u^*) = \inf_u \{\langle u, u^* \rangle - K(u, x^*)\}.$$

Moreover,

$$K(u, x^*) = \langle Fu, x^* \rangle = \langle u, F^*x^* \rangle$$

if $u \in \text{ri} (\text{dom } F)$ or if $x^* \in \text{ri} (\text{dom } F^*)$.

PROOF. First we shall show that each equivalence class of closed concave-convex functions is contained in a unique $\Omega(F)$. Then we shall show that the functions in $\Omega(F)$ are equivalent and have all the properties claimed. This will establish the theorem.

Let K be any closed concave-convex function on $R^m \times R^n$. Then

$$\mathrm{cl}_1 \, (\mathrm{cl}_2 \, K) = \mathrm{cl}_1 \, K, \qquad \mathrm{cl}_2 \, (\mathrm{cl}_1 \, K) = \mathrm{cl}_2 \, K,$$

so by Corollary 33.3.1 there exists a unique closed convex bifunction F such that

$$(\mathrm{cl}_2 \, K)(u, x^*) = \langle Fu, x^* \rangle, \qquad (\mathrm{cl}_1 \, K)(u, x^*) = \langle u, F^*x^* \rangle.$$

Inasmuch as

$$\mathrm{cl}_2 \, K \leq K \leq \mathrm{cl}_1 \, K,$$

K must belong to $\Omega(F)$. Furthermore, if L is any concave-convex function equivalent to K, we have

$$\mathrm{cl}_2 \, K = \mathrm{cl}_2 \, L \leq L \leq \mathrm{cl}_1 \, L = \mathrm{cl}_1 \, K,$$

and therefore L too must belong to $\Omega(F)$.

Now let K be an arbitrary member of $\Omega(F)$. By Theorem 33.2,

$$\mathrm{cl}_1 \, \bar{K} = \mathrm{cl}_1 \, \mathrm{cl}_1 \, \underline{K} = \mathrm{cl}_1 \, \underline{K} = \bar{K},$$

$$\mathrm{cl}_2 \, \underline{K} = \mathrm{cl}_2 \, \mathrm{cl}_2 \, \bar{K} = \mathrm{cl}_2 \, \bar{K} = \underline{K}.$$

Since $\underline{K} \leq K \leq \bar{K}$, this implies that

$$\mathrm{cl}_1 \, K = \bar{K}, \qquad \mathrm{cl}_2 \, K = \underline{K},$$

and consequently

$$\mathrm{cl}_1 \, \mathrm{cl}_2 \, K = \mathrm{cl}_1 \, K, \qquad \mathrm{cl}_2 \, \mathrm{cl}_1 \, K = \mathrm{cl}_2 \, K.$$

Thus K is closed and equivalent to $\underline{K}$ and $\bar{K}$. Since the convex function $K(u, \cdot)$ has $\underline{K}(u, \cdot)$ as its closure, we have

$$K(u, \cdot)^* = \underline{K}(u, \cdot)^* = Fu, \quad \forall u.$$

In particular, it follows that $u \notin \mathrm{dom}_1 \, K$ if and only if Fu is the constant function $+\infty$, i.e. $u \notin \mathrm{dom} \, F$. Similarly,

$$K(\cdot, x^*)^* = \bar{K}(\cdot, x^*)^* = F^*x^*, \quad \forall x^*,$$

and we have $x^* \notin \mathrm{dom}_2 \, K$ if and only if F^*x^* is the constant function $-\infty$, i.e. $x^* \notin \mathrm{dom} \, F^*$. This proves that

$$\mathrm{dom}_1 \, K \times \mathrm{dom}_2 \, K = \mathrm{dom} \, F \times \mathrm{dom} \, F^*$$

and that the formulas for F and F^* in terms of K are valid. The last assertion of the theorem is justified by Corollary 33.2.1. ‖

COROLLARY 34.2.1. *Let K be a closed saddle-function on $R^m \times R^n$, and let L be a saddle-function equivalent to K. Then $\mathrm{dom} \, L = \mathrm{dom} \, K$, and one has $L(u, v) = K(u, v)$ whenever $u \in \mathrm{ri} \, (\mathrm{dom}_1 \, K)$ or $v \in \mathrm{ri} \, (\mathrm{dom}_2 \, K)$.*

COROLLARY 34.2.2. *A lower closed or upper closed (or fully closed) saddle-function is in particular closed. Each equivalence class of closed saddle-functions contains a unique lower closed function (the least member of the class) and a unique upper closed function (the greatest member of the class).*

PROOF. By Theorem 33.3. ‖

For a closed convex bifunction F, the class $\Omega(F)$ in Theorem 34.2 consists of all the concave-convex functions equivalent to the function

$$(u, x^*) \to \langle Fu, x^* \rangle.$$

We define $\Omega(G)$ for a closed *concave* bifunction G as the class of all concave-convex functions equivalent to the function

$$(x^*, u) \to \langle x^*, Gu \rangle$$

(so that under the Ω notation we are always speaking of concave-convex functions rather than convex-concave functions). Thus, for the concave adjoint F^* of a closed convex F, $\Omega(F^*)$ consists of all the concave-convex functions equivalent to

$$(u, x^*) \to \langle v, F^*x^* \rangle.$$

By Theorem 34.2, these are precisely the concave-convex functions K such that

$$\langle Fu, x^* \rangle \leq K(u, x^*) \leq \langle u, F^*x^* \rangle, \forall u, \forall x^*,$$

and we have

$$\Omega(F^*) = \Omega(F).$$

The latter formula might be regarded as the "true" analogue of the formula

$$\langle Au, x^* \rangle = \langle u, A^*x^* \rangle$$

defining the adjoint of a linear transformation A.

How is the properness of the saddle-functions in $\Omega(F)$ related to the properness of F? If F is proper, F^* is proper by Theorem 30.1, so that dom $F \neq \emptyset$ and dom $F^* \neq \emptyset$. Then every $K \in \Omega(F)$ is proper, because dom $K \neq \emptyset$ by Theorem 34.2. On the other hand, if F is a closed convex bifunction which is not proper, the graph function of F must be identically $+\infty$ or identically $-\infty$. In the first case, we have

$$\langle Fu, x^* \rangle = \langle u, F^*x^* \rangle = -\infty, \forall u, \forall x^*,$$

while in the second case

$$\langle Fu, x^* \rangle = \langle u, F^*x^* \rangle = +\infty, \forall u, \forall x^*.$$

The following conclusion may be drawn.

COROLLARY 34.2.3.   *The only improper closed saddle-functions on $R^m \times R^n$ are the constant functions $+\infty$ and $-\infty$ (which are not equivalent).*

The structure of certain equivalence classes of proper closed saddle-functions can also be described without further ado.

COROLLARY 34.2.4.   *Let C and D be non-empty closed convex sets in R^m and R^n, respectively, and let K be a finite continuous concave-convex function on $C \times D$. Let Ω be the class of all concave-convex extensions of K to $R^m \times R^n$ satisfying*

$$K(u, v) = \begin{cases} +\infty & \text{if} \quad u \in C, \quad v \notin D, \\ -\infty & \text{if} \quad u \notin C, \quad v \in D. \end{cases}$$

(The lower simple extension of K and the upper simple extension of K are the least and greatest members of Ω, respectively.) Then Ω is an equivalence class of proper closed saddle-functions.

PROOF.   This is immediate from Theorem 34.2 and Corollary 33.3.3.   ‖

We shall now show that the structure of general equivalence classes of proper closed saddle-functions is only slightly more complicated.

THEOREM 34.3.   *Let K be a proper concave-convex function on $R^m \times R^n$. Let $C = \mathrm{dom}_1 K$ and $D = \mathrm{dom}_2 K$. In order that K be closed, it is necessary and sufficient that K have the following properties:*

(a) *For each $u \in \mathrm{ri}\, C$, $K(u, \cdot)$ is a closed proper convex function with effective domain D.*

(b) *For each $u \in (C \setminus \mathrm{ri}\, C)$, $K(u, \cdot)$ is a proper convex function whose effective domain lies between D and $\mathrm{cl}\, D$.*

(c) *For each $u \notin C$, $K(u, \cdot)$ is an improper convex function which has the value $-\infty$ throughout $\mathrm{ri}\, D$ (throughout D itself if actually $u \notin \mathrm{cl}\, C$).*

(d) *For each $v \in \mathrm{ri}\, D$, $K(\cdot, v)$ is a closed proper concave function with effective domain C.*

(e) *For each $v \in (D \setminus \mathrm{ri}\, D)$, $K(\cdot, v)$ is a proper concave function whose effective domain lies between C and $\mathrm{cl}\, C$.*

(f) *For each $v \notin D$, $K(\cdot, v)$ is an improper concave function which has the value $+\infty$ throughout $\mathrm{ri}\, C$ (throughout C itself if $v \notin \mathrm{cl}\, D$).*

PROOF.   Assume that K is closed. Let F be the unique closed proper convex bifunction from R^m to R^n such that $K \in \Omega(F)$ as in Theorem 34.2. Then $C = \mathrm{dom}\, F$ and $D = \mathrm{dom}\, F^*$. Let $\underline{K}(u, v) = \langle Fu, v \rangle$ and $\bar{K}(u, v) = \langle u, F^*v \rangle$. We have

$$\underline{K}(u, v) > -\infty, \forall v, \quad \text{if} \quad u \in C,$$
$$\underline{K}(u, v) = -\infty, \forall v, \quad \text{if} \quad u \notin C,$$
$$\bar{K}(u, v) < +\infty, \forall u, \quad \text{if} \quad v \in D,$$
$$\bar{K}(u, v) = +\infty, \forall u, \quad \text{if} \quad v \notin D.$$

For each u, the convex function $K(u, \cdot)$ lies between the convex function $\bar{K}(u, \cdot)$ and the closure of $K(u, \cdot)$, all three functions coinciding when $u \in \text{ri } C$ (Theorem 34.2). The $\pm \infty$ relations imply that, for each $u \in C$, $\bar{K}(u, \cdot)$ has D as its effective domain and $\underline{K}(u, \cdot)$ is proper. Properties (a) and (b) are immediate from this and the basic properties of the closure operation for convex functions. The proof of (d) and (e) is similar. Properties (c) and (f) are trivial consequences of (a), (b), (d) and (e).

Conversely, assume that K has properties (a) through (f). By (a) we have

$$(\text{cl}_2 K)(u, v) = K(u, v), \forall v,$$

when $u \in \text{ri } C$. On the other hand, by (c)

$$(\text{cl}_2 K)(u, v) = -\infty, \forall v,$$

when $u \notin C$. Hence, for each $v \notin D$, the concave functions $(\text{cl}_2 K)(\cdot, v)$ and $K(\cdot, v)$ are both improper with the value $+\infty$ on ri C. For each $v \in D$, $(\text{cl}_2 K)(\cdot, v)$ and $K(\cdot, v)$ are proper and their effective domains have the same relative interior, namely ri C, on which they coincide. It follows that $(\text{cl}_2 K)(\cdot, v)$ and $K(\cdot, v)$ have the same (concave) closure for every v, i.e. $\text{cl}_1 \text{cl}_2 K = \text{cl}_1 K$. By a parallel argument, $\text{cl}_2 \text{cl}_1 K = \text{cl}_2 K$. Thus K is closed. ‖

The restriction of a saddle-function K to dom K is a certain finite saddle-function on a product of convex sets, as we have already noted. The restriction of K to

$$\text{ri } (\text{dom } K) = \text{ri } (\text{dom}_1 K) \times \text{ri } (\text{dom}_2 K)$$

will be called the *kernel* of K.

THEOREM 34.4. *Two closed proper concave-convex functions on $R^m \times R^n$ are equivalent if and only if they have the same kernel.*

PROOF. Let K and L be closed proper concave-convex functions on $R^m \times R^n$. If L is equivalent to K, then L has the same kernel as K by Corollary 34.2.1. Conversely, suppose that L has the same kernel as K. Then, in particular, the effective domains of K and L have the same relative interior. Let

$$C' = \text{ri } (\text{dom}_1 K) = \text{ri } (\text{dom}_1 L),$$
$$D' = \text{ri } (\text{dom}_2 K) = \text{ri } (\text{dom}_2 L).$$

Property (a) of Theorem 34.3 asserts that, for each $u \in C'$ the convex function $K(u, \cdot)$ is closed and has D' as the relative interior of its effective domain; likewise $L(u, \cdot)$. Moreover $K(u, \cdot)$ and $L(u, \cdot)$ agree on D' when $u \in C'$, because K and L have the same kernel. Since a closed convex function is uniquely determined by its values on the relative interior of its effective domain, it follows that $K(u, \cdot)$ and $L(u, \cdot)$ agree throughout R^n when $u \in C'$. In particular, $\mathrm{dom}_2 K$ and $\mathrm{dom}_2 L$ must be the same convex set D by property (a) of Theorem 34.3. The agreement of $K(u, \cdot)$ and $L(u, \cdot)$ when $u \in C'$ can be expressed another way: the concave functions $K(\cdot, v)$ and $L(\cdot, v)$ agree on C' for each $v \in R^n$. By properties (d), (e) and (f) of Theorem 34.3, $K(\cdot, v)$ and $L(\cdot, v)$ are proper and have C' as the relative interior of their effective domains when $v \in D$, whereas both are improper with the value $+\infty$ throughout C' when $v \notin D$. Thus $K(\cdot, v)$ and $L(\cdot, v)$ must have the same (concave) closure for each $v \in R^n$, i.e. $\mathrm{cl}_1 K = \mathrm{cl}_1 L$. By a parallel argument, $\mathrm{cl}_2 K = \mathrm{cl}_2 L$. Thus L is equivalent to K. ‖

According to Theorem 34.4, each equivalence class of closed proper saddle-functions has a uniquely determined kernel, and the correspondence between equivalence classes and kernels is one-to-one. Each kernel is a finite saddle-function on a non-empty product of relatively open convex sets. Is *every* function of the latter sort the kernel of some equivalence class of closed proper saddle-functions? The answer is yes. To prove this, we need to examine more closely the lower and upper closure operations.

A concave-convex function K on $R^m \times R^n$ will be said to be *simple* if the effective domain of the convex function $K(u, \cdot)$ is contained in $\mathrm{cl}\,(\mathrm{dom}_2 K)$ for every $u \in \mathrm{ri}\,(\mathrm{dom}_1 K)$, and the effective domain of the concave function $K(\cdot, v)$ is contained in $\mathrm{cl}\,(\mathrm{dom}_1 K)$ for every $v \in \mathrm{ri}\,(\mathrm{dom}_2 K)$.

The most important examples of simple saddle-functions, for our purposes, are the lower and upper simple extensions of finite saddle-functions on convex sets $C \times D$. Every closed proper saddle-function is simple by Theorem 34.3. The reader can show, as an exercise, that every saddle-function of the form

$$K(u, x^*) = \langle Fu, x^* \rangle$$

(F a convex or concave bifunction from R^m to R^n) is simple. It can be shown further that every saddle-function whose effective domain has a non-empty interior is simple. An example of a concave-convex function which is *not* simple is a function on $R \times R$ already encountered:

$$K(u, v) = \begin{cases} +\infty & \text{if } uv > 0, \\ 0 & \text{if } uv = 0, \\ -\infty & \text{if } uv < 0 \end{cases}$$

THEOREM 34.5.   *Let K be any proper concave-convex function on $R^m \times R^n$ which is simple. The lower closure $\mathrm{cl}_2\,\mathrm{cl}_1 K$ and upper closure $\mathrm{cl}_1\,\mathrm{cl}_2 K$ of K are then equivalent, and*

$$\mathrm{cl}_2\,\mathrm{cl}_1 K \le \mathrm{cl}_1\,\mathrm{cl}_2 K.$$

The concave-convex functions between $\text{cl}_2 \text{cl}_1 K$ *and* $\text{cl}_1 \text{cl}_2 K$ *form an equivalence class of closed proper concave-convex functions having the same kernel as* K.

PROOF. We shall demonstrate first that $\text{cl}_2 K$ is simple and has the same kernel as K. By the definition of $\text{dom}_1 K$ and $\text{dom}_2 K$, when $u \notin \text{dom}_1 K$ the convex function $K(u, \cdot)$ has the value $-\infty$ somewhere, whereas when $u \in \text{dom}_1 K$ the effective domain of $K(u, \cdot)$ includes the non-empty set $\text{dom}_2 K$ and $K(u, \cdot)$ is proper. Thus $(\text{cl}_2 K)(u, \cdot)$ is the constant function $-\infty$ when $u \notin \text{dom}_1 K$, whereas when $u \in \text{dom}_1 K$ it is again a proper convex function whose effective domain includes $\text{dom}_2 K$. This shows that

$$\text{dom}_1 (\text{cl}_2 K) = \text{dom}_1 K,$$
$$\text{dom}_2 (\text{cl}_2 K) \supset \text{dom}_2 K,$$

and in fact that $\text{dom}_1 K$ is the effective domain of every one of the concave functions $(\text{cl}_2 K)(\cdot, v)$. Since K is simple, for each $u \in \text{ri} (\text{dom}_1 K)$ the convex function $(\text{cl}_2 K)(u, \cdot)$ actually agrees with $K(u, \cdot)$ on $\text{ri} (\text{dom}_2 K)$, and its effective domain is contained in $\text{cl} (\text{dom}_2 K)$. Therefore

$$\text{dom}_2 (\text{cl}_2 K) \subset \text{cl} (\text{dom}_2 K),$$

and $\text{dom}_2 (\text{cl}_2 K)$ has the same relative interior and closure as $\text{dom}_2 K$. It follows that $\text{cl}_2 K$ is simple, and that the relative interior of its effective domain is the same as $\text{ri} (\text{dom} K)$. The kernels of $\text{cl}_2 K$ and K are the same, because $\text{cl}_2 K$ agrees with K on $\text{ri} (\text{dom} K)$. This proves that the operation cl_2 preserves the class of simple proper concave-convex functions and their kernels. By a parallel argument, cl_1 has this property too. Therefore $\text{cl}_2 \text{cl}_1 K$ and $\text{cl}_1 \text{cl}_2 K$ must be simple proper concave-convex functions having the same kernel as K. Since $\text{cl}_2 \text{cl}_1 K$ is lower closed and $\text{cl}_1 \text{cl}_2 K$ is upper closed (Theorem 34.1), these two functions are in particular closed (Corollary 34.2.2), and by Theorem 34.4 they must be equivalent. The saddle-functions equivalent to $\text{cl}_2 \text{cl}_1 K$ and $\text{cl}_1 \text{cl}_2 K$ likewise have the same kernel as K by Theorem 34.4. The equivalence class contains a unique lower closed $\underline{K}$ and upper closed $\bar{K}$ with $\underline{K} \leq \bar{K}$, and it consists of the concave-convex functions between $\underline{K}$ and $\bar{K}$ (Theorem 34.2). The functions $\underline{K}$ and $\bar{K}$ must be $\text{cl}_2 \text{cl}_1 K$ and $\text{cl}_1 \text{cl}_2 K$, respectively. ‖

COROLLARY 34.5.1. *Let C and D be non-empty convex sets in R^m and R^n respectively, and let K be a finite concave-convex function on $C \times D$. Then there exists one and only one equivalence class of closed proper concave-convex functions on $R^m \times R^n$ having as its kernel the restriction of K to the relative interior of $C \times D$.*

PROOF. To see that an equivalence class with this kernel exists, one need only apply the theorem to the lower (or upper) simple extension of K to all of $R^m \times R^n$. The class is unique by Theorem 34.4. ‖

Continuity and Differentiability of Saddle-Functions

The purpose of this section is to show how the main results about regularity properties of convex functions, such as continuity and differentiability, can be extended to saddle-functions. The continuity and convergence theorems in §10 will be dealt with first.

THEOREM 35.1. *Let C and D be relatively open convex sets in R^m and R^n, respectively, and let K be a finite concave-convex function on $C \times D$. Then K is continuous relative to $C \times D$. In fact, K is Lipschitzian on every closed bounded subset of $C \times D$.*

PROOF. It suffices to show that K is Lipschitzian on $S \times T$, where S and T are arbitrary closed bounded subsets of C and D, respectively. By Theorem 10.1, $K(u, v)$ is in any case continuous in $u \in C$ for each $v \in D$ and continuous in $v \in D$ for each $u \in C$. The collection of concave functions $K(\cdot, v)$, $v \in T$, is therefore pointwise bounded on C, and hence by Theorem 10.6 it is equi-Lipschitzian on S. Thus there exists a non-negative real number α_1 such that

$$|K(u', v) - K(u, v)| \leq \alpha_1 |u' - u|, \quad \forall u', u \in S, \quad \forall v \in T.$$

At the same time, the collection of convex functions $K(u, \cdot)$, $u \in S$, is pointwise bounded on T, so that there exists a non-negative real number α_2 such that

$$|K(u, v') - K(u, v)| \leq \alpha_2 |v' - v|, \quad \forall v', v \in T, \quad \forall u \in S.$$

Let $\alpha = 2(\alpha_1 + \alpha_2)$. Given any two points (u, v) and (u', v') in $S \times T$, we have

$$|K(u', v') - K(u, v)| \leq |K(u', v') - K(u, v')| + |K(u, v') - K(u, v)|$$
$$\leq \alpha_1 |u' - u| + \alpha_2 |v' - v| \leq \alpha(|u' - u| + |v' - v|)/2$$
$$\leq \alpha(|u' - u|^2 + |v' - v|^2)^{1/2} = \alpha |(u', v') - (u, v)|.$$

Thus K is Lipschitzian as claimed. ‖

THEOREM 35.2. *Let C and D be relatively open convex sets in R^m and R^n, respectively, and let $\{K_i \mid i \in I\}$ be a collection of finite concave-convex functions on $C \times D$. Suppose there exist subsets C' and D' of C and D, respectively, such that*

$$\operatorname{conv}(\operatorname{cl}(C' \times D')) \supset C \times D$$

and $\{K_i \mid i \in I\}$ is pointwise bounded on $C' \times D'$. Then, relative to every closed bounded subset of $C \times D$, $\{K_i \mid i \in I\}$ is uniformly bounded and equi-Lipschitzian.

PROOF. It suffices to consider a closed bounded subset of $C \times D$ having the form $S \times T$. For each $u \in C'$, the collection of convex functions $\{K_i(u, \cdot) \mid i \in I\}$ is pointwise bounded on D' and hence uniformly bounded on T by Theorem 10.6. The collection of concave functions

$$\{K_i(\cdot, v) \mid i \in I, v \in T\}$$

is therefore pointwise bounded on C', so that by Theorem 10.6 it is uniformly bounded on S and there exists a non-negative real number α_1 such that

$$|K_i(u', v) - K_i(u, v)| \leq \alpha_1 |u' - u|, \qquad \forall u', u \in S, \qquad \forall v \in T, \qquad \forall i \in I.$$

By a parallel argument, there exists a non-negative real number α_2 such that

$$|K_i(u, v') - K_i(u, v)| \leq \alpha_2 |v' - v|, \qquad \forall v', v \in T, \qquad \forall u \in S, \qquad \forall i \in I.$$

Then for any (u, v) and (u', v') in $S \times T$ we have

$$|K_i(u', v') - K_i(u, v)| \leq \alpha |(u', v') - (u, v)|, \qquad \forall i \in I,$$

where $\alpha = 2(\alpha_1 + \alpha_2)$, by the calculation in the proof of preceding theorem. ‖

THEOREM 35.3. *Let C and D be relatively open convex sets in R^m and R^n, respectively, and let T be any locally compact topological space. Let K be a real-valued function on $C \times D \times T$ such that $K(u, v, t)$ is concave in u for each v and t, convex in v for each u and t, and continuous in t for each u and v. Then K is continuous on $C \times D \times T$, i.e. jointly continuous in u, v and t.*

The conclusion remains valid if the assumption about continuity in t is weakened to the following: there exist dense subsets C' and D' of C and D, respectively, such that $K(u, v, \cdot)$ is a continuous function on T for each $(u, v) \in C' \times D'$.

PROOF. The same as the proof fo Theorem 10.7, except for changes of notation. Theorem 35.2 is invoked in place of Theorem 10.6. ‖

THEOREM 35.4. *Let C and D be relatively open convex sets in R^m and R^n, respectively, and let $K_1, K_2, \ldots$, be a sequence of finite concave-convex functions on $C \times D$. Suppose that, for each (u, v) in a certain dense subset $C' \times D'$ of $C \times D$, the limit of $K_1(u, v), K_2(u, v), \ldots$, exists and is finite. The limit then exists for every $(u, v) \in C \times D$, and the function K, where*

$$K(u, v) = \lim_{i \to \infty} K_i(u, v),$$

is finite and concave-convex on $C \times D$. Moreover, the sequence $K_1, K_2, \ldots$, converges to K uniformly on each closed bounded subset of $C \times D$.

PROOF. The same as the proof of Theorem 10.8, except for changes of notation. Again Theorem 35.2 is invoked in place of Theorem 10.6. ‖

THEOREM 35.5. *Let C and D be relatively open convex sets in R^m and R^n respectively, and let $K_1, K_2, \ldots$, be a sequence of finite concave-convex functions on $C \times D$. Suppose that, for every (u, v) in a certain dense subset $C' \times D'$ of $C \times D$, the sequence $K_1(u, v), K_2(u, v), \ldots$, is bounded. Then there exists a subsequence of $K_1, K_2, \ldots$, which converges uniformly on closed bounded subsets of $C \times D$ to some finite concave-convex function K.*

PROOF. An imitation of the proof of Theorem 10.9, with Theorem 35.4 invoked instead of Theorem 10.8. ‖

We turn now to results about directional derivatives and subgradients of saddle-functions.

Let K be a saddle-function on $R^m \times R^n$, and let (u, v) be a point where K is finite. The (one-sided) *directional derivative* of K at (u, v) with respect to (u', v') is, of course, defined to be the limit

$$K'(u, v; u', v') = \lim_{\lambda \downarrow 0} [K(u + \lambda u', v + \lambda v') - K(u, v)]/\lambda,$$

if this limit exists. The directional derivatives

$$K'(u, v; u', 0) = \lim_{\lambda \downarrow 0} [K(u + \lambda u', v) - K(u, v)]/\lambda,$$

$$K'(u, v; 0, v') = \lim_{\lambda \downarrow 0} [K(u, v + \lambda v') - K(u, v)]/\lambda,$$

certainly exist by Theorem 23.1, but the existence of $K(u, v; u', v')$ is problematical. In most of what follows, we shall restrict ourselves, for the sake of simplicity, to the study of directional derivatives at interior points of dom K.

THEOREM 35.6. *Let K be a concave-convex function on $R^m \times R^n$, and let $C \times D$ be an open convex set on which K is finite. Then, for each $(u, v) \in C \times D$, $K'(u, v; u', v')$ exists and is a finite positively homogeneous concave-convex function of (u', v') on $R^m \times R^n$. In fact,*

$$K'(u, v; u', v') = K'(u, v; u', 0) + K'(u, v; 0, v').$$

PROOF. We know from §23 that, for each $u \in C$ and $v \in D$, $K'(u, v; 0, v')$ is a finite positively homogeneous convex function of v' and $K'(u, v; u', 0)$ is a finite positively homogeneous concave function of u'. The properties claimed for $K'(u, v; u', v')$ follow therefore from the equation in the theorem, and only this equation needs to be established. We shall show that

$$\limsup_{\lambda \downarrow 0} \, [K(u + \lambda u', v + \lambda v') - K(u, v)]/\lambda$$

$$\leq K'(u, v; u', 0) + K'(u, v; 0, v').$$

By a dual argument, we will have

$$\liminf_{\lambda \downarrow 0} \, [K(u + \lambda u', v + \lambda v') - K(u, v)]/\lambda$$

$$\geq K'(u, v; u', 0) + K'(u, v; 0, v'),$$

and the existence of $K'(u, v; u', v')$ and the equality in the theorem will both be proved. The difference quotient

$$[K(u + \lambda u', v + \lambda v') - K(u, v)]/\lambda$$

can be expressed as

$$([K(u + \lambda u', v) - K(u, v)]/\lambda)$$

$$+ ([K(u + \lambda u', v + \lambda v') - K(u + \lambda u', v)]/\lambda),$$

where the first quotient has limit $K'(u, v; u', 0)$ as $\lambda \downarrow 0$. We must show that

$$\limsup_{\lambda \downarrow 0} \, [K(u + \lambda u', v + \lambda v') - K(u + \lambda u', v)]/\lambda \leq K'(u, v; 0, v').$$

Given any $\mu > K'(u, v; 0, v')$, there exists an $\alpha > 0$ such that

$$\mu > [K(u, v + \alpha v') - K(u, v)]/\alpha.$$

Since K is continuous on $C \times D$ by Theorem 35.1, for all sufficiently small values of λ, $0 < \lambda < \alpha$, we have

$$\mu > [K(u + \lambda u', v + \alpha v') - K(u + \lambda u', v)]/\alpha$$

$$\geq [K(u + \lambda u', v + \lambda v') - K(u + \lambda u', v)]/\lambda.$$

The "lim sup" of the latter quotient thus cannot exceed μ, and the result follows. ‖

The directional derivatives of saddle-functions correspond to certain "subgradients," much as in the case of purely convex or concave functions. Given any concave-convex function K on $R^m \times R^n$, we define

$$\partial_1 K(u, v) = \partial_u K(u, v)$$

to be the set of all subgradients of the concave function $K(\cdot, v)$ at u, i.e. the set of all vectors $u^* \in R^m$ such that

$$K(u', v) \le K(u, v) + \langle u^*, u' - u \rangle, \qquad \forall u' \in R^m.$$

Similarly, we define

$$\partial_2 K(u, v) = \partial_v K(u, v)$$

to be the set of all subgradients of the convex function $K(u, \cdot)$ at v, i.e. the set of all vectors $v^* \in R^n$ such that

$$K(u, v') \ge K(u, v) + \langle v^*, v' - v \rangle, \qquad \forall v' \in R^n.$$

The elements (u^*, v^*) of the set

$$\partial K(u, v) = \partial_1 K(u, v) \times \partial_2 K(u, v)$$

are then defined to be the *subgradients* of K at (u, v), and the multivalued mapping

$$\partial K: (u, v) \to \partial K(u, v)$$

is called the *subdifferential* of K.

Note that $\partial K(u, v)$ is a (possibly empty) closed convex subset of $R^m \times R^n$ for each $(u, v) \in R^m \times R^n$. By Theorem 23.2, if (u, v) is any point where K is finite, the closure of the convex function

$$u' \to -K'(u, v; -u', 0)$$

is the support function of $\partial_1 K(u, v)$, while the closure of the convex function

$$v' \to K'(u, v; 0, v')$$

is the support function of $\partial_2 K(u, v)$. If K is proper and (u, v) is an interior point of dom K, then u is an interior point of dom $K(\cdot, v)$ and v is an interior point of dom $K(u, \cdot)$, so that $\partial_1 K(u, v)$ and $\partial_2 K(u, v)$ are non-empty closed bounded convex sets by Theorem 23.4 and

$$K'(u, v; u', 0) = \inf \{ \langle u^*, u' \rangle \mid u^* \in \partial_1 K(u, v) \},$$

$$K'(u, v; 0, v') = \sup \{ \langle v^*, v' \rangle \mid v^* \in \partial_2 K(u, v) \}.$$

It follows in this case from Theorem 35.6 that $K'(u, v; u', v')$ is a "minimax" of the function

$$(u^*, v^*) \to \langle u^*, u' \rangle + \langle v^*, v' \rangle$$

over $\partial K(u, v)$.

The theorems which follow concern the continuity and single-valuedness of ∂K. Certain other general results about ∂K will be stated in §37.

THEOREM 35.7. *Let K be a concave-convex function on $R^m \times R^n$, and let $C \times D$ be an open convex set on which K is finite. Let $K_1, K_2, \ldots$, be a sequence of concave-convex functions finite on $C \times D$ and converging pointwise on $C \times D$ to K. Let $(u, v) \in C \times D$, and let $(u_i, v_i), i = 1, 2, \ldots$, be a sequence in $C \times D$ converging to (u, v). Then*

$$\liminf_{i \to \infty} K_i'(u_i, v_i; u', 0) \geq K'(u, v; u', 0), \qquad \forall u' \in R^m,$$

$$\limsup_{i \to \infty} K_i'(u_i, v_i; 0, v') \leq K'(u, v; 0, v'), \qquad \forall v' \in R^n.$$

Moreover, given any $\varepsilon > 0$, there exists an index i_0 such that

$$\partial K_i(u_i, v_i) \subset \partial K(u, v) + \varepsilon B, \qquad \forall i \geq i_0,$$

where B is the Euclidean unit ball of $R^m \times R^n = R^{m+n}$.

PROOF. All of this is immediate from Theorem 24.5 and the continuity properties of $K(u, v)$. ‖

COROLLARY 35.7.1. *Let $C \times D$ be an open convex set in $R^m \times R^n$, and let K be a concave-convex function finite on $C \times D$. Then, for each u', $K'(u, v; u', 0)$ is a lower semi-continuous function of (u, v) on $C \times D$ and, for each v', $K'(u, v; 0, v')$ is an upper semi-continuous function of (u, v) on $C \times D$. Moreover, given any (u, v) in $C \times D$ and any $\varepsilon > 0$, there exists a $\delta > 0$ such that*

$$\partial K(x, y) \subset \partial K(u, v) + \varepsilon B, \qquad \forall (x, y) \in [(u, v) + \delta B]$$

(where B is the Euclidean unit ball).

PROOF. Take $K_i = K$ for all indices i. ‖

THEOREM 35.8. *Let K be a concave-convex function $R^m \times R^n$, and let (u, v) be a point where K is finite. If K is differentiable at (u, v), then $\nabla K(u, v)$ is the unique subgradient of K at (u, v). Conversely, if K has a unique subgradient at (u, v), then K is differentiable at x.*

PROOF. By definition, K has a unique subgradient at (u, v) if and only if the convex function $K(u, \cdot)$ has a unique subgradient at v and the concave function $K(\cdot, v)$ has a unique subgradient at u. This situation is equivalent to K being differentiable in its concave and convex arguments separately at (u, v), according to Theorem 25.1. The only question is whether separate differentiability implies joint differentiability, i.e. whether

$$\lim_{(u',v')\to 0} \frac{K(u + u', v + v') - K(u, v) - \langle u^*, u' \rangle - \langle v^*, v' \rangle}{(|u'|^2 + |v'|^2)^{1/2}} = 0$$

when $u^* = \nabla_1 K(u, v)$ and $v^* = \nabla_2 K(u, v)$. This can be established by the following argument similar to the one in Theorem 25.1. For each $\lambda > 0$, let h_λ be the concave-convex function on $R^m \times R^n$ defined by

$$h_\lambda(x, y) = [K(u + \lambda x, v + \lambda y) - K(u, v) - \lambda \langle u^*, x \rangle - \lambda \langle v^*, y \rangle]/\lambda.$$

Assuming that K is separately differentiable at (u, v), one has in particular

$$\lim_{\lambda \downarrow 0} h_\lambda(x, y) = 0, \quad \forall x, \forall y.$$

It follows from this by Theorem 35.4 that, as λ decreases to 0, the functions h_λ must actually converge to 0 uniformly on all bounded sets. Thus, given any $\varepsilon > 0$, there exists a $\delta > 0$ such that, when $0 < \lambda \leq \delta$, one has $|h_\lambda(x, y)| \leq \varepsilon$ for all (x, y) with $(|x|^2 + |y|^2)^{1/2} \leq 1$. Then

$$\left| \frac{K(u + u', v + v') - K(u, v) - <u^*, u'> - <v^*, v'>}{(|u'|^2 + |v'|^2)^{1/2}} \right| \leq \varepsilon$$

for every (u', v') such that

$$0 < (|u'|^2 + |v'|^2)^{1/2} \leq \delta,$$

as is seen by taking $\lambda = (|u'|^2 + |v'|^2)^{1/2}$ and $(x, y) = \lambda^{-1}(u', v')$. Since, given any $\varepsilon > 0$, there exists a $\delta > 0$ with this property, K is jointly differentiable at (u, v) as claimed. ‖

COROLLARY 35.8.1. *Let K be a concave-convex function on $R^m \times R^n$, and let (u, v) be a point at which K is finite. A necessary and sufficient condition for K to be differentiable at (u, v) is that K be finite on a neighborhood of (u, v) and the directional derivative function $K'(u, v; \cdot, \cdot)$ be linear. Moreover, this condition is satisfied if merely the $m + n$ two-sided partial derivatives of K exist and are finite.*

PROOF. This follows by Theorem 25.2. ‖

THEOREM 35.9. *Let $C \times D$ be an open convex set in $R^m \times R^n$, and let K be a concave-convex function finite on $C \times D$. Let E be the subset of $C \times D$ where K is differentiable. Then E is dense in $C \times D$. In fact the complement of E in $C \times D$ is a set of measure zero. The gradient mapping ∇K is continuous from E to $R^m \times R^n$.*

PROOF. Let E_p be the subset of $C \times D$ where the finite two-sided partial derivative of K with respect to the pth of its $m + n$ real arguments exists. Since $E = E_1 \cap \cdots \cap E_{m+n}$ by the preceding corollary, it suffices to prove that the complement of E_p in $C \times D$ is of measure zero and that the partial derivative corresponding to E_p is continuous on E_p. For simplicity of exposition, we shall limit ourselves to the case where $p = m + n$. Let $e = (0, \ldots, 0, 1) \in R^n$. The set E_{m+n} consists of the points $(u, v) \in C \times D$ such that

$$-K'(u, v; 0, -e) = K'(u, v; 0, e).$$

Since $K'(u, v; 0, e)$ and $K'(u, v; 0, -e)$ are both upper semi-continuous functions of (u, v) by Corollary 35.7.1, the $(m + n)$th partial derivative is simultaneously upper and lower semi-continuous on E_{m+n}, i.e. it is continuous. For $k = 1, 2, \ldots$, let

$$S_k = \{(u, v) \in C \times D \mid K'(u, v; 0, e) + K'(u, v; 0, -e) \geq 1/k\}.$$

Since in general
$$-K'(u, v; 0, -e) \leq K'(u, v; 0, e),$$

the complement of E_{m+n} in $C \times D$ is the union of the sets $S_1, S_2, \ldots$, each of which is closed by the upper semi-continuity of K'. Thus E_{m+n} is measurable. For a given point (u, v), the values of λ such that $(u, v + \lambda e) \in S_k$ are those where the right derivative of the convex function

$$h(\lambda) = K(u, v + \lambda e)$$

jumps by at least as much as $1/k$. Since the right derivative of a convex function is non-decreasing, there can be only finitely many jumps as large as $1/k$ in any bounded interval of λ values. Thus, for a given k, each line parallel to the $(m + n)$th coordinate axis has at most finitely many points of S_k in any bounded interval and therefore meets S_k in a set of measure zero. It follows that S_k itself is of measure zero, and hence that the complement of E_{m+n} in $C \times D$ is of measure zero. ‖

THEOREM 35.10. *Let $C \times D$ be an open convex set in $R^m \times R^n$, and let K be a finite differentiable concave-convex function on $C \times D$. Let K_1, $K_2, \ldots$, be a sequence of finite differentiable concave-convex functions on $C \times D$ such that $\lim_{i \to \infty} K_i(u, v) = K(u, v)$ for every $(u, v) \in C \times D$. Then*

$$\lim_{i \to \infty} \nabla K_i(u, v) = \nabla K(u, v), \qquad \forall (u, v) \in C \times D.$$

In fact, the mappings ∇K_i converge uniformly to ∇K on all closed bounded subsets of $C \times D$.

PROOF. It is enough to prove the convergence for each of the $m + n$ partial derivatives, and this can be done exactly as in the case of Theorem

25.7. The citations of Theorems 10.8 and 25.5 in the proof of Theorem 25.7 are replaced by citations of Theorems 35.4 and 35.9. ‖

It suffices actually, in the hypothesis of Theorem 35.10, if $K_i(u, v)$ converges to $K(u, v)$ for every (u, v) in a certain dense subset $C' \times D'$ of $C \times D$. This implies by Theorem 35.4 and the continuity of finite saddle-functions on $C \times D$ that $K_i(u, v)$ converges to $K(u, v)$ for every $(u, v) \in C \times D$.

Minimax Problems

Minimax theory treats a class of extremum problems which involve, not simply minimization or maximization, but a combination of both. Let C and D be arbitrary non-empty sets, and let K be a function from $C \times D$ to $[-\infty, +\infty]$. For each $u \in C$, one can take the infimum of $K(u, v)$ over $v \in D$ and then take the supremum of this infimum as a function on C. The quantity so obtained is

$$\sup_{u \in C} \inf_{v \in D} K(u, v).$$

On the other hand, for each $v \in D$ one can take the supremum of $K(u, v)$ over $u \in C$ and then take the infimum of this supremum as a function on D. This forms

$$\inf_{v \in D} \sup_{u \in C} K(u, v).$$

If the "sup inf" and "inf sup" are equal, the common value is called the *minimax* or the *saddle-value* of K (with respect to maximizing over C and minimizing over D).

One of the tasks of minimax theory is to furnish conditions under which the saddle-value exists and is attained in some suitable sense. In general, of course, the "sup inf" and the "inf sup" might not be equal, but a certain inequality is at least satisfied.

LEMMA 36.1. *If K is any function from a non-empty product set $C \times D$ to $[-\infty, +\infty]$, then*

$$\sup_{u \in C} \inf_{v \in D} K(u, v) \leq \inf_{v \in D} \sup_{u \in C} K(u, v).$$

PROOF. Let $f(u) = \inf \{K(u, v) \mid v \in D\}$ for each $u \in C$, and let

$$\alpha = \sup_{u \in C} \inf_{v \in D} K(u, v).$$

For each $v \in D$, one has $K(u, v) \geq f(u)$ for every $u \in C$, and consequently

$$\sup_{u \in C} K(u, v) \geq \sup_{u \in C} f(u) = \alpha.$$

379

Since this relation holds for every $v \in D$, one has

$$\inf_{v \in D} \sup_{u \in C} K(u, v) \geq \alpha,$$

and the lemma is proved. ‖

It is not entirely obvious what one should mean by a saddle-value being "attained." The proper concept is that of a saddle-point. By definition, a point $(\bar{u}, \bar{v})$ is a *saddle-point* of K with respect to maximizing over C and minimizing over D if $(\bar{u}, \bar{v}) \in C \times D$ and

$$K(u, \bar{v}) \leq K(\bar{u}, \bar{v}) \leq K(\bar{u}, v), \qquad \forall u \in C, \qquad \forall v \in D.$$

This means that the function $K(\bar{u}, \cdot)$ attains its infimum over D at $\bar{v}$, while $K(\cdot, \bar{v})$ attains its supremum over C at $\bar{u}$. The relationship between saddle-points and saddle-values is as follows.

LEMMA 36.2. *Let K be any function from a non-empty product set $C \times D$ to $[-\infty, +\infty]$. A point $(\bar{u}, \bar{v})$ is a saddle-point of K (with respect to maximizing over C and minimizing over D) if and only if the supremum in the expression*

$$\sup_{u \in C} \inf_{v \in D} K(u, v)$$

is attained at $\bar{u}$, the infimum in the expression

$$\inf_{v \in D} \sup_{u \in C} K(u, v)$$

is attained at $\bar{v}$, and these two extrema are equal. If $(\bar{u}, \bar{v})$ is a saddle-point, the saddle-value of K is $K(\bar{u}, \bar{v})$.

PROOF. If $(\bar{u}, \bar{v})$ is a saddle-point, we have

$$K(\bar{u}, \bar{v}) = \inf_{v \in D} K(\bar{u}, v) \leq \sup_{u \in C} \inf_{v \in D} K(u, v),$$

$$K(\bar{u}, \bar{v}) = \sup_{u \in C} K(u, \bar{v}) \geq \inf_{v \in D} \sup_{u \in C} K(u, v).$$

In view of the inequality in Lemma 36.1, these quantities must actually all be equal, so the three conditions in the lemma are satisfied. Conversely, if these conditions are satisfied, the saddle-value α of K exists, and one has

$$\sup_{u \in C} K(u, \bar{v}) = \alpha = \inf_{v \in D} K(\bar{u}, v),$$

where the supremum is at least as great as $K(\bar{u}, \bar{v})$ and the infimum is no greater than $K(\bar{u}, \bar{v})$. Thus $\alpha = K(\bar{u}, \bar{v})$, and $(\bar{u}, \bar{v})$ is a saddle-point. ‖

There is an elementary heuristic interpretation of saddle-values and saddle-points which is worth knowing for the sake of motivation. Given K on $C \times D$, we may think of the following game for two players I and II. At each play of the game, I selects a point u of C and II selects a point v of D. The players make their choices known to each other simultaneously, and at that time II must pay $K(u, v)$ units of money to I. (A negative $K(u, v)$ corresponds to a positive payment from I to II instead of from II to I.) For each $u \in C$, inf $\{K(u, v) \mid v \in D\}$ is the amount of winnings which I can *guarantee* for himself in selecting u. The highest amount of winnings which player I can guarantee in this way is

$$\sup_{u \in C} \inf_{v \in D} K(u, v).$$

A point $\bar{u}$ for which the supremum is attained is an optimal strategy for I (according to the von Neumann minimax principle). On the other hand, consider the game from the point of view of II. For each $v \in D$, sup $\{K(u, v) \mid u \in C\}$ is the most that II can possibly lose if he selects v. Thus

$$\inf_{v \in D} \sup_{u \in C} K(u, v)$$

is the lowest ceiling which II can put on his losses. A point $\bar{v}$ for which the infimum is attained is an optimal strategy for II.

When the lowest ceiling to the losses of II coincides with the highest floor to the winnings of I, the common level is the saddle-value of K. A saddle-point represents an "equilibrium" choice of points for I and II, in the sense that neither player can gain any advantage by unilaterally altering his choice.

As already observed, the problem of minimizing a real-valued function f over a subset S of R^n can be expressed conveniently as the problem of minimizing f over all of R^n, if one defines $f(x)$ to be $+\infty$ for every $x \notin S$. A similar technical device is useful in the study of minimax problems.

Let C and D be non-empty sets in R^m and R^n, respectively, and let K be a real-valued function on $C \times D$. Suppose that one extends K beyond $C \times D$ by setting

$$K(u, v) = \begin{cases} +\infty & \text{if} \quad u \in C, v \notin D, \\ -\infty & \text{if} \quad u \notin C, v \in D, \\ \text{any value in } [-\infty, +\infty] & \text{if} \quad u \notin C, v \notin D. \end{cases}$$

Then obviously

$$\inf_{v \in R^n} K(u, v) = \inf_{v \in D} K(u, v) < +\infty, \qquad \forall u \in R^m,$$

where the infima are $-\infty$ if $u \notin C$, and hence

$$\sup_{u \in R^m} \inf_{v \in R^n} K(u, v) = \sup_{u \in C} \inf_{v \in D} K(u, v).$$

Similarly,

$$\sup_{u \in R^m} K(u, v) = \sup_{u \in C} K(u, v) > -\infty, \qquad \forall v \in R^n,$$

where the suprema are $+\infty$ if $v \notin D$, and hence

$$\inf_{v \in R^n} \sup_{u \in R^m} K(u, v) = \inf_{v \in D} \sup_{u \in C} K(u, v).$$

In particular, if either the saddle-value of K with respect to $R^m \times R^n$ or the saddle-value of K with respect to $C \times D$ exists, then both exist and are equal. Furthermore, the saddle-points of K with respect to $R^m \times R^n$ are the same as the saddle-points of K with respect to $C \times D$ (if any). Indeed, according to what we have just established, $(\bar{u}, \bar{v})$ satisfies the condition

$$\sup_{u \in R^m} K(u, \bar{v}) = K(\bar{u}, \bar{v}) = \inf_{v \in R^n} K(\bar{u}, v)$$

if and only if it satisfies

$$\sup_{u \in C} K(u, \bar{v}) = K(\bar{u}, \bar{v}) = \inf_{v \in D} K(\bar{u}, v),$$

in which case one necessarily has $(\bar{u}, \bar{v}) \in C \times D$. (If $\bar{u}$ were not in C, the infimum would be $-\infty$ and hence could not possibly equal the supremum. Similarly, if $\bar{v}$ were not in D the supremum would be $+\infty$ and could not equal the infimum.)

In what follows, we shall be concerned only with saddle-values and saddle-points of concave-convex (or convex-concave) functions on products of convex sets. It is to be understood always that the minimization takes place in the convex argument of the function, and that the maximization takes place in the concave argument. The observations above allow us to reduce almost everything to the case of concave-convex functions defined on all of $R^m \times R^n$. The closedness of such functions is imposed as a natural (and essentially constructive) regularity condition.

In general, minimax problems for closed proper saddle-functions on $R^m \times R^n$ correspond in the following way to minimax problems for certain finite saddle-functions on convex product sets.

THEOREM 36.3. *Let K be a closed proper concave-convex function on $R^m \times R^n$, and let $C = \mathrm{dom}_1 K$ and $D = \mathrm{dom}_2 K$. Then*

$$\sup_{u \in R^m} \inf_{v \in R^n} K(u, v) = \sup_{u \in C} \inf_{v \in D} K(u, v),$$

$$\inf_{v \in R^n} \sup_{u \in R^m} K(u, v) = \inf_{v \in D} \sup_{u \in C} K(u, v).$$

The saddle-value and saddle-points of K with respect to $R^m \times R^n$ are the same as those with respect to $C \times D$.

PROOF. For a convex function f on R^n, one has

$$\inf \{f(v) \mid v \in R^n\} = \inf \{f(v) \mid v \in D\}$$

for any set D containing ri (dom f) (Corollary 7.3.1). Similarly in the case of the supremum of a concave function. The domain relations in Theorem 34.3 imply therefore that

$$\inf_{v \in R^n} K(u, v) = \inf_{v \in D} K(u, v) < + \infty, \quad \forall u \in R^m,$$

$$\sup_{u \in R^m} K(u, v) = \sup_{u \in C} K(u, v) > - \infty, \quad \forall v \in R^n,$$

where the infima are $- \infty$ if $u \notin C$ and the suprema are $+ \infty$ if $v \notin D$. The desired conclusions follow from these facts exactly as in the discussion preceding the theorem. ∥

COROLLARY 36.3.1. *Let K be a closed proper saddle-function on $R^m \times R^n$. If K has a saddle-point, this saddle-point lies in* dom K, *and the saddle-value of K is finite.*

PROOF. Let $(\bar{u}, \bar{v})$ be a saddle-point of K (with respect to $R^m \times R^n$). By the theorem, $(\bar{u}, \bar{v})$ is also a saddle-point with respect to the set $C \times D =$ dom K, so that $(\bar{u}, \bar{v}) \in C \times D$. The saddle-value of K is $K(\bar{u}, \bar{v})$ by Lemma 36.2, and this is finite because K is finite on dom K. ∥

In particular, according to Theorem 36.3, the minimax theory of closed proper saddle-functions on $R^m \times R^n$ includes as a special case (in view of Corollary 34.2.4) the minimax theory of continuous finite saddle-functions defined on non-empty products $C \times D$, where C is a closed convex set in R^m and D is a closed convex set in R^n.

Minimax problems for saddle-functions on $R^m \times R^n$ really correspond to equivalence classes of saddle-functions, rather than to individual saddle-functions:

THEOREM 36.4. *Equivalent saddle-functions on $R^m \times R^n$ have the same saddle-value and saddle-points (if any).*

PROOF. Let K and K' be equivalent concave-convex functions on $R^m \times R^n$. By the definition of equivalence, $\text{cl}_1 K = \text{cl}_1 K'$ and $\text{cl}_2 K = \text{cl}_2 K'$. Two convex functions with the same closure have the same infimum, and two concave functions with the same closure have the same supremum. Thus

$$\inf_v K(u, v) = \inf_v K'(u, v), \forall u,$$

$$\sup_u K(u, v) = \sup_u K'(u, v), \forall v.$$

But the saddle-values and saddle-points of K and K' depend only on these infimum and supremum functions. ∥

According to the above, the natural objects of concave-convex minimax theory are the *equivalence classes of closed proper concave-convex functions on $R^m \times R^n$*, each class corresponding to a single "regularized" saddle-point problem. We shall now show that there is a one-to-one correspondence, in terms of Lagrangians, between such saddle-point problems and the generalized convex programs associated with closed proper convex bifunctions.

If F is any bifunction from R^m to R^n, the *inverse* of F is defined to be the bifunction

$$F_*:x \to F_*x:u \to (F_*x)(u)$$

from R^n to R^m given by

$$(F_*x)(u) = -(Fu)(x), \qquad \forall x \in R^n, \qquad \forall u \in R^m.$$

Note that F_* is concave if F is convex, and vice versa. This notion of "inverse" generalizes the one for single-valued or multivalued mappings in the sense that, if F is the $+\infty$ indicator bifunction of a mapping A from R^m to R^n, then F_* is the $-\infty$ indicator bifunction of A^{-1}.

The inverse operation $F \to F_*$ clearly preserves closedness and properness of convex or concave bifunctions, and it is involutory, i.e.

$$(F_*)_* = F.$$

Moreover, the inverse operation commutes with the adjoint operation for convex and concave bifunctions:

$$(F_*)^* = (F^*)_*.$$

To see this, suppose that F is convex, so that F_* is concave. Then by definition $(F_*)^*$ is given by

$$((F_*)^*u^*)(x^*) = \sup_{u,\, x} \{(F_*x)(u) - \langle u, u^* \rangle + \langle x, x^* \rangle\}$$

$$= \sup_{u,\, x} \{-(Fu)(x) - \langle u, u^* \rangle + \langle x, x^* \rangle\}$$

$$= -\inf_{u,\, x} \{(Fu)(x) - \langle x, x^* \rangle + \langle u, u^* \rangle\}$$

$$= -(F^*x^*)(u^*) = ((F^*)_*u^*)(x^*).$$

(The argument is the same if F is concave instead of convex, except that "inf" and "sup" are interchanged.)

The relation $(F_*)^* = (F^*)_*$ may be regarded as a generalization of the fact that

$$(A^{-1})^* = (A^*)^{-1}$$

for any non-singular linear transformation A.

One may simply write F_*^* in place of $(F_*)^*$ or $(F^*)_*$. Of course, if F is a convex bifunction from R^m to R^n, then F_*^* is likewise a convex bifunction from R^m to R^n, and

$$(F_*^*)_*^* = F^{**} = \text{cl } F.$$

By definition, the *Lagrangian* of the convex program (P) associated with a convex bifunction F from R^m to R^n is the function L on $R^m \times R^n$ given by

$$L(u^*, x) = \inf_u \{\langle u^*, u \rangle + (Fu)(x)\}.$$

In terms of F_*, the formula for L becomes

$$L(u^*, x) = \inf_u \{\langle u^*, u \rangle - (F_*x)(u)\}$$
$$= \langle u^*, F_*x \rangle.$$

Therefore L is a concave-convex function on $R^m \times R^n$ by Theorem 33.1, and we have the following characterization.

THEOREM 36.5. *In order that L be the Lagrangian of a convex program (P) associated with a closed convex bifunction F from R^m to R^n, it is necessary and sufficient that L be an upper closed concave-convex function on $R^m \times R^n$.*

PROOF. This is immediate from Theorem 33.3. ‖

Given any upper closed concave-convex function L on $R^m \times R^n$, the unique "closed" convex program (P) having L as its Lagrangian is easily determined from the correspondences in §33. Indeed, (P) is the convex program associated with F, where F is the closed convex bifunction such that F_*x is the (concave) conjugate of $L(\cdot, x)$ for each x, i.e.

$$(Fu)(x) = -\inf_{u^*} \{\langle u^*, u \rangle - L(u^*, x)\}$$
$$= \sup_{u^*} \{L(u^*, x) - \langle u^*, u \rangle\}.$$

The objective function in (P) is thus the convex function

$$\sup_{u^*} L(u^*, \cdot),$$

and the optimal value in (P) is

$$\inf_x \sup_{u^*} L(u^*, x).$$

At the same time, the adjoint of F is given by

$$(F^*x^*)(u^*) = \inf_u \inf_x \{(Fu)(x) - \langle x^*, x \rangle + \langle u^*, u \rangle\}$$
$$= \inf_x \{L(u^*, x) - \langle x^*, x \rangle\},$$

so that the objective function in the concave program (P^*) dual to (P) is the concave function

$$\inf_x L(\cdot, x),$$

and the optimal value in (P^*) is

$$\sup_{u^*} \inf_x L(u^*, x).$$

The saddle-points of L correspond to optimal solutions and Kuhn–Tucker vectors for (P) and (P^*), as explained in Theorem 29.3 and Theorem 30.5.

Each equivalence class of closed proper concave-convex functions on $R^m \times R^n$ contains a unique upper closed function L (Corollary 34.2.2). Thus the general "regularized" saddle-point problems which we arrived at by the natural considerations of minimax theory turn out to be precisely the Lagrangian problems corresponding to the (generalized) "closed proper" convex programs.

It follows that the main results of (concave-convex) minimax theory concerning existence of saddle-values and saddle-points will essentially be corollaries to theorems already proved in §29 and §30. These results will be presented in §37 in terms of a conjugacy correspondence for saddle-functions.

Since the Lagrangian L of a convex program (P) is a concave-convex function, the theory of subgradients can be used to characterize the saddle-points of L. The condition

$$L(u^*, \bar{x}) \leq L(\bar{u}^*, \bar{x}) \leq L(\bar{u}^*, x), \forall u^*, \forall x,$$

holds if and only if the convex function $L(\bar{u}^*, \cdot)$ achieves its minimum at $\bar{x}$, i.e.

$$0 \in \partial_2 L(\bar{u}^*, \bar{x}),$$

and the concave function $L(\cdot, \bar{x})$ achieves its maximum at $\bar{u}^*$, i.e.

$$0 \in \partial_1 L(\bar{u}^*, \bar{x}).$$

But, by the definition in §35,

$$\partial L(\bar{u}^*, \bar{x}) = \partial_1 L(\bar{u}^*, \bar{x}) \times \partial_2 L(\bar{u}^*, \bar{x}).$$

Thus $(\bar{u}^*, \bar{x})$ is a saddle-point of L if and only if

$$(0, 0) \in \partial L(\bar{u}^*, \bar{x}).$$

The latter relation will be called the *Kuhn–Tucker condition* for (P). It reduces to the Kuhn–Tucker conditions in Theorem 28.3 when (P) is an ordinary convex program, as is seen simply from the calculus of subgradients. On the other hand, it reduces to the Kuhn–Tucker conditions in Theorem 31.3 when (P) is a convex program of the type in Theorem 31.2, since the Lagrangian of (P) in that case is of the form

$$L(u^*, x) = \inf_u \{\langle u^*, u \rangle + f(x) - g(Ax + u)\}$$

$$= \begin{cases} f(x) + g^*(u^*) - \langle u^*, Ax \rangle & \text{if } x \in \text{dom } f, \\ +\infty & \text{if } x \notin \text{dom } f. \end{cases}$$

The general Kuhn–Tucker Theorem for convex programs (Corollary 29.3.1) may be restated as:

THEOREM 36.6. *Let* (P) *be the convex program associated with a closed proper convex bifunction* F *from* R^m *to* R^n. *Assume that* (P) *is strongly (or strictly) consistent, or that* (P) *is polyhedral and consistent. In order that a given vector* $\bar{x} \in R^n$ *be an optimal solution to* (P), *it is necessary and sufficient that there exist a vector* $\bar{u}^* \in R^m$ *such that*

$$(0, 0) \in \partial L(\bar{u}^*, \bar{x}),$$

where L *is the Lagrangian of* (P). *The vectors* $\bar{u}^*$ *satisfying this condition for a given* $\bar{x}$ *(if any) are precisely the Kuhn–Tucker vectors for* (P).

Conjugate Saddle-Functions and Minimax Theorems

Questions about saddle-values and saddle-points of concave-convex functions can be reduced essentially to questions about (generalized) convex programs and their associated Lagrangian problems, as has been shown in §36. The main existence theorems will be presented here in terms of a conjugacy correspondence among concave-convex functions, much as the main theorems concerning the minimum of a convex function were presented in §27 in terms of the conjugacy correspondence for convex functions.

The notion of the conjugate of a saddle-function is derived from properties of the inverse operation for convex bifunctions, which was introduced in the preceding section. Thus, as it turns out, the inverse operation is the natural foundation for minimax theory, just as the adjoint operation for convex bifunctions was the natural foundation for the duality theory of convex programs in §30.

If F is any convex bifunction from R^m to R^n, the inverse F_* of F is a concave bifunction from R^n to R^m, and hence $\langle u^*, F_* x \rangle$ is a concave-convex function of (u^*, x) on $R^m \times R^n$ (Theorem 33.1). How is $\langle u^*, F_* x \rangle$ related to $\langle Fu, x^* \rangle$, which similarly is concave-convex in (u, x^*)? By definition,

$$\langle u^*, F_* x \rangle = (F_* x)^*(u^*)$$
$$= \inf_u \{\langle u, u^* \rangle - (F_* x)(u)\}$$
$$= \inf_u \{\langle u, u^* \rangle + (Fu)(x)\}.$$

If F is closed (or merely image-closed), we have

$$(Fu)(x) = \sup_{x^*} \{\langle x, x^* \rangle - \langle Fu, x^* \rangle\}$$

by Corollary 33.1.2 and consequently

$$\langle u^*, F_* x \rangle = \inf_u \sup_{x^*} \{\langle u, u^* \rangle + \langle x, x^* \rangle - \langle Fu, x^* \rangle\}.$$

This reasoning, applied also to the bifunctions F^* and F_*^*, leads to the following basic result.

THEOREM 37.1.   *Let F be a closed convex bifunction from R^m to R^n, and let K be any one of the closed concave-convex functions in the equivalence class $\Omega(F)$ corresponding to F, i.e. any concave-convex function on $R^m \times R^n$ such that*

$$\langle Fu, x^* \rangle \leq K(u, x^*) \leq \langle u, F^*x^* \rangle, \forall u, \forall x^*.$$

Then, for every $u^* \in R^m$ and $x \in R^n$,

$$\inf_{u} \sup_{x^*} \{\langle u, u^* \rangle + \langle x, x^* \rangle - K(u, x^*)\} = \langle u^*, F_*x \rangle,$$

$$\sup_{x^*} \inf_{u} \{\langle u, u^* \rangle + \langle x, x^* \rangle - K(u, x^*)\} = \langle F_*^*u^*, x \rangle.$$

On the other hand, let K^* be any one of the closed concave-convex functions in the equivalence class $\Omega(F_*)$ corresponding to F_*, i.e. any concave-convex function on $R^m \times R^n$ such that

$$\langle F_*^*u^*, x \rangle \leq K^*(u^*, x) \leq \langle u^*, F_*x \rangle, \forall u^*, \forall x.$$

Then, for every $u \in R^m$ and every $x^* \in R^n$,

$$\inf_{u^*} \sup_{x} \{\langle u, u^* \rangle + \langle x, x^* \rangle - K^*(u^*, x)\} = \langle u, F^*x^* \rangle,$$

$$\sup_{x} \inf_{u^*} \{\langle u, u^* \rangle + \langle x, x^* \rangle - K^*(u^*, x)\} = \langle Fu, x^* \rangle.$$

PROOF.   This is immediate from the definition of the inverse operation and the properties of the equivalence class $\Omega(F)$ in Theorem 34.2.   ‖

The saddle-function correspondence in Theorem 37.1 can be regarded as a generalization of the conjugacy correspondences for convex or concave functions. Let K be any concave-convex function on $R^m \times R^n$. For each $u^* \in R^m$ and $v^* \in R^n$,

$$\langle u, u^* \rangle + \langle v, v^* \rangle - K(u, v)$$

is a convex-concave function of (u, v), the sort of function one naturally minimizes in u and maximizes in v. We define the *lower conjugate* $\underline{K}^*$ of K by

$$\underline{K}^*(u^*, v^*) = \sup_{v} \inf_{u} \{\langle u, u^* \rangle + \langle v, v^* \rangle - K(u, v)\}$$

and the *upper conjugate* $\bar{K}^*$ of K by

$$\bar{K}^*(u^*, v^*) = \inf_{u} \sup_{v} \{\langle u, u^* \rangle + \langle v, v^* \rangle - K(u, v)\}.$$

Of course, $\underline{K}^* \leq \bar{K}^*$ by Lemma 36.1.

COROLLARY 37.1.1.   *Let K be any closed concave-convex function on $R^m \times R^n$. The lower conjugate $\underline{K}^*$ of K is then a lower closed concave-convex function on $R^m \times R^n$, and the upper conjugate $\bar{K}^*$ of K is an upper closed concave-convex function on $R^m \times R^n$. Moreover, $\underline{K}^*$ and $\bar{K}^*$ are equivalent, and they depend only on the equivalence class containing K. If K^* is any closed concave-convex function equivalent to $\underline{K}^*$ and $\bar{K}^*$, the lower and upper conjugates of K^* are in turn equivalent to K.*

PROOF.   By Theorem 34.2, the equivalence classes in Theorem 37.1 are the most general equivalence classes of closed concave-convex functions. The fact that $\underline{K}^*$ is lower closed and $\bar{K}^*$ is upper closed is deduced by applying Theorem 33.3 to F^* and F_*^*, since for $K \in \Omega(F)$ one has

$$\underline{K}^*(u^*, v^*) = \langle F_*^* u^*, v^* \rangle,$$

$$\bar{K}^*(u^*, v^*) = \langle u^*, F_* v^* \rangle,$$

by Theorem 37.1. ‖

Any saddle-function K^* which is equivalent to both the lower and upper conjugates of a given saddle-function K will simply be called a *conjugate* of K. In this terminology, Corollary 37.1.1 describes a conjugacy correspondence among closed saddle-functions which is symmetric and one-to-one *up to equivalence*. The constant functions $+\infty$ and $-\infty$ on $R^m \times R^n$ are closed saddle-functions conjugate to each other; since these are the only improper closed saddle-functions, a saddle-function conjugate to a closed proper saddle-function must be proper. In general, the equivalence class conjugate to the equivalence class $\Omega(F)$, where F is a closed convex or concave bifunction, is $\Omega(F_*)$ according to Theorem 37.1.

The importance of Corollary 37.1.1 for minimax theory is that it reduces the possible discrepancies between "sup inf" and "inf sup" to the possible discrepancies between saddle-functions which are equivalent to each other. The fact that sometimes

$$\text{sup inf} \neq \text{inf sup}$$

is thus (by the results in §34) precisely dual to the fact that in general

$$cl_2\, cl_1 \neq cl_1\, cl_2,$$

and the peculiar non-uniqueness of closures and infinity-valued extensions of saddle-functions turns out to have a natural dual significance.

COROLLARY 37.1.2.   *The lower and upper conjugates $\underline{K}^*$ and $\bar{K}^*$ of a closed proper saddle-function K have the structural properties in Theorem 34.3 with respect to a certain non-empty convex product set $C^* \times D^*$ (the effective domain of both $\underline{K}^*$ and $\bar{K}^*$), and they satisfy the relations*

$$cl_1\, \underline{K}^* = \bar{K}^*, \qquad cl_2\, \bar{K}^* = \underline{K}^*.$$

In particular, one has

$$\underline{K}^*(u^*, v^*) = \bar{K}^*(u^*, v^*)$$

if either $u^* \in \text{ri } C^*$ *or* $v^* \in \text{ri } D^*$.

PROOF. By Theorem 34.2 and Theorem 34.3. ‖

Since by definition

$$\underline{K}^*(0, 0) = \sup_v \inf_u \{\langle u, 0 \rangle + \langle v, 0 \rangle - K(u, v)\},$$

$$\bar{K}^*(0, 0) = \inf_u \sup_v \{\langle u, 0 \rangle + \langle v, 0 \rangle - K(u, v)\}.$$

one has

$$\inf_v \sup_u K(u, v) = -\underline{K}^*(0, 0),$$

$$\sup_u \inf_v K(u, v) = -\bar{K}^*(0, 0).$$

The existence of the saddle-value of K depends therefore on the position of $(0, 0)$ relative to

$$C^* \times D^* = \text{dom } \underline{K}^* = \text{dom } \bar{K}^*.$$

In particular, we have:

COROLLARY 37.1.3. *Let K be a closed proper concave-convex function on $R^m \times R^n$, and let $C^* \times D^*$ be the common effective domain of the concave-convex functions conjugate to K. If either* ri C^* *contains the origin of R^m or* ri D^* *contains the origin of R^n, then*

$$\inf_v \sup_u K(u, v) = \sup_u \inf_v K(u, v).$$

If both conditions hold, this saddle-value must be finite.

In order to get the most use out of the minimax criterion in Corollary 37.1.3, we need a direct characterization of the sets C^* and D^* in terms of K. This is provided by the next theorem.

THEOREM 37.2. *Let K be a closed proper concave-convex function on $R^m \times R^n$ with effective domain $C \times D$. Let $C^* \times D^*$ be the common effective domain of the concave-convex functions K^* conjugate to K. The support functions of C^* and D^* are then given by the formulas*

$$\delta^*(w \mid D^*) = \sup_{u \in \text{ri } C} \sup_{v \in D} \{K(u, v + w) - K(u, v)\},$$

$$-\delta^*(-z \mid C^*) = \inf_{v \in \text{ri } D} \inf_{u \in C} \{K(u + z, v) - K(u, v)\}.$$

PROOF. Let F be the unique closed proper convex bifunction from R^m to R^n such that

$$(\text{cl}_2 K)(u, v) = \langle Fu, v \rangle, \quad \forall u, \forall v$$

(see Theorem 34.2). We have $C = \text{dom } F$. Since the equivalence class of saddle-functions conjugate to K corresponds to F_*, we have $D^* = \text{dom } F_*$. Let G be the effective domain of the graph function of F, i.e.

$$G = \{(u, x) \mid (Fu)(x) < +\infty\}.$$

We have

$$D^* = \{x \mid \exists u, (u, x) \in G\} = \bigcup \{\text{dom } Fu \mid u \in C\}.$$

In fact, by Theorem 6.8

$$\text{ri } D^* = \{x \mid \exists u, (u, x) \in \text{ri } G\} = \bigcup \{\text{ri } (\text{dom } Fu) \mid u \in \text{ri } C\}.$$

Therefore

$$\delta^*(w \mid D^*) = \sup \{\langle x, w \rangle \mid x \in \text{ri } D^*\}$$

$$= \sup \{\langle x, w \rangle \mid x \in \text{ri } (\text{dom } Fu), u \in \text{ri } C\}$$

$$= \sup \{\delta^*(w \mid \text{dom } Fu) \mid u \in \text{ri } C\}.$$

On the other hand, for each $u \in \text{ri } C$, $K(u, \cdot)$ is a closed proper convex function with effective domain D (Theorem 34.3), and hence it agrees with $(\text{cl}_2 K)(u, \cdot)$, which is the conjugate of the closed proper convex function Fu. The support function of dom Fu is the recession function of the conjugate of Fu, according to Theorem 13.3. Thus, for each $u \in \text{ri } C$, $\delta^*(\cdot \mid \text{dom } Fu)$ is the recession function of $K(u, \cdot)$, and we have

$$\delta^*(w \mid \text{dom } Fu) = \sup \{K(u, v + w) - K(u, v) \mid v \in D\}$$

by the first recession function formula in Theorem 8.5. This proves the formula for $\delta^*(\cdot \mid D^*)$. The proof of the formula for $\delta^*(\cdot \mid C^*)$ is similar. ‖

COROLLARY 37.2.1.   *In the notation of the theorem, one has* $0 \in \text{int } D^*$ *if and only if the convex functions* $K(u, \cdot)$ *for* $u \in \text{ri } C$ *have no common direction of recession. Similarly, one has* $0 \in \text{int } C^*$ *if and only if the convex functions* $-K(\cdot, v)$ *for* $v \in \text{ri } D$ *have no common direction of recession.*

PROOF.   One has $0 \notin \text{int } D^*$ if and only if there exists a vector $w \neq 0$ such that $\delta^*(w \mid D^*) \leq 0$, i.e. (according to the preceding proof)

$$K(u, v + w) - K(u, v) \leq 0, \qquad \forall v \in D, \qquad \forall u \in \text{ri } C.$$

Since the effective domain of $K(u, \cdot)$ is D for every $u \in \text{ri } C$, the latter conditions means that w belongs to the recession cone of $K(u, \cdot)$ for every $u \in \text{ri } C$. The proof of the other part of the corollary is analogous. ‖

The main theorem about the existence of saddle-values may now be stated.

THEOREM 37.3.   *Let K be a closed proper concave-convex function on $R^m \times R^n$ with effective domain $C \times D$. Then either of the following* *conditions implies that the saddle-value of K exists. If both conditions hold, the saddle-value must be finite.*

(a) *The convex functions $K(u, \cdot)$ for $u \in \mathrm{ri}\ C$ have no common direction of recession.*

(b) *The convex functions $-K(\cdot, v)$ for $v \in \mathrm{ri}\ D$ have no common direction of recession.*

PROOF. This simply combines Corollary 37.1.3 and Corollary 37.2.1. ‖

COROLLARY 37.3.1. *Let K be a closed proper concave-convex function on $R^m \times R^n$ with effective domain $C \times D$. If either C or D is bounded, the saddle-value of K exists.*

PROOF. The effective domain of $K(u, \cdot)$ is D for every $u \in \mathrm{ri}\ C$ by Theorem 34.3, so condition (a) is fulfilled when D is bounded. Similarly, condition (b) is fulfilled when C is bounded. ‖

The saddle-value of K with respect to $R^m \times R^n$ in Theorem 37.3 and Corollary 37.3.1 is of course the same as the saddle-value of K with respect to $C \times D$, as explained in §36. To emphasize this, we state as a special case:

COROLLARY 37.3.2. *Let C and D be non-empty closed convex sets in R^m and R^n, respectively, and let K be a continuous finite concave-convex function on $C \times D$. If either C or D is bounded, one has*

$$\inf_{v \in D} \sup_{u \in C} K(u, v) = \sup_{u \in C} \inf_{v \in D} K(u, v).$$

PROOF. Apply the preceding corollary to the lower (or upper) simple extension of K to all of $R^m \times R^n$, which is a closed proper concave-convex function with effective domain $C \times D$ by Corollary 34.2.4. ‖

We shall see below that, when both conditions hold in Theorem 37.3 a saddle-point actually exists. This result will be obtained from properties of the subdifferential mappings ∂K defined in §35, where $\partial K(u, v)$ is the closed convex set

$$\partial_1 K(u, v) \times \partial_2 K(u, v)$$

for each u and v, and

$$\mathrm{dom}\ \partial K = \{(u, v) \mid \partial K(u, v) \neq \emptyset\}.$$

THEOREM 37.4. *Let K be a concave-convex function on $R^m \times R^n$. For each (u, v), $\partial K(u, v)$ consists of the pairs (u^*, v^*) such that the concave-convex function $K - \langle \cdot, u^* \rangle - \langle \cdot, v^* \rangle$ has (u, v) as a saddle-point. If K is closed and proper, one has*

$$\mathrm{ri}\ (\mathrm{dom}\ K) \subset \mathrm{dom}\ \partial K \subset \mathrm{dom}\ K.$$

PROOF. The sets $\partial_1 K(u, v)$ and $\partial_2 K(u, v)$ are closed and convex in R^m and R^n respectively, so $\partial K(u, v)$ is closed and convex. By definition,

(u^*, v^*) belongs to $\partial K(u, v)$ if and only if

$$K(u', v) - \langle u', u^* \rangle \leq K(u, v) - \langle u, u^* \rangle, \forall u',$$

$$K(u, v') - \langle v', v^* \rangle \geq K(u, v) - \langle v, v^* \rangle, \forall v'.$$

Setting $K_0 = K - \langle \cdot, u^* \rangle - \langle \cdot, v^* \rangle$, we can express these inequalities as the condition that

$$K_0(u', v) \leq K_0(u, v) \leq K_0(u, v'), \forall u', \forall v',$$

which means that (u, v) is a saddle-point of K_0. Assume now that K is closed and proper. Then K_0 is closed and proper, and dom $K_0 = $ dom K. The saddle-points of K_0 all lie in dom K_0, according to Corollary 36.3.1. The condition $(u^*, v^*) \in \partial K(u, v)$ therefore implies that $(u, v) \in $ dom K. In other words, dom ∂K is included in dom K. On the other hand, suppose that

$$(u, v) \in \text{ri (dom } K) = \text{ri (dom}_1 K) \times \text{ri (dom}_2 K).$$

Then v is in the relative interior of the effective domain of the convex function $K(u, \cdot)$ (Theorem 34.3), and consequently $K(u, \cdot)$ has at least one subgradient at v (Theorem 23.4). Thus $\partial_2 K(u, v) \neq \emptyset$. Similarly $\partial_1 K(u, v) \neq \emptyset$, so that $\partial K(u, v) \neq \emptyset$. ‖

COROLLARY 37.4.1. *If K and L are equivalent saddle-functions on $R^m \times R^n$ then $\partial K = \partial L$. Moreover, the values of K and L agree on the set* dom $\partial K = $ dom ∂L.

PROOF. For any (u^*, v^*), the saddle-functions

$$K_0(u, v) = K(u, v) - \langle u, u^* \rangle - \langle v, v^* \rangle,$$

$$L_0(u, v) = L(u, v) - \langle u, u^* \rangle - \langle v, v^* \rangle,$$

are equivalent like K and L. According to the theorem, one has $(u^*, v^*) \in \partial K(u, v)$ if and only if (u, v) is a saddle-point of K_0, in which case the saddle-value of K_0 is of course $K_0(u, v)$. Since equivalent saddle-functions have the same saddle-value and saddle-points if any (Theorem 36.4), one has $(u^*, v^*) \in \partial K(u, v)$ if and only if $(u^*, v^*) \in \partial L(u, v)$, in which case $K(u, v) = L(u, v)$. ‖

The subdifferential of a saddle-function depends only on the equivalence class containing the saddle-function, by Corollary 37.4.1. Thus we may speak of the subdifferential of an equivalence class.

Of course, the equivalence classes of closed proper concave-convex functions on $R^m \times R^n$ correspond one-to-one with the closed proper convex bifunctions from R^m to R^n (Theorem 34.2), and the latter correspond one-to-one with the closed proper convex functions on R^{m+n}. The next theorem describes how subdifferentials behave under these correspondences. It asserts in particular that, for saddle-function equivalence classes which are conjugate to each other, the associated subdifferentials are the inverses of each other in the sense of multivalued mappings, just as in the case of the subdifferentials of purely convex functions in Corollary 23.5.1.

THEOREM 37.5. *Let K be a closed proper concave-convex function on $R^m \times R^n$, and let K^* be one of the equivalent concave-convex functions conjugate to K. Let F be the (unique) closed proper convex bifunction from R^m to R^n such that $(\text{cl}_2 K)(u, v) = \langle Fu, v \rangle$, and let f be the graph function of F on R^{m+n}, i.e.*

$$f(u, v^*) = \sup_v \{\langle v, v^* \rangle - K(u, v)\}.$$

Then the following conditions on (u, v) and (u^*, v^*) are equivalent:

(a) $(u^*, v^*) \in \partial K(u, v)$;

(b) $(u, v) \in \partial K^*(u^*, v^*)$;

(c) $(-u^*, v) \in \partial f(u, v^*)$;

(d) $(Fu)(v^*) - \langle v, v^* \rangle = (F^*v)(u^*) - \langle u, u^* \rangle$.

PROOF. We show first that (a) implies (d). By definition, $v^* \in \partial_2 K(u, v)$ if and only if the supremum of the function $\langle \cdot, v^* \rangle - K(u, \cdot)$ on R^n is attained at v. This supremum is $(Fu)(v^*)$, since the convex function Fu is conjugate to the closure of the convex function $K(u, \cdot)$. Thus $v^* \in \partial_2 K(u, v)$ if and only if

$$\langle v, v^* \rangle - K(u, v) = (Fu)(v^*).$$

By a dual argument, $u^* \in \partial_1 K(u, v)$ if and only if

$$\langle u, u^* \rangle - K(u, v) = (F^*v)(u^*).$$

Thus $(u^*, v^*) \in \partial K(u, v)$ if and only if

$$\langle v, v^* \rangle - (Fu)(v^*) = K(u, v) = \langle u, u^* \rangle - (F^*v)(u^*).$$

This condition implies (d), and the reverse implication is also true because of the general inequality

$$\langle v, v^* \rangle - (Fu)(v^*) \leq \langle Fu, v \rangle \leq K(u, v)$$

$$\leq \langle u, F^*v \rangle \leq \langle u, u^* \rangle - (F^*v)(u^*).$$

Therefore (a) is equivalent to (d). Since K^* corresponds to the inverse bifunctions F_*^* and F_* in the same way that K corresponds to F and F^* (i.e. as in Theorem 37.1), it follows that (b) is equivalent to the condition

$$(F_*^* u^*)(v) - \langle v, v^* \rangle = (F_* v^*)(u) - \langle u, u^* \rangle.$$

This is identical to (d), because

$$(F_*v^*)(u) = -(Fu)(v^*)$$

$$(F_*^*u^*)(v) = -(F^*v)(u^*).$$

We have

$$(F^*v)(u^*) = \inf_{u,v^*} \{(Fu)(v^*) - \langle v, v^* \rangle + (u, u^*)\}$$

$$= -\sup_{u,v^*} \{\langle u, -u^* \rangle + \langle v, v^* \rangle - f(u, v^*)\} = -f^*(-u^*, v)$$

by definition, so that (d) can also be expressed by

$$f(u, v^*) + f^*(-u^*, v) = \langle u, -u^* \rangle + \langle v, v^* \rangle.$$

This condition is equivalent to (c) by Theorem 23.5.   ‖

The equivalence of (a) and (b) in Theorem 37.5 means, according to Theorem 37.4, that the concave-convex function $K - \langle \cdot, u^* \rangle - \langle \cdot, v^* \rangle$ has a saddle-point at (u, v) if and only if the concave-convex function $K^* - \langle u, \cdot \rangle - \langle v, \cdot \rangle$ has a saddle-point at (u^*, v^*).

The equivalence of (a) and (c) shows that the multivalued mappings which are the subdifferentials of closed proper saddle-functions K can be obtained in a simple way by "partial inversion" of the subdifferentials of closed proper convex functions f. The results which have been established about the geometric nature of the mappings ∂f therefore yield results about the mappings ∂K.

COROLLARY 37.5.1.   *If K is a closed proper concave-convex function on $R^m \times R^n$, the graph of ∂K is closed, and it is homeomorphic to $R^m \times R^n$ under the mapping*

$$(u, v, u^*, v^*) \to (u - u^*, v + v^*).$$

PROOF.   This is immediate from Theorem 24.4 and Corollary 31.5.1.   ‖

COROLLARY 37.5.2.   *If K is a closed proper concave-convex function on $R^m \times R^n$, the mapping*

$$\rho: (u, v) \to \{(-u^*, v^*) \mid (u^*, v^*) \in \partial K(u, v)\}$$

is a maximal monotone mapping from $R^m \times R^n$ to $R^m \times R^n$. In particular, if K is everywhere finite and differentiable,

$$(u, v) \to (-\nabla_1 K(u, v), \nabla_2 K(u, v))$$

is a maximal monotone mapping.

PROOF.   By the theorem, one has

$$(u^*, v^*) \in \rho(u, v)$$

if and only if

$$(u^*, v) \in \partial f(u, v^*),$$

where f is a certain closed proper convex function. The maximal monotonicity of ρ follows from the maximal monotonicity of ∂f (Corollary 31.5.2). ‖

For the purpose of studying the existence of saddle-points, the following corollary is the most important fact embodied in Theorem 37.5.

COROLLARY 37.5.3. *Let K be a closed proper saddle-function on $R^m \times R^n$, and let K^* be one of the equivalent saddle-functions conjugate to K. Then $\partial K^*(0, 0)$ is the set of saddle-points of K. The saddle-points of K thus form a closed convex product set in $R^m \times R^n$, and a saddle-point exists if and only if*

$$(0, 0) \in \text{dom } \partial K^*.$$

In particular, K has a saddle-point if

$$(0, 0) \in \text{ri (dom } K^*).$$

PROOF. We have $(u, v) \in \partial K^*(0, 0)$ if and only if $(0, 0) \in \partial K(u, v)$, i.e. if and only if (u, v) is a saddle-point of K. Apply Theorem 37.4 to K^*. ‖

To get an existence theorem for saddle-points, we need only translate the condition $(0, 0) \in \text{ri (dom } K^*)$ in Corollary 37.5.3 into a convenient condition on K itself. This can easily be done using the formula in Theorem 37.2 for the support functions of $\text{dom}_1 K^*$ and $\text{dom}_2 K^*$ in terms of K. For the sake of simplicity, we shall only state the general saddle-point theorem which corresponds to the condition

$$(0, 0) \in \text{int (dom } K^*) = \text{int (dom}_1 K^*) \times \text{int (dom}_2 K^*).$$

THEOREM 37.6. *Let K be a closed proper concave-convex function on $R^m \times R^n$ with effective domain $C \times D$. If conditions* (a) *and* (b) *of Theorem 37.3 are both satisfied, K has a saddle-point (which necessarily lies in $C \times D$).*

PROOF. The hypothesis implies by Corollary 37.2.1 that $0 \in \text{int (dom}_1 K^*)$ and $0 \in \text{int (dom}_2 K^*)$. In this case K has a saddle-point by Corollary 37.5.3. ‖

COROLLARY 37.6.1. *Let K be a closed proper concave-convex function on $R^m \times R^n$ with effective domain $C \times D$. If C and D are bounded, K has a saddle-point and a finite saddle-value.*

PROOF. As for Corollary 37.3.1. ‖

COROLLARY 37.6.2. *Let C and D be non-empty closed bounded convex sets in R^m and R^n, respectively, and let K be a continuous finite concave-convex function on $C \times D$. Then K has a saddle-point with respect to*

$C \times D$, *i.e. there exists some $\bar{u} \in C$ and $\bar{v} \in D$ such that*

$$K(u, \bar{v}) \leq K(\bar{u}, \bar{v}) \leq K(\bar{u}, v), \qquad \forall u \in C, \qquad \forall v \in D.$$

PROOF.   As for Corollary 37.3.2.   ‖

More generally, given a finite saddle-function K on a non-empty *relatively open* convex product set $C_0 \times D_0$ in $R^m \times R^n$, one can always extend K to get a closed proper saddle-function on $R^m \times R^n$ such that

$$C_0 \times D_0 \subseteq \mathrm{dom}\, K \subseteq \mathrm{cl}\, (C_0 \times D_0)$$

(Corollary 34.5.1). If $C_0 \times D_0$ is bounded, the extended K has a saddle-point with respect to $C \times D = \mathrm{dom}\, K$ by Corollary 37.6.1 and Theorem 36.3.

Part VIII · Convex Algebra

The Algebra of Bifunctions

The adjoint and inverse operations for convex and concave bifunctions generalize the adjoint and inverse operations for linear transformations in the following sense, as has already been pointed out. Let A be a linear transformation from R^m to R^n, and let F be the convex indicator bifunction of A, i.e. the closed proper convex bifunction from R^m to R^n defined by

$$(Fu)(x) = \delta(x \mid Au) = \begin{cases} 0 & \text{if } x = Au, \\ +\infty & \text{if } x \neq Au. \end{cases}$$

The adjoint F^* of F is then the concave indicator bifunction of the adjoint linear transformation A^*,

$$(F^*x^*)(u^*) = -\delta(u^* \mid A^*x^*) = \begin{cases} 0 & \text{if } u^* = A^*x^*, \\ -\infty & \text{if } u^* \neq A^*x^*, \end{cases}$$

and we have

$$\langle Fu, x^* \rangle = \langle Au, x^* \rangle = \langle u, A^*x^* \rangle = \langle u, F^*x^* \rangle.$$

If A is nonsingular, the inverse F_* of F is the concave indicator bifunction of A^{-1}, and F_*^* is the convex indicator bifunction of $(A^*)^{-1} = (A^{-1})^*$.

Our purpose here will be to show how other familiar operations of linear algebra, such as addition and multiplication of linear transformations, can be generalized in a natural way to bifunctions, and to explain the behavior of these generalized operations with respect to taking adjoints.

Given any proper convex bifunctions F_1 and F_2 from R^m to R^n, we define the bifunction $F = F_1 \square F_2$ from R^m to R^n by infimal convolution of F_1u and F_2u for each u, i.e.

$$(Fu)(x) = (F_1u \square F_2u)(x) = \inf_y \{(F_1u)(x - y) + (F_2u)(y)\}.$$

This operation generalizes the addition of linear transformations, in the sense that $F_1 \square F_2$ is the convex indicator bifunction of $A_1 + A_2$ when F_1 and F_2 are the convex indicator bifunctions of linear transformations

A_1 and A_2, respectively. For the concave bifunctions, the operation $\square$ is defined in the same way, except that the infimum is replaced by a supremum.

THEOREM 38.1. *Let F_1 and F_2 be proper convex bifunctions from R^m to R^n. Then $F_1 \square F_2$ is a convex bifunction from R^m to R^n, and*

$$\text{dom}\,(F_1 \square F_2) = \text{dom}\,F_1 \cap \text{dom}\,F_2.$$

Furthermore, one has

$$\langle (F_1 \square F_2)u, x^* \rangle = \langle F_1 u, x^* \rangle + \langle F_2 u, x^* \rangle, \forall u, \forall x^*,$$

if one sets $\infty - \infty = -\infty + \infty = -\infty$. (Similarly for concave bifunctions, but with $\infty - \infty = -\infty + \infty = +\infty$.)

PROOF. The graph function of $F_1 \square F_2$ is obtained by partial infimal convolution of the graph functions of F_1 and F_2, which are proper convex functions, and hence it is a convex function. Thus $F_1 \square F_2$ is a convex bifunction. If u belongs to dom $F_1 \cap$ dom F_2, both $F_1 u$ and $F_2 u$ are proper convex functions on R^n. Then $(F_1 \square F_2)u$ is not identically $+\infty$, so that $u \in \text{dom}\,(F_1 \square F_2)$, and by Theorem 16.4

$$((F_1 \square F_2)u)^* = (F_1 u \square F_2 u)^* = (F_1 u)^* + (F_2 u)^*.$$

In inner product notation, this relation is

$$\langle (F_1 \square F_2)u, \cdot \rangle = \langle F_1 u, \cdot \rangle + \langle F_2 u, \cdot \rangle.$$

If u does not belong to dom $F_1 \cap$ dom F_2, one of the functions $F_1 u$ and $F_2 u$ is identically $+\infty$. Then $(F_1 \square F_2)u$ is identically $+\infty$, implying $u \notin \text{dom}\,(F_1 \square F_2)$. The function $\langle (F_1 \square F_2)u, \cdot \rangle$ is then identically $-\infty$, as is one of the functions $\langle F_1 u, \cdot \rangle$ or $\langle F_2 u, \cdot \rangle$, so that the inner product equation in the theorem holds if one takes $\infty - \infty$ to be $-\infty$. ∥

The operation $\square$ is a commutative, associative operation in the class of convex bifunctions from R^m to R^n to the extent that it is defined, inasmuch as infimal convolution of convex functions has these properties. One can extend $\square$ to improper bifunctions using the "geometric" definition of infimal convolution of improper convex functions given in §5. The class of all convex bifunctions from R^m to R^n is then a commutative semigroup under $\square$, with the indicator of the zero linear transformation as the identity element.

The next theorem generalizes the familiar formula

$$(A_1 + A_2)^* = A_1^* + A_2^*$$

for linear transformations to the case of bifunctions.

THEOREM 38.2. *Let F_1 and F_2 be proper convex bifunctions from R^m to R^n. If* ri (dom F_1) *and* ri (dom F_2) *have a point in common, one has*

$$(F_1 \square F_2)^* = F_1^* \square F_2^*.$$

PROOF. For any x^*, $(F_1 \square F_2)^* x^*$ is the conjugate of the concave function $\langle \cdot, (F_1 \square F_2)^* x^* \rangle$, since $(F_1 \square F_2)^*$, being the adjoint of the convex bifunction $F_1 \square F_2$, is closed (Theorem 30.1) and hence in particular image-closed. On the other hand, $\langle \cdot, (F_1 \square F_2)^* x^* \rangle$ is the closure of the concave function

$$u \to \langle (F_1 \square F_2)u, x^* \rangle$$

(Theorem 33.2). It follows by the formula in Theorem 38.1 that $(F_1 \square F_2)^* x^*$ is the conjugate of g, where

$$g(u) = \langle F_1 u, x^* \rangle + \langle F_2 u, x^* \rangle$$

with $\infty - \infty = -\infty$. Now (by Theorem 33.1 and the fact that the conjugate of $F_i u$ has the value $-\infty$ at x^* if and only if $F_i u$ is identically $+\infty$, i.e. $u \notin$ dom F_i) the concave functions

$$g_1(u) = \langle F_1 u, x^* \rangle, \qquad g_2(u) = \langle F_2 u, x^* \rangle$$

have dom F_1 and dom F_2 as their effective domains, respectively, and these sets have a relative interior point in common by hypothesis. If

$$x^* \in \text{dom} \, (F_1^* \square F_2^*) = \text{dom} \, F_1^* \cap \text{dom} \, F_2^*,$$

g_1 and g_2 nowhere assume the value $+\infty$, and we have

$$g^* = (g_1 + g_2)^* = g_1^* \square g_2^*$$

by (the concave version of) Theorem 16.4. Since $g_1^* = F_1^* x^*$ and $g_2^* = F_2^* x^*$, this relation says that

$$(F_1 \square F_2)^* x^* = (F_1^* \square F_2^*) x^*.$$

If x^* is not in dom $(F_1^* \square F_2^*)$, then $(F_1^* \square F_2^*) x^*$ and one of the concave functions $F_1^* x^*$ or $F_2^* x^*$ must be identically $-\infty$. One of the functions g_1 or g_2 must then take on $+\infty$ somewhere. A concave function which takes on $+\infty$ must actually have $+\infty$ throughout the relative interior of its effective domain (Theorem 7.2), so in this case g likewise takes on $+\infty$ somewhere. Then $(F_1 \square F_2)^* x^* = g^*$ is identically $-\infty$ and coincides again with $(F_1^* \square F_2^*) x^*$. ‖

COROLLARY 38.2.1. *Let F_1 and F_2 be closed proper convex bifunctions from R^m to R^n. If* ri (dom F_1^*) *and* ri (dom F_2^*) *have a point in common, then $F_1 \square F_2$ is closed and*

$$(F_1 \square F_2)^* = \text{cl} \, (F_1^* \square F_2^*).$$

PROOF. The relative interior condition implies that

$$(F_1^* \square F_2^*)^* = F_1^{**} \square F_2^{**}.$$

But $F_1^{**} = F_1$ and $F_2^{**} = F_2$, since F_1 and F_2 are closed. The adjoint of a convex or concave bifunction is always closed, so it follows that $F_1 \square F_2$ is closed and

$$(F_1 \square F_2)^* = (F_1^* \square F_2^*)^{**} = \text{cl} \, (F_1^* \square F_2^*)$$

(Theorem 30.1). ‖

In general, of course, $F_1 \square F_2$ need not be closed. One does, however, have

$$(\text{cl} \, (F_1 \square F_2))u = \text{cl} \, (F_1 u \square F_2 u)$$

by Theorem 29.4 for each u in the relative interior of dom $(F_1 \square F_2)$, and hence in particular for each

$$u \in \text{ri} \, (\text{dom} \, F_1) \cap \text{ri} \, (\text{dom} \, F_2).$$

Scalar multiples $F\lambda$ are defined for $\lambda > 0$ by the formula $(F\lambda)u = (Fu)\lambda$, i.e.

$$((F\lambda)u)(x) = \lambda(Fu)(\lambda^{-1}x).$$

This corresponds to scalar multiplication of linear transformations: if F is the convex indicator bifunction of the linear transformation A, then $F\lambda$ is the convex indicator bifunction of λA.

THEOREM 38.3. *Let F be a convex bifunction from R^m to R^n, and let $\lambda > 0$. Then $F\lambda$ is a convex bifunction, closed or proper according as F itself is closed or proper, and one has*

$$\langle (F\lambda)u, x^* \rangle = \lambda \langle Fu, x^* \rangle, \, \forall u, \forall x^*.$$

Moreover $(F\lambda)^* = F^*\lambda$.

PROOF. Let f be the graph function of F, i.e. $f(u, x) = (Fu)(x)$. The epigraph of the graph function of $F\lambda$ is the image of epi f under the one-to-one linear transformation

$$(u, x, \mu) \rightarrow (u, \lambda x, \lambda \mu)$$

from R^{m+n+1} onto itself, so the graph function of $F\lambda$ is convex, and so forth. The inner product formula follows from the fact that

$$((Fu)\lambda)^* = \lambda(Fu)^*, \, \forall u$$

(Theorem 16.1). The bifunctions $(F\lambda)^*$ and $F^*\lambda$ are both closed (Theorem 30.1), and by applying the inner product formula just established we have

$$\langle u, (F\lambda)^*x^* \rangle = \text{cl}_u \, \langle (F\lambda)u, x^* \rangle = \lambda \, \text{cl}_u \, \langle Fu, x^* \rangle$$
$$= \lambda \langle u, F^*x^* \rangle = \langle u, (F^*\lambda)x^* \rangle$$

(Theorem 33.2). Therefore $(F\lambda)^* = F^*\lambda$ (Theorem 33.3). ‖

Let F be a proper convex bifunction from R^m to R^n. Given any convex function f on R^m which does not take on $-\infty$, we define the function Ff, the *image of f under F*, by the formula

$$(Ff)(x) = \inf_u \{f(u) + (Fu)(x)\} = \inf (f - F_*x).$$

(Analogously when f and F are concave instead of convex.) If F is the indicator of a linear transformation A, then $Ff = Af$.

THEOREM 38.4. *Let F be a proper convex bifunction from R^m to R^n, and let f be a proper convex function on R^m. Then Ff is a convex function on R^n. If the sets* ri $(\mathrm{dom} f)$ *and* ri $(\mathrm{dom} F)$ *have a point in common, one has*

$$(Ff)^* = F_*^* f^*$$

and the infimum in the definition of $(F_*^* f^*)(x^*)$ is attained for each x^*.

PROOF. Let $h(u, x) = f(u) + (Fu)(x)$. Then h is a convex function on R^{m+n}, and Ff is the image of h under the projection $(u, x) \to x$. Hence Ff is convex (Theorem 5.7). For any $x^* \in R^n$, we have

$$(Ff)^*(x^*) = \sup_x \{\langle x, x^* \rangle - \inf_u \{f(u) + (Fu)(x)\}\}$$

$$= \sup_{u,x} \{\langle x, x^* \rangle - (Fu)(x) - f(u)\}.$$

The concave function

$$g(u) = \langle Fu, x^* \rangle = \sup_x \{\langle x, x^* \rangle - (Fu)(x)\}$$

has dom F as its effective domain and $F^* x^*$ as its conjugate. Assume that ri $(\mathrm{dom} f)$ meets ri $(\mathrm{dom} g)$ = ri $(\mathrm{dom} F)$. If $x^* \in \mathrm{dom} F^*$, g is proper and by Fenchel's Duality Theorem (or more exactly by the result obtained when both sides of the equation in Theorem 31.1 are multiplied by -1) we have

$$(Ff)^*(x^*) = \sup_u \{g(u) - f(u)\} = \inf_{u^*} \{f^*(u^*) - g^*(u^*)\}$$

$$= \inf_{u^*} \{f^*(u^*) + (F_*^* u^*)(x^*)\} = (F_*^* f^*)(x^*),$$

where the infimum is attained. On the other hand, if $x^* \notin \mathrm{dom} F^*$ the concave function g is improper and hence identically $+\infty$ on the relative interior of its effective domain. Then $(Ff)^*(x^*)$ must be $+\infty$. At the same time, $F^* x^*$ is the constant function $-\infty$, so that the infimum defining $(F_*^* f^*)(x^*)$ is $+\infty$ and trivially attained. ‖

COROLLARY 38.4.1. *Let F be a closed proper convex bifunction from R^m to R^n, and let f be a closed proper convex function on R^m. If* ri $(\mathrm{dom} f^*)$ *meets* ri $(\mathrm{dom} F_*^*)$, *then Ff is closed, and the infimum in the definition of $(Ff)(x)$ is attained for each x. Moreover, then $(Ff)^* = \mathrm{cl}\, (F_*^* f^*)$.*

PROOF. We have $(F_*^*)_*^* = F$ and $f^{**} = f$. Apply the theorem to F_*^* and f^*. ∥

The operation of taking the image of a function under a bifunction suggests the natural way to define the *product* of two bifunctions. Let F be a proper convex bifunction from R^m to R^n, and let G be a proper convex bifunction from R^n to R^p. We define the bifunction GF from R^m to R^p by

$$(GF)u = G(Fu),$$

or in other words

$$((GF)u)(y) = \inf_x \{(Fu)(x) + (Gx)(y)\} = \inf \{Fu - G_*y\}.$$

When F and G are concave, one takes the supremum instead of the infimum. Obviously

$$(GF)_* = F_*G_*.$$

Observe that, when F and G are the convex indicator bifunctions of linear transformations A and B respectively, GF is the indicator of BA.

THEOREM 38.5. *Let F be a proper convex bifunction from R^m to R^n, and let G be a proper convex bifunction from R^n to R^p. Then GF is a convex bifunction from R^m to R^p. If* ri (dom F_*) *and* ri (dom G) *have a point in common, one has*

$$(GF)^* = F^*G^*,$$

and the supremum in the definition of $((F^*G^*)y^*)(u^*)$ is attained for each $u^* \in R^m$ and $y^* \in R^p$.

PROOF. Let

$$h(u, x, y) = (Fu)(x) + (Gx)(y).$$

Then h is a convex function on R^{m+n+p}. The graph function of GF is the image of h under the linear transformation $(u, x, y) \to (u, y)$, and hence it is convex (Theorem 5.7). For any $u^* \in R^m$ and $y^* \in R^p$, we have

$$((GF)^*y^*)(u^*) = \inf_{u,y} \inf_x \{(Fu)(x) + (Gx)(y) - \langle y, y^* \rangle + \langle u, u^* \rangle\}$$

$$= \inf_{x,u,y} \{\langle u, u^* \rangle - (F_*x)(u) - \langle y, y^* \rangle + (Gx)(y)\}.$$

The concave function

$$g(x) = \langle Gx, y^* \rangle = \sup_y \{\langle y, y^* \rangle - (Gx)(y)\}$$

has dom G as its effective domain and G^*y^* as its conjugate. The convex function

$$f(x) = \langle u^*, F_*x \rangle = \inf_u \{\langle u, u^* \rangle - (F_*x)(u)\}$$

has dom F_* as its effective domain and $F_*^*u^*$ as its conjugate. Assume that dom G and dom F_* have a relative interior point in common. If $y^* \in$ dom G^* and $u^* \in$ dom F_*^*, g and f are proper and by Fenchel's Duality Theorem we have

$$((GF)^*y^*)(u^*) = \inf_x \{f(x) - g(x)\} = \sup_{x^*} \{g^*(x^*) - f^*(x^*)\}$$

$$= \sup_{x^*} \{(G^*y^*)(x^*) + (F^*x^*)(u^*)\} = ((F^*G^*)y^*)(u^*),$$

where the supremum is attained.

If $y^* \notin$ dom G^*, g is improper and hence identically $+\infty$ on the relative interior of its effective domain. Then, for any x in ri (dom F_*) $\cap$ ri (dom G) and any u such that $(F_*x)(u)$ is finite, we have

$$\inf_y \{\langle u, u^* \rangle - (F_*x)(u) - \langle y, y^* \rangle + (Gx)(y)\} = -\infty$$

and hence $((GF)^*y^*)(u^*) = -\infty$. At the same time, $y^* \notin$ dom G^* implies that G^*y^* is the constant function $-\infty$, so that the supremum defining $((F^*G^*)y^*)(u^*)$ is $-\infty$ and trivially attained. Thus

$$((GF)^*y^*)(u^*) = ((F^*G^*)y^*)(u^*)$$

whenever $y^* \notin$ dom G^*. A similar argument covers the case where $u^* \notin$ dom F_*^*. ‖

COROLLARY 38.5.1. *Let F be a closed proper convex bifunction from R^m to R^n, and let G be a closed proper convex bifunction from R^n to R^p. If ri (dom F^*) and ri (dom G_*^*) have a point in common, then GF is closed and the infimum in the definition of $((GF)u)(y)$ is always attained. Moreover, then $(GF)^* = $ cl (F^*G^*).*

PROOF. Apply the theorem to F^* and G^*. Since F and G are closed, we have $F^{**} = F$, $G^{**} = G$, and hence $(F^*G^*)^* = GF$. As the adjoint of something, GF is closed. ‖

The convex set dom F_* in Theorem 38.5 is, of course, the image of the effective domain of the graph function $f(u, x) = (Fu)(x)$ under the projection $(u, x) \to x$ (whereas dom F is the image under $(u, x) \to u$). Thus dom F_* is the union of all the sets dom Fu. Moreover ri (dom F_*) is the image of ri (dom f) under $(u, x) \to x$ (Theorem 6.6), so

$$\text{ri (dom } F_*) = \bigcup \{\text{ri (dom } Fu) \,|\, u \in \text{ri (dom } F)\}$$

by Theorem 6.8. The condition in Theorem 38.5 that ri (dom F_*) and ri (dom G) have a point in common can therefore be stated equivalently as the condition that there exist some $u \in$ ri (dom F) such that ri (dom Fu) meets ri (dom G). It is not difficult to show that, when such vectors u exist, they form ri (dom (GF)). Of course, in general, dom GF itself consists of the vectors $u \in$ dom F such that dom Fu meets dom G.

Multiplication of convex bifunctions is plainly associative to the extent that it is defined, i.e. one has

$$H(GF) = (HG)F$$

when F, G, H, GF and HG are proper. The associative law is valid even for improper convex bifunctions, if one extends the definition of GF to the improper case simply by invoking the rule $\infty - \infty = +\infty$ to interpret $(Fu)(x) + (Gx)(y)$ where necessary. Under the extended definition of multiplication, the class of all convex bifunctions from R^n to itself is a (non-commutative) semigroup having as identity element the indicator of the identity linear transformation. Products $F^2 = FF$, $F^3 = FFF, \ldots$, may be studied, etc.

A more general notion of inner product is helpful in describing the properties of expressions like $\langle (GF)u, y^* \rangle$. Let f be a proper convex function on R^n, and let g be a proper concave function on R^n. Let $C = \operatorname{dom} f$ and $D = \operatorname{dom} g$. If the quantity

$$\sup_{x \in C} \inf_{y \in D} \{\langle x, y \rangle - f(x) - g(y)\} = \sup_{x} \{g^*(x) - f(x)\}$$

and the quantity

$$\inf_{y \in D} \sup_{x \in C} \{\langle x, y \rangle - f(x) - g(y)\} = \inf_{y} \{f^*(y) - g(y)\}$$

are equal, we call the common extremum the *inner product* of f and g and denote it by $\langle f, g \rangle$. (If the quantities are not equal, $\langle f, g \rangle$ is undefined.) According to Fenchel's Duality Theorem, $\langle f, g \rangle$ exists in particular when g is closed and ri (dom f) meets ri (dom g^*), or when f is closed and ri (dom g) meets ri (dom f^*). For instance, a simple condition which ensures the existence of $\langle f, g \rangle$ is that f and g be closed and either C or D be bounded. (If C is bounded, dom f^* is all of R^n by Corollary 13.3.1. Similarly, if D is bounded, dom g^* is all of R^n.)

When f and g are the indicators of points a and b, i.e. $f(x) = \delta(x \mid a)$ and $g(y) = -\delta(y \mid b)$, one has $\langle f, g \rangle$ equal to the ordinary inner product $\langle a, b \rangle$.

The new definition of inner product agrees with the notation $\langle f, x^* \rangle$ for $f^*(x^*)$ introduced in §33, in the sense that $\langle f, g \rangle = \langle f, x^* \rangle$ when $g(y) = -\delta(y \mid x^*)$.

The two extrema in the definition of $\langle f, g \rangle$ can equally well be expressed as

$$\sup \{g^*(x) - f(x) \mid x \in (\operatorname{dom} g^* \cap \operatorname{dom} f)\},$$

$$\inf \{f^*(y) - g(y) \mid y \in (\operatorname{dom} f^* \cap \operatorname{dom} g)\}.$$

These expressions are unambiguous even when f or g is improper, and they therefore allow us to extend the definition of $\langle f, g \rangle$ to the case of improper convex and concave functions. It can be verified without difficulty that the relative interior conditions cited above also suffice for the existence of $\langle f, g \rangle$ in the improper case.

LEMMA 38.6. *Let f be a convex function on R^n, and let g be a concave function on R^n. If $\langle f, g \rangle$ exists, then $\langle f^*, g^* \rangle$ exists and*

$$\langle f^*, g^* \rangle = -\langle f, g \rangle.$$

Moreover, then $\langle \mathrm{cl}\, f, \mathrm{cl}\, g \rangle$ exists and coincides with $\langle f, g \rangle$.

PROOF. Given any vector x such that $f^{**}(x) < +\infty$ and $g^*(x) > -\infty$, and given any vector y such that $g^{**}(y) > -\infty$ and $f^*(y) < +\infty$, we have

$$+\infty > f^{**}(x) + f^*(y) \geq \langle x, y \rangle \geq g^*(x) + g^{**}(y) > -\infty$$

by Fenchel's Inequality, and consequently

$$f^{**}(x) - g^*(x) \geq g^{**}(y) - f^*(y).$$

Since $f^{**} \leq f$ and $g^{**} \geq g$, it follows that

$$\inf \{f(x) - g^*(x) \mid x \in (\mathrm{dom}\, f \cap \mathrm{dom}\, g^*)\}$$

$$\geq \inf \{f^{**}(x) - g^*(x) \mid x \in (\mathrm{dom}\, f^{**} \cap \mathrm{dom}\, g^*)\}$$

$$\geq \sup \{g^{**}(y) - f^*(y) \mid y \in (\mathrm{dom}\, g^{**} \cap \mathrm{dom}\, f^*)\}$$

$$\geq \sup \{g(y) - f^*(y) \mid y \in (\mathrm{dom}\, g \cap \mathrm{dom}\, f^*)\}.$$

The two middle extrema give $\langle f^*, g^* \rangle$ when they are equal. If $\langle f, g \rangle$ exists, the first and last extrema equal $-\langle f, g \rangle$, so that all four extrema coincide. The existence of $\langle f^*, g^* \rangle$ implies in turn that the inner product $\langle f^{**}, g^{**} \rangle = \langle \mathrm{cl}\, f, \mathrm{cl}\, g \rangle$ exists and equals $-\langle f^*, g^* \rangle$. Thus $\langle \mathrm{cl}\, f, \mathrm{cl}\, g \rangle = \langle f, g \rangle$. ‖

THEOREM 38.7. *Let F be a proper convex bifunction from R^m to R^n. Let f be a proper convex function on R^m, and let g be a proper concave function on R^n. Assume there exists at least one u in ri (dom f) $\cap$ ri (dom F) such that ri (dom Fu) meets ri (dom g). Then the following equation holds (where in particular all four inner products exist):*

$$\langle Ff, g^* \rangle = \langle f, F^*g^* \rangle = -\langle f^*, F_*g \rangle = -\langle F_*^*f^*, g \rangle.$$

PROOF. Let C and D be the convex sets defined by

$$C = \{(u, x) \mid u \in R^m, x \in R^n, (Fu)(x) < +\infty\},$$

$$D = \{(u, x) \mid u \in R^m, x \in R^n, f(u) < +\infty\}.$$

The image of $C \cap D$ under the linear transformation $(u, x) \to x$ is dom Ff, and hence the image of ri $(C \cap D)$ is ri (dom Ff) (Theorem 6.6). We have

$$\text{ri } C = \{(u, x) \mid u \in \text{ri (dom } F), \, x \in \text{ri (dom } Fu)\}$$

according to Theorem 6.8. Since ri (dom f) $\cap$ ri (dom F) is non-empty by hypothesis, ri $C \cap$ ri D is non-empty, so that

$$\text{ri } (C \cap D) = \text{ri } C \cap \text{ri } D$$

(Theorem 6.5). Thus ri (dom Ff) is the image of ri $C \cap$ ri D under $(u, x) \to x$, and we have

$$\text{ri (dom } Ff) = \bigcup \{\text{ri (dom } Fu) \mid u \in \text{ri (dom } f) \cap \text{ri (dom } F)\}.$$

The latter set meets ri (dom g) by hypothesis. Hence ri (dom Ff) meets ri (dom g^{**}) and $\langle Ff, g^* \rangle$ exists. Similarly, ri (dom $(Ff)^{**}$) meets ri (dom g) and $\langle (Ff)^*, g \rangle$ exists. By Theorem 38.4, $(Ff)^* = F_*^* f^*$. Therefore $\langle F_*^* f^*, g \rangle$ exists, and from Lemma 38.6 we have

$$\langle F_*^* f^*, g \rangle = \langle (Ff)^*, g \rangle = -\langle \text{cl } (Ff), g^* \rangle = -\langle Ff, g^* \rangle.$$

A dual argument can now be applied to the inverse bifunction F_*. The formula

$$\text{ri } C = \{(u, x) \mid x \in \text{ri (dom } F_*), \, u \in \text{ri (dom } F_* x)\}$$

holds by Theorem 6.8. Our hypothesis about relative interiors can therefore be expressed equivalently as follows: there exists at least one x in ri (dom g) $\cap$ ri (dom F_*) such that ri (dom $F_* x$) meets ri (dom f). Of course, $(F_*)_*^* = F^*$. Thus by the reasoning above $\langle f^*, F_* g \rangle$ exists, $\langle f, F^* g^* \rangle$ exists and

$$\langle f, F^* g^* \rangle = -\langle f^*, F_* g \rangle.$$

By definition,

$$\langle f, F^* g^* \rangle = \inf \{f^*(u^*) - (F^* g^*)(u^*) \mid u^* \in (\text{dom } f^* \cap \text{dom } F^* g^*)\}$$

$$= \inf_{u^*, x^*} \{f^*(u^*) - g^*(x^*) - (F^* x^*)(u^*)\}.$$

On the other hand,

$$\langle F_*^* f^*, g \rangle = \sup \{g^*(x^*) - (F_*^* f^*)(x^*) \mid x^* \in (\text{dom } g^* \cap \text{dom } F_*^* f^*)\}$$

$$= \sup_{x^*, u^*} \{g^*(x^*) - f^*(u^*) - (F_*^* u^*)(x^*)\}.$$

Therefore

$$\langle F_*^* f^*, g \rangle = -\langle f, F^* g^* \rangle$$

and the proof is complete. $\|$

CoROLLARY 38.7.1.   *Let F be a proper convex bifunction from R^m to R^n. Let f be a proper convex function on R^m such that ri (dom f) meets* ri (dom F). *Then* $\langle f, F^*x^* \rangle$ *exists for every* $x^* \in R^n$, *and*

$$\langle Ff, x^* \rangle = \langle f, F^*x^* \rangle.$$

PROOF. Given any x^*, apply the theorem with $g = \langle \cdot, x^* \rangle$. ‖

Corollary 38.7.2. *Let* F *be a proper convex bifunction from* R^m *to* R^n, *and let* G *be a proper convex bifunction from* R^n *to* R^p. *Assume that* ri (dom F_*) *and* ri (dom G) *have a point in common. Then, for each* $u \in$ ri (dom GF), $\langle Fu, G^*y^* \rangle$ *exists for every* $y^* \in R^p$ *and*

$$\langle GFu, y^* \rangle = \langle Fu, G^*y^* \rangle = \langle u, F^*G^*y^* \rangle.$$

PROOF. The first and last of these inner products are equal simply by Corollary 33.2.1, since $F^*G^* = (GF)^*$ by Theorem 38.5. On the other hand, the first equality is valid by the preceding corollary if ri (dom Fu) meets ri (dom G). It suffices therefore to show that

$$\text{ri (dom } GF) = \{u \in \text{ri (dom } F) \mid \text{ri (dom } Fu) \cap \text{ri (dom } G) \neq \emptyset\}$$

under the hypothesis that

$$\text{ri (dom } F_*) \cap \text{ri (dom } G) \neq \emptyset.$$

We leave this to the reader as a pithy exercise in the calculus of relative interiors (cf. the remarks following Corollary 38.5.1). ‖

The results above take on an especially nice form when the bifunctions are co-finite. A convex (or concave) bifunction F from R^m to R^n is said to be *co-finite* if, for every $u \in R^m$, the convex (or concave) function Fu is co-finite (i.e. closed, proper and without any non-vertical half-lines in its epigraph). This condition implies that dom $F = R^m$, and that F is closed and proper (Theorem 29.4).

Since the co-finite convex (or concave) functions are precisely the conjugates of the finite convex (or concave) functions (Corollary 13.3.1), F (closed) is co-finite if and only if $\langle Fu, x^* \rangle$ is finite for all u and x^*. The co-finite convex (or concave) bifunctions F from R^m to R^n are thus in one-to-one correspondence with the finite saddle-functions on $R^m \times R^n$ (Corollary 33.1.2). It follows that the adjoint F^* of a co-finite F is co-finite and satisfies

$$\langle Fu, x^* \rangle = \langle u, F^*x^* \rangle, \ \forall u, \forall x^*$$

(Corollary 33.2.1). It follows further that a closed convex bifunction F from R^m to R^n is co-finite if and only if dom $F = R^m$ and dom $F^* = R^n$ (Theorem 34.2).

The indicator bifunctions of linear transformations from R^m to R^n are, of course, examples of co-finite bifunctions. They correspond to the saddle-functions which are *bilinear* functions on $R^m \times R^n$.

If F_1 and F_2 are co-finite convex bifunctions from R^m to R^n, then $F_1 \square F_2$ is co-finite and

$$(F_1 \square F_2)^* = F_1^* \square F_2^*.$$

This is immediate from Corollary 38.2.1 and the inner product formula in Theorem 38.1. The operation $F \to F\lambda$, $\lambda > 0$, likewise preserves co-finiteness.

If F is a co-finite convex bifunction from R^m to R^n and G is a co-finite convex bifunction from R^n to R^p, then GF is a co-finite convex bifunction from R^m to R^p and

$$(GF)^* = F^*G^*$$

(Theorem 38.5 and Corollary 38.5.1). The co-finite convex bifunctions from R^n to itself thus form a non-commutative semigroup under multiplication.

The inner product equation

$$\langle Ff, g^* \rangle = \langle f, F^*g^* \rangle$$

is always valid when F, f and g^* are co-finite (Theorem 38.7).

Most of the results in this section can be sharpened in the case of *polyhedral* bifunctions. However, we shall leave this to the reader.

Convex Processes

The notion of a convex process is intermediate between that of a linear transformation and that of a convex bifunction. Convex processes form an algebra of multivalued mappings with many interesting duality properties. These properties can be deduced from theorems already established for bifunctions, which they help to illuminate.

A *convex process* from R^m to R^n is a multivalued mapping $A: u \to Au$ such that

(a) $A(u_1 + u_2) \supset Au_1 + Au_2, \forall u_1, \forall u_2,$

(b) $A(\lambda u) = \lambda Au, \forall u, \forall \lambda > 0,$

(c) $0 \in A0.$

Condition (c) means that the set

$$\text{graph } A = \{(u, x) \mid u \in R^m, x \in R^n, x \in Au\}$$

contains the origin of R^{m+n}. Condition (a) is equivalent to the condition that

$$(u_1, x_1) + (u_2, x_2) \in \text{graph } A,$$

$$\forall (u_1, x_1) \in \text{graph } A, \qquad \forall (u_2, x_2) \in \text{graph } A,$$

while (b) is equivalent to

$$\lambda(u, x) \in \text{graph } A, \qquad \forall (u, x) \in \text{graph } A, \qquad \forall \lambda > 0.$$

Thus a multivalued mapping A from R^m to R^n is a convex process if and only if its graph is a non-empty subset of R^{m+n} closed under addition and non-negative scalar multiplication, i.e. a convex cone in R^{m+n} containing the origin.

Various elementary properties of convex processes are immediate from the definition. If A is a convex process from R^m to R^n, Au is a convex set in R^n. The set $A0$ is a convex cone containing the origin, and it consists precisely of the vectors y such that

$$Au + y \subset Au, \forall u.$$

The *domain* of A, which is defined of course by

$$\text{dom } A = \{u \mid Au \neq \emptyset\},$$

is a convex cone in R^m containing the origin, and the *range* of A,

$$\text{range } A = \bigcup \{Au \mid u \in R^m\},$$

is a convex cone in R^n containing the origin.
  The *inverse* A^{-1} of A, where

$$A^{-1}x = \{u \mid x \in Au\}, \forall x,$$

is a convex process from R^n to R^m such that

$$\text{dom } A^{-1} = \text{range } A$$
$$\text{range } A^{-1} = \text{dom } A.$$

The convex cone $A^{-1}0$ consists of the vectors v with the property that

$$A(u + v) \supset Au, \forall u.$$

If A is a single-valued mapping with dom $A = R^m$, condition (a) in the definition of "convex process" reduces simply to

$$A(u_1 + u_2) = Au_1 + Au_2.$$

Linear transformations are therefore special cases of convex processes. They are the only convex processes A such that Au is a non-empty bounded set for every u, as the following theorem shows.

THEOREM 39.1. *If A is a convex process from R^m to R^n such that* dom $A = R^m$ *and $A0$ is bounded, then A is a linear transformation.*

PROOF. Since $A0$ is a convex cone containing the origin, boundedness implies that $A0$ consists of the origin alone. The relation

$$Au + A(-u) \subset A0$$

then implies that Au consists of a single vector (also denoted by Au) for each u, and $A(-u) = -Au$. In this case $A(u_1 + u_2) = Au_1 + Au_2$ for every u_1 and u_2 as pointed out above, and $A(\lambda u) = \lambda Au$ for every $\lambda \in R$ by condition (b) in the definition of "convex process." Thus A is linear. ‖
  A good example of a convex process which is *not* a linear transformation is the mapping A defined by

$$Au = \begin{cases} \{x \mid x \leq Bu\} & \text{if } u \geq 0, \\ \emptyset & \text{if } u \ngeq 0, \end{cases}$$

where B is a linear transformation from R^m to R^n. Note that

$$A^{-1}x = \{u \mid u \geq 0, Bu \geq x\}, \forall x.$$

The convex process A^{-1} thus expresses the dependence upon x of the set of solutions to a certain linear inequality system having x as its "vector of constants."

A convex process is said to be *polyhedral* if its graph is a polyhedral convex cone. The convex processes A and A^{-1} in the preceding paragraph are polyhedral, as are of course all linear transformations. The results below which involve conditions on closures and relative interiors can be stated much more simply in the case where the convex processes are polyhedral, although we shall not pursue this point.

Let A be a convex process from R^m to R^n. The closure of the graph of A is a convex cone in R^{m+n} containing the origin, and hence it is the graph of a certain convex process. We call this convex process the *closure* of A and denote it by cl A. Obviously $x \in (\text{cl } A)u$ if and only if there exist sequences $u_1, u_2, \ldots$, and $x_1, x_2, \ldots$, such that u_i converges to u, $x_i \in Au_i$ and x_i converges to x. We say that A is *closed* if cl $A = A$. Clearly, cl A is itself closed, and

$$\text{cl } (A^{-1}) = (\text{cl } A)^{-1}.$$

If A is a closed convex process, all the sets Au for $u \in \text{dom } A$ are closed, and they all have the same recession cone, namely $A0$. (The latter is apparent from the fact that the sets Au correspond to the "parallel" cross-sections

$$L_u \cap \text{graph } A, \qquad u \in R^m,$$

where L_u is the affine set in R^{m+n} consisting of all of the pairs (u, x), $x \in R^n$. The recession cone of $L_u \cap (\text{graph } A)$ is L_0 for every u by Corollary 8.3.3.)

Scalar multiples λA of a convex process A from R^m to R^n are defined for every $\lambda \in R$ by

$$(\lambda A)u = \lambda(Au).$$

These scalar multiples are obviously convex processes.

If A and B are convex processes from R^m to R^n, the sum $A + B$ is defined by

$$(A + B)u = Au + Bu.$$

It follows immediately from the definitions that $A + B$ is another convex process, and

$$\text{dom } (A + B) = \text{dom } A \cap \text{dom } B.$$

Addition is a commutative, associative operation under which the collection of all convex processes from R^m to R^n is a semigroup with an identity element (the zero linear transformation).

If C is a convex set in R^m and A is a convex process from R^m to R^n, the image of C under A is defined as

$$AC = \bigcup \{Au \mid u \in C\}.$$

This image is a convex set in R^n, because, for $0 < \lambda < 1$,

$$(1 - \lambda)AC + \lambda AC = A((1 - \lambda)C)$$

$$+ A(\lambda C) \subseteq A((1 - \lambda)C + \lambda C) \subseteq AC.$$

The image Af of a convex function f on R^m under a convex process A from R^m to R^n is defined by

$$(Af)(x) = \inf \{f(u) \mid u \in A^{-1}x\}.$$

It is easy to verify that Af is a convex function on R^n.

The product BA of a convex process A from R^m to R^n and a convex process B from R^n to R^p is defined by

$$(BA)u = B(Au) = \bigcup \{Bx \mid x \in Au\}.$$

One has

$$BA(u_1 + u_2) \supset B(Au_1 + Au_2) \supset BAu_1 + BAu_2,$$

$$BA(\lambda u) = B(\lambda Au) = \lambda(BAu), \; \forall \lambda > 0,$$

$$0 \in B(A0) = (BA)0,$$

so that BA is a convex process from R^m to R^p. Clearly

$$(BA)^{-1} = A^{-1}B^{-1}.$$

Note that $A^{-1}A$ is generally a multivalued mapping and not just the identity transformation. Multiplication of convex processes is an associative operation. The collection of all convex processes from R^n to itself is a (non-commutative) semigroup under multiplication, with the identity linear transformation I acting as the identity element.

The distributive law does not generally hold between addition and multiplication. Instead one has *distributive inequalities*:

$$A(A_1 + A_2) \supset AA_1 + AA_2,$$

$$(A_1 + A_2)A \subseteq A_1A + A_2A.$$

Inclusion here is in the sense of graphs, i.e. in the sense that $A \supset B$ if and only if $Au \supset Bu$ for every u.

The collection of all convex processes from R^m to R^n is, of course, a complete lattice under the partial ordering defined by inclusion (inasmuch as the collection of all convex cones containing the origin in R^{m+n} is a complete lattice under inclusion).

In order to develop a sound duality theory for convex processes, one needs to introduce a concept of orientation which reflects the convexity-concavity dualism in the theory of bifunctions. A convex set C can be treated as a special case of a convex function by identifying it with its convex indicator function $\delta(\cdot \mid C)$, or as a special case of a concave function by identifying it with $-\delta(\cdot \mid C)$. When the first identification is made we speak of C as having a *supremum orientation* and define

$$\langle C, x^* \rangle = \langle x^*, C \rangle = \sup \{\langle x, x^* \rangle \mid x \in C\}, \forall x^*,$$

while when the second identification is made we speak of C as having an *infimum orientation* and define

$$\langle C, x^* \rangle = \langle x^*, C \rangle = \inf \{\langle x, x^* \rangle \mid x \in C\}, \forall x^*.$$

(Strictly speaking, we should say that an oriented convex set is a pair consisting of a convex set and one of the words "supremum" or "infimum." This word is the "orientation" of the set, and it specifies how the set is to be manipulated in various formulas below.) For a supremum oriented convex set, $\langle C, \cdot \rangle$ is the support function of C, the convex conjugate of $\delta(\cdot \mid C)$, while for an infimum oriented convex set one has

$$\langle C, x^* \rangle = -\delta^*(-x^* \mid C),$$

i.e. $\langle C, \cdot \rangle$ is the concave conjugate of $-\delta(\cdot \mid C)$.

A *supremum oriented convex process* is a convex process A with Au supremum oriented for every u; similarly for an *infimum oriented convex process*. The inverse of an oriented convex process is given the opposite orientation. The sum or product, etc., of convex processes with like orientation is given this same orientation. (Only sums and products of convex processes with like orientation are considered below.)

The *indicator bifunction* F of a supremum oriented convex process A from R^m to R^n is the bifunction from R^m to R^n defined by

$$(Fu)(x) = \delta(x \mid Au).$$

Clearly F is convex and proper, since the graph function of F is the indicator function of a non-empty convex cone in R^{m+n}, namely the graph of A. Also, F is closed if and only if A is closed. One has

$$\operatorname{dom} F = \operatorname{dom} A.$$

If A is infimum oriented, instead of supremum oriented, the indicator bifunction of A is concave, instead of convex, and it is defined by

$$(Fu)(x) = -\delta(x \mid Au).$$

The algebraic operations for convex processes correspond to operations introduced for bifunctions in the preceding section. For example, if A_1 and A_2 are supremum oriented convex processes from R^m to R^n with indicator bifunctions F_1 and F_2 respectively, then the indicator bifunction of $A_1 + A_2$ is $F_1 \square F_2$. If A is a supremum oriented convex process from

R^m to R^n with indicator bifunction F, and B is a supremum oriented convex process from R^n to R^p with indicator function G, then the indicator function of BA is GF.

Adjoints of convex processes can be defined unambiguously by consistent use of orientations. Given a supremum oriented convex process A from R^m to R^n, we take the *adjoint* of A to be the infimum oriented mapping A^* (actually a convex process, as will be shown below) defined by

$$A^*x^* = \{u^* \mid \langle u, u^* \rangle \geq \langle x, x^* \rangle, \forall x \in Au, \forall u\}$$
$$= \{u^* \mid \langle u, u^* \rangle \geq \langle Au, x^* \rangle, \forall u\}.$$

The adjoint of an infimum oriented convex process is defined in the same way, except that it is supremum oriented and the inequality in the definition is reversed. Evidently

$$(A^*)^{-1} = (A^{-1})^*.$$

Note that, when A is a linear transformation, the adjoint of A as a convex process (given either orientation) is the adjoint linear transformation. Indeed, the condition

$$\langle u, u^* \rangle \geq \langle Au, x^* \rangle, \forall u,$$

implies that

$$\langle u, u^* \rangle = \langle Au, x^* \rangle, \forall u,$$

i.e. that u^* is the image of x^* under the adjoint linear transformation.

THEOREM 39.2. *Let A be an oriented convex process from R^m to R^n. Then A^* is a closed convex process from R^n to R^m having the opposite orientation, and $A^{**} = \text{cl } A$. The adjoint of the indicator bifunction of A is the indicator bifunction of A^*.*

PROOF. Suppose that A is supremum oriented. Let $K = \text{graph } A$, and let f be the graph function of the indicator bifunction F of A, i.e. $f = \delta(\cdot \mid K)$. The conjugate f^* of f is $\delta(\cdot \mid K^\circ)$, where K° is the polar of K (see §14), whereas

$$(F^*x^*)(u^*) = -f^*(-u^*, x^*)$$

by definition. The graph of A^* consists of the vectors $z^* = (u^*, x^*)$ in R^{m+n} such that $\langle z, \bar{z}^* \rangle \leq 0$ for every z in the graph of A, where $\bar{z}^* = (-u^*, x^*)$. Thus

$$\text{graph } A^* = \{(u^*, x^*) \mid (-u^*, x^*) \in K^\circ\}.$$

It follows that the graph of A^* is a closed convex cone in R^{m+n} containing the origin, and hence that A^* is a closed convex process. In fact, we have

$$(F^*x^*)(u^*) = -\delta(u^* \mid A^*x^*).$$

Thus F^* is the indicator bifunction of A^*.

The case of an infimum oriented convex process is argued similarly. The relation $A^{**} = \text{cl } A$ follows from $F^{**} = \text{cl } F$ (Theorem 30.1), or equivalently from $K^{\infty\infty} = \text{cl } K$. ‖

If A is an oriented convex process with indicator bifunction F, we have

$$\langle Au, x^* \rangle = \langle Fu, x^* \rangle, \; \forall u, \forall x^*,$$

by definition. The general theorems about inner products involving convex and concave bifunctions can be specialized in this way to theorems about inner products involving convex processes.

THEOREM 39.3. *If A is a supremum oriented convex process from R^m to R^n, then $\langle Au, x^* \rangle$ is a positively homogeneous closed convex function of x^* for each u and a positively homogeneous concave function of u for each x^*. Likewise when A is infimum oriented, except that then convexity and concavity are reversed. In either case,*

$$\langle u, A^*x^* \rangle = \text{cl}_u \langle Au, x^* \rangle.$$

If A is closed, one also has

$$\langle Au, x^* \rangle = \text{cl}_{x^*} \langle u, A^*x^* \rangle.$$

Indeed, if A is closed,

$$\langle Au, x^* \rangle = \langle u, A^*x^* \rangle$$

whenever $u \in \text{ri (dom } A)$ or $x^* \in \text{ri (dom } A^*)$.

PROOF. The positive homogeneity in x^* follows from the fact that $\langle Au, \cdot \rangle$ is the support function of A, while the positive homogeneity in u follows from condition (b) in the definition of "convex process." Everything else is just a special case of Theorem 33.1, Theorem 33.2 and Corollary 33.2.1. ‖

THEOREM 39.4. *The relations*

$$K(u, x^*) = \langle Au, x^* \rangle,$$

$$Au = \{x \mid \langle x, x^* \rangle \leq K(u, x^*), \forall x^*\},$$

define a one-to-one correspondence between the lower closed concave-convex functions K on $R^m \times R^n$ such that $K(0, 0) = 0$ and

$$K(\lambda u, x^*) = \lambda K(u, x^*) = K(u, \lambda x^*), \quad \forall \lambda > 0, \forall u, \forall x^*,$$

and the supremum oriented closed convex processes A from R^m to R^n. (Similarly for upper closed convex-concave functions and infimum oriented convex processes.)

PROOF. This specializes Theorem 33.3. It is an easy exercise to show that F is the indicator bifunction of a convex process in Theorem 33.3 if and only if K has the additional properties cited here. ‖

We would like to emphasize that the relation

$$\langle Au, x^* \rangle = \langle u, A^*x^* \rangle,$$

which "usually" holds according to the last statement in Theorem 39.3, expresses a duality between two extremum problems, as has already been pointed out in the more general context of bifunction following Corollary 33.2.2. If A is a supremum oriented convex process, A^* is infimum oriented and for each fixed u and x^* we have

$$\langle Au, x^* \rangle = \sup \{\langle x, x^* \rangle \mid x \in Au\},$$

$$\langle u, A^*x^* \rangle = \inf \{\langle u, u^* \rangle \mid u^* \in A^*x^*\}.$$

Thus $\langle Au, x^* \rangle$ is obtained by maximizing the linear function $\langle \cdot, x^* \rangle$ over a certain convex set Au, whereas $\langle u, A^*x^* \rangle$ is obtained by minimizing the linear function $\langle u, \cdot \rangle$ over a certain convex set A^*x^*. If A is polyhedral, implying by Theorems 39.2 and 30.1 that A^* too is polyhedral, the sets Au and A^*x^* are polyhedral, as is easy to see, so that these two extremum problems can be expressed by linear programs.

For instance, suppose as in the example described earlier in this section that

$$Au = \begin{cases} \{x \mid x \leq Bu\} & \text{if } u \geq 0, \\ \emptyset & \text{if } u \not\geq 0, \end{cases}$$

$$A^{-1}x = \{u \mid u \geq 0, Bu \geq x\},$$

where B is a linear transformation from R^m to R^n. Let A be supremum oriented, so that A^{-1} is infimum oriented. We then have

$$\langle Au, x^* \rangle = \begin{cases} \langle Bu, x^* \rangle & \text{if } u \geq 0, x^* \geq 0, \\ +\infty & \text{if } u \geq 0, x^* \not\geq 0, \\ -\infty & \text{if } u \not\geq 0. \end{cases}$$

Closing $\langle Au, x^* \rangle$ as a concave function of u for each x^* yields $\langle u, A^*x^* \rangle$ according to Theorem 39.3. Thus

$$\langle u, A^*x^* \rangle = \begin{cases} \langle u, B^*x^* \rangle & \text{if } u \geq 0, x^* \geq 0, \\ -\infty & \text{if } u \not\geq 0, x^* \geq 0, \\ +\infty & \text{if } x^* \not\geq 0. \end{cases}$$

It follows that

$$A^*x^* = \begin{cases} \{u^* \mid u^* \geq B^*x^*\} & \text{if} \quad x^* \geq 0, \\ \emptyset & \text{if} \quad x^* \not\geq 0, \end{cases}$$

$$(A^{-1})^*u^* = A^{*-1}u^* = \{x^* \mid x^* \geq 0, B^*x^* \leq u^*\},$$

where A^* is infimum oriented and $(A^{-1})^* = A^{*-1}$ is supremum oriented. For each fixed x and u^*, we have

$$\langle u^*, A^{-1}x \rangle = \inf \{\langle u, u^* \rangle \mid u \geq 0, Bu \geq x\},$$

$$\langle (A^{-1})^*u^*, x \rangle = \sup \{\langle x, x^* \rangle \mid x^* \geq 0, B^*x^* \leq u^*\}.$$

The fact that these two extrema are usually equal has already been encountered following the proof of Theorem 30.4 as the Gale–Kuhn–Tucker Duality Theorem for linear programs.

Results about sums and products of convex processes can be obtained simply by specializing the results in the preceding section to indicator bifunctions.

THEOREM 39.5. *Let A_1 and A_2 be convex processes from R^m to R^n with the same orientation. If* ri (dom A_1) *and* ri (dom A_2) *have a point in common, one has*

$$(A_1 + A_2)^* = A_1^* + A_2^*.$$

If A_1 and A_2 are closed and ri (dom A_1^*) *and* ri (dom A_2^*) *have a point in common, then $A_1 + A_2$ is closed and $(A_1 + A_2)^*$ is the closure of $A_1^* + A_2^*$.*

PROOF. This is a special case of Theorem 38.2 and Corollary 38.2.1. ‖

THEOREM 39.6. *For any oriented convex process A, one has $(\lambda A)^* = \lambda A^*$ for every $\lambda > 0$.*

PROOF. This is a special case of Theorem 38.3. ‖

THEOREM 39.7. *Let A be a supremum oriented convex process from R^m to R^n, and let f be a proper convex function on R^m. If* ri (dom f) *meets* ri (dom A), *one has*

$$(Af)^* = A^{*-1}f^*,$$

and the infimum in the definition of $(A^{*-1}f^*)(x^*)$ is attained for each x^*.

If A and f are closed and ri (dom f^*) *meets* ri (dom A^{*-1}), *then Af is closed and the infimum in the definition of $(Af)(x)$ is attained for each x. Moreover, then $(Af)^*$ is the closure of $A^{*-1}f^*$.*

PROOF. This specializes Theorem 38.4 and Corollary 38.4.1. ‖

COROLLARY 39.7.1. *Let A be a closed convex process from R^m to R^n, and let C be a non-empty closed convex set in R^m. If no non-zero vector in*

$A^{-1}0$ belongs to the recession cone of C (which is true in particular if C is bounded), then AC is closed in R^n.

PROOF. Make A supremum oriented, and apply the theorem with $f = \delta(\cdot \mid C)$. The set $K = \text{dom} f^*$ is the barrier cone of C, and its polar is the recession cone of C (Corollary 14.2.1). If the convex cones K and

$$\text{dom } A^{*-1} = \text{range } A^*$$

have relative interior points in common, Af is closed by the theorem, and since Af is the indicator function of AC it follows that AC is closed. On the other hand, if these cones had no relative interior points in common they could be separated properly by some hyperplane. Thus there would exist some non-zero $v \in R^m$ such that $\langle v, u^* \rangle \leq 0$ for every $u^* \in K$ and $\langle v, u^* \rangle \geq 0$ for every u^* in the range of A^*. Then $v \in K^\circ$ and (inasmuch as $A^{**} = A$) $v \in A^{-1}0$. But this case is excluded by hypothesis. ‖

THEOREM 39.8. *Let A be a convex process from R^m to R^n, let B be a convex process from R^n to R^p, and let A and B have the same orientation. If $\text{ri (range } A)$ meets $\text{ri (dom } B)$, one has*

$$(BA)^* = A^*B^*.$$

If A and B are closed and $\text{ri (range } B^*)$ meets $\text{ri (dom } A^*)$, then BA is closed and $(BA)^*$ is the closure of A^*B^*.

PROOF. This specializes Theorem 38.5 and Corollary 38.5.1. ‖

Let C and D be non-empty convex sets in R^n such that C is supremum oriented and D is infimum oriented. If the quantity

$$\sup_{x \in C} \inf_{y \in D} \langle x, y \rangle = \sup_{x \in C} \langle x, D \rangle$$

and the quantity

$$\inf_{y \in D} \sup_{x \in C} \langle x, y \rangle = \inf_{y \in D} \langle C, y \rangle$$

are equal, we call this the *inner product* of C and D and denote it by $\langle C, D \rangle$ (or by $\langle D, C \rangle$). This definition agrees with that of the inner product of a convex and concave function in §38, in the sense that $\langle C, D \rangle = \langle f, g \rangle$ when $f = \delta(\cdot \mid C)$ and $g = -\delta(\cdot \mid D)$. Note that $\langle C, D \rangle$ always exists when C and D are both closed and either C or D is bounded (Corollary 37.3.2).

If h is a proper concave function on R^n, we naturally define $\langle C, h \rangle = \langle f, h \rangle$, where $f = \delta(\cdot \mid C)$. Thus

$$\langle C, h \rangle = \sup \{h^*(x) \mid x \in C\} = \inf_y \{\langle C, y \rangle - h(y)\}$$

when these two extrema are equal, and otherwise $\langle C, h \rangle$ is undefined. Similarly, if h is a proper convex function on R^n the inner product of h

with the infimum oriented set D is defined as

$$\langle h, D \rangle = \inf \{ h^*(y) \mid y \in D \} = \sup_x \{ \langle x, D \rangle - h(x) \}$$

when these extrema are equal, and otherwise it is undefined.

If C and C' are supremum oriented non-empty convex sets in R^n and D and D' are infimum oriented non-empty convex sets in R^n, the following laws hold (to the extent that all the inner products in question exist and $\infty - \infty$ is not involved):

$$\langle \lambda C, D \rangle = \lambda \langle C, D \rangle = \langle C, \lambda D \rangle, \qquad \forall \lambda > 0,$$
$$\langle C + C', D \rangle \geq \langle C, D \rangle + \langle C', D \rangle,$$
$$\langle C + C', y \rangle = \langle C, y \rangle + \langle C', y \rangle, \qquad \forall y \in R^n,$$
$$\langle C, D + D' \rangle \leq \langle C, D \rangle + \langle C, D' \rangle,$$
$$\langle x, D + D' \rangle = \langle x, D \rangle + \langle x, D' \rangle, \qquad \forall x \in R^n.$$

These laws are all elementary consequences of the definitions.

Using this expanded inner product notation, Theorem 38.7 and its corollaries can be specialized in the obvious way to the case of oriented convex processes and oriented convex sets.

The set of all convex processes A from R^n to itself is a non-commutative semigroup under multiplication, and it includes the semigroup of all linear transformations from R^n to itself. The structure of an individual A may be analyzed in terms of the powers $A^2, A^3, \ldots$, and more generally in terms of convex processes given by expressions such as $A - \lambda I$ or

$$I + \alpha_1 A + \alpha_2 A^2 + \cdots + \alpha_k A^k.$$

Eigensets of A, i.e. convex sets C such that

$$AC = \lambda C$$

for some λ, may also be studied.

Comments and References

Part I: Basic Concepts

The foundations of the general theory of convex sets and functions were laid around the turn of the century, chiefly by Minkowski [1, 2]. For a survey of the progress of the subject up to 1933, at least in those aspects pertinent to geometry, see the book of Bonnesen–Fenchel [1]. For the history of the role of convex functions in the theory of inequalities up to 1948, consult Beckenbach [1].

Modern expositions of convexity in R^n have been written by Fenchel [2], Eggleston [1], Berge [1] and Valentine [1], among others. Valentine's book treats infinite-dimensional spaces to some extent, as well as R^n; material on infinite-dimensional convexity can also be found in almost any text on functional analysis, such as Bourbaki [1]. The 1967 lecture notes of Moreau [17] provide an excellent reference for the theory of convex functions in topological vector spaces of arbitrary dimension; the reader should turn to these notes for generalizations of various results about conjugate convex functions which have been presented in this book in the finite-dimensional case only.

The matrix representations of affine sets referred to as Tucker representations in §1 have been used extensively by Tucker [4, 5, 6] in developing the theory of linear programs.

Our approach to the theory of convex functions in §4 and §5 is based on that of Fenchel [2], except that Fenchel handled everything in terms of pairs (C, f) rather than infinity-valued functions. In particular, the idea of identifying convex sets with certain "degenerate" convex functions (their indicator functions) originates with Fenchel, as does the important operation of infimal convolution. For discussions of the arithmetic of $\pm \infty$ and of infimal convolution with respect to convex cones, see Moreau [3, 6, 7, 8].

Part II: Topological Properties

The results about relative interiors of convex sets in §6 are almost all classical; see especially the paper of Steinitz [1]. The theory of the closure operation for convex functions in §7 is due to Fenchel [1, 2].

Unbounded convex sets were first studied systematically by Steinitz [1], who proved most of the basic facts about recession cones, such as Theorem 8.3. Recession cones have been used to prove closedness theorems, such as those in §9, by Fenchel [2] and later by Choquet [1] and Rockafellar [1, 6]. Theorems about the closedness of sums and projections of convex sets have also been established by Klee [10] and Gale–Klee [1].

Theorems 10.2 and 10.3 stem from Gale–Klee–Rockafellar [1], but all the other continuity and convergence theorems in §10 must be regarded as classical, even though some of them do not seem to have been stated explicitly anywhere in the literature. Similar theorems in a more geometric formulation appear in the theory of convex surfaces; see Bonnesen–Fenchel [1], Alexandroff [2] and Busemann [1]. The continuity of various operations with respect to convergence of convex sets and functions has been studied recently by Wijsman [1, 2] and Walkup–Wets [1].

Part III: Duality Correspondences

Separation theorems were first investigated by Minkowski. The traditional proofs in R^n rely on nearest-point arguments; see for instance the exposition of Botts [1]. The approach taken in §11, however, is the approach typical of functional analysis, where Theorem 11.2 corresponds to the Hahn–Banach Theorem. Theorem 11.3 was first proved in its full generality by Fenchel [2]. For other results on separation, both in finite- and infinite-dimensional spaces, we refer the reader to the definitive papers of Klee [1, 2, 3, 4, 5, 15].

The general conjugacy correspondence for convex functions was discovered by Fenchel [1], although conjugates of functions of a single-variable were considered earlier by Mandelbrojt [1]. Monotone conjugacy on R has a long history beginning with the work of Young [1] and Birnbaum–Orlicz [1]; see the book of Krasnosel'skii–Rutickii [1]. Monotone conjugacy of n-dimensional non-decreasing *concave* functions has been studied by Bellman–Karush [3]. See Moreau [2, 4] and Brøndsted [1] for the generalization of Fenchel's correspondence to infinite-dimensional spaces.

Support functions, originally defined by Minkowski for bounded convex sets, have been studied for general convex sets by Fenchel [1, 2] and in infinite-dimensional spaces by Hörmander [1]. Theorems 13.3 and 13.5 are from Rockafellar [1, 6]; however, see Klee [7] for an earlier proof that f^* is finite everywhere when dom f is bounded. Theorem 13.4 has been proved by Brøndsted [1].

Steinitz [1] invented the polarity correspondence for convex cones, but polars of bounded convex sets were considered earlier by Minkowski, as were the correspondences between gauge functions and support functions described in §15, at least in the case of dual norms. For the theorems relating polarity to conjugacy, see Rockafellar [6]. Theorem 14.7 should be credited to Moreau [9, 11]; special cases are also known in the theory of Orlicz spaces, see Krasnosel'skii–Rutickii [1]. Theorem 15.3 in its general form comes from Aggeri–Lescarret [1], but Corollary 15.3.2 was developed earlier in terms of the Legendre transformation by Lorch [1]. The general polarity correspondence for non-negative convex functions vanishing at the origin is defined here for the first time.

The duality results in §16 are virtually all contained in the lecture notes of Fenchel [2]. The duality between infimal convolution and addition has been used by Bellman–Karush [1, 2, 3, 4, 5] to solve certain recursive functional relations.

Part IV: Representation and Inequalities

For an account of Carathéodory's Theorem and some of its extensions, see the 1965 monograph of Reay [1].

Our presentation of the theory of extremal structure of convex sets in §18 is based on the work of Klee [2, 6, 7, 8]. The fact that a compact convex set in R^n is the convex hull of its extreme points (Corollary 18.5.1) was first proved by Minkowski. More famous, however, is an infinite-dimensional generalization by Krein–Milman [1], to the effect that a compact convex set in a locally convex Hausdorff topological vector space is the *closure* of the convex hull of its extreme points. Related results for convex functions have been formulated by Aggeri [1] and Brøndsted [2]. Theorems 18.6 and 18.7 were first established for bounded convex sets by Straszewicz [1].

Theorem 19.1 is a celebrated result attributable primarily to Minkowski [1] and Weyl [1]. The early history of polyhedral convexity can be found in the book of Motzkin [1]. As excellent sources for further information about polyhedral convexity, we recommend Grünbaum [1], Klee [8, 13] and the 1956 collection of papers edited by Kuhn–Tucker [2]. Theorem 20.1 is new. Theorems 20.2 and 20.3 seem to be stated here for the first time, but a broader result, from which Theorem 20.2 could be deduced, has been proved by Klee [15, Theorem 4(i)]. Theorems 20.4 and 20.5 are classical.

Theorem 21.1 and its proof are due to Fan–Glicksberg–Hoffman [1]. A different proof of Theorem 21.2 in the case where $C = R^n$ is given in the book of Berge–Ghouila-Houri [1]. Fenchel [2] originated the version of Helly's Theorem involving recession cones which we have stated as Corollary 21.3.3, as well as Theorem 21.3 itself in the special case where C

and the effective domains of the functions f_i have no common direction of recession. For an earlier proof of Theorem 21.3 in the case where C is compact, see Bohnenblust–Karlin–Shapley [1]. Theorems 21.4 and 21.5 stem from Rockafellar [4]. Theorem 21.6 is one of the forms of Helly's Theorem due to Helly [1] himself. A thorough review of the literature on Helly's Theorem up to 1963 has been put together by Danzer–Grünbaum–Klee [1]. Some further results about infinite systems of inequalities may be found in papers of Fan [3, 4].

For other expositions of the classical results in §22 about the consistency of linear inequalities, along with historical comments, we refer the reader to Tucker [2], and also to Kuhn [1]. Theorem 22.6 is a recent outgrowth of graph-theoretic investigations and should be credited mainly to Camion [1], although the proof is based on an earlier argument of Ghouila-Houri. Consult Rockafellar [13] for a different proof of Theorem 22.6 and an explanation of the relationship with earlier results of Minty [1] concerning flows in networks.

Part V: Differential Theory

The existence of one-sided derivatives of convex functions was noted as early as 1893 by Stoltz [1]. The properties of such derivatives received much attention in the beginning decades of this century in connection with the theory of convex bodies and convex surfaces; cf. Bonnesen–Fenchel [1], Alexandroff [2] and Busemann [1]. Most of the results in §24 and §25 concerning differentiability and differential continuity or convergence may be said to date from this period, although it is difficult to give explicit references apart from Fenchel's 1951 exposition, the older context being one of geometry rather than analysis. The explicit development of the theory of multivalued subdifferential mappings is comparatively recent; the reader should refer to Moreau [16, 17] for a general review of the literature.

Theorems 23.1 through 23.5 are essentially contained in the lecture notes of Fenchel [2] (and to a certain extent in Bonnesen–Fenchel [1]). Theorems 23.6, 23.8, 23.9, 24.8 and 24.9 are from Rockafellar [1, 7], while Theorems 24.6 and 25.6 are new. Complete non-decreasing curves in R^2 were first studied in their full generality by Minty [1].

The nature of the set of points where a convex function is not differentiable is known in much greater detail than indicated in Theorem 25.5; see Anderson–Klee [1]. Much is also known about the second derivative of convex functions; see Alexandroff [1], Busemann–Feller [1] and Busemann [1].

The relationship between the Legendre transformation and conjugacy was noted by Fenchel [1]. Some classical applications of the Legendre transformation are described in Courant–Hilbert [1].

Part VI: Constrained Extremum Problems

The constrained minimization of convex functions is a subject which has attracted a great deal of attention since about 1950. For some of the computational aspects, see Dantzig [1], Goldstein [1] and Wolfe [2, 3]. For some of the applications to mathematical economics, see Karlin [1].

The theory of ordinary convex programs is historically an outgrowth of the paper of Kuhn–Tucker [1]. Although Lagrange multiplier conditions closely related to the Kuhn–Tucker conditions were derived earlier for general (differentiable) inequality constraints by John [1], it was Kuhn and Tucker who discovered the connection between Lagrange multipliers and saddle-points and who focused attention on the role of convexity. This is why we have called the special Lagrange multiplier values corresponding to saddle-values of Lagrangian functions *Kuhn–Tucker coefficients*. (In most of the literature, the term "Lagrange multiplier" is used, not only as we have used it to refer to the coefficients λ_i as variables, but also, perhaps confusingly, to refer to the particular values of these variables which satisfy certain relations such as the Kuhn–Tucker conditions. In non-convex programming such values of the λ_i's do not necessarily correspond to saddle-points of the Lagrangian. On the other hand, Kuhn–Tucker coefficients are well-defined even in programs in which, due to the lack of an optimal solution x, the Kuhn–Tucker conditions cannot be satisfied.)

The original theorems of Kuhn and Tucker relied on the differential calculus, but it was foreseen by those authors and soon verified by others that, in the case of convex functions, gradient conditions could be replaced by something not involving differentiability. Slater [1] seems to have been the first to substitute for the constraint qualification condition of Kuhn–Tucker [1] a hypothesis like the one in Theorem 28.2 about the existence of a feasible solution satisfying the inequality constraints with strict inequality. Theorem 28.2 (and hence the Kuhn–Tucker Theorem) has previously been proved by Fan–Glicksberg–Hoffman [1] in the case where there are no equality constraints, and by Uzawa (see Arrow–Hurwicz–Uzawa [1, p. 36]) in the case where there are (linear) equality constraints, C is the non-negative orthant of R^n, and the functions f_i are all finite throughout R^n.

The decomposition principle was first discovered in the case of linear programs by Dantzig and Wolfe; see Dantzig [1] for a thorough treatment in that case. Our more general exposition is based to some extent on Falk [1].

The theory of generalized convex programs in §29 has never been presented before, but it owes very much to a paper of Gale [1] in which, in effect, generalized convex programs are considered in the context of economics, and Theorem 29.1 and some of its corollaries are demonstrated (although not in terms of "perturbations" or bifunctions). The Lagrangian theory in §29 and the general duality theory in §30 are both new. However, duality has a long history in the study of ordinary convex programs and other types of problems, such as those in §31.

The basic duality result which has served as a model for all subsequent developments is the theorem of Gale–Kuhn–Tucker [1] for linear programs, discovered around 1948. The duality theorem of Fenchel [2] in §31 dates from 1951. Duals of ordinary convex programs have been defined in terms of the differential calculus by Dorn [2] (linear constraints), Dennis [1] (linear constraints) and Wolfe [1]. Wolfe's dual problem, which has stimulated work of Huard [1, 2], Mangasarian [1] and many others, corresponds in our notation to maximizing

$$f_0(x) + v_1^* f_1(x) + \cdots + v_m^* f_m(x)$$

in x and u^* subject to $u^* \geq 0$ and

$$\nabla f_0(x) + v_1^* \nabla f_1(x) + \cdots + v_m^* \nabla f_m(x) = 0.$$

The connection between this and the dual program (P^*) in §30 is explained following Corollary 30.5.1. A closely related generalization of Wolfe's dual problem has been given by Falk [1]. In the logarithmic example at the very end of §30, program (R) is equivalent to the standard "geometric program" of Duffin–Peterson [1], whereas the dual program (R^*) is the so-called general chemical equilibrium problem when $n_0 = 1$; see Duffin–Peterson–Zener [1, Appendix C] and the references given there.

A general duality theory in which constrained minimization or maximization problems are derived from Lagrangian minimax problems, rather than the other way around, has been developed by Dantzig–Eisenberg–Cottle [1], Stoer [1, 2] and Mangasarian–Ponstein [1]. It can be shown that the pairs of mutually dual problems considered by these authors can essentially be expressed in the form

(I)　　minimize $\varphi(x) = \sup \{L(u^*, x) \,|\, u^* \in A\}$ subject to $x \in B_0$,

(II)　　maximize $\psi(u^*) = \inf \{L(u^*, x) \,|\, x \in B\}$ subject to $u^* \in A_0$,

where A and B are given non-empty closed convex sets in R^m and R^n, respectively, L is a given continuous real-valued concave-convex function on $A \times B$ satisfying certain regularity conditions, and A_0 and B_0 are certain subsets of A and B (e.g. the sets of points for which the supremum in (I) and the infimum in (II) are attained, respectively). Such pairs of problems can be viewed as restricted versions of the problems in §30 according to the discussion in §36 following Theorem 36.5.

The original version of Fenchel's Duality Theorem did not include the final assertion of Theorem 31.1 concerning polyhedral convexity. The extensions of the theorem to take advantage of polyhedral convexity, and to include a linear transformation A as in Theorem 31.2, were carried out by Rockafellar [1, 2, 9]; see also Berge–Ghouila-Houri [1] and Eisenberg [1] for special cases.

As remarked, Corollary 31.4.1 yields the Gale–Kuhn–Tucker theorem for linear programs when f is taken to be a partial affine function, as can be seen by giving f any Tucker representation. The various Tucker representations which are possible correspond to the various "tableaus" which may be encountered in the course of solving a given linear program by the well-known simplex algorithm of Dantzig. Similarly, it can be shown that Corollary 31.4.1 yields the duality theorem of Cottle [1] for quadratic programs when f is taken to be a partial quadratic function; cf. Rockafellar [12]. For some additional duality results which may be viewed as special cases of Corollary 31.4.2, although they are developed in terms of the Legendre transformation rather than Fenchel's conjugacy operation, see Dennis [1] and Duffin [2]. Corollary 31.4.2 can be sharpened in the case where f is separable, a very important case for many applications, e.g. to extremum problems involving flows and potentials in networks; see Minty [1], Camion [2], Rockafellar [10].

The theory of proximations, including Theorem 31.5 and its corollaries, has been developed by Moreau [13].

Theorem 32.3 may be found in Hirsch–Hoffman [1]; see also Bauer [1].

Part VII: Saddle-Functions and Minimax Theory

Proofs of most of the results in §33 and §34 have already been given elsewhere by Rockafellar [3, 12], but not in terms of bifunctions. The results in §35 are new, as are Theorems 36.5, 36.6, 37.2 and Corollaries 37.5.1, 37.5.2.

Minimax theorems have been investigated by many authors, starting with von Neumann; in particular, the result stated as Corollary 37.6.2 was first proved by Kakutani [1]. For an excellent summary of the literature up to 1958, see Sion [2]. The sharpest results described in the Sion paper generally require something less than the concavity-convexity of $K(u, v)$ but require the compactness of C or D. In contrast, Theorems 37.3

and 37.6 (which come from Rockafellar [3]; see also Moreau [12]) require concavity-convexity but something less than compactness.

For the original development of the conjugacy correspondence for saddle-functions, see Rockafellar [3, 12].

Part VIII: Convex Algebra

The theory in §38 and §39 is new. However, see Rockafellar [14] for a generalization of some of the theory of non-negative matrices to a special class of convex processes arising in mathematical economics.

Bibliography

J.-C. Aggeri
[1] "Les fonctions convexes continue et le théorème de Krein-Milman," *C.R. Acad. Sci. Paris 262* (1966), 229–232.

J.-C. Aggeri and C. Lescarret
[1] "Fonctions convexes duales associées à une couple d'ensembles mutuellement polaires," *C.R. Acad. Sci. Paris 260* (1965), 6011–6014.
[2] "Sur une application de la théorie de la sous-différentiabilité à des fonctions convexes duales associées à un couple d'ensembles mutuellement polaires," Séminaire de Mathématiques, Faculté des Sciences, Université de Montpellier (1965).

A. D. Alexandroff
[1] "Almost everywhere existence of the second differential of a convex function and some properties of convex surfaces connected with it," *Leningrad State Univ. Ann., Math. Ser. 6* (1939), 3–35 (Russian).
[2] *The Inner Geometry of Convex Surfaces*, Moscow, 1948 (Russian). German translation: Berlin, 1955.

R. D. Anderson and V. L. Klee
[1] "Convex functions and upper semi-continuous collections," *Duke Math. J. 19* (1952), 349–357.

K. J. Arrow and L. Hurwicz
[1] "Reduction of constrained maxima to saddle-point problems," in *Proceedings of the Third Berkeley Symposium on Mathematical Statistics and Probability*, J. Neyman, ed., Univ. of California Press, 1956, Vol. V, 1–20.

K. J. Arrow, L. Hurwicz, and H. Uzawa
[1] *Studies in Linear and Nonlinear Programming*, Stanford University Press, 1958.
[2] "Constraint qualifications in maximization problems," *Naval Res. Logist. Quart. 8* (1961), 175–191.

E. Asplund
[1] "Positivity of duality mappings," *Bull. Amer. Math. Soc. 73* (1967), 200–203.
[2] "Fréchet differentiability of convex functions," *Acta Math. 121* (1968), 31–48.

E. Asplund and R. T. Rockafellar

[1] "Gradients of convex functions," *Trans. Amer. Math. Soc. 139* (1969), 443–467.

H. Bauer

[1] "Minimalstellen von Funktionen und Extremalpunkte," *Arch. Math. 9* (1958), 389–393.

[2] "Minimalstellen von Funktionen und Extremalpunkte. II," *Arch. Math. 11* (1960), 200–205.

E. F. Beckenbach

[1] "Convex functions," *Bull. Amer. Math. Soc. 54* (1948), 439–460.

E. F. Beckenbach and R. Bellman

[1] *Inequalities*, Springer, Berlin, 1961.

R. Bellman and W. Karush

[1] "On a new functional transform in analysis: the maximum transform," *Bull. Amer. Math. Soc. 67* (1961), 501–503.

[2] "On the maximum transform and semigroups of transformations," *Bull. Amer. Math. Soc. 68* (1962), 516–518.

[3] "Mathematical programming and the maximum transform," *J. Soc. Indust. Appl. Math. 1* (1962), 550–567.

[4] "On the maximum transform," *J. Math. Anal. Appl. 6* (1963), 67–74.

[5] "Functional equations in the theory of dynamic programming XII: an application of the maximum transform," *J. Math. Anal. Appl. 6* (1963), 155–157.

A. Ben-Israel

[1] "Notes on linear inequalities, I: the intersection of the non-negative orthant with complementary orthogonal subspaces," *J. Math. Anal. Appl. 10* (1964) 303–314.

C. Berge

[1] *Espaces Topologiques*, Paris, 1959.

[2] "Sur une propriété combinatoire des ensembles convexes," *C.R. Acad. Sci. Paris 248* (1959), 2698.

C. Berge and A. Ghouila-Houri

[1] *Programmes, Jeux et Réseaux de Transport*, Dunod, Paris, 1962.

Z. Birnbaum and W. Orlicz

[1] "Über die Verallgemeinerung des Begriffes der zueinander konjugierten Potenzen," *Studia Math. 3* (1931), 1–67.

H. F. Bohnenblust, S. Karlin, and L. S. Shapley

[1] "Games with continuous pay-off," in *Annals of Mathematics Studies*, No. 24 (1950), 181–192.

T. Bonnesen and W. Fenchel
[1] *Theorie der konvexen Körper*, Springer, Berlin, 1934.

T. Botts
[1] "Convex sets," *Amer. Math. Soc. Monthly 49* (1942) 527–535.

N. Bourbaki
[1] *Espaces Vectoriels Topologiques* I, II. Hermann, Paris, 1953 and 1955.

A. Brøndsted
[1] "Conjugate convex functions in topological vector spaces," *Mat.-Fys. Medd. Dansk. Vid. Selsk. 34* (1964), No. 2, 1–26.
[2] "Milman's theorem for convex functions," *Math. Scand. 19* (1966), 5–10.

A. Brøndsted and R. T. Rockafellar
[1] "On the subdifferentiability of convex functions," *Proc. Amer. Math. Soc. 16* (1965), 605–611.

F. E. Browder
[1] "On a theorem of Beurling and Livingston," *Canad. J. Math. 17* (1965), 367–372.
[2] "Multivalued monotone non-linear mappings in Banach spaces," *Trans. Amer. Math. Soc. 118* (1965), 338–351.

H. Busemann
[1] *Convex Surfaces*, Interscience, New York, 1958.

H. Busemann and W. Feller
[1] "Krümmungseigenschaften konvexer Flächen," *Acta Math. 66* (1935), 1–47.

P. Camion
[1] "Modules unimodularies," *J. Comb. Theory 4* (1968), 301–362.
[2] "Application d'une généralisation du lemme de Minty a une problème d'infimum de fonction convexe," *Cahiers Centre Res. Op. 7* (1965), 230–247.

C. Carathéodory
[1] "Über den Variabilitätsbereich der Fourier'schen Konstanten von positiven harmonischen Funktionen," *Rend. Circ. Mat. Palermo 32* (1911), 193–217.

G. Choquet
[1] "Ensembles et cônes convexes faiblement complets," *C.R. Acad. Sci. Paris 254* (1962), 1908–1910.

G. Choquet, H. Corson, and V. L. Klee
[1] "Exposed points of convex sets," *Pacific J. Math. 16* (1966), 33–43.

R. W. Cottle
[1] "Symmetric dual quadratic programs," *Quart. Appl. Math. 21* (1963), 237.

R. Courant and D. Hilbert
[1] *Methods of Mathematical Physics*, Vol. I, Berlin, 1937; Vol. II, New York, 1962.

G. B. Dantzig
[1] *Linear Programming and Extensions*, Princeton University Press, 1963.

G. B. Dantzig, J. Folkman, and N. Shapiro
[1] "On the continuity of the minimum set of a continuous function," *J. Math. Anal. Appl. 17* (1967), 519–548.

G. Dantzig, E. Eisenberg, and R. W. Cottle
[1] "Symmetric dual nonlinear programs," *Pacific J. Math. 15* (1965), 809–812.

L. Danzer, B. Grünbaum, and V. L. Klee
[1] "Helly's theorem and its relatives," in *Convexity*, V. L. Klee, ed., Proceedings of Symposia in Pure Mathematics, Vol. VII, American Mathematical Society, 1963, 101–180.

C. Davis
[1] "All convex invariant functions of hermitian matrices," *Arch. Math. 8* (1957), 276–278.

J. B. Dennis
[1] *Mathematical Programming and Electrical Networks*, Technology Press, Cambridge, Mass., 1959.

U. Dieter
[1] "Dual extremum problems in locally convex linear spaces," in *Proceedings of the Colloquium on Convexity, Copenhagen, 1965*, W. Fenchel, ed., Copenhagen, Matematisk Institut, 1967, 185–201.
[2] "Dualität bei konvexen Optimierungs- (Programmierungs-) Aufgaben," *Unternehmensforschung 9* (1965), 91–111.
[3] "Optimierungsaufgaben in topologischen Vektorräumen I: Dualitätstheorie," *Z. Wahrscheinlichkeitstheorie verw. Geb. 5* (1966), 89–117.

L. L. Dines
[1] "On convexity," *Amer. Math. Monthly 45* (1938), 199–209.

W. S. Dorn
[1] "Duality in quadratic programming," *Quart. Appl. Math. 18* (1960), 155–162.
[2] "A duality theorem for convex programs," *IBM J. Res. Develop. 4* (1960), 407–413.
[3] "Self-dual quadratic programs," *J. Soc. Indust. Appl. Math. 9* (1961), 51–54.
[4] "On Lagrange multipliers and inequalities," *Operations Res. 9* (1961), 95–104.

R. J. Duffin
[1] "Infinite programs," in *Annals of Mathematics Studies*, No. 38 (1956), 157–170.
[2] "Dual programs and minimum cost," *J. Soc. Indust. Appl. Math. 10* (1962), 119–124.

R. J. Duffin and E. L. Peterson
[1] "Duality theory for geometric programming," *S.I.A.M. J. Appl. Math. 14* (1966), 1307–1349.

R. J. Duffin, E. L. Peterson, and C. Zener
[1] *Geometric Programming—Theory and Application*, Wiley, New York, 1967.

H. G. Eggleston
[1] *Convexity*, Cambridge Univ., 1958.

E. Eisenberg
[1] "Duality in homogeneous programming," *Proc. Amer. Math. Soc. 12* (1961), 783–787.
[2] "Supports of a convex function," *Bull. Amer. Math. Soc. 68* (1962), 192–195.

J. E. Falk
[1] "Lagrange multipliers and nonlinear programming," *J. Math. Anal. Appl. 19* (1967), 141–159.

Ky Fan
[1] "Fixed-point and minimax theorems in locally convex topological linear spaces," *Proc. Natl. Acad. Sci. U.S. 38* (1952), 121–126.
[2] "Minimax theorems," *Proc. Natl. Acad. Sci. U.S. 39* (1953), 42–47.
[3] "On systems of linear inequalities," in *Annals of Mathematics Studies*, No. 38 (1956), 99–156.
[4] "Existence theorems and extreme solutions for inequalities concerning convex functions or linear transformations," *Math. Z. 68* (1957), 205–216.

[5] "On the equilibrium value of a system of convex and concave functions," *Math. Z. 70* (1958), 271–280.

[6] "On the Krein-Milman theorem," in *Convexity*, V. L. Klee, ed., Proceedings of Symposia in Pure Mathematics, Vol. VII, American Mathematical Society, 1963, 211–219.

[7] "Sur une théorème minimax," *C.R. Acad. Sci. Paris 259* (1964), 3925–3928.

[8] "A generalization of the Alaoglu-Bourbaki theorem and its applications," *Math. Z. 88* (1965), 48–60.

[9] "Sets with convex sections," *Proceedings of the Colloquium on Convexity, Copenhagen, 1965*, W. Fenchel, ed., Copenhagen, Matematisk Institut, 1967, 72–77.

Ky Fan, I. Glicksberg, and A. J. Hoffman

[1] "Systems of inequalities involving convex functions," *Proc. Amer. Math. Soc. 8* (1957), 617–622.

J. Farkas

[1] "Über die Theorie der einfachen Ungleichungen," *J. Math. 124* (1902), 1–24.

W. Fenchel

[1] "On conjugate convex functions," *Canad. J. Math. 1* (1949), 73–77.

[2] "Convex Cones, Sets and Functions," mimeographed lecture notes, Princeton University, 1951.

[3] "A remark on convex sets and polarity," *Medd. Lunds Univ. Mat. Sem.* (Supplementband, 1952), 82–89.

[4] "Über konvexe Funktionen mit vorgeschriebenen Niveaumannigfaltigkeiten," *Math. Z. 63* (1956), 496–506.

[5] (editor), *Proceedings of the Colloquium on Convexity, Copenhagen, 1965*, Copenhagen, Matematisk Institut, 1967.

M. Frank and P. Wolfe

[1] "An algorithm for quadratic programming," *Naval Res. Logist. Quart. 3* (1956), 95–110.

D. Gale

[1] "A geometric duality theorem with economic application," *Rev. Econ. Studies 34* (1967), 19–24.

D. Gale and V. L. Klee

[1] "Continuous convex sets," *Math. Scand. 7* (1959), 379–391.

D. Gale, V. L. Klee, and R. T. Rockafellar

[1] "Convex functions on convex polytopes," *Proc. Amer. Math. Soc. 19* (1968), 867–873.

D. Gale, H. W. Kuhn, and A. W. Tucker
[1] "Linear programming and the theory of games," in *Activity Analysis of Production and Allocation*, T. C. Koopmans, ed., Wiley, New York, 1951.

M. Gerstenhaber
[1] "Theory of convex polyhedral cones," in *Activity Analysis of Production and Allocation*, T. C. Koopmans, ed., Wiley, New York, 1951, 298–316.

A. Ghouila-Houri
[1] "Sur l'étude combinatoire des familles de convexes," *C.R. Acad. Sci. Paris 252* (1961), 494.

A. J. Goldman and A. W. Tucker
[1] "Theory of linear programming" in *Annals of Mathematics Studies*, No. 38 (1956), 53–98.

A. A. Goldstein
[1] *Constructive Real Analysis*, Harper and Row, New York, 1967.

B. Grünbaum
[1] *Convex Polytopes*, Wiley, New York, 1967.

M. Guignard
[1] "Conditions d'optimalité et dualité en programmation mathématique," thèse (Univ. de Lille, 1967).

M. A. Hanson
[1] "A duality theorem in nonlinear programming with nonlinear constraints," *Austral. J. Statist. 3* (1961), 64–72.

E. Helly
[1] "Über Systeme linearer Gleichungen mit unendlich vielen Unbekannten," *Monatschr. Math. Phys. 31* (1921), 60–91.

W. M. Hirsch and A. J. Hoffman
[1] "Extreme varieties, concave functions, and the fixed charge problem," *Comm. Pure Appl. Math. XIV* (1961), 355–369.

L. Hörmander
[1] "Sur la fonction d'appui des ensembles convexes dans une espace localement convexe," *Arkiv för Mat. 3* (1954), 181–186.

P. Huard
[1] "Dual programs," *IBM J. Res. Develop. 6* (1962), 137–139.
[2] "Dual programs," in *Recent Advances in Math. Programming*, R. L. Graves and P. Wolfe, eds., McGraw-Hill, New York, 1963, 55–62.

440

J. L. W. V. Jensen
[1] "Om konvexe Funktioner og Uligheder mellem Middelvaerdier," *Nyt Tidsskr. Math. 16B* (1905), 49–69.
[2] "Sur les fonctions convexes et les inegalités entre les valeurs moyennes," *Acta Math. 30* (1906), 175–193.

F. John
[1] "Extremum problems with inequalities as subsidiary conditions," in *Studies and Essays, Courant Anniversary Volume*, Interscience, New York, 1948, 187–204.

W. L. Jones
[1] "On conjugate functionals," dissertation (Columbia University, 1960).

R. I. Kachurovskii
[1] "On monotone operators and convex functionals," *Uspekhi 15* (1960), 213–215 (Russian).

S. Kakutani
[1] "A generalization of Brouwer's fixed point theorem," *Duke Math. J. 8* (1941), 457–459.

S. Karlin
[1] *Mathematical Methods and Theory in Games, Programming and Economics*, Vol. I, McGraw-Hill, New York, 1960.

V. L. Klee
[1] "Convex sets in linear spaces," *Duke Math. J. 18* (1951), 443–466.
[2] "Convex sets in linear spaces, II," *Duke Math. J. 18* (1951), 875–883.
[3] "Convex sets in linear spaces, III," *Duke Math. J. 20* (1953), 105–112.
[4] "Separation properties for convex cones," *Proc. Amer. Math. Soc. 6* (1955), 313–318.
[5] "Strict separation of convex sets," *Proc. Amer. Math. Soc. 7* (1956), 735–737.
[6] "Extremal structure of convex sets," *Arch. Math. 8* (1957), 234–240.
[7] "Extremal structure of convex sets, II," *Math. Z. 69* (1958), 90–104.
[8] "Some characterizations of convex polyhedra," *Acta Math. 102* (1959), 79–107.
[9] "Polyhedral sections of convex bodies," *Acta Math. 103* (1960), 243–267.
[10] "Asymptotes and projections of convex sets," *Math. Scand. 8* (1960), 356–362.
[11] (editor), *Convexity*, Proceedings of Symposia in Pure Mathematics, Vol. VII, American Mathematical Society, 1963.

[12] "Infinite-dimensional intersection theorems," in *Convexity*, V. L. Klee, ed., Proceedings of Symposia in Pure Math., Vol. VII, American Mathematical Society, 1963, 349–360.

[13] "Convex polytopes and linear programming," in *Proceedings of the IBM Scientific Computing Symposium on Combinatorial Problems*, Yorktown Heights, 1964.

[14] "Asymptotes of convex bodies," *Math. Scand. 20* (1967), 89–90.

[15] "Maximal separation theorems for convex sets," *Trans. Amer. Math. Soc., 134* (1968) 133–148.

H. Kneser

[1] "Sur une théorème fondamentale de la théorie des jeux," *C.R. Acad. Sci. Paris 234* (1952), 2418–2420.

M. A. Krasnosel'skii and Ya. B. Rutickii

[1] *Convex Functions and Orlicz Spaces*, Noordhoff, Groningen, 1961.

M. Krein and D. Milman

[1] "On the extreme points of regularly convex sets," *Studia Math. 9* (1940), 133–138.

K. S. Kretchmer

[1] "Programmes in paired spaces," *Canad. J. Math. 13* (1961), 221–238.

H. W. Kuhn

[1] "Solvability and consistency for linear equations and inequalities," *Amer. Math. Monthly 63* (1956), 217–232.

H. W. Kuhn and A. W. Tucker

[1] "Nonlinear programming," in *Proceedings of the Second Berkeley Symposium on Mathematical Statistics and Probability*, Univ. of California Press, Berkeley, 1951, 481–492.

[2] (editors), *Linear Inequalities and Related Systems*, Annals of Mathematics Studies, No. 38 (1956).

F. Lannér

[1] "On convex bodies with at least one point in common," *Medd. Lunds Univ. Mat. Sem. 5* (1943), 1–10.

C. Lescarret

[1] "Sur la sous-différentiabilité d'une somme de fonctionelles convexes semi-continues inférieurement," *C.R. Acad. Sci. Paris 262* (1966), 443–446.

E. R. Lorch

[1] "Differentiable inequalities and the theory of convex bodies," *Trans. Amer. Math. Soc. 71* (1951), 243–266.

442 BIBLIOGRAPHY

S. Mandelbrojt
[1] "Sur les fonctions convexes," *C.R. Acad. Sci. Paris 209* (1939), 977–978.

O. L. Mangasarian
[1] "Duality in nonlinear programming," *Quart. Appl. Math. 20* (1962), 300–302.
[2] "Pseudo-convex functions," *S.I.A.M. J. Control 3* (1965), 281–290.

O. L. Mangasarian and J. Ponstein
[1] "Minimax and duality in nonlinear programming," *J. Math. Anal. Appl. 11* (1965), 504–518.

H. Minkowski
[1] *Geometrie der Zahlen*, Teubner, Leipzig, 1910.
[2] *Theorie der Konvexen Körper, Insbesondere Begründung ihres Oberflächenbegriffs*, Gesammelte Abhandlungen II, Leipzig, 1911.

G. J. Minty
[1] "Monotone networks," *Proc. Roy. Soc. London (Ser. A) 257* (1960), 194–212.
[2] "On the monotonicity of the gradient of a convex function," *Pacific J. Math. 14* (1964), 243–247.

J.-J. Moreau
[1] "Décomposition orthogonale d'un espace hilbertien selon deux cônes mutuellement polaires," *C.R. Acad. Sci. Paris 255* (1962), 238–240.
[2] "Fonctions convexes en dualité," multigraph, Séminaires de Mathématiques, Faculté des Sciences, Université de Montpellier (1962).
[3] "Inf-convolution," multigraph, Séminaires de Mathématiques, Faculté des Sciences, Université de Montpellier (1962).
[4] "Fonctions duales et points proximaux dans un espace hilbertien," *C.R. Acad. Sci. Paris, 255* (1962), 2897–2899.
[5] "Propriétés des applications prox," *C.R. Acad. Sci. Paris 256* (1963), 1069–1071.
[6] "Inf-convolution des fonctions numériques sur un espace vectoriel," *C.R. Acad. Sci. Paris 256* (1963), 5047–5049.
[7] "Fonctions à valeurs dans $[-\infty, +\infty]$; notions algebraiques," Séminaires de Mathématiques, Faculté des Sciences, Université de Montpellier (1963).
[8] "Remarques sur les fonctions à valeurs dans $[-\infty, +\infty]$ définies sur on demi-groupe," *C.R. Acad. Sci. Paris 257* (1963), 3107–3109.
[9] "Étude locale d'une fonctionelle convexe," multigraph, Séminaires de Mathématiques, Faculté des Sciences, Université de Montpellier (1963).

[10] "Fonctionelles sous-differentiables," *C.R. Acad. Sci. Paris 257* (1963), 4117–4119.

[11] "Sur la fonction polaire d'une fonction semi-continue supérieurment," *C.R. Acad. Sci. Paris 258* (1964), 1128–1131.

[12] "Théorèmes 'inf-sup'," *C.R. Acad. Sci. Paris 258* (1964), 2720–2722.

[13] "Proximité et dualité dans un espace hilbertien," *Bull. Soc. Math. France 93* (1965), 273–299.

[14] "Semi-continuité de sous-gradient d'une fonctionelle," *C.R. Acad. Sci. Paris 260* (1965), 1067–1070.

[15] "Convexity and duality," in *Functional Analysis and Optimization*, E. R. Caianello, ed., Academic Press, New York, 1966, 145–169.

[16] "Sous-differentiabilité," in *Proceedings of the Colloquium on Convexity, Copenhagen, 1965*, W. Fenchel, ed., Copenhagen, Matematisk Institut, 1967, 185–201.

[17] "Fonctionelles Convexes," lecture notes, Séminaire "Equations aux dérivées partielles," Collège de France, 1966.

T. Motzkin

[1] *Beiträge zur Theorie der linearen Ungleichungen*, Azriel, Jerusalem, 1936.

J. von Neumann

[1] "Zur Theorie der Gesellschaftsspiele," *Math. Ann. 100* (1928), 295–320.

H. Nikaidô

[1] "On von Neumann's minimax theorem," *Pacific J. Math. 4* (1954), 65–72.

T. Popoviciu

[1] *Les Fonctions Convexes*, Hermann, Paris, 1945.

J. R. Reay

[1] *Generalizations of a Theorem of Carathéodory*, Amer. Math. Soc. Memoir No. 54 (1965).

R. T. Rockafellar

[1] Convex Functions and Dual Extremum Problems, thesis, Harvard, 1963.

[2] "Duality theorems for convex functions," *Bull. Amer. Math. Soc. 70* (1964), 189–192.

[3] "Minimax theorems and conjugate saddle-functions," *Math. Scand. 14* (1964), 151–173.

[4] "Helly's theorem and minima of convex functions," *Duke Math. J. 32* (1965), 381–398.

[5] "An extension of Fenchel's duality theorem for convex functions," *Duke Math. J. 33* (1966), 81–90.

[6] "Level sets and continuity of conjugate convex functions," *Trans. Amer. Math. Soc. 123* (1966), 46–63.

[7] "Characterization of the subdifferentials of convex functions," *Pacific J. Math. 17* (1966), 497–510. (A correction to the maximality proof given in this paper appears in [17].)

[8] "Conjugates and Legendre transforms of convex functions," *Canad. J. Math. 19* (1967), 200–205.

[9] "Duality and stability in extremum problems involving convex functions," *Pacific J. Math. 21* (1967), 167–187.

[10] "Convex programming and systems of elementary monotonic relations," *J. Math. Anal. Appl. 19* (1967), 543–564.

[11] "Integrals which are convex functions," *Pacific J. Math. 24* (1968), 867–873.

[12] "A general correspondence between dual minimax problems and convex programs," *Pacific J. Math. 25* (1968), 597–611.

[13] "The elementary vectors of a subspace of R^n," in *Combinatorial Mathematics and Its Applications*, R. C. Bose and T. A. Dowling, eds., University of North Carolina Press, 1969, 104–127.

[14] *Monotone Processes of Convex and Concave Type*, Amer. Math. Soc. Memoir No. 77 (1967).

[15] "Duality in nonlinear programming," in *Mathematics of the Decision Sciences, Part 1*, Lectures in Applied Mathematics, Vol. 11, American Mathematical Society, 1968, 401–422.

[16] "Monotone operators associated with saddle-functions and minimax problems," in *Nonlinear Functional Analysis*, Proceedings of Symposia in Pure Mathematics, American Mathematical Society, 1969.

[17] "On the maximal monotonicity of subdifferential mappings," *Pacific J. Math.*, to appear.

L. Sandgren
[1] "On convex cones," *Math. Scand. 2* (1954), 19–28.

F. W. Sinden
[1] "Duality in convex programming and in projective space," *J. Soc. Indust. Appl. Math. 11* (1963), 535–552.

M. Sion
[1] "Existence de cols pour les fonctions quasi-convexes et semi-continues," *C.R. Acad. Sci. Paris 244* (1957), 2120–2123.

[2] "On general minimax theorems," *Pacific J. Math. 8* (1958), 171–176.

M. Slater
[1] "Lagrange multipliers revisited: a contribution to non-linear programming," Cowles Commission Discussion Paper, Math. 403 (1950).

E. Steinitz
[1] "Bedingt konvergente Reihen und konvexe Systeme, I, II, III," *J. Math. 143* (1913), 128–175; *144* (1914), 1–40; *146* (1916), 1–52.

J. Stoer
[1] "Duality in nonlinear programming and the minimax theorem," *Numer. Math. 5* (1963), 371–379.
[2] "Über einen Dualitätsatz der nichtlinearen Programmierung," *Numer. Math. 6* (1964), 55–58.

J. J. Stoker
[1] "Unbounded convex sets," *Amer. J. Math. 62* (1940), 165–179.

O. Stolz
[1] *Grundzüge der Differential- und Integralrechnung*, Vol. I, Teubner, Leipzig, 1893.

M. H. Stone
[1] "Convexity," mimeographed lecture notes, U. of Chicago, 1946.

S. Straszewicz
[1] "Über exponierte Punkte abgeschlossener Punktmengen," *Fund. Math. 24* (1935), 139–143.

A. W. Tucker
[1] "Extensions of theorems of Farkas and Steimke," *Bull. Amer. Math. Soc. 56* (1950), 57.
[2] "Dual systems of homogeneous linear relations," in *Annals of Mathematics Studies*, No. 38 (1956), 53–97.
[3] "Linear and nonlinear programming," *Operations Res. 5* (1957), 244–257.
[4] "A combinatorial equivalence of matrices," in *Combinatorial Analysis*, R. Bellman and M. Hall, eds., Proceedings of Symposia in Applied Mathematics, Vol. X, American Mathematical Society, 1960, 129–134.
[5] "Combinatorial theory underlying linear programs," in *Recent Advances in Mathematical Programming* (L. Graves and P. Wolfe, eds.), McGraw-Hill, New York, 1963.
[6] "Pivotal Algebra," mimeographed lecture notes compiled by T. D. Parsons (Princeton University, 1965).

F. A. Valentine
[1] *Convex Sets*, McGraw-Hill, New York, 1964.
[2] "The dual cone and Helly type theorems," in *Convexity*, V. L. Klee, ed., Proceedings of Symposia in Pure Mathematics, Vol. VII, American Mathematical Society, 1963, 473–494.

R. M. Van Slyke and R. J.-B. Wets
[1] "A duality theory for abstract mathematical programs with applications to optimal control theory," *J. Math. Anal. Appl. 22* (1968), 679–706.

D. W. Walkup and R. J.-B. Wets
[1] "Continuity of some convex-cone-valued mappings," *Proc. Amer. Math. Soc. 18* (1967), 229–235.

H. Weyl
[1] "Elementare Theorie der konvexen Polyeder," *Commentarii Math. Helvetici 7* (1935), 290–306.

A. Whinston
[1] "Some applications of the conjugate functions theory to duality," in *Nonlinear Programming*, J. Abadie, ed., North-Holland, Amsterdam, 1967, 75–96.

R. A. Wijsman
[1] "Convergence of sequences of convex sets, cones and functions," *Bull. Amer. Math. Soc. 70* (1964), 186–188.
[2] "Convergence of sequences of convex sets, cones and functions, II," *Trans. Amer. Math. Soc. 123* (1966), 32–45.

P. Wolfe
[1] "A duality theorem for nonlinear programming," *Quart. Appl. Math. 19* (1961), 239–244.
[2] "Methods of nonlinear programming," Chap. 10 of *Recent Advances in Mathematical Programming*, R. L. Graves and P. Wolfe, eds., McGraw-Hill, 1963.
[3] "Methods of nonlinear programming," Chap. 6 of *Nonlinear Programming*, J. Abadie, ed., North-Holland, Amsterdam, 1967.

W. H. Young
[1] "On classes of summable functions and their Fourier series," *Proc. Royal Soc. (A) 87* (1912), 225–229.

Index

www.ingramcontent.com/pod-product-compliance
Ingram Content Group UK Ltd.
Pitfield, Milton Keynes, MK11 3LW, UK
UKHW020220060325
455916UK00012B/201

9 780691 015866